A Century
of Mathematics
in America

Part II

HISTORY OF MATHEMATICS
Volume 2

A Century of Mathematics in America

Part II

Edited by Peter Duren
with the assistance of Richard A. Askey
Uta C. Merzbach

American Mathematical Society • Providence, Rhode Island

Library of Congress Cataloging-in-Publication Data

A century of mathematics in America. Part II.
  (History of mathematics; v. 1-)
  1. Mathematics—United States—History—20th century. I. Askey, Richard, 1933-   .
II. Duren, Peter L., 1935-   . III. Merzbach, Uta C., 1933-   .
QA27.U5C46 1988   510′.973   88-22155
  ISBN 0-8218-0124-4 (v. 1)
  ISBN 0-8218-0130-9 (v. 2)

Contents

Historical Surveys

Preface

Part I of *A Century of Mathematics in America*, published for the Centennial celebration of the American Mathematical Society in August 1988, featured a collection of "autobiographically oriented historical articles" by senior American mathematicians. In Part II the emphasis now shifts to histories of mathematical activity at some of the major academic institutions in America: Harvard, Yale, Chicago, Princeton, Stanford, Berkeley, and NYU. The selection of institutions was governed to some extent by the willingness of qualified people to put aside other obligations and produce these historical accounts on relatively short notice. In most cases the primary accounts are supplemented by reprinted articles of recent vintage, and by older materials.

In addition to the articles on mathematical centers, this volume reprints the proceedings of a very unusual conference held at Princeton in 1946 to mark the Bicentennial of the University. The world war had just ended, mathematicians had returned to their university positions, and large numbers of veterans were beginning or resuming graduate work. It was a good time to take stock of open problems and to try to chart the future course of research. Many of the leading mathematicians of the day assembled at Princeton to hold informal discussions on the state of the art in each of nine broad areas of mathematics. Written versions of the discussions were recorded in the proceedings, which unfortunately were not widely circulated at the time. In reading the proceedings today, one cannot fail to see how different mathematics was in 1946, how far matters have progressed over the last forty years. The picture is further clarified by expert commentaries on the Princeton discussions, prepared in 1988 by current leaders in the same fields of research.

The volume concludes with articles describing several aspects of America's mathematical past. Among them are retrospective analyses of political and social currents which have affected the mathematical scene in America, discussions of some major mathematicians and their work, and accounts of the historical development of certain areas of mathematics.

On behalf of the mathematical community, the editors would like to thank the writers for their superb contributions. Their heroic efforts and friendly spirit of cooperation are most appreciated. The editors are also grateful to Albert Tucker for his enthusiastic moral support of the historical project and for some very good suggestions, including the reprinting of the Princeton proceedings. Once again, the editors want to acknowledge the close participation of Mary Lane, Director of Publication at the AMS, in all of the detailed work which went into the formation of this volume. Finally, a special word of appreciation to Donna Harmon and Michael Saitas for their efficient handling of many administrative details, and to all of the AMS production staff, whose expertise will be readily apparent.

Peter Duren
Richard Askey
Uta Merzbach

A Century of Mathematics in America

Part II

Garrett Birkhoff has had a lifelong connection with Harvard mathematics. He was an infant when his father, the famous mathematician G. D. Birkhoff, joined the Harvard faculty. He has had a long academic career at Harvard: A.B. in 1932, Society of Fellows in 1933–1936, and a faculty appointment from 1936 until his retirement in 1981. His research has ranged widely through algebra, lattice theory, hydrodynamics, differential equations, scientific computing, and history of mathematics. Among his many publications are books on lattice theory and hydrodynamics, and the pioneering textbook A Survey of Modern Algebra, *written jointly with S. Mac Lane. He has served as president of SIAM and is a member of the National Academy of Sciences.*

Mathematics at Harvard, 1836–1944

GARRETT BIRKHOFF

0. Outline

As my contribution to the history of mathematics in America, I decided to write a connected account of mathematical activity at Harvard from 1836 (Harvard's bicentennial) to the present day. During that time, many mathematicians at Harvard have tried to respond constructively to the challenges and opportunities confronting them in a rapidly changing world.

This essay reviews what might be called the *indigenous* period, lasting through World War II, during which most members of the Harvard mathematical faculty had also studied there. Indeed, as will be explained in §§1–3 below, mathematical activity at Harvard was dominated by Benjamin Peirce and his students in the first half of this period.

Then, from 1890 until around 1920, while our country was becoming a great power economically, basic mathematical research of high quality, mostly in traditional areas of analysis and theoretical celestial mechanics, was carried on by several faculty members. This is the theme of §§4–7. Finally, I will review some mathematical developments at Harvard in the quarter-century 1920–44, during which mathematics flourished there (and at Princeton) as well as anywhere in the world.

3

Whereas §§1–13 of my account are based on reading and hearsay, much of §§14–20 reviews events since 1928, when I entered Harvard as a freshman, and expresses my own first-hand impressions, mellowed by time. Throughout, I will pay attention not only to "core" mathematics, but also to "applied" mathematics, including mathematical physics, mathematical logic, statistics, and computer science. I will also round out the picture by giving occasional glimpses into aspects of the contemporary scientific and human environment which have influenced "mathematics at Harvard". Profound thanks are due to Clark Elliott and I. Bernard Cohen at Harvard, and to Uta Merzbach and the other editors of this volume, for many valuable suggestions and criticisms of earlier drafts.

1. BENJAMIN PEIRCE[1]

In 1836, mathematics at Harvard was about to undergo a major transition. For a century, all Harvard College students had been introduced to the infinitesimal calculus and the elements of physics and astronomy by the Hollis Professor of Mathematics and Natural Philosophy. Since 1806, the Hollis Professor had been John Farrar (1779–1853), who had accepted the job after it had been declined by Nathaniel Bowditch (1773–1838),[2] a native of Salem.

Bowditch's connection with mathematics at Harvard was truly unique. Having had to leave school at the age of 10 to help his father as a cooper, Bowditch had been almost entirely *self-educated*. After teaching himself Latin and reading Newton's *Principia*, he sailed on ships as supercargo on four round trips to the East Indies. He then published *the* most widely used book on the science of navigation, *The New American Practical Navigator*, before becoming an executive actuary for a series of insurance companies.

After awarding Bowditch an honorary M.A. in 1802, Harvard offered him the Hollis Professorship of Mathematics and Natural Philosophy in 1806. Imagine Harvard offering a professorship today to someone who had never gone to high school or college! Though greatly honored, Bowditch declined because he could not raise his growing family properly on the salary offered ($1200/yr.), and remained an actuarial executive. A prominent member of Boston's American Academy of Arts and Sciences, he did however stay active in Harvard affairs.

In American scientific circles, Bowditch became most famous through his translation of Laplace's *Mécanique Céleste*, with copious notes explaining many sketchy derivations in the original. He did most of the work on this about the same time that Robert Adrain showed that Laplace's value of $1/338$ for the earth's eccentricity $(b - a)/a$ should be $1/316$. Bowditch decided (correctly) that it is more nearly $1/300$.[3]

Bowditch's scientific interests were shared by a much younger Salem native, Benjamin Peirce (1809–1880). Peirce had become friendly at the Salem

Private Grammar School with Nathaniel's son, Henry Ingersoll Bowditch, who acquainted his father with Benjamin's skill in and love for mathematics, and Peirce is reputed to have discussed mathematics and its applications with Nathaniel Bowditch from his boyhood. In 1823, Benjamin's father became Harvard's librarian;[4] in 1826, Bowditch became a member of the Harvard Corporation (its governing board). By then, the Peirce and Bowditch sons were fellow students in Harvard College, and their families had moved to Boston.

There Peirce's chief mentor was Farrar. For more than 20 years, Farrar had been steadily improving the quality of instruction in mathematics, physics, and astronomy by making translations of outstanding 18th century French textbooks available under the title "Natural Philosophy for the Students at Cambridge in New England". Actually, his own undergraduate thesis of 1803 had contained a calculation of the solar eclipse which would be visible in New England in 1814. Although very few Harvard seniors could do this today, it was not an unusual feat at that time.

By 1829, Nathaniel Bowditch had become affluent enough to undertake the final editing and publication of Laplace's *Mécanique Céleste* at his own expense, and young Peirce was enthusiastically assisting him in this task. Peirce must have found it even more exhilarating to participate in criticizing Laplace's masterpiece than to predict a future eclipse!

In 1831, Peirce was made tutor in mathematics at Harvard College, and in 1833 he was appointed University Professor of Mathematics and Natural Philosophy. As a result, *two* of the nine members of the 1836 Harvard College faculty bore almost identical titles. In the same year, he married Sara Mills of Northampton, whose father Elijah Hunt Mills had been a U.S. senator [DAB]. They had five children, of whom two would have an important influence on mathematics at Harvard, as we shall see.

By 1835, still only 26, Peirce had authored seven booklets of Harvard course notes, ranging from "plane geometry" to "mechanics and astronom".[5] Moreover Farrar, whose health was failing, had engaged another able recent student, Joseph Lovering (1813–92), to share the teaching load as instructor. In 1836, Farrar resigned because of poor health, and Lovering succeeded to his professorship two years later. For the next 44 years, Benjamin Peirce and Joseph Lovering would cooperate as Harvard's senior professors of mathematics, astronomy, and physics.

Nathaniel Bowditch died in 1838, and his place on the Harvard Corporation was taken by John A. Lowell (1798–1881), a wealthy textile industry financier. By coincidence, in 1836 (Harvard's bicentennial year) his cousin John Lowell, Jr. had left $250,000 to endow a series of public lectures, with John A. as sole trustee of the Lowell Institute which would pay for them. Lowell (A.B. 1815) must have also studied with Farrar. Moreover by another coincidence, having entered Harvard at 13, John A. Lowell had lived as a freshman in the house of President Kirkland, whose resignation in 1829 after a deficit of $4,000 in a budget of $30,000, student unrest, and a slight stroke had been due to pressure from Bowditch![6] As a final coincidence, by 1836 Lowell's tutor Edward Everett had become governor of Massachusetts, in which capacity he inaugurated the Lowell Lectures in 1840.

In 1842, Peirce was named Perkins Professor of Astronomy and Mathematics, a newly endowed professorship. By that time, qualified Harvard students devoted two years of study to Peirce's book *Curves and Functions*, for which he had prepared notes. Ambitious seniors might progress to Poisson's *Mécanique Analytique*, which he would replace in 1855 with his own textbook, *A System of Analytical Mechanics*. This was fittingly dedicated to "My master in science, NATHANIEL BOWDITCH, the father of American Geometry".

Meanwhile, Lovering was becoming famous locally as a teacher of physics and a scholar. His course on "electricity and magnetism" had advanced well beyond Farrar, and over the years, he would give no less than nine series of Lowell Lectures [NAS #6, 327–44].[7]

In 1842–3, Peirce and Lovering also founded a quarterly journal called *Cambridge Miscellany of Mathematics, Physics, and Astronomy*, but it did not attract enough subscribers to continue after four issues. They also taught Thomas Hill '43, who was awarded the Scott Medal by the Franklin Institute for an astronomical instrument he invented as an undergraduate. He later wrote two mathematical textbooks while a clergyman, and became Harvard's president from 1862 to 1868 [DAB 20, 547–8].

2. PEIRCE REACHES OUT

During his lifetime Peirce was without question the leading American mathematical astronomer. In 1844, perhaps partly stimulated by a brilliant comet in 1843, "The Harvard Observatory was founded on its present site ... by a public subscription, filled largely by the merchant shipowners of Boston" [Mor, p. 292]. For decades, William C. Bond (1789–1859) had been advising Harvard about observational astronomy, and he may well have helped Hill with his instrument. In any event, Bond finally became the salaried director of the new Observatory, where he was succeeded by his son George (Harvard '45). Peirce and Lovering both collaborated effectively with William Bond in interpreting data.[8] Pursuing further the methods he had learned from Laplace's *Mécanique Céleste*, Peirce also analyzed critically Leverrier's successful prediction of the new planet Neptune, first observed in 1846.

Meanwhile, our government was beginning to play an important role in promoting science. At about this time, the Secretary of the Navy appointed

Lieutenant (later Admiral) Charles Henry Davis head of the Nautical Almanac Office. Like Peirce, Davis had gone to Harvard and married a daughter of Senator Mills. Because of Peirce and the Harvard Observatory, he decided to locate the Nautical Almanac Office in Cambridge, where Peirce served from 1849 to 1867 as Consulting Astronomer, supervising for several years the preparation (in Cambridge) of the *American Ephemeris and Nautical Almanac*, our main government publication on astronomy. This activity attracted to Cambridge such outstanding experts in celestial mechanics as Simon Newcomb (1835–1909) and G.W. Hill (1838–1914), who later became the third and fourth presidents of the American Mathematical Society.[9] Others attracted there were John M. van Vleck, an early AMS vice-president, and John D. Runkle, founder of a short-lived but important *Mathematical Monthly*.

Similarly, Alexander Dallas Bache [DAB 1, 461–2], after getting the Smithsonian Institution organized in 1846 with Joseph Henry as its first head [EB 20, 698–700], became head of the U.S. Coast Survey. Bache appointed Peirce's former student B. A. Gould[10] as head of the Coast Survey's longitude office and Gould, who was Harvard president Josiah Quincy's son-in-law, decided to locate his headquarters at the Harvard Observatory and also use Peirce as scientific adviser. Peirce acted in this capacity from 1852 to 1874, aided by Lovering's selfless cooperation, succeeding Bache as superintendent of the Coast Survey for the last eight of these years.[11] In both of these roles, Peirce showed that he could not only *apply* mathematics very effectively (see [Pei, p. 12]); he was also a creative organizer and persuasive promoter.

Indeed, from the 1840s on, Peirce was reaching out in many directions. Thus, he became president of the newly founded American Association for the Advancement of Science in 1853. He was also active locally in Harvard's Lawrence Scientific School (LSS) during its early years.

The Lawrence Scientific School. The LSS is best understood as an early attempt to promote graduate education in pure and applied science, including mathematics. Established when Harvard's president was Edward Everett, John A. Lowell's former tutor, and its treasurer was Lowell's then affluent business associate Samuel Eliot, the LSS was named for Abbott Lawrence, another New England textile magnate who had been persuaded to give $50,000 to make its establishment possible in 1847. By then, the cloudy academic concepts of "natural philosophy" and "natural history" were becoming articulated into more clearly defined "sciences" such as astronomy, geology, physics, chemistry, zoology, and botany. Correspondingly, the LSS was broadly conceived as a center where college graduates and other qualified aspirants could receive advanced instruction in these sciences and engineering. Its early scientifically minded graduates included not only several notable "applied" mathematicians such as Newcomb and Runkle, but also the classmates Edward

Pickering and John Trowbridge. John Trowbridge would later join and ulti-mately replace Joseph Lovering as Harvard's chief physicist, while Pickering would become director of the Harvard Observatory and an eminent astro-physicist. Most important for mathematics at Harvard, Trowbridge would have as his "first research student in magnetism" B. O. Peirce.

The leading "pure" scientists on the LSS faculty were Peirce, the botanist Asa Gray, the anatomists Jeffries Wyman, and the German-educated Swiss naturalist Louis Agassiz (1807–73), already internationally famous when he joined the faculty of the LSS as professor of zoology and geology in 1847.

The original broad conception of the Lawrence Scientific School was best exemplified by Agassiz. Like Benjamin Peirce, Agassiz expected students to think for themselves; but unlike Peirce, he was a brilliant lecturer, who soon "stole the show" from Gray and Peirce at the LSS. Although he did not influence mathematics directly, we shall see that three of his students *indirectly* influenced the mathematical sciences at Harvard: his son Alexander (1835–1912); the palaeontologist and geologist Nathaniel Shaler; and the eminent psychologist and philosopher William James.

The Lazzaroni.   Like Alexander Bache, Louis Agassiz and Benjamin Peirce were members of an influential group of eminent American scientists who, calling themselves the "Lazzaroni", tried to promote research. However, very few tuition-paying LSS students had research ambitions. The majority of them studied civil *engineering* under Henry Lawrence Eustis ('38), who had taught at West Point [Mor, p. 414] before coming to Harvard. Most of the rest studied *chemistry*, until 1863 under Eben Horsford, whose very "applied" interests made it appropriate to assign his Rumford Professorship to the LSS.

Peirce's students.   Peirce had many outstanding students. Among them may be included Thomas Hill, B. A. Gould, John Runkle, and Simon New-comb. Partly through his roles in the Coast Survey, the Harvard Observa-tory, and the *Nautical Almanac*, he was responsible for making Harvard our nation's leading research center in the mathematical sciences in the years 1845–65.

Although most of his students "apprehended imperfectly what Professor Peirce was saying", he also was "a very inspiring and stimulating teacher" for those eager to learn. We know this from the vivid account of his teach-ing style [Pei, pp. 1–4] written by Charles William Eliot (1834–1926), sev-enty years after taking (from 1849 to 1853) the courses in mathematics and physics taught by Peirce and Lovering. The same courses were also taken by Eliot's classmate, Benjamin's eldest son James Mills Peirce (1834–1906). Af-ter graduating, these two classmates taught mathematics together from 1854 to 1858, collaborating in a daring and timely educational reform. At the time, Harvard students were examined orally by state-appointed overseers whose duty it was to make sure that standards were being maintained. "Offended by the dubious expertness and obvious absenteeism of the Overseers, ... the young tutors [Eliot and J. M. Peirce] obtained permission to substitute written examinations, which they graded themselves" [HH, p. 15].

After 1858, J. M. Peirce tried his hand at the ministry, while Eliot increasingly concentrated his efforts on chemistry (his favorite subject) and university administration. Josiah Parsons Cooke ('48), the largely self-taught Erving Professor of Chemistry, had shared chemicals with Eliot when the latter was still an undergraduate [HH, p. 11]. Then in 1858, these two friends successfully proposed a new course in chemistry [HH, p. 16], in which students performed laboratory exercises "probably for the first time". Eliot demonstrated remarkable administrative skill; at one point, he was acting dean of the LSS and in charge of the chemistry laboratory. In these roles, he proposed a thorough revision in the program for the S.B. in chemistry, based on a "firm grounding in chemical and mathematical fundamentals". He then served with Dean Eustis of that School and Louis Agassiz on a committee appointed to revamp the school's curriculum as a whole [HH, pp. 23–25]. However Eliot's zeal for order and discipline antagonized the more informal Agassiz, and Eliot's reformist ideas were rejected. After losing out to the more research-oriented "Lazzarone" Wolcott Gibbs in the competition for the Rumford professorship [HH, pp. 25–27], in spite of J. A. Lowell's support of his candidacy, Eliot left Harvard (with his family) for two years of study in France and Germany.

MIT. Academic job opportunities in "applied" science in the Boston area were improved by the founding there of the Massachusetts Institute of Technology (MIT). Since the mid-1840s, Henry and then his brother William Rogers, "one of those accomplished general scientists who matured before the age of specialization", had been lobbying with J. A. Lowell and others for the benefits of polytechnical education, and in 1862 William Rogers became its first president. Peirce's student John Runkle joined its new faculty in 1865 as professor of mathematics and analytical mechanics, and later became the second president of MIT.[12] Recognizing Eliot's many skills, Rogers soon also invited him to go to MIT, and Eliot accepted [HH, pp. 34–7]. There Eliot wrote with F. H. Storer, an LSS graduate and his earlier collaborator and hiking companion, a landmark *Manual of Inorganic Chemistry*.

3. Peirce's Golden Years

When the Civil War broke out in 1861, J. M. Peirce was a minister in Charleston, South Carolina. Benjamin promptly had James made an assistant professor of mathematics at Harvard, to help him carry the teaching load. Benjamin's brilliant but undisciplined younger son Charles Sanders

Peirce (1839–1914), after being tutored at home, had graduated from Harvard two years earlier without distinction. Nevertheless, Benjamin secured for him a position in the U.S. Coast Survey exempting Charles from military service. As we shall see, this benign nepotism proved to be very fruitful.

A year later, his former student Thomas Hill became Harvard's president, having briefly (and reluctantly) been president of Antioch College. At Harvard, Hill promoted the "elective system", encouraging students to decide between various courses of study. He also initiated series of university lectures which, like the LSS, constituted a step toward the provision of graduate instruction. Peirce participated in this effort most years during the 1860s, lecturing on abstruse mathematics with religious fervor.

From the 1850s on, Peirce had largely freed himself from the drudgery of teaching algebra, geometry, and trigonometry. Moreover, whereas Lovering's courses continued to be required, Peirce's were all optional (electives), and were taken by relatively few students. Peirce was frequently "lecturing on his favorite subject, Hamilton's new calculus of quaternions" [Pei, p. 6], to W. E. Byerly ('71) among others. In 1873 Byerly was persuaded by Peirce to write a doctoral thesis on "The Heat of the Sun". In it, he calculated the total energy of the sun, under the assumption (common at the time) that this was gravitational.[13] The calculation only required using the calculus and elementary thermodynamics. Nevertheless, Byerly became Harvard's first Ph.D., and a pillar of Harvard's teaching staff until his retirement in 1912!

Benjamin Peirce had long been interested in Hamilton's quaternions $a + b\underline{i} + c\underline{j} + d\underline{k}$; moreover Chapter X of his *Analytical Mechanics* (1855) contained a masterful chapter on 'functional determinants' of $n \times n$ matrices.[14] During the Civil War, Benjamin and Charles became interested in generalizations of quaternions to *linear associative algebras*. After lecturing several times to his fellow members of the National Academy of Sciences on this subject, Benjamin published his main results in 1870 in a privately printed paper.[15] This contained the now classic "Peirce decomposition"

$$x = exe + ex(1 - e) + (1 - e)xe + (1 - e)x(1 - e)$$

with respect to any "idempotent" e satisfying $ee = e$.

An 1881 sequel, published posthumously by J. J. Sylvester in the newly founded *American Journal of Mathematics* (**4**, 97–229), contained numerous addenda by Charles. Most important was Appendix III, where Charles proved that the only division algebras of finite order over the real field **R** are **R** itself, the complex field **C**, and the real quaternions. This very fundamental theorem had been proved just three years earlier by the German mathematician Frobenius.[16]

Charles Peirce also worked with his father in improving the scientific instrumentation of the U.S. "Coast Survey", which by 1880 was surveying the entire United States! But most relevant to mathematics at Harvard, and most distinguished, was his unpaid role as a logician and philosopher. An active member of the Metaphysical Club presided over by Chauncey Wright ('52), another of Benjamin's ex-students who made his living as a computer for the *Nautical Almanac*, Charles gave a brilliant talk there proposing a new philosophical doctrine of "pragmatism". He also published a series of highly original papers on the then new algebra of relations. Although surely "imperfectly apprehended" by most of his contemporaries, these contributions earned him election in 1877 to the National Academy of Sciences, which had been founded 12 years earlier by the Lazzaroni.

In his last years, increasingly absorbed with quaternions, Benjamin Peirce's unique teaching personality influenced other notable students. These included Harvard's future president, Abbott Lawrence Lowell, and his brilliant brother, the astronomer Percival. They were grandsons of both Abbott Lawrence and John A. Lowell! To appreciate the situation, one must realize that the two grandfathers of the Lowell brothers were John A. Lowell, the trustee of the Lowell Institute, and Abbott Lawrence, for whom Harvard's Lawrence Scientific School (see §2) had been named. From 1869 to 1933, the presidents of Harvard would be former students of Benjamin Peirce!

Two others were B. O. Peirce, a distant cousin of Benjamin's who would succeed Lovering (and Farrar) as Hollis Professor of Mathematics and Natural Philosophy; and Arnold Chace, later chancellor of Brown University. Peirce's lectures inspired both A. L. Lowell and Chace to publish papers in which "quaternions" (now called vectors) were applied to geometry. Moreover both men would describe in [Pei] fifty years later, as would Byerly, how Peirce influenced their thinking. Still others were W. E. Story, who went on to get a Ph.D. at Leipzig, and W. I. Stringham (see [S-G]). Story became president of the 1893 International Mathematical Congress in Chicago, and himself supervised 12 doctoral theses including that of Solomon Lefschetz [CMA, p. 201].

Benjamin Peirce's funeral must have been a very impressive affair. His pallbearers fittingly included Harvard's president C. W. Eliot and ex-president Thomas Hill; Simon Newcomb, J. J. Sylvester, and Joseph Lovering; his famous fellow students and lifelong medical friends Henry Bowditch and Oliver Wendell Holmes (the "Autocrat of the Breakfast Table"); and the new superintendent of the Coast Survey, C. P. Patterson.

4. Eliot Takes Hold[17]

When Thomas Hill resigned from the presidency of Harvard in 1868, the Corporation (with J. A. Lowell in the lead) recommended that Eliot be his successor, and the overseers were persuaded to accept their nomination in 1869. After becoming president, Eliot immediately tried (unsuccessfully) to merge MIT with the LSS [Mor, p. 418].[18] At the same time, he tried to build up the series of university lectures, inaugurated by his predecessor, into a viable graduate school. It is interesting that for the course in philosophy his choice of lecturers included Ralph Waldo Emerson and Charles Sanders Peirce; in 1870–71 thirty-five courses of university lectures were offered; but the scheme "failed hopelessly" [Mor, p. 453].

To remedy the situation, Eliot then created a graduate department, with his classmate James Mills Peirce as secretary of its guiding academic council. This was authorized to give M.A. and Ph.D. degrees, such as those given to Byerly and Trowbridge. At the same time, Eliot transferred his former rival Wolcott Gibbs from being dean of the LSS to the physics department, and chemistry from the LSS to Harvard College, where his former mentor Josiah Parsons Cooke was in charge. A year earlier, Eliot had proposed an elementary course in chemistry, to be taught partly in the laboratory. This gradually became immensely popular, partly because of its emphasis on the chemistry of such familiar phenomena as photography [Mor, p. 260].

By 1886, all members of the LSS scientific faculty had transferred to Harvard College. Moreover undergraduates wanting to study engineering had no incentive for enrolling in the LSS rather than Harvard College. As a result of this, and competition from MIT and other institutions, there was a steady decline in the LSS enrollment, until only 14 students enrolled in 1886! Its four-year programme in "mathematics, physics, and astronomy", inherited from the days of Benjamin Peirce, had had no takers at all for many years, and was wisely replaced in 1888 by a programme in electrical engineering.

Meanwhile, the new graduate department itself was struggling. After 1876, the Johns Hopkins University attracted many of the best graduate students. Two of them would later be prominent members of the Harvard faculty: Edwin Hall in physics and Josiah Royce in philosophy. During its entire lifetime (1872–90), the graduate department awarded only five doctorates in mathematics, including those of Byerly and F. N. Cole (the latter earned in Germany, see §5).

To remedy the situation, Eliot made a second administrative reorganization in 1890. From it, the graduate department emerged as the graduate school; the LSS engineering faculty joined the Harvard College faculty in a new Faculty of Arts and Sciences. J. M. Peirce's title was changed from secretary of the graduate department to dean of the Graduate School of Arts and Sciences, and the economist Charles F. Dunbar ('51) was made dean of the Faculty of Arts and Sciences.

Additional programmes of study were introduced into the LSS, and Nathaniel Shaler made its new dean. This was a brilliant choice: the enrollment in the LSS increased to over 500 by 1900, and Shaler's own Geology 4 became one of Harvard College's most popular courses. Shaler was also allowed to assume management of the mining companies of his friend and neighbor, the aging inventor and mining tycoon Gordon McKay [HH, pp. 213–15], Shaler persuaded McKay in 1903 to bequeath his fortune to the school, where it supports the bulk of Harvard's program in "applied" mathematics to this day!

Mention should also be made of the appointment in 1902 of the self-educated, British-born scientist Arthur E. Kennelly (1851–1939). Joint discoverer of the upper altitude "Kennelly–Heaviside layer" which reflects radio waves, Kennelly had been Edison's assistant for 13 years and president of the American Institute of Electric Engineers. In his thoughtful biography of Kennelly [NAS 22: 83–119], Vannevar Bush describes how Kennelly's career spanned the entire development of electrical engineering to 1939. Bush and my father [GDB III, 734–8] both emphasize that Kennelly revolutionized the mathematical theory of alternating current (a.c.) circuits by utilizing the complex exponential function. Curiously, this major application is still rarely explained in mathematics courses in our country, at Harvard or elsewhere!

For further information about the changes I have outlined, and other interpretations of the conflicting philosophies of scientific education which motivated them, I refer you to [HH] and especially [Love]. The latter document was written by James Lee Love, who taught mathematics under the auspices of the Lawrence Scientific School from 1890 to 1906, when the LSS was renamed the Graduate School of Applied Science. Officially affiliated with Harvard until 1911, Love returned to Burlington, North Carolina, in 1918 to become president of the Gastonia Cotton Manufacturing Company. Reorganized as Burlington Mills, this became one of our largest textile companies. During these years, Love donated $50,000 to the William Byerly Book Fund.

5. A Decade of Transition

When Benjamin Peirce died, his son James had been ably assisting him in teaching Harvard undergraduates for more than 20 years. Byerly had joined them in 1876. At the time, B. O. Peirce was still studying physics and mathematics with John Trowbridge and Benjamin Peirce, but he became an instructor in mathematics in 1881, assistant professor of mathematics and physics in 1884, and Hollis Professor of Mathematics and Natural Philosophy (succeeding Lovering) in 1888. In the decade following Benjamin Peirce's death, the triumvirate consisting of J. M. Peirce (1833–1906), W. E. Byerly (1850–1934), and B. O. Peirce (1855–1913) would be Harvard's principal mathematics teachers.

As a mathematician, J. M. Peirce has been aptly described as an "understudy" to his more creative father [Mor, p. 249]. However, "to no one, excepting always President Eliot, [was] the Graduate School so indebted" for "the promotion of graduate instruction" [Mor, p. 455]. Moreover his teaching, unlike that of his father, seems to have been popular and easily comprehended. In the 1880s, he and Byerly began giving in alternate years Harvard's first higher geometry course (Mathematics 3) with the title "Modern methods in geometry – determinants". Otherwise, his advanced teaching covered mainly topics of algebra and geometry in which Benjamin and C. S. Peirce had done research, such as "quaternions", "linear associative algebra", and "the algebra of logic".

While James Peirce was administering graduate degrees at Harvard as secretary of the academic council, Byerly was cooperating most effectively in making mathematics courses better understood by undergraduates. His *Differential Calculus* (1879), his *Integral Calculus* (1881), and his revised and abridged edition of Chauvenet's *Geometry* (1887), presumably the text for Math. 3, were widely adopted in other American colleges and universities.[19]

In 1883–4, Byerly and B. O. Peirce introduced a truly innovative course in mathematical physics (or "applied mathematics") which has been taught at Harvard in suitably modified form ever since. Half of this course (taught by Byerly) dealt with the expansion of "arbitrary functions" in *Fourier Series and Spherical Harmonics*, this last being the title of a book he wrote in 1893. The other half treated potential theory, and Peirce wrote for it a book, *Newtonian Potential Function*, published in three editions (1884, 1893, 1902). Like Byerly's other books, they were among the most influential and advanced American texts of their time.

B. O. Peirce was an able and scholarly, if traditional, mathematical physicist. A brilliant undergraduate physics major, his "masterly" later physical research was mostly empirical. Although it was highly respected for its thoroughness, and Peirce became president of the American Physical Society in 1913, it lay in "the unexciting fields of magnetism and the thermal conduction of non-metallic substances". His main mathematical legacy consisted in his text for Mathematics 10, and his *Table of Integrals...*, originally written as a supplement to Byerly's *Integral Calculus*. This was still being used at Harvard when I was an undergraduate, but such tables may soon be superseded by packages of carefully written, debugged, and documented computer programs like Macsyma.

In short, Harvard's three professors of mathematics regarded their profession as that of *teaching* reasonably advanced mathematics in an understandable way. Their success in this can be judged by the quality of their students, who included M. W. Haskell, Arthur Gordon Webster, who became president of the American Physical Society in 1903, Frank N. Cole, W. F. Osgood, and Maxime Bôcher. In 1888, when the AMS was founded, two of them had inherited the titles of Benjamin Peirce and Lovering; only Byerly had the simple title "professor of mathematics".[21]

C. S. Peirce. When his father died, C. S. Peirce (1839–1914) was at the zenith of his professional career. From 1879 to 1884, he was a lecturer at Johns Hopkins as well as a well-paid and highly respected employee of the Coast Survey (cf. [SMA, pp. 13–20]). While there, he discovered the fundamental connection between Boolean algebra and what are today called "partially ordered sets" (cf. American J. Math. **3** (1880), 15–57), thus fore-shadowing the "Dualgruppen" of Dedekind ("Verbände" or *lattices* in today's terminology). Unfortunately, in describing this connection, he erroneously claimed that the distributive law $a(b \vee c) = ab \vee ac$ necessarily relates least upper bounds $x \vee y$ and greatest lower bounds xy.

Indeed, the 1880s were a disastrous decade for C. S. Peirce. His lectures at the Johns Hopkins Graduate School were not popular; his personality was eccentric; and his appointment there was not renewed after Sylvester returned to England. He also lost his job with the Coast Survey soon after 1890. Although he continued to influence philosophy at Harvard (see §8), he never again held a job with any kind of tenure. An early member of the New York Mathematical Society, his brilliant turns of speech continued to enliven its meetings [CMA, pp. 15–16], but he was not taken seriously.

6. OSGOOD AND BÔCHER[22]

By 1888, when the American Mathematical Society (AMS) was founded (in New York), a new era in mathematics at Harvard was dawning. Frank Nelson Cole (Harvard '82) had returned three years earlier after "two years under Klein at Leipzig" [Arc, p. 100]. "Aglow with enthusiasm, he gave courses in modern higher algebra, and in the theory of functions of a complex variable, geometrically treated, as in Klein's famous course of lectures at Leipzig." His "truly inspiring" lectures were attended by two undergraduates, W. F. Osgood (1864–1943) and Maxime Bôcher (1867–1918), "as well as by nearly all members of the department," including Professors J. M. Peirce, B. O. Peirce, and W. E. Byerly.

After graduating, Osgood and Bôcher followed Cole's example and went to Germany to study with Felix Klein, who had by then moved to Göttingen.[23] After earning Ph.D. degrees (Osgood in 1890, Bôcher with especial distinction in 1891), both men joined the expanding Harvard staff as instructors for three years. Inspired by the example of Göttingen under Klein, they spearheaded a revolution in mathematics at Harvard, where they continued to serve as assistant professors for another decade before becoming full professors (Osgood in 1903, Bôcher in 1904). All this took place in the heyday of the Eliot regime, under the benevolent but mathematically nominal leadership of the two Peirces and Byerly.

The most conspicuous feature of the revolution resulting from the appointments of Bôcher and Osgood was a sudden increase in research activity. By

1900, Osgood had published 21 papers (six in German), while Bôcher had published 30 in addition to a book *On the series expansions of potential theory*, and a survey article on "Boundary value problems of ordinary differential equations" for Klein's burgeoning *Enzyklopädie der Mathematischen Wissenschaften*, both in German. Moreover Bôcher and Yale's James Pierpont had given the first AMS Colloquium Lectures in 1896, to an audience of 13, while Osgood and A. G. Webster (a Lawrence Scientific School alumnus) had given the second, in 1898.[24]

Similar revolutions had taken place in the 1890s at other leading American universities. Most important of these was at the newly founded University of Chicago, where the chairman of its mathematics department, E. H. Moore, was inspiring a series of Ph.D. candidates [LAM, §3]. Under the leadership of H. B. Fine, who had been stimulated by Sylvester's student G. B. Halsted, Princeton would blossom somewhat later. Meanwhile, Cole had become a professor at Columbia, secretary of the AMS, and editor of its *Bulletin* (cf. [Arc, Ch. V]). The Cole prize in algebra is named for him.

Thus it was most appropriate for Osgood, Bôcher, and Pierpont to cooperate with E. H. Moore (1862–1932) of Chicago in making the promotion of mathematical research the central concern of the AMS. Feeling "the great need of a journal in which original investigations might be published" [Arc, p. 56], these men succeeded in establishing the *Transactions Amer. Math. Soc.* [Arc, Ch. V]. From 1900 on, this new periodical supplemented the *American Journal of Mathematics*, complete control over which Simon Newcomb was unwilling to relinquish. The *Annals of Mathematics* was meanwhile being published at Harvard from 1899 to 1911, with Bôcher as chief editor. Primarily designed for "graduate students who are not yet in a position to read the more technical journals", this also "contained some articles ... suitable for undergraduates."

Harvard continued to educate many mathematically talented students during the years 1890–1905, including most notably J. L. Coolidge, E. V. Huntington, and E. B. Wilson, all for four years; and for shorter periods E.R. Hedrick ('97–'99), Oswald Veblen ('99–'00), and G. D. Birkhoff ('03–'05) . At the same time, there was a great improvement in the quality and quantity of advanced courses designed "primarily for graduate students", but taken also by a few outstanding undergraduates. By 1905, the tradition of Benjamin Peirce had finally been supplanted by new courses stressing new concepts, mostly imported from Germany and Paris; in 1906 J. M. Peirce died.

By that time, Harvard's graduate enrollment had increased mightily. From 28 students in 1872, when Eliot had appointed J. M. Peirce secretary of his new "graduate department", it had grown to 250 when Peirce resigned as dean of Harvard's "graduate school", to become dean of the entire Faculty of Arts and Sciences. A key transition had occurred in 1890, when the graduate "department" was renamed a "school", and the Harvard catalog first divided all courses into three tiers: "primarily for undergraduates", "for undergraduates and graduates", and "primarily for graduates", as it still does.

The Peirces and Byerly had explained to their students many of the methods of Fourier, Poisson, Dirichlet, Hamilton, and Thomson and Tait's *Principles of Natural Philosophy* (1867). However, they had largely ignored the advances in rigor due to Cauchy, Riemann, and Weierstass. For example, Byerly's *Integral Calculus* of 1881 still defined a definite integral vaguely as "the limit of a sum of infinitesimals", although Cauchy–Moigno's *Leçons de Calcul Integral* had already defined integrals as limits of sums $\sum f(x_i)\Delta x_i$, and sketched a proof of the fundamental theorem of the calculus in 1844, while in 1883 volume 2 of Jordan's *Cours d'Analyse* would even define *uniform* continuity.

The key graduate course (Mathematics 13) on functions of a *complex variable* became modernized gradually. Under J. M. Peirce, it had been a modest course based on Briot and Bouquet's *Fonctions Elliptiques*. In 1891–92, Osgood followed this with a more specialized course on elliptic functions as such, and the next year with another treating abelian integrals, while Bôcher gave a course on "functions defined by differential equations", in the spirit of Poincaré. Then, from 1893 to 1899, Bôcher developed Mathematics 13 into the basic full course on complex analysis that it would remain for the next half-century, introducing students to many ideas of Cauchy, Riemann, and Weierstrass. Then, beginning in 1895, he and Osgood supplemented Mathematics 13 with a half-course on "infinite series and products" (Mathematics 12) which treated *uniform* convergence. By 1896, Osgood had written a pamphlet *Introduction to Infinite Series* covering its contents.

In his moving account of "The life and services of Maxime Bôcher" (*Bull. Amer. Math. Soc.* **25** (1919), 337–50) Osgood has described Bôcher's lucid lecture style, and how much Bôcher contributed to his own masterly treatise *Funktionentheorie* (1907), which became the standard advanced text on the subject on both sides of the Atlantic. (Weaker souls, whose mathematical sophistication or German was not up to this level, could settle for Goursat-Hedrick.) Osgood's other authoritative articles on complex function theory, written for the *Enzyklopädie der Mathematischen Wissenschaften* and as Colloquium Lectures,[24] established him as America's leading figure in classical complex analysis.

On a more elementary level, Osgood wrote several widely used textbooks beginning with an *Introduction to Infinite Series* (1897). Ten years later, his *Differential and Integral Calculus* appeared, with acknowledgement of its debt to Professors B. O. Peirce and Byerly. There one finds stated, for the first time in a Harvard textbook, a (partial) "fundamental theorem of the calculus". These were followed by his *Plane and Solid Analytic Geometry* with W. C. Graustein (1921), his *Introduction to the Calculus* (1922), and his

Advanced Calculus (1925), the last three of which were standard fare for Harvard undergraduates until around 1940. Osgood also served for many years on national and international commissions for the teaching of mathematics.

Less systematic than Osgood, Bôcher was more inspiring as a lecturer and thesis adviser. As an analyst, his main work concerned expansions in Sturm–Liouville series (including Fourier series) associated with the partial differential equations of mathematical physics (after "separating variables"). His *Introduction to the Study of Integral Equations* (1909, 1914) and his *Leçons sur les Méthodes de Sturm* ... (1913–14) were influential pioneer monographs. Like Bôcher's papers which preceded them, they established clearly and rigorously by classical methods[24a] precise interpretations of many basic formulas concerned with potential theory and orthogonal expansions (Mathematics 10a and Mathematics 10b).

Several of Bôcher's Ph.D. students had very distinguished careers, most notable among them being G. C. Evans, who in the 1930s would pilot the mathematics department of the University of California at Berkeley to the level of preeminence that it has maintained ever since. Others were D. R. Curtiss (Northwestern University), Tomlinson Fort (Georgia Tech), and L. R. Ford (Rice Institute).

By 1900, the presence of Osgood, Bôcher, Byerly, and B. O. Peirce had made Harvard very strong in analysis. Moreover this strength was increased in 1898 by the addition to its faculty of Charles Leonard Bouton (1860–1922), who had just written a Ph.D. thesis with Sophus Lie.[25] However, it was clear that advanced instruction in other areas of mathematics, mostly given before 1900 by J. M. Peirce, needed to be rejuvenated by new ideas.

The first major step in building up a balanced curriculum was taken by Bôcher. In the 1890s, he had given with Byerly in alternate years Harvard's first higher geometry course (Mathematics 3) with the title "Modern methods in geometry – determinants". Then, in 1902–3, he inaugurated a new version of Mathematics 3, entitled "Modern geometry and modern algebra", with a very different outline leading up to "the fundamental conceptions in the theory of invariants." The algebraic component of this course matured into Bôcher's book, *Introduction to Higher Algebra* (1907), in which §26 on "sets, systems, and groups" expresses modern algebraic ideas. This book would introduce a generation of American students to linear algebra, polynomial algebra, and the theory of elementary divisors. But to build higher courses on this foundation, without losing strength in analysis, would require new faculty members.

7. COOLIDGE AND HUNTINGTON

Harvard's course offerings in higher geometry were revitalized in the first decades of this century by the addition to its faculty of Julian Lowell Coolidge

(1873–1954). After graduating from Harvard (*summa cum laude*) and Balliol College in Oxford, Coolidge taught for three years at the Groton School before returning to Harvard. At Groton, he began a lifelong friendship with Franklin Roosevelt, which illustrates his concern with the *human* side of education (see §15). Indeed, somewhat like his great-great-grandfather Thomas Jefferson, our "most mathematical president", Coolidge was unusually many-sided.[26]

From 1900 on, Coolidge gave in rotation a series of lively and informative graduate courses on such topics as the geometry of position, non-Euclidean geometry, algebraic plane curves, and line geometry. After he had spent two years (1902–4) in Europe and written a Ph.D. thesis under the guidance of Eduard Study and Corrado Segre, these courses became more authoritative. In time, the contents of four of them would be published as books on *Non-Euclidean Geometry* (1909), *The Circle and the Sphere* (1916), *The Geometry of the Complex Domain* (1924), and *Algebraic Plane Curves* (1931).

In 1909–10, Coolidge also initiated a half-course on probability (Mathematics 9), whose contents were expanded into his readable and timely *Introduction to Mathematical Probability* (1925), soon translated into German (Teubner, 1927). Coolidge's informal and lively expository style is well illustrated by his 1909 paper on "The Gambler's Ruin".[27] This concludes by reminding the reader of "the disagreeable effect on most of humanity of anything which refers, even in the slightest degree, to mathematical reasoning or calculation." The preceding books were all published by the Clarendon Press in Oxford, as would be his later historical books (see §19). These later books reflect an interest that began showing itself in the 1920s, when he wrote thoughtful accounts of the history of mathematics at Harvard such as [JLC] and [Mor, Ch. XV] which have helped me greatly in preparing this paper.

A vivid lecturer himself, Coolidge always viewed research and scholarly publication as the last of four major responsibilities of a university faculty member. In his words [JLC, p. 355], these responsibilities were:

1. To inject the elements of mathematical knowledge into a large number of frequently ill informed pupils, the numbers running up to 500 each year. Mathematical knowledge for these people has come to mean more and more the calculus.

2. To provide a large body of instruction in the standard topics for a College degree in mathematics. In practice this is the one of the four which it is hardest to maintain.

3. To prepare a number of really advanced students to take the doctor's degree, and become university teachers and productive scholars. The number of these men slowly increased [at Harvard] from one in two or three years, to three or four a year.

4. To contribute fruitfully to mathematical science by individual research.

Coolidge's sprightly wit and his leadership as an educator led to his election as president of the Mathematical Association (MAA) of America in the mid-1920s, during which he also headed a successful fund drive of the American Mathematical Society [Arc, pp. 30–32].

An important Harvard contemporary of Coolidge was Edward Vermilye Huntington (1874–1952). After completing graduate studies on the *foundations* of mathematics in Germany, he began a long career of down-to-earth teaching, at first under the auspices of the Lawrence Scientific School. Concurrently, he quickly established a national reputation for clear thinking by definitive research papers on postulate systems for groups, fields, and Boolean algebra. These are classics, as is his lucid monograph on *The Continuum and Other Types of Serial Order* (Harvard University Press 1906; 2d ed., 1917).

From 1907–8 on, he gave biennially a course (Mathematics 27) on "Fundamental Concepts of Mathematics", cross-listed by the philosophy deptartment (see the end of §8), which introduced students to abstract mathematics. He also became coauthor in 1911 (with Dickson, Veblen, Bliss, and others) of the thought-provoking survey *Fundamental Concepts of Modern Mathematics* (J. W. Young, ed.); 2d ed. 1916. This survey still introduced mathematics concentrators to 20th century axiomatic mathematics when I began teaching, 25 years later. It is interesting to compare this book with Bôcher's address on "The Fundamental Conceptions and Methods of Mathematics" (*Bull. Amer. Math. Soc.* **11** (1904), 11–35), and with §26 of his *Introduction to Higher Algebra*.

In the 1920s, Huntington broadened his interests. Four years after making "mathematics and statistics" the subject of his retiring presidential address to the MAA (*Amer. Math. Monthly* **26** (1919), 421–35), he began teaching *statistics* in Harvard's Faculty of Arts and Sciences. Offered initially in 1923 as a replacement to a course on *interpolation and approximation* given earlier (primarily for actuaries) by Bôcher and L. R. Ford, it was given biennially from 1928 on as a companion to the course on probability for which Coolidge wrote his book.

Finally, as a related sideline, he invented in 1921 a method of proportions for calculating how many representatives in the U.S. Congress each state is entitled to, on the basis of its population.[28] This method successfully avoids the "Alabama paradox" and the "population paradox" that had flawed the methods previously in use. Adopted by Congress in 1943, it has been used successfully by our government ever since.

8. PASSING ON THE TORCH

As I tried to explain in §5, the mathematics courses above freshman level offered at Harvard in the 1870s and 1880s could be classified into two main groups: (i) courses on the calculus and its applications in the tradition of Benjamin Peirce's texts (including his *Analytical Mechanics*), designed to make books on classical mathematical physics (Poisson, Fourier, Maxwell) readable, and (ii) courses on topics in algebra and geometry related to the later research of Benjamin and C. S. Peirce. Broadly speaking, Byerly and B. O. Peirce revitalized the courses in the first group with their new Mathematics 10, while J. M. Peirce made comprehensible those of the second. It was primarily J. M. Peirce's courses that Coolidge and Huntington replaced, giving them new content and new emphases.

The first major change in the mathematics courses at Harvard initiated by Bôcher and Osgood concerned Mathematics 13 and its new sequels, and these changes bear a clear imprint of the ideas of Riemann, Weierstrass, and Felix Klein, who had "passed the torch" to his enthusiastic young American students. We have already discussed this change in §6.

The emphasis on "the theory of invariants" in Bôcher's revitalized Mathematics 3 and his *Introduction to Higher Algebra* (cf. §6) also reflects Felix Klein's influence, while the emphasis on "elementary divisors" clearly stems from Weierstrass. It is much harder to trace the evolution of ideas about the *foundations* of mathematics. In §11 of his article in the *Ann. of Math.* **6** (1905), 151–89, Huntington clearly anticipated the modern concepts of *relational structure* and *algebraic structure*, as defined by Bourbaki, far more clearly than Bôcher had in his 1904 article on "The Fundamental Conceptions and Methods of Mathematics", and probably influenced §26 of Bôcher's *Introduction to Higher Algebra*. However, it would be hard to establish clearly the influence of this pioneer work. Indeed, although supremely important for human culture, the evolution of basic ideas is nearly impossible to trace reliably, because each new recipient of an idea tends to modify it before "passing it on".

C. S. Peirce, conclusion. This principle is illustrated by the evolution of two major ideas of C. S. Peirce: his philosophical concept of "pragmatism", and his ideas about the algebra of logic. Both of these ideas were transmitted at Harvard primarily through members of its philosophy department, as we shall see.

The idea of *pragmatism* was apparently first suggested in a brilliant philosophical lecture given by C. S. Peirce at Chauncey Wright's Metaphysical Club in the 1870s. In this lecture, Peirce claimed that the human mind created *ideas* in order to consider the effects of pursuing different courses of action. This lecture deeply impressed William James (1842–1910), whose 1895 *Principles of Psychology* was a major landmark in that subject [EB 12, 1863–5]. During our Civil War, James had studied anatomy at the Lawrence Scientific School and Harvard Medical School, inspired by Jeffries Wyman and Louis Agassiz. After spending the years 1872–76 as an instructor in physiology at

Harvard College, and twenty more years in preparing his famous book, James
turned to philosophy and religion.

In 1906, James finally applied Peirce's idea to a broad range of philosophical problems in his Lowell Lectures on "*Pragmatism...*", published in
book form. In turn, James' lectures and writings on psychology and "pragmatism" strongly influenced John Dewey (1859–1952), whose philosophy dominated the teaching of elementary mathematics in our country during the
first half of this century [EB 7, 346–7]. It is significant that the last three
chapters of Bertrand Russell's *History of Western Philosophy* are devoted to
William James, John Dewey, and the "philosophy of logical analysis" underlying mathematics, as Russell saw it.

Peirce's concern with *logic* overlapped that of Huntington with postulate
theory. Actually, C. S. Peirce was a visiting lecturer in philosophy at Harvard
and a Lowell lecturer on logic in Boston in 1903, and Huntington's article
on the "algebra of logic" in the *Trans. Amer. Math. Soc.* **5** (1904), 288–309,
contains a deferential reference to Peirce's 1880 article on the same subject,
and a letter from Peirce which totally misrepresents the facts, and shows how
far he had slipped since 1881. The facts are as follows.

Never analyzed critically at Harvard, Peirce's pioneer papers on the algebra of relations and his 1881 article basing Boolean algebra on the concept
of partial order inspired the German logician Ernst Schröder. First in his
Operationskreis des Logikkalkuls, and then in his three volume *Algebra der
Logik* (1890–95), Schröder made a systematic study of Peirce's papers. In
turn, these books stimulated Richard Dedekind to investigate the concept of
a "Dualgruppe" (*lattice*; see §16), in two pioneer papers which were ignored
at the time.

Although Huntington did impart to Harvard students many of the other
fundamental concepts of Dedekind, Cantor, Peano and Hilbert, transmitting
them in his course Mathematics 27 and to readers of the books cited in §7,
he paid little attention, if any, to this work of Schröder and Dedekind.

Indeed, it was primarily through Josiah Royce that the ideas of C. S. Peirce
had any influence at Harvard. Royce, whose interests were many-sided, made
logic the central theme of his courses. In turn, he influenced H. M. Sheffer
(A.B. '05) and C. I. Lewis (A.B. '06), two distinguished logicians who wrote
Ph.D. theses with Royce and later became members of the Harvard philosophy department (see §12).

Royce also influenced Norbert Wiener, who wrote a Ph.D. thesis comparing
Schröder's algebra of relations with that of Whitehead and Russell at Harvard
in 1913, and later became one of America's most famous mathematicians.
Indeed, an examination of the first 332 pages of Wiener's *Collected Works*

(MIT Press, 1976) shows that until 1920 he felt primarily affiliated with Harvard's philosophy department.

9. FROM ELIOT TO LOWELL

As the preceding discussion indicates, great advances were made at Harvard in mathematical teaching and research during Eliot's tenure as president (1869–1909). However, besides many ambitious mathematical courses, Harvard also offered in 1900 a number of very popular 'gut' courses. After 30 years of President Eliot's unstructured "free elective" system, it became possible to get an A.B. from Harvard in three years with relatively little effort. Moreover, whereas athletic excellence was greatly admired by students, scholastic excellence was not. Someone who worked hard at his studies might be called a "greasy grind", and a social cleavage had developed between "the men who studied and those who played".[29]

Abbott Lawrence Lowell, who himself became the world's leading authority on British government without attending graduate school,[30] had in 1887 drawn attention "to the importance of making the undergraduate work out ... a rational system of choosing his electives ... [with] the benefit of the experience of the faculty" [Low, p. 11]. Fifteen years later, he spearheaded in 1901–2 a faculty committee whose purpose was to reinstate intellectual achievement as the main objective of undergraduate education ([Yeo, Ch. V], [Mor, xlv–xlvi]). After six more years of continuing faculty discussions in which Osgood and Bôcher were both active [Yeo, pp. 77–78], and many votes, Eliot appointed in 1908 a committee selected by Lowell "to consider how the tests for rank and scholarly distinction in Harvard College can be made a more generally recognized measure of intellectual power" [Yeo, p. 80]. In 1909 Lowell succeeded Eliot as president at the age of 52.

In his inaugural address [Mor, pp. lxxix–lxxxviii], Lowell outlined his plan of concentration and distribution, stating that a college graduate should "know a little of everything and something well" [Low, p. 40]. Having in mind the examples of Oxford and Cambridge Universities, he also proposed creating residential halls (at first for freshmen) to foster social integration. I shall discuss the fruition of these and other educational reforms of Lowell's in §12 below. His ideas have been expressed very clearly by himself and by Henry Yeomans,[31] his colleague in the government department and frequent companion in later life. For the moment, I shall describe only some major changes in undergraduate mathematics at Harvard which he encouraged, that took place during the years 1906–29.

Calculus instruction. During its lifetime (1847–1906), the Lawrence Scientific School had shared in the teaching of elementary mathematics at Harvard. In 1910, during its transition into a graduate school of engineering (completed in 1919), this responsibility was turned over to the mathematics department, doubling the latter's elementary teaching load. At the time, "nine-tenths of all living [Harvard] graduates who took an interest in mathematics at college got their inspiration from Mathematics C," which then covered only analytic geometry through the conic sections.

This seemed deplorable to Lowell, who knew that the calculus, its extensions to differential equations, differential geometry, and function theory, and its applications to celestial mechanics, physics, and engineering, had dominated the development of mathematics ever since 1675. Aware of this domination, he sometimes identified the phonetic alphabet, the Hindu–Arabic decimal notation for numbers, symbolic algebra, and the calculus, as the four most impressive inventions of the human mind.

Lowell soon persuaded the faculty to require each undergraduate to take for "distribution" at least one course in mathematics or philosophy, presumably to develop power in abstract thinking. Through the visiting committee of the Harvard mathematics department (see below), he also encouraged devoting substantial time in Mathematics C to the calculus. Within a decade, "half of the Freshman course was devoted to the subject [of the calculus], and in 1922 the Faculty of Arts and Sciences, through the President's deciding vote, passed a motion that no mathematics course where the calculus was not taught would be counted for distribution" [Mor, p. 255]. This change was followed by steadily increasing emphasis (at Harvard) on the calculus and its applications, until "In 1925–26, 327 young men, just out of secondary school, were receiving a half-year of instruction in the differential calculus" [Mor, p. 255].

Visiting Committees. Since 1890, the activities of each Harvard department have been reviewed by a benevolent visiting committee, which reports triennially to the board of overseers. Beginning in 1906, Lowell's brother-in-law William Lowell Putnam played a leading role on the visiting committee of the mathematics department, and in 1912, Lowell invited George Emlen Roosevelt, a first cousin of Franklin Delano Roosevelt, to join it as well.[32] Both men had been outstanding mathematics students, and their 1913 report with George Leverett and Philip Stockton contained "the important suggestion that the bulk of freshmen be taught in small sections" [Mor, p. 254].

This new plan allowed an increasing number of able graduate students in mathematics to be self-supporting by teaching elementary courses (based on Osgood's texts). For example, during the years 1927–40, S. S. Cairns, G. A. Hedlund, G. Baley Price, C. B. Morrey, T. F. Cope, J. S. Frame, D. C. Lewis, Sumner Myers, J. H. Curtiss, Walter Leighton, Arthur Sard, John W. Calkin, Ralph Boas, Herbert Robbins, R. F. Clippinger, Lynn Loomis, Philip Whitman, and Maurice Heins served in this role. At the same time, a few outstanding new Ph.D.'s were invited to participate in Harvard's research environment by becoming Benjamin Peirce instructors. Among these, one may mention John Gergen, W. Seidel, Magnus Hestenes, Saunders Mac Lane, Holbrook MacNeille, Everett Pitcher, Israel Halperin, John Green, Leon Alaoglu, and W. J. Pettis in the decade preceding World War II.

Besides giving benign and wise advice, the visiting committees of the mathematics department established and financed for many decades a departmental library, where for at least seventy years the bulk of reading in advanced mathematics has taken place. Among the many grateful users of this library should be recorded George Yale Sosnow. More than 60 years after studying mathematics in it around 1920, he left $300,000 in his will to endow its expansion and permanent maintenance.

10. George David Birkhoff

A major influence on mathematics at Harvard from 1912 until his death was my father, George David Birkhoff (1884–1944). His personality and mathematical work have been masterfully analyzed by Marston Morse in [GDB, vol. I, xxiii–lvii], reprinted from *Bull. Amer. Math. Soc.* **52** (1946), 357–91. Moreover I have already sketched some more personal aspects of his career in [LAM, §7 and §§14–15]. Therefore, I will concentrate here on his roles at Harvard.

When my father entered Harvard as a junior in 1903, he had already been thinking creatively about geometry and number theory for nearly a decade. According to his friend, H. S. Vandiver [Van, p. 272] "he rediscovered the lunes of Hippocrates when he was ten years old". In this connection, I still recall him showing my sister and me how to draw them with a compass (see Fig. 1) when I was about nine, joining the tips of these lunes with a regular hexagon, and mentioning that with ingenuity, one could construct regular pentagons by analogous methods. By age 15, he had solved the problem (proposed in the *Amer. Math. Monthly*) of proving that any triangle with two equal angle bisectors is isosceles.

Before entering Harvard, he had proved (with Vandiver) that every integer $a^n - b^n$ $(n > 2)$ except $63 = 2^6 - 1^6$ has a prime divisor p which does not divide $a^k - b^k$ for any proper divisor k of n. He had also reduced the question of the existence of solutions of $x^m y^n + y^m z^n + z^m x^n = 0$ (m, n not both even) to the Fermat problem of finding nontrivial solutions of $u^t + v^t + w^t = 0$, where $t = m^2 - mn + n^2$. Indeed, he had already begun his career as a research mathematician when he entered the University of Chicago in 1902. There he soon began a lifelong friendship with Oswald Veblen, a graduate student who had received an A.B. from Harvard (his second) two years earlier.[33]

I have outlined in [LAM, §7] some high points of my father's career during the final "formative years" in Cambridge, Chicago, Madison, and Princeton that preceded his return to Cambridge. He himself has described with feeling, in [GDB, vol. III, pp. 274–5], his intellectual debt to E. H. Moore, Bolza,

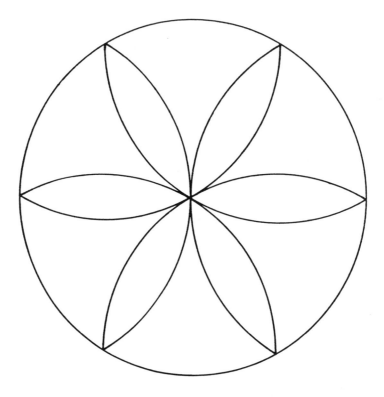

FIGURE 1

and Bôcher, thanking Bôcher "for his suggestions, for his remarkable critical insight, and his unfailing interest in the often crude mathematical ideas which I presented". It was presumably under the stimulus of Bôcher (and perhaps Osgood) that he wrote his first substantial paper (*Trans. Amer. Math. Soc.* 7 (1906), 107–36), entitled "General mean value and remainder theorems". The questions raised and partially answered in this are still the subject of active research.[34] Moreover his 1907 Ph.D. thesis, on expansion theorems generalizing Sturm–Liouville series, was also stimulated by Bôcher's ideas about such expansions, at least as much as by those of his thesis adviser, E. H. Moore, about integral equations.

Return to Harvard.   As Veblen has written [GDB, p. xvii], my father's return in 1912 as a faculty member to Harvard, "the most stable academic environment then available in this country," marked "the end of the formative period of his career". He had just become internationally famous for his proof of Poincaré's last geometric theorem. Moreover Bôcher had devoted much of his invited address that summer at the International Mathematical

Congress in Cambridge, England, to explaining the importance and depth of my father's work on boundary value problems for ordinary differential equations. Equally remarkable, my father had been chosen to review for the *Bull. Amer. Math. Soc.* (**17**, pp. 14–28) the "New Haven Colloquium Lectures" given by his official thesis supervisor, E. H. Moore, and Moore's distinguished Chicago colleagues E. J. Wilczynski and Max Mason.

It is therefore not surprising that, in his first year as a Harvard assistant professor, he and Osgood led a seminar in analysis for research students, or that he remained one of the two leaders of this seminar until 1921. By that time, it "centered around those branches of analysis which are related to mathematical physics". This statement reflected interest in the theory of relativity (see §11). It may seem more surprising that the reports of the visiting committee of 1912 and 1913 took no note of this unique addition to Harvard's faculty, until one remembers that their main concern was with the mathematical *education* of typical undergraduates!

1912 as a milestone. By coincidence, 1912 also bisects the time interval from 1836 to 1988, and so is a half-way mark in this narrative. It can also be viewed as a milestone marking the transition from primary emphasis on mathematical *education* at Harvard to primary emphasis on *research*. Since Byerly retired and B. O. Peirce died in 1913, it also marks the end of Benjamin Peirce's influence on mathematics at Harvard. Finally, since I was one year old at the time, it serves as a convenient reminder that all the changes that I will recall took place during two human life spans.

During the next two decades, G. D. Birkhoff would supervise the Ph.D. theses of a remarkable series of graduate students. These included Joseph Slepian (inventor of the magnetron), Marston Morse, H. J. Ettlinger, J. L. Walsh, R. E. Langer, Carl Garabedian (father of Paul), D. V. Widder, H. W. Brinkmann, Bernard Koopman, Marshall Stone, C. B. Morrey, D. C. Lewis, G. Baley Price, and Hassler Whitney. Four of them (Morse, Walsh, Stone, and Morrey) would become AMS presidents.

In retrospect, my father's role in bringing *topology* to Harvard (as Veblen did to Princeton), at a time just after L. E. J. Brouwer had proved some of its most basic theorems rigorously, seems to me especially remarkable. So does his early introduction to Harvard of *functional analysis*, through his 1922 paper with O. D. Kellogg on "Invariant points in function space", his probable influence on Stone and Koopman, and his "pointwise ergodic theorem" of 1931. But deepest was probably his creative research on the *dynamical systems* of celestial mechanics. It was to present this research that he was made AMS colloquium lecturer in 1920, and to honor it that he was awarded the first Bôcher prize in 1922.

It is interesting to consider my father's related work on celestial mechanics as a continuation of the tradition of Bowditch and Benjamin Peirce, which

George David Birkhoff

was carried on by Hill and Newcomb, and after them by E. W. Brown at Yale. Brown became a president of the AMS, and my father was happy to teach his course on celestial mechanics one year in the early 1920s, and to coauthor with him, Henry Norris Russell, and A. O. Lorchner a Natural Research Council Bulletin (#4) on "Celestial Mechanics". This document provides a very readable account of the status of the theory from an astronomical standpoint as of 1922, including the impact of Henri Poincaré's *Méthodes nouvelles de la Mécanique céleste*.[35]

Although my father's lectures were not always perfectly organized or models of clarity, his contagious enthusiasm for new mathematical ideas stimulated students at all levels to enjoy thinking mathematically. He also enjoyed considering all kinds of situations and phenomena from a mathematical standpoint, an aspect of his scientific personality that I shall take up next.

11. MATHEMATICAL PHYSICS

Among research mathematicians, my father will be longest remembered for his contributions to the theory of dynamical systems (including his ergodic theorem), and his work on linear ordinary differential and difference equations. These were admirably reviewed by Marston Morse in [GDB, I, pp. xv–xlix; *Bull. Amer. Math. Soc.* **52**, 357–83], and it would make little sense for me to discuss them further here. At Harvard, however, there were very few who could appreciate these deep researches, and so from 1920 on, my father's ideas about mathematical physics and the *philosophy* of science aroused much more interest. These were also the themes of his invited addresses at plenary sessions of the International Mathematical Congresses of 1928 and 1936, and of most of his public lectures. Accordingly, I shall concentrate below on these aspects of his work (cf. Parts V and VI of Morse's review).

Relativity. Of all my father's "outside" interests, the most durable concerned Einstein's special and general theories of *relativity*. Unfortunately, it is also this interest that has been least reliably analyzed. Thus Morse's review suggests that it began in 1922, whereas in fact his 1911 review of Poincaré's Göttingen lectures concludes with a discussion of "the new mechanics" of Einstein's special theory of relativity (cf. [GDB, III, pp. 193–4] and *Bull. Amer. Math. Soc.* **17**, pp. 193–4). Moreover, he had touched on these theories and discussed "The significance of dynamics for general scientific thought" at length in his 1920 colloquium lectures,[36] before initiating in 1921–22 an "intermediate level" course on "space, time, and relativity" (Mathematics 16) having second-year calculus as its only prerequisite. He promptly wrote (with the cooperation of Rudolph Langer) a text for this course, entitled *Relativity and Modern Physics* (Harvard University Press,

1923, 1927). In 1922, he also gave a series of public Lowell lectures on relativity. Two years later, he gave a similar series at U.C.L.A. (then called "the Southern Branch of the University of California"), and edited them into a book entitled *The Origin, Nature, and Influence of Relativity* (Macmillan, 1925). It was not until 1927 that he finally published in book form his deep AMS colloquium lectures, in a book *Dynamical Systems*, which omitted many of these topics which he had presented orally seven years earlier.

Bridgman, Kemble, van Vleck. My father's interest in relativity and the philosophy of science was shared by his friend and contemporary Percy W. Bridgman (1882–1961). (Bridgman's notes of 1903–4 on B. O. Peirce's Mathematics 10 are still in the Harvard archives, and it seems likely that my father attended the same lectures.) Bridgman would get the Nobel prize 25 years later for his ingenious experiments on the "physics of high pressure", his own research specialty, but in the 1920s he amused himself by writing *the* classic book on *Dimensional Analysis* (1922, 1931), by giving a half-course on "electron theory and relativity", and writing a thought-provoking book on *The Logic of Modern Physics* (1927). The central philosophical idea of this book, that concepts should be examined *operationally*, in terms of how they relate to actual experiments, is reminiscent of the *pragmatism* of William James and C. S. Peirce.

In 1916, Bridgman had supervised a doctoral thesis on "Infra-red absorption spectra" by Edwin C. Kemble which (as was required by the physics department at that time) included a report on *experiments* made to confirm its theoretical conclusions. Five years later, Kemble supervised the thesis of John H. van Vleck (1899–1980), grandson of Benjamin Peirce's student John M. van Vleck and son of the twelfth AMS president E. B. van Vleck. This thesis, entitled "A critical study of possible models of the Helium atom", is a case study of the unsatisfactory state of quantum mechanics at that time.

Quantum mechanics. However, in 1926, Schrödinger's equations finally provided satisfactory mathematical foundations for *non*relativistic mechanics, shifting the main focus of mathematical physics from relativity to atomic physics. In that same year, my father began trying to correlate his relativistic concept of an elastic "perfect fluid", having a "disturbance velocity equal to that of light at all densities" [GDB, II, 737–63 and 876–86], with the spectrum of monatomic hydrogen, usually derived from Schrödinger's non-relativistic wave equation. Although this work was awarded an AAAS prize in 1927, of greater permanent value was probably his later use of the theory of *asymptotic series* to reinterpret the *WKB*-approximations of quantum mechanics, which yield classical particle mechanics in the limiting case of very short wave length (ibid., pp. 837–56). Related ideas about quantum mechanics also constituted the theme of his address at the 1936 International Congress in Oslo [GDB II, 857–75].

In the meantime, Kemble had taught me most of what I know about quantum mechanics. Far more important, he had just about completed his 1937 book, *The Fundamental Principles of Quantum Mechanics*. The preface of this book mentions his "distress" at "the tendency to gloss over the numerous mathematical uncertainties and pitfalls which abound in the subject", and his own "consistent emphasis on the operational point of view".

Like Kemble, van Vleck (Harvard Ph.D., 1922) made *non*-relativistic quantum mechanics his main analytical tool; but unlike Kemble, he attached little importance to its mathematical rigor. Instead, he *applied* it so effectively to models of magnetism that he was awarded a Nobel prize around 1970. As a junior fellow in 1934, I audited his half-course (Mathematics 39) on "group theory and quantum mechanics", and was startled by his use of the convenient assumption that every matrix is similar to a diagonal matrix.* The courses (Mathematics 40) on the "differential equations of wave mechanics" given in alternate years through 1940 by my father, must have had a very different flavor.

12. Philosophy; Mathematical Logic

From his philosophical analysis of the concepts of space and time, my father also gradually developed radical ideas about how high school geometry should be taught. His public lectures on relativity had included (in Chapter II, on "the nature of space and time") a system of eight postulates for *plane geometry*, of which the first two concern measurement. They assert that length and angle are *measurable quantities* (magnitudes, or real numbers), measurable by "ruler and protractor". Whereas Euclid had devoted his *axioms* to properties of such "quantities", my father saw no good reason why high school students should not use them freely.

A decade later, he proposed a reduced system of *four* postulates for plane geometry, including besides these measurement postulates only two: the existence of a unique straight line through any two points, and the proportionality of the lengths of the sides of any two triangles ABC and $A'B'C'$ having equal corresponding interior angles. His presentation to the National Council of Teachers of Mathematics two years earlier had included a fifth postulate: that "All straight angles have the same measure, $180°$."[37] This presentation was coauthored by Ralph Beatley of Harvard's Graduate School of Education, and their ideas expanded into an innovative textbook on *Basic Geometry* (Scott, Foresman, 1940, 1941).

Less innovative analogous texts on high-school physics and chemistry, coauthored by N. Henry Black of Harvard's Education School with Harvey

---

*Of course, every *finite group* of complex matrices is similar to a group of *unitary* matrices.

Davis and James Conant,[38] respectively, had been widely adopted. However, perhaps because it came out just before World War II, the book by G. D. Birkhoff and Beatley never achieved comparable success.

Expanded from 4 (or 5) postulates to 23, and from 293 pages to 578, G. D. Birkhoff's idea of allowing high-school students to assume that *real numbers* express measurements of distance and angles was developed by E. E. Moise and F. L. Downs, Jr. into a commercially successful text *Geometry* (Addison-Wesley, 1964).

Aesthetic Measure. According to Veblen, my father "was already speculating on the possibility of a mathematical theory of music, and indeed of art in general, when he was in Princeton" (in 1909–12). At the core of his speculations was the formula

$$(1) \qquad M = f(O/C), \quad O = \sum O_i, \quad C = \sum C_j,$$

where the O_i are pleasing, suitably weighted elements of *order*, the C_j suitably weighted elements of *complexity*, intended to express the effort required to "take in" the given art object, and M is the resulting *aesthetic measure* (or "value"). Attempts were made to quantify (1) by David Prall at Harvard and others, through psychological measurements; those interested in aesthetics should read G. D. Birkhoff's book *Aesthetic Measure* (Harvard University Press, 1933). See also his papers reprinted in [GDB, III, pp. 288–307, 320–34, 382–536, and 755–838], the first of which constitutes his invited address at the 1928 International Congress in Bologna.

Of my father's last five papers (##199–203 in [GDB, vol. iii, p. 897]), one is concerned with quaternions and refers to Benjamin and C. S. Peirce; a second with axioms for one-dimensional "geometries"; and a third with generalizing Boolean algebra. His enthusiasm for analyzing basic mathematical structures and recognizing their interrelations never flagged.

Like the relativistic theory of gravitation in flat space–time which was his dominant interest in the last years of his life (see §20), these speculative contributions are less highly appreciated by most professional mathematicians today than his technical work on dynamical systems. However, they made him more interesting to the undergraduates in his classes, his tutees, and his colleagues on the Harvard faculty. In particular, they contributed substantially to his popularity as dean of the faculty, and to the high esteem in which he was held by President Lowell and the Putnam family.[39] They must have also influenced his election as president of the American Association for the Advancement of Science.

A. N. Whitehead. My father's ventures into mathematical physics, the foundations of geometry, and mathematical aesthetics were comparable to the ventures into relativity, the foundations of mathematics, and mathematical logic of A. N. Whitehead, who joined Harvard's philosophy department in

1924. The Whiteheads lived two floors above my parents at 984 Memorial Drive, and were very congenial with them.

The situation had changed greatly since 1910, when Josiah Royce was the only Harvard philosopher who found technical mathematics interesting, and (perhaps because of William James) Harvard's courses in *psychology* were given under the auspices of the philosophy department. In the 1920s and 1930s, not only Whitehead, but also C. I. Lewis (author of the *Survey of Symbolic Logic*) and H. M. Sheffer of the philosophy department (cf. §8) were important mathematical logicians. Moreover Huntington's course Mathematics 27 on "Fundamental concepts..." (cf. §7) was cross-listed for credit in philosophy, and there was even a joint field of concentration in mathematics and philosophy.

In the 1920s, mathematical logic was a bridge connecting mathematics and philosophy, making the former seem more human and the latter more substantial. Whitehead and Russell's monumental *Principia Mathematica* was considered in the English-speaking world to have revolutionized the foundations of mathematics, reducing its principles to rules governing the mechanical manipulation of symbols. In particular, its claim to have made axioms "either unnecessary or demonstrable" was widely accepted by both mathematicians and philosophers.[40]

In the following decade, Gödel and Turing would revolutionize ideas about the role and significance of mathematical logic; the Association for Symbolic Logic would be formed; and the subject would gradually become detached from the rest of mathematics, concentrating more and more on its own internal problems. However, the addition of W.V. Quine to the Harvard philosophical faculty, and the presence in Cambridge of Alfred Tarski for several years, continued to stimulate fruitful interchanges of ideas until long after World War II.

13. Postwar Recruitment

The retirement of Byerly in 1913 and the death of B. O. Peirce in 1914, together with the deaths of Bôcher and G. M. Green, and the departure of Dunham Jackson after six years as secretary in 1919,[41] created a serious void in Harvard mathematics. This void was filled slowly, at first (in 1920) by Oliver D. Kellogg (1878–1932), and William C. Graustein (1897–1942), who had earned Ph.D.'s in Germany before the war with Hilbert and Study, respectively. Then came Joseph L. Walsh (1895–1973) in 1921, and (after H. W. Brinkmann in 1925) H. Marston Morse (1892–1977) in 1926. Both Walsh's and Morse's Ph.D. theses had been supervised by my father.[42] Like Osgood, Bôcher, Coolidge, Huntington, and Dunham Jackson, Graustein (A.B. 1910) and Walsh (S.B. 1916) had been Harvard undergraduates.

Kellogg immediately modernized and infused new life into Mathematics 10a ("potential theory"), took on the teaching of Mathematics 4 (mechanics), and joined my father in running the seminar in analysis. The 1921–22 department pamphlet announced that in that seminar, "the topics assigned will centre about those branches of analysis which are related to mathematical physics". This statement was repeated for two more years, during the first of which Kellogg and Einar Hille (as B.P. Instructor) directed the seminar, a fact which confirms my impression, described in §12, that in those years it was relativity theory and not "dynamical systems" that seemed most exciting at Harvard.

Graustein was an extremely clear lecturer and writer. His and Osgood's *Analytic Geometry*, and his texts for Mathematics 3 (*Introduction to Higher Geometry*, 1930) and Mathematics 22 (*Differential Geometry*, 1935), were models of careful exposition. Combined with Coolidge's lively lectures and more informal texts on special topics, they made geometry second only to analysis in popularity at Harvard during the years 1920–36.

Walsh concentrated on analysis. In 1924–5, he expanded Osgood's half-course Mathematics 12 on infinite series, which had remained static for 30 years, into a full course on "functions of a real variable" which included the Lebesgue integral. He also soon invented "Walsh functions",[43] and became an authority on the approximation of complex and harmonic functions. His interest in this area may have been stimulated by Dunham Jackson, who had done distinguished work in approximation theory fifteen years earlier (see *Trans. AMS* **12**). Most striking was Walsh's result that, in any bounded simply connected domain with boundary C, every harmonic function is the limit of a sequence of harmonic polynomials which converges *uniformly* on any closed set interior to C (*Bull. Amer. Math. Soc.* **35** (1929), 499–544).

Brinkmann came from Stanford, where H. F. Blichfeldt had interested him in group representations. A year's post-doctoral stay in Göttingen with Emmy Noether had not converted him to the axiomatic approach. A brilliant and versatile lecturer, his graduate courses were mostly on algebra and number theory, in which he interested J. S. Frame and Joel Brenner, see [Bre]. However, he also gave a course on "mathematical methods of the quantum theory" with Marshall Stone in 1929–30.

Morse applied variational and topological ideas related to those of my father (and of Poincaré before him). Just before he came to Harvard, he had derived the celebrated derived Morse inequalities (*Trans. Amer. Math. Soc.* **27** (1925), 345–96). The main fruit of his Harvard years was his 1934 Colloquium volume, *Calculus of Variations in the Large*. The foreword of this volume describes admirably its connection with earlier ideas and results of Poincaré, Bôcher, my father, and my father's Ph.D. student Ettlinger.[44]

14. MY UNDERGRADUATE MATHEMATICS COURSES

Most of my own undergraduate courses in mathematics were taught by these relatively new members of the Harvard staff, and by two other thesis students of my father: H. W. Brinkmann and Hassler Whitney, who joined the Harvard mathematics staff in the later 1920s (Whitney, Simon Newcomb's grandson, as a graduate student). It may be of interest to record my own youthful impressions of their teaching and writing styles.[45]

In this connection, I should repeat that whereas my description of mathematical developments at Harvard before 1928 has been based largely on reading, hearsay, and reflection, from then on it will be based primarily on my own impressions during fifteen years of slowly increasing maturity.

Shortly after joining my parents in Paris in the summer of 1928, my father ordered me to "learn the calculus" from a second-hand French text which he picked up in a bookstall along the Seine. Later that summer, after explaining to me Fermat's "method of infinite descent", he challenged me to prove that there were no (least) positive integers satisfying $x^4 + y^4 = z^4$. After making substantial progress, I lost heart, and felt ashamed when he showed me how to complete the proof in two or three more steps.

The next fall, I was fortunate in being taught second-year calculus as a freshman by Morse and Whitney. Their lectures made the *theory* of the calculus interesting and intuitively clear; especially fascinating to me was their construction of a twice-differentiable function $U(x, y)$ for which $U_{yx} \neq U_{xy}$. The daily exercises from Osgood's text gave the needed manipulative skill in problem solving. Likewise, the clarity of Osgood and Graustein made it easy and pleasant to learn from their *Analytic Geometry*, in "tutorial" reading (see §15), not only the reduction of conics to canonical form, but also the theory of determinants.

I learned the essentials of analytic mechanics (Mathematics 4) from Kellogg concurrently. In his lectures Kellogg explained how to reduce systems of forces to canonical form, and derived the conservation laws for systems of particles acting on each other by equal and opposite "internal" forces. His presentation of Newton's solution of the two-body problem opened my eyes to the beauty and logic of celestial mechanics, and reinforced my interest in the calculus and the elementary theory of differential equations. In an unsolicited course paper, I also tried my hand at applying conservation laws to deduce the effect of spin on the bouncing of a tennis ball (I had played tennis with Kellogg, for several years a next door neighbor), as a function of its coefficient of (Coulomb) friction and its "coefficient of restitution". I was also delighted to learn the mathematical explanation of the "center of percussion" of a baseball bat.

That spring my father gave me a short informal lecture on the crucial difference between *pointwise* and *uniform* convergence of a sequence of functions, and then challenged me to prove that any uniform limit of a sequence of continuous functions is continuous. After I wrote out a proof (in two or three hours), he seemed satisfied. In any event, he encouraged me to take as a sophomore the graduate course on functions of a complex variable (Mathematics 13) from Walsh, omitting Mathematics 12. This was only 15 months after I had begun learning the calculus.

This was surely my most inspiring course. Walsh had a dramatic way of presenting delicate proofs, lowering his voice more and more as he approached the key point, which he would make in a whisper. Each week we were assigned theorems to prove as homework. As I was to learn decades later, our correctors were J. S. Frame, who became a distinguished mathematician, and Harry Blackmun, now a justice on the U. S. Supreme Court. They did their job most ably, conscientiously checking my homemade proofs, which often differed from those of the rest of the class. What a privilege it was!

Concurrently, I took advanced calculus (Mathematics 5) from Brinkmann, who drilled a large class on triple integration, the beta and gamma functions and many other topics. He made concise and elegant formula derivations into an art form, leaving little room for student initiative. Osgood's *Advanced Calculus* supplemented Brinkmann's lectures admirably, by including an explanation of how to express the antiderivative $\int R(x, \sqrt{Q(x)})dx$ of any rational function of x and the square root of a quadratic function $Q(x)$ in elementary terms, good introductions to the wave and Laplace equations, etc. Through Brinkmann's lectures and Osgood's book, I acquired a deep respect for the power of the calculus, which I have always enjoyed trying to transmit to students.

My junior year, I took half-courses on the calculus of variations (Mathematics 15) from Morse, on differential geometry (Mathematics 22a) from Graustein, and on ordinary differential equations (Mathematics 32) from my father. Morse's imaginative presentation again made me conscious of many subtleties, especially the sufficient conditions required to prove (from considerations of 'fields of extremals') that solutions of the Euler–Lagrange equations are actually maxima or minima. Graustein, on the other hand, explained details of proofs so carefully that there was little left for students to think about by themselves. I preferred my father's lecture style, which included a digression on the three 'crucial effects' of the general theory of relativity, and a challenge to classify qualitatively the solutions of the autonomous DE $\ddot{x} = F(x)$. (He did not suggest using the Poincaré phase plane.)

He mentioned in class the fact (first proved by Picard) that one cannot reduce to quadratures the solution of

$$(14.1) \qquad U'' + p(x)U' + q(x)U = 0.$$

Not knowing anything then about solvable groups or Lie groups, I was skeptical and wasted many hours in trying to find a formula for solving (14.1) by quadratures.

Finally, in my senior year, I took a half-course on potential theory (Mathematics 10a) with Kellogg. There I found the concept of a harmonic function and Green's theorems exciting, but was bored by elaborate formulas for expanding functions in Legendre polynomials. I also took one on "analysis situs" (combinatorial topology) with Morse, which was an unmitigated joy, however, especially because of its classification of bounded surfaces, proofs of the topological invariance of Betti numbers, etc. The reduction of rectangular matrices of 0s and 1s to canonical form under row equivalence was another stimulating experience.

15. Harvard Undergraduate Education: 1928–32

My own undergraduate career was strongly influenced by the philosophy of education developed by President Lowell. Lowell was an "elitist", who believed that excellence was fostered by competition, and best developed through a combination of drill, periodic written examinations, oral discussions with experts, and the writing of original essays of variable length. He also believed that *breadth* should be balanced by *depth*, and that *originality* was a precious gift which could not be taught. His primary educational aim was to foster intellectual and human development through his system of concentration and distribution of courses, tutorial discussions, "general examinations", and senior honors theses. I believe that he wanted Harvard to train public-spirited leaders with clear vision, who could think hard, straight, and deep. This philosophy dominated Harvard undergraduate education, at least for mathematics concentrators, from 1928 to 1942, during which I observed its effects at first hand.

During eleven academic years, 1927–38, I slept in a dormitory, ate most meals with students in dining halls (from 1936 to 1938 as a tutor), usually participated in athletics during the afternoon, and studied in the evening. From 1929 on, my primary aim was to achieve excellence as a mathematician, and I think the Harvard educational environment of those years was ideal for that purpose also. After 1931, I continued to think about mathematics during summers, if somewhat less systematically.

I think I have already said enough about individual mathematics courses at Harvard. All my teachers impressed me as trying hard to communicate to a mix of students a mature view of the subject they were teaching and (equally important) as being themselves deeply interested in it. Of even greater value was the encouragement I got in tutorial to do guided reading and 'creative' thinking about mathematics and a few of its applications. These efforts were tactfully monitored by leading mathematicians, who were surely conscious of my limitations and slowly decreasing immaturity, and communicated their evaluations to my father.

My first tutorial assignment was to learn about (real) linear algebra and solid analytic geometry as a freshman by reading the book by Osgood and Graustein. In Walsh's Mathematics 13, I spent the spring reading period of my sophomore year on a much more advanced topic: figuring out how to reconstruct any doubly periodic function without essential singularities from the array of its poles. In a junior course by G. W. Pierce on "Electric oscillations and electric waves", I wrote a term paper on the refraction and reflection of electromagnetic waves by a plane interface separating two media having different dielectric constants and magnetic permeabilities, and presented my results as one of the speakers at a physics seminar the following fall. I surely learned more from giving my talk than the audience did from hearing it!

From the middle of that year on, my main tutorial efforts were devoted to planning and writing a senior honors thesis, for which endeavor approximately one-fourth of my time was officially left free. My tutorial reading for this began with Hausdorff's *Mengenlehre* (first ed.) and de la Vallée-Poussin's beautiful *Cours d'Analyse Infinitésimale*, from which I learned the foundations of set-theoretic topology and the theory of the Lebesgue integral, respectively. In retrospect, I can see that this reading and my father's oral examination on "uniform convergence" essentially covered the content of Mathematics 12 on "functions of a real variable" (cf. [Tex, p. 16]). My acquaintance with general topology was broadened by reading Fréchet's Thesis (1906), which introduced me to function spaces, and his book *Les Espaces Abstraits*. It was also deepened by reading the fundamental papers of Urysohn, Alexandroff, Niemytski, and Tychonoff (*Math. Ann.*, vols. 92–95). I was fascinated by Caratheodory's paper "on the linear measure of sets" and Hausdorff's fractional-dimensional measure, so brilliantly applied to *fractals* by Benoit Mandelbrot in recent decades. This reading was guided and monitored by Marston Morse; like all faculty members, his official duties included talking with each of his 'tutees' for about an hour every two weeks.

By that time, Lowell's ambition of establishing "houses" at Harvard similar to the Colleges of Oxford and Cambridge had also come to fruition, and I became a member of Lowell House, of which Coolidge was the dedicated "master". Brinkmann was a resident tutor in mathematics, and J. S. Frame a resident graduate student. My mathematical tutorials with Morse were supplemented by occasional casual chats on a variety of subjects with these and other friendly tutors, as well as (naturally) with my father. The Coolidges tried to set a tone of good manners by entertaining suitably clad undergraduates in their tastefully furnished residence.

Hours of study in Lowell House were relieved by lighter moments. One of these involved a humorous letter from President Roosevelt to Coolidge, which ended "··· do you remember your first day's class at Groton? You stood up at the blackboard — announced to the class that a straight line is the shortest distance between two points — and then tried to draw one. All I can say is that I, too, have never been able to draw a straight line. I am sure you shared my joy when Einstein proved there ain't no such thing as a straight line!"

As a senior in Lowell House, I wrote a rambling 80 page thesis centering around what would today be called *multisets* (but which I called "counted point-sets"), such as might arise from a parametrically defined rectifiable curve $x(s)$ allowed to recross itself any number of times. Not taking the hint from the fact that pencilled comments by the official thesis reader ended on page 41, I submitted all 80 pages for publication in the AMS *Transactions*, and was shocked when a kindly letter from J. D. Tamarkin explained why it could not be accepted![46]

In the comfortable and well-stocked Lowell House library, I became acquainted with the difficulty of defining "probability" rigorously. But above all, in the one room departmental library funded by the visiting committee, I discovered Miller, Blichfeldt and Dickson's book on finite groups, and soon became fascinated by the problem of determining *all* groups of given finite order. There I also saw Klein's *Enzyklopädie der Mathematischen Wissenschaften* with its awe-inspiring multivolume review of mathematics as a whole. After finishing my honors thesis, which touched on fractional-dimensional measure, I decided to see what was known about it. To my horror, I found everything I knew compressed into two pages, in which a large fraction of the space was devoted to references! Although profoundly impressed, I decided not to allow myself to be overawed.

Among nearly contemporary Harvard undergraduates, I suspect that Joseph Doob, Arthur Sard, Joel Brenner, Angus Taylor, and Herbert Robbins were profiting similarly from their Harvard undergraduate education; human minds are at their most receptive during the years from 17 to 22. Although expert professorial guidance is doubtless most beneficial when given to students planning an academic career, and conditions today are very different from those of the 1930s, I think it would be hard to improve on my mathematical education![47] It prepared me well for a year as a research student at Cambridge University (see §16), after which I was ready to carry on three years of free research in Harvard's Society of Fellows (see §17).

As a junior fellow, I ate regularly in Lowell House for three more years with undergraduates, a handful of resident Law School students, and tutors. I then participated actively for two more years in dormitory life, as faculty instructor and senior tutor of Lowell House, trying to live up to the ideals of intellectual communication from which I had myself benefited so much.

In retrospect, although I had pleasant human relations with my prewar undergraduate tutees, I fear I gave some of them an overdose of mathematical ideology. They decided (no doubt rightly) that mathematics as I presented it was simply not their 'dish of tea'! As senior tutor, I was more popular for being otherwise a normal and gregarious human being, and top man on the Lowell House squash team (#1), than for being inspiring mathematically.

Thesis topics. To be a good mathematics thesis adviser at any level, one should be acquainted with a variety of interesting possible thesis topics, and the mathematical thinking processes of a variety of students. At Harvard, a substantial fraction of theses in the 1930s dealt with such simple topics as relating the vibrating string and Fourier series to musical scales and harmony; there was (and is) a Wister Prize for excellence in "mathematics and music". My tutee Russell ("Rusty") Greenhood, later a financial officer at the Massachusetts General Hospital, got a prize for his thesis on "The χ^2 test and goodness of fit", a statistical topic about which I knew nothing. He may have discussed his thesis with Huntington, but all students were encouraged to work independently. Generally speaking, prospective research mathematicians chose advanced thesis topics in very *pure* mathematics. Thus Harry Pollard (A.B. '40, Ph.D. '42) wrote an impressive undergraduate thesis on the Riemann zeta function, which may be found in the Harvard archives.

16. Harvard, Yale, and Oxbridge

Harvard and Yale have often been considered as American (New England?) counterparts of Cambridge and Oxford universities in "old" England. Actually, it was John Harvard of Emmanuel College, Cambridge, who gave to Harvard its first endowment. More relevant to this account, the House Plan at Harvard and the College Plan at Yale (both endowed by Yale's Edward Harkness) were modelled on the educational traditions that had (in 1930) been evolving at "Oxbridge" for centuries. Moreover, since the time of Newton, Cambridge had been one of the world's greatest centers of mathematics and physics, and I formed as a senior the ambition of becoming a graduate student there.

Fortunately for me, Lady Julia Henry endowed in 1932 four choice fellowships, one to be awarded by each of these four universities, to support a year's study across the ocean. President Lowell in person interviewed the candidates applying at Harvard, of whom I was one. He asked me two questions: (i) was I more interested in theoretical or applied mathematics? (ii) since most candidates seemed to want to go to Cambridge, would I accept a fellowship at Oxford? Thinking that being a theorist sounded more distinguished than being a problem-solver, I replied that my interests were theoretical. Moreover I knew of no famous mathematicians or physicists at Oxford, and stated that I would try to find another means of getting to Cambridge.

As a research student interested in quantum mechanics, I attended Dirac's lectures and was given R. H. Fowler as adviser during my first term. Like Widder two years later [CMA, p. 82], but far less mature, I also attended the brilliant lectures given by Hardy in each of three terms, and sampled several other lecture courses. When I first met Hardy, he asked me how my father was progressing with his theory of esthetics. I told him with pride that my father's book *Aesthetic Measure* had just appeared. His only comment was: "Good! Now he can get back to *real* mathematics". I was shocked by his lack of appreciation!

The Julia Henry Fellow from Yale was the mathematician Marshall Hall, who has since done outstanding work in combinatorial theory. We compared impressions concerning the system of Tripos Examinations used at Cambridge to rank students, for which Cambridge students were prepared by their tutors. We agreed that Cambridge students were better trained than we, but thought that the paces they were put through took much of the bloom off their originality!

My course with E. C. Kemble at Harvard had left me with the mistaken impression that quantum mechanics was concerned with solving the Schrödinger equation in a physical universe containing only atomic nuclei and electrons. Dirac's lectures were much more speculative, and it was not until I heard Carl Anderson lecture on the newly discovered *positron* in the spring of 1933 that I realized that Dirac's lectures were concerned with a much broader concept of quantum mechanics than that postulated by Schrödinger's equations.

In the meantime, I had decided to concentrate on finite *group theory*, and was transferred to Philip Hall as adviser. By that spring, I had rediscovered lattices (the "Dualgruppen" of Dedekind; see §8), which had also been independently rediscovered a few years earlier by Fritz Klein who called them "Verbände". Recognizing their widespread occurrence in "modern algebra" and point-set topology, I wrote a paper giving "a number of interesting applications" of what I called "lattice theory", and wrote my father about them. He mentioned my results to Oystein Ore at Yale, who had taught algebra to both Marshall Hall and Saunders Mac Lane. Ore immediately recalled Dedekind's prior work, and soon a major renaissance of the subject was under way. This has been ably described by H. Mehrtens in his book, *Die Entstehung der Verbände*, cf. also [GB, Part I].

In retrospect, I think that I was very lucky that Emmy Noether, Artin, and other leading German algebraists had not taken up Dedekind's "Dualgruppe" concept before 1932. As it was, by 1934 Ore had rediscovered the idea of C. S. Peirce (see §8), of defining lattices as *partially ordered sets*, and by 1935 he had done a far more professional job than I in applying them to determine the *structure of algebras* — and especially that of "groups with operators" (e.g., vector spaces, rings, and modules). However, by that time (in continuing correspondence with Philip Hall) I had applied lattices to projective geometry, Whitney's "matroids", the logic of quantum mechanics (with von Neumann), and set-theoretic topology, as well as to what is now called *universal algebra*, so that my self-confidence was never shattered!

17. THE SOCIETY OF FELLOWS

Our modern Ph.D. degree requirements were originally designed in Germany to *train* young scholars in the art of advancing knowledge. The German emphasis was on discipline, and Ph.D. advisers might well use candidates as assistants to further their own research. Having never "earned" a Ph.D. by serving as a research apprentice himself, Lowell was always skeptical of its value for the very best minds, somewhat as William James once decried "The Ph.D. Octopus". Throughout his academic career, Lowell kept trying to imagine the most stimulating and congenial environment in which a select group of the most able and original recent college graduates could be *free* to develop their own ideas [Yeo, Ch. XXXII]. In his last decade as Harvard's president, he discussed what this environment should be with the physiologist L. J. Henderson and the mathematician-turned-philosopher A. N. Whitehead, among others.

As successful models for such a select group, these very innovative men analyzed the traditions of the prize fellows of Trinity and Kings Colleges at Cambridge University, of All Souls College at Oxford, and of the Fondation Thiers in Paris. They decided that a group of about 24 young men (a natural social unit), appointed for a three year term (with possible reappointment for a second term), dining once a week with mature creative scholars called senior fellows, and lunching together as a group twice a week, would provide a good environment. The only other stated requirement was negative: "not to be a candidate for a degree" while a junior fellow.

The proper attitude of such a junior fellow was defined in the following noble "Hippocratic Oath of the Scholar" [SoF, p. 31], read each year before the first dinner:

> 'You have been selected as a member of this Society for your personal prospect of serious achievement in your chosen field, and your promise of notable contributions to knowledge and thought. That promise you must redeem with your whole intellectual and moral force.
>
> You will practice the virtues, and avoid the snares, of the scholar. You will be courteous to your elders who have explored to the point from which you may advance; and helpful to your juniors who will progress farther by reason of your labors. Your aim will be knowledge and wisdom, not the reflected glamour of fame. You will

not accept credit that is due to another, or harbor jealousy of an explorer who is more fortunate.

You will seek not a near, but a distant, objective, and you will not be satisfied with what you may have done. All that you may achieve or discover you will regard as a fragment of a larger pattern, which from his separate approach every true scholar is striving to descry.

To these things, in joining the Society of Fellows, you dedicate yourself.'

Some months later, we were informed frankly that if one out of every four of us had an outstanding career, the senior fellows would feel that their enterprise had been very successful.

Like all institutions, Harvard's Society of Fellows has changed with the times. Thus junior fellows may now be women, and may use their work to fulfill departmental Ph.D. requirements. But the ceremony of reading the preceding statement to new junior fellows at their first dinner in the Society's rooms has not changed.

As a junior fellow, I was so absorbed in developing my own ideas and in exploring the literature relating to them (especially abstract algebra, set-theoretic topology, and Banach spaces), that I attended only two Harvard courses or seminars. I had studied in 1932–33 Stone's famous *Linear Transformations in Hilbert Space*, one of the three books that established functional analysis (the study of operators on "function spaces") as a major area of mathematics.[48] Moreover, Whitney was rapidly becoming famous as a topologist with highly original ideas. Therefore, I audited Stone's course (Mathematics 12) on the theory of real functions, which he ran as a seminar, in 1933–34, and I participated actively in Whitney's seminar on topological groups in 1935–36.

I also attended the weekly colloquia. At an early one of these, Stone announced his theorem that every Boolean algebra is isomorphic to a field of sets. Having proved the previous spring that every distributive lattice was isomorphic to a ring of sets, I became quite excited. He went on to prove much deeper results in the next few years, while I kept on exploring the mathematical literature for other examples of lattices.

There were five mathematical junior fellows during the years 1933–44: John Oxtoby, Stan Ulam, Lynn Loomis, Creighton Buck, and myself. In addition, the mathematical logician W. V. Quine was (like me) among the first six selected, as was the noted psychologist B. F. Skinner. Like many other junior fellows, the last four of the six just named joined the Harvard faculty, where their influence would be felt for decades. But that is another story!

Ulam and Oxtoby.   Instead, I will take up here the accomplishments of Ulam and Oxtoby through 1944. Most important was their proof that, in the sense of (Baire) category theory, almost every measure-preserving home-omorphism of any "regularly connected" polyhedron of dimension $r \geq 2$ is metrically transitive. As they observed in their paper,[49] "the effect of the ergodic theorem was to replace the ergodic hypothesis (of Ehrenfest) by the hypothesis of metric transitivity (of Birkhoff)". Philosophically, therefore, they in effect showed that Hamiltonian systems should almost surely satisfy the ergodic theorem. This constituted a notable modern extension of the tradition of Lagrange, Laplace, Poincaré, and G. D. Birkhoff.

During World War II, like von Neumann (but full-time), Ulam worked at Los Alamos. There he is credited as having conceived, independently of Edward Teller, the basic idea underlying the H-bomb developed some years later.

Two other junior fellows of the same vintage who applied mathematics to important *physical* problems after leaving Harvard were John Bardeen and James Fisk. After joining the Bell Telephone Laboratories in 1938, Bardeen went on to win *two* Nobel prizes. Fisk became briefly director of research of the Atomic Energy Commission after the war, and finally vice president in charge of research at the Bell Telephone Labs. I hope that these few examples will suggest the wisdom and timeliness of the plan worked out by Lowell, Whitehead, Henderson, and others, and endowed by Lowell's own fortune. Of the first fifty junior fellows, no less than six (Bardeen, Fisk, W. V. Quine, Paul Samuelson, B. F. Skinner, and E. Bright Wilson) have received honorary degrees from Harvard!

The Putnam Competition.   The aim of Lowell and his brother-in-law William Lowell Putnam, to restore undergraduate admiration for intellectual excellence (see §7), was given a permanent national impetus in 1938 with the administration of the first Putnam Competition by the Mathematical Associ-ation of America. For a description of the establishment of this competition, in which George D. Birkhoff played a major role, and its subsequent history to 1965, I refer you to the *Amer. Math. Monthly* **72** (1965), 469–83. Of the five prewar Putnam Fellowship winners, Irving Kaplansky is current director of the NSF funded Mathematical Sciences Research Institute in Berkeley, after a long career as a leading American algebraist; he and Andrew Gleason have recently been presidents of the AMS; while Richard Arens and Harvey Cohn have also had distinguished and productive research careers. All of them except Gleason (who joined the U.S. Navy as a code breaker in 1942) contributed through their teaching to the mathematical vitality of Harvard in the years 1938–44!

18. FOUR NOTABLE MEETINGS

I shall now turn to some impressions of the moods of, and Harvard's participation in, four notable meetings that took place in the late 1930s: the International Topological Congress in Moscow in 1935; the International Mathematical Congress in Oslo and Harvard's Tercentenary in 1936; and the Semicentennial meeting of the AMS in 1938.

Lefschetz was a major organizer of the 1935 Congress in Moscow. He, von Neumann, Alexander and Tucker went to it from Princeton; Hassler Whitney, Marshall Stone, David Widder (informally) and I from Harvard. Whitney's paper [CAM, pp. 97–118] describes the fruitfulness for topology of this Congress, an event which Widder also mentions [CAM, p. 82]. For me, it provided a marvellous opportunity to get first-hand impressions of the thinking of many mathematicians whose work I admired, above all Kolmogoroff, but also Alexandroff and Pontrjagin.

Widder, Stone, and I met in Helsinki, just before the Congress, whence we took a wood-fired train to Leningrad. There we were greeted at the station by L. Kantorovich and an official Cadillac. By protocol, he took a street-car to his home, where he had invited us for tea, while we were driven there in the Cadillac. I was astonished! I would have been even more astonished had I realized that within two years I would be studying the work of Kantorovich on vector lattices (and that of Freudenthal, also at the Congress); that 20 years later I would be admiring his book with V.I. Krylov on *Approximation Methods of Higher Analysis*; or that in about 30 years he would get a Nobel prize for inventing the simplex method of linear programming, discovered independently by George Dantzig in our country somewhat later.[50]

Marshall Stone, infinitely more worldly wise than I, reported privately that evening Kantorovich's disaffection with the Stalin regime. I was astonished for the third time, having assumed that all well-placed Soviet citizens supported their government. Many of my other naive suppositions were corrected in Moscow.

For example, when I expressed to Kolmogoroff my admiration for his *Grundbegriffe der Wahrscheinlichkeitsrechnung*, he remarked that he considered it only an introduction to Khinchine's deeper *Asymptotische Gesetze der Wahrscheinlichkeitsrechnung*. The algebraist Kurosh and I made a limited exchange of opinions in German, and I also met at the Congress I. Gelfand, who would get an honorary degree at Harvard 50 years later!

Above all, I was impressed by the crowding and poverty I saw in Moscow (the famine had just ended a year earlier), and the inaccessibility of government officials behind the Kremlin walls.

At the International Mathematical Congress in Oslo a year later, I was dazzled by the depth and erudition of the invited speakers, and the panorama of fascinating areas of research that their talks opened up. I was permitted to present three short talks (Marcel Riesz gave four!), and there seemed to be an adequate supply of listeners for all the talks presented. Paul Erdös gave one talk, and he must have been the only speaker who did not wear a necktie!

Naturally, I was pleased that the two Fields medallists (Lars Ahlfors and Jesse Douglas) were both from Cambridge, Massachusetts, and delighted that the 1940 International Congress was scheduled to be held at Harvard, with my father as Honorary President! I was also impressed by the efficient organization for the *Zentralblatt* of reviews of mathematical papers displayed by Otto Neugebauer (cf. [LAM, §21]). This convinced me of the desirability of transplanting his reviewing system to AMS auspices, if funds could be found to cover the initial cost. Of course, this was accomplished three years later.

On both my 1935 and 1936 trips to Europe, I stopped off in Hamburg to see Artin in Hamburg. In 1935, I also stopped off in Berlin to meet Erhard Schmidt and my future colleague Richard Brauer and his brother Alfred. Near Hamburg in 1936, the constant drone of military airplanes made me suddenly very conscious of the menace of Hitler's campaign of rearmament!

The serene atmosphere of Harvard's Tercentenary celebration that September was a welcome contrast, and I naturally went to the invited mathematical lectures. Among them, Hardy's famous lecture on Ramanujan was most popular.[51] It did not bother me that the technical content of the others was over my head, and I dare say over the heads of the vast majority of the large audiences present!

The summer meeting of the AMS was held at Harvard in conjunction with this Tercentenary; its description in the *Bull. Amer. Math. Soc.* (**42**, 761–76) states that: "Among the more than one thousand persons attending the meetings ..., approximately eight hundred registered, of whom 443 are members of the Society". What a contrast with the Harvard of John Farrar and Nathaniel Bowditch, a hundred years earlier!

A fourth notable mathematical meeting celebrated the Golden Jubilee of the AMS at Columbia University in September, 1938. It was to celebrate this anniversary that R. C. Archibald wrote the historical review [Arc] on which I have drawn so heavily, here and in [LAM], and that my father surveyed "Fifty years of American mathematics" from his contemporary standpoint.

The meeting honored Thomas Scott Fiske of Columbia, who had by then attended 164 of the 352 AMS meetings that had taken place. (Of these 352 meetings, 221 had been held at Columbia.) A review of the occasion was published in the *Bull. Amer. Math. Soc.* **45** (1939), 1–51, including Fiske's reminiscence that, in the early days of the AMS, C. S. Peirce was "equally brilliant, whether under the influence of liquor or otherwise, and his company was prized ... so he was never dropped ... even though he was unable to pay his dues."

19. ANOTHER DECADE OF TRANSITION

In §12 and §13, I recalled the mathematical activity in physics and philosophy at Harvard through 1940. I shall now give some impressions of the main themes of research and teaching of the Harvard mathematics department from 1930 through 1943.

During these years, it was above all G. D. Birkhoff who acted as a magnet attracting graduate students to Harvard. After getting an honorary degree from Harvard in 1933, he served as dean of the faculty under President Conant from 1934 to 1938, meanwhile being showered with honorary degrees and elected a member of the newly founded Pontifical Academy. He directed the theses of C. B. Morrey, D. C. Lewis, G. Baley Price, Hassler Whitney, and 12 other Harvard Ph.D.'s after 1930. In 1935, he wrote with Magnus Hestenes an important series of papers on natural isoperimetric conditions in the calculus of variations, and throughout the 1930s he wrote highly original sequels to his earlier papers on dynamical systems, the four color theorem, etc., while continuing to lecture to varied audiences also on relativity, his ideas about quantum mechanics, and his philosophy of science.

Meanwhile, Walsh and Widder pursued their special areas of research in classical analysis, Walsh publishing many papers as well as a monograph on "Approximation by polynomials in the complex domain" in the tradition of Montel, Widder his well-known *Laplace Transform*. Variety within classical analysis and its applications was provided at Harvard by Walsh and Widder. For example, Joseph Doob and Lynn Loomis wrote theses with Walsh, while Ralph Boas and Harry Pollard wrote theses with Widder during these years. While Ahlfors was there (from 1935 to 1938), Harvard's national leadership in classical analysis was even more pronounced, being further strengthened by the presence of Wiener in neighboring MIT.

Coolidge, Graustein, and Huntington continued to give well-attended courses on geometry and axiomatic foundations, keeping these subjects very much alive at Harvard. In particular, Coolidge gave a series of Lowell lectures on the history of geometry, while Graustein published occasional papers on differential geometry, and served as editor of the *Transactions Amer. Math. Soc.* from 1936 until his death in 1941. In his role of associate dean,

Graustein also worked out a detailed "Graustein plan" which metered skillfully the tenure positions available in each department of the Faculty of Arts and Sciences, aimed at achieving a roughly uniform age distribution.

Former Presidents of the Society at Harvard University, September 1936
Left to right: White, Fiske, Bliss, Osgood, Coble, Dickson, and Birkhoff

Moreover, every department member performed capably and conscientiously his teaching and tutorial duties, undergraduate honors being "based on the quality of the student's work in his courses, on his thesis, and on the general examination" (the latter a less sophisticated version of the Cambridge Tripos).

New trends.   However, this seeming emphasis on classical mathematics was deceptive. By 1935, Kellogg had died, Osgood had retired, and Morse had gone to the Institute for Advanced Study at Princeton. Their places were taken by Marshall Stone, Hassler Whitney, Saunders Mac Lane, and myself. (I recall that like Walsh and Widder, Stone and Whitney were Ph.D. students of G. D. Birkhoff.) Stone, already famous as a functional analyst, was concentrating on Boolean algebra and its relation to topology. Whitney was founding the theories of differentiable manifolds and sphere bundles [CMA, pp. 109–117]. Mac Lane was exhibiting great versatility and expository skill in papers on algebra and graph theory.

Before 1936, when I became a faculty instructor after attending all the four "notable meetings" described in §17, I had never taught a class. I realized that my survival at Harvard depended on my success in interesting freshmen in the calculus, and was most grateful for the common sense advice given by Ralph Beatley regarding pitfalls to be avoided. "Teach the student, not just the subject", and "face the class, not the backboard" were two of his aphorisms. All new instructors were "visited" by experienced teachers, who reported candidly on what they witnessed at department meetings, usually with humor. I was visited by Coolidge, and became so unnerved that I splintered a pointer while sliding a blackboard down. I survived the test, and became a colleague of Stone, Whitney, and Mac Lane. Thus, after 1938, the four youngest members of the Harvard mathematical faculty were primarily interested in functional analysis, topology, and abstract algebra. In addition, Quine had introduced a new full graduate course in mathematical logic (Mathematics 19). This treated general "deductive systems", thus going far beyond Huntington's half-course on "fundamental concepts".

I am happy to say that Stone (Harvard '22), Whitney, and Mac Lane are still active, while both Widder and Beatley (Harvard '13) are in good health. Stone recently managed the AMS conference honoring von Neumann, while Whitney, Mac Lane, and Widder are fellow contributors to the series of volumes in which this report is being published.

David Widder and I were put in charge of the Harvard Colloquium in the years 1936–40. My father and Norbert Wiener usually sat side by side in the front row, and made lively comments on almost every lecture. C. R. Adams and Tamarkin often drove up from Brown to attend the colloquium, bringing graduate students with them. My role included shopping conscientiously for good cookie bargains for these convivial and sociable affairs, where tea was served by a faculty wife. Most interesting for me were the talks by Ore, von

Neumann, and Menger on lattice theory, then my central research interest. In 1938, these three participated in the first AMS Symposium on lattice theory (see *Bull. Amer. Math. Soc.* **44** (1938), 793–837), with Stone, Stone's thesis student Holbrook MacNeille, who would become the first executive director of the AMS, and myself. Two years later the AMS published the first edition of my book *Lattice Theory*.

Beginning in 1937–38, Mac Lane and I taught alternately a new undergraduate full course on algebra (Mathematics 6), which immediately became very popular. I began the course with sets and ended with groups; in the second year, my students included Loomis, Mackey, and Philip Whitman. The next year, Mac Lane began with groups and ended with sets; his students included Irving Kaplansky. After amicable but sometimes intense discussions, we settled on the sequence of topics presented in our *Survey of Modern Algebra* (Macmillan, 1941). In it and in our course, we systematically correlated rigorous axiomatic foundations with elementary *applications* to number theory, the theory of equations, geometry, and logic.

20. END OF AN ERA

Meanwhile, war clouds were getting more and more threatening! Germany and Russia invaded and absorbed Poland in 1939, and the International Mathematical Congress scheduled to be held at Harvard was postponed indefinitely. After the fall of France in the spring of 1940, Germany's invasion of Russia, and Pearl Harbor, it became clear that our country would have to devote all its strength to winning a war against totalitarian tyranny.

It was clear to me that our war effort was unlikely to be helped by any of the beautiful ideas about "modern" algebra, topology, and functional analysis that had fascinated me since 1932, and so from 1942 until the war ended, I concentrated my research efforts on more relevant topics. Most interesting of these scientifically was trying to predict the underwater trajectories of air-launched torpedoes, a problem on which I worked with Norman Levinson and Lynn Loomis, a study in which my father also took an interest. I believe that our work freed naval research workers in the Bureau of Ordnance to concentrate on more urgent and immediate tasks.

George D. Birkhoff. During these years, my father continued to think about natural philosophy, much as Simon Newcomb and C. S. Peirce had. He lectured on a broad range of topics at the Rice Institute, and also in South America and Mexico, where he and my mother were good will ambassadors cooperating in Nelson Rockefeller's effort to promote hemispheric solidarity against Hitler.

My father finally succeeded in constructing a relativistic model of gravitation which was invariant under the Lorentz group, yet predicted the "three crucial effects" whose explanation had previously required Einstein's *general* theory of relativity. Because it assumed Minkowski's four-dimensional *flat* space-time, the model also accommodated electromagnetic phenomena such as the relativistic motion of particles in electron and proton accelerators.[52]

The exploration of this theory and other ideas he had talked about provided an important stimulus to the development of the National University of Mexico into a significant research center. The honorary degree that I received there in 1955, as well as my honorary membership in the Academy of Sciences in Lima, were in large part tributes to his influence on the two oldest universities in the Western Hemisphere.

The department pamphlet of 1942–43. In spite of the war, the pamphlet of the Harvard mathematics department for 1942–43 gives the illusion of a balance of mathematical activities that had been fairly constant for nearly a decade. Although President Conant had gone to Washington to run the National Defense Research Council with Vannevar Bush, he had left intact the plan of undergraduate education worked out by Lowell.

Perhaps suggestive of future trends, Beatley was in charge of three sections of freshman calculus, Chuck (C. E.) Rickart (then a B.P.) of two; only Whitney's and Mac Lane's sections were taught by tenured research faculty members. Stone, Kaplansky, and I taught second-year calculus; of the three of us, Kaplansky was the most popular teacher. Advanced calculus was taught by Whitney and my father, geometry by Coolidge, and undergraduate algebra (Mathematics 6) by Ed Hewitt. Real and complex analysis (our main introductory graduate courses) were taught by Loomis and Widder, respectively; ordinary differential equations (a full course) by my father; and mechanics by van Vleck. Graustein had died, but differential geometry was taught by Kaplansky; topology was taught by Mac Lane. Widder's student Harry Pollard and I taught Mathematics 10b and 10a, respectively.

Applied mathematics. The only "applied" touch visible in this 1942–43 pamphlet was my changed wording for the description of Mathematics 10a: I announced that it would treat "the computation of [potential] fields in special cases of importance in physics and airfoil theory", and that "In 1942–43, analogous problems for compressible non-viscous flow will also be treated, and emphasis ... put on airfoil theory and air resistance to bullets". Also, two courses in "mechanics" were listed: Mathematics 4 to be taught by van Vleck, and Mathematics 8 by Kemble. Actually, van Vleck and Dean Westergard of the Engineering School had agreed with me that we should teach Mathematics 4 (= Engineering Science 6) in rotation. When my turn came, John Tate (in naval uniform) was in the class.

Moreover, appreciation for "applied" mathematics as such was reviving in the Harvard Engineering School, with whose faculty I was getting acquainted as part of my "continuing education". Though they did not worry about Weierstrassian rigor, let alone Cantorian set theory or symbolic logic, Richard von Mises[53] and my friend Howard Emmons knew infinitely more about *real* flows around airfoils than I. Associated with von Mises were his coauthor Philipp Frank, by then primarily interested in the philosophy of science, and Stefan Bergman of "kernel function" fame, as well as Hilda Geiringer von Mises at Wheaton and Will Prager at Brown. After emigrating together from Berlin to Istanbul to escape Hitler, all of these distinguished mathematicians had come to New England,[54] greatly enhancing its role in *Continuum Mechanics*, including especially the mathematical analysis of fluid motions, elastic vibrations, and plastic deformations.

But most important for the post-war era, the Gordon McKay bequest of 1903, which Nathaniel Shaler had labored so hard to secure for Harvard, was about to become available. In addition, the 1940 bequest of $125,000, given by Professor A. E. Kennelly because "the great subject of mathematics applied to electric engineering, together with its study and teaching, have throughout my life been an inspiration in my work", was being used to pay the salary of Howard Aiken, while he worked at IBM on the development of a *programmable computer*. Harvard was getting ready for the dawn of the computer age!

NOTES

Further Supplementary Notes and references for this essay, identified by letters, will be deposited in the Harvard Archives.

[1] For Peirce's career and influence, see [Pei] and [DAB 14, 393–7].

[2] See John Pickering's *Eulogy of Nathaniel Bowditch*, Little Brown, 1868; [DAB 2, 496–8]; [EB 4, p. 31], and [Bow, vol. 1, pp. 1–165].

[3] See p. 69 of Pickering's *Eulogy*. The accepted value today is $(a - b)/a = 1/297$.

[4] Benjamin Peirce senior also wrote a notable history of Harvard, recording the many benefactions made to it before the American Revolution.

[5] See [Cat, 1835].

[6] See [TCH, p. 220] and [Qui]. Kirkland was succeeded by Josiah Quincy, who would be followed in 1846 by Edward Everett.

[7] For Lovering's scientific biography, by B. O. Peirce, see [NAS 2: 327–44]. He was president of the American Academy from 1880 to 1892.

[8] For William Bond's biography, see [DAB 2, 434–5]. His son George succeeded him as director of the Harvard Observatory. For more information,

see *The Harvard College Observatory: the first four directorships, 1839-1919*, by Bessie Z. Jones and Lyle G. Boyd, Harvard University Press, 1971.

[9]See Simon Newcomb's autobiography, *Reminiscences of an Astronomer* for colorful details about his life, and [DAB 13, 452–5] for a biographical survey. Hill's first substantial paper was published in Runkle's *Mathematical Monthly*. For his later work, see [NAS 8: 275–309], by E.W. Brown, and [DAB 9, 32–3].

[10]See [DAB 7, 447–9] for biographies of Gould (Harvard '44), who founded the *Astronomical Journal*, and his father of the same name.

[11][DAB 14, 393-7]. As superintendent, he received $4000/yr, which must have doubled his salary.

[12]Runkle was MIT President from 1870 to 1878.

[13]For the model used, see Newcomb's *Popular Astronomy*, 5th ed., Part IV, Ch. III. Until nuclear energy was discovered, the source of the sun's energy was a mystery. W. E. Story was Byerly's classmate.

[14]These are associated with systems of linear DE's of the form $dx_i/dt = \Sigma a_{ij}(t)x_j$. Hamilton had discovered quaternions in 1843, while Cayley's famous paper on matrices was published in 1853.

[15]U.S. government employees helped to prepare Peirce's manuscript for lithographing.

[16]Crelle's *J. für Math.* **84** (1878), 1–68.

[17]See [HH, p. 42], Eliot's article on "The New Education" in the *Atlantic Monthly* **23** (1869), expresses Eliot's opinions *before* he became president; his inaugural address is reprinted in [Mor, pp. lix–lxxviii].

[18]See *When MIT was Boston Tech.*, by Samuel C. Prescott, MIT Press, 1954.

[20]Byerly was also active in promoting Radcliffe (Harvard's "Female Annex"), where Byerly Hall is named for him; see [DAB, Suppl., pp. 145–6]. Elizabeth Cary Agassiz was its president. For the Radcliffe story, see [HH, pp. 193–7].

[21]Cf. [S-G, p. 69]. Oliver Wendell Holmes Sr. wittily observed that professorial chairs in "astronomy and mathematics" and "geology and zoology", like those of Louis Agassiz and his classmate Benjamin Peirce, should be called "settees, not chairs".

[22]See §7 (pp. 32-4) of my article in [Tar, pp. 25–78], and pp. 293–5 of my father's article in [AMS, pp. 270–315], reprinted in [GDB, vol. iii, pp. 605–52]. A biography of Osgood by J. L. Walsh will be included in this volume. For "The Scientific Work of Maxime Bôcher", see my father's article in the *Bull. AMS* **25** (1919), 197–215, reprinted in [GDB, vol. iii, pp. 227–45].

[23]For Klein's great influence on American mathematics, see the Index of [Arc]; also [Tar, pp. 30–32], and §10 of my article with M. K. Bennett in Wm. Aspray and Philip Kitcher (eds.), *History and Philosophy of Modern Mathematics*, University of Minnesota Press, 1988.

[24]*Bull. Amer. Math. Soc.* **5** (1898), 59–87, and vol. 7 of the AMS Colloquium Publications (1914).

[25]See Bouton's Obituary in *Bull. Amer. Math. Soc.* **28** (1922), 123–4.

[26]For an appreciative account of Coolidge's career, see the Obituary by D. J. Struik in the *Amer. Math. Monthly* **62** (1955), 669–82. Ref. 60 there to a biography of Graustein by Coolidge seems not to exist.

[27]*Ann. of Math.* **10** (1909), 181–92.

[28]Senate Document # 304 (41 pp.), U.S. Printing Office, 1940. See also EVH in *Quart. Amer. Statist. Assn.* (1921), 859–70, and *Trans. Amer. Math. Soc.* **30** (1928), 85–110.

[29][Yeo, p. 67]. Owen Wister's book *Philosophy Four* gives an amusing description of the "Zeitgeist" at Harvard in those years.

[30]Lowell's *The Government of England* and (his friend) James Bryce's *American Commonwealth* were the leading books on these two important subjects. See [Yeo, p. 111]. Lord Bryce, when British ambassador to the United States, gave Lowell's manuscript a helpful critical reading.

[31]In the two volumes [Low] and [Yeo].

[32]Cf. [LAM, §12]. For many years, the Putnams graciously hosted dinner meetings of the visiting committee, to which all the members of the mathematics department were invited.

[33]See [GDB, pp. xv–xxi] for Veblen's recollections and appraisal of my father's work. The grandson of a Norwegian immigrant, Veblen had graduated at 18 from the University of Iowa before going to Harvard. See [Arc, pp. 206–18], for biographies of Veblen and my father.

[34]See G. G. Lorentz, K. Jetter, and S. D. Riemenschneider, *Birkhoff Interpolation*, Addison-Wesley, 1983.

[35]This classic is currently being republished by the American Physical Society in translated form, prefaced by an excellent historical introduction by Daniel Goroff.

[36]The outline of these (*Bull. Amer. Math. Soc.* **27**, 67–69) includes many topics of general interest that were not included in the printed volume. These include from the first lecture: (7) methods of computation and their validity, (8) relativistic dynamics, and (9) dissipative systems. The last lecture was entitled "The significance of dynamical systems for general scientific theory", and dealt with (1) the dynamical model in physics, (2) modern cosmogony and dynamics, (3) dynamics and biological thought, and (4) dynamics and philosophical speculation. My father's interest in relativity presumably dates from a course he took with A. A. Michelson at Chicago around 1900; see his review "Books on relativity", *Bull. Amer. Math. Soc.* **28** (1922), 213–21.

[37]Cf. [GDB, III, pp. 365–81], reprinted from the *Ann. of Math.* **33** (1932), 329–45, and the *Fifth Yearbook* (1930) of the NCTM.

[38]Harvey Davis, after teaching mathematics (as a graduate student), physics, and engineering [Mor, p. 430] at Harvard, became president of the Stevens Institute of Technology. Conant, of course, was President Lowell's successor at Harvard.

[39]Mrs. William Lowell Putnam lent her summer home to the Birkhoffs during the summer of 1927; see also §17.

[40]See [Whi], in which pp. 125–65 contain an essay by Quine on "Whitehead and the rise of modern logic".

[41]Dunham Jackson had been Secretary of the Division since 1913. Other losses were: the differential geometer Gabriel Marcus Green (cf. *Bull. Amer. Math. Soc.* **26**, pp. 1–13), and Leonard Bouton (in 1921).

[42]Actually, Walsh had asked Osgood to supervise his thesis, but Osgood declined. Like Coolidge and Huntington ('95), Graustein ('10) and Walsh ('16) had both been Harvard undergraduates.

[43]Closely related to Haar functions, these would prove very useful for signal processings in the 1970s.

[44]For a charming description of Morse and his contributions, see Raoul Bott, *Bull. Amer. Math. Soc.* (*N.S.*) **3** (1980), 907–50.

[45]The majority of students, not being interested in a mathematical career, presumably had very different impressions.

[46]Though original, my ideas were not new. Tamarkin kindly softened the blow by writing that my paper "showed promise". Six months later, I published a revised and very condensed paper containing my sharpest results in *Bull. Amer. Math. Soc.* **39** (1933), 601–7.

[47]In Stone's words [Tex, p. 15], "the Harvard of my student days could not have offered more opportunity or encouragement to a student eager for study and learning."

[48]The others were von Neumann's *Mathematische Grundlagen der Quantenmechanik* and Banach's *Théorie des Opérations Linéaires*; cf. *Historia Math.* **11** (1984), 258–321.

[49]*Ann. of Math.* **42** (1941), 874–920. See also Ulam's charming *Adventures of a Mathematician* (Scribners, 1976) for other aspects of his life.

[50]For the story of the independent discoveries of linear programming by Kantorovich (1939), Frank Hitchcock (1941), T. C. Koopmans ($\sim$ 1944), and G. B. Dantzig (1946), see Robert Dorfman, *Ann. Hist. Comput.* **6** (1984), 283–95.

[51]Published in the *Amer. Math. Monthly* **44** (1937), 137–55, and as Ch. I of Hardy's book *Ramanujan* (Cambridge University Press, 1940).

[52]See [GDB, pp. 920–83], and the article by Carlos Graef Fernandez in pp. 167–89 of the AMS Symposium *Orbit Theory* (G. Birkhoff and R. E. Langer, eds.), Amer. Math. Soc., 1959.

[53]In a very different way, von Mises' book *Probability, Statistics and Truth* was a famous contribution to the foundations of probability theory, which are shaky because sequential frequencies are not countably additive.

[54]Minkowski's son-in-law Reinhold Rüdenberg had also come from the University of Berlin to Harvard, while Hans Reissner had come to MIT.

REFERENCES

[Ahl] Lars V. Ahlfors, *Collected Papers*, vol. 1. Birkhäuser, Boston, 1982.

[AMM] *American Mathematical Monthly*, published by the Math. Assn. of America.

[AMS] *Semicentennial Addresses*, American Mathematical Society Centennial Publications, vol. II, Amer. Math. Soc., 1938; Arno Press reprint, 1980.

[Arc] Raymond Clare Archibald, *A Semicentennial History of the American Mathematical Society*, Amer. Math. Soc., 1940.

[Bow] Nathaniel Bowditch, translator and annotator, "The *Mécanique Céleste* of the Marquis de La Place", 4 vols., Boston, 1829-39. References in this paper will be to the Chelsea reprint (1963).

[Bre] Joel L. Brenner, "Student days — 1930", *Amer. Math. Monthly* **86** (1979), 350-6.

[Cat] *Harvard College Catalogues*, published annually since 1835. Copies on display in Harvard Archives.

[CMA] *A Century of Mathematics in America*, Peter Duren et al., (eds.), vol. 1, Amer. Math. Soc., 1988.

[DAB] *Dictionary of American Biography*, Allen Johnson and Dumas Malone (eds.), Scribners, 1928-1936.

[DMP] Harvard Division of Mathematics Pamphlets from 1891 through 1941. Copies in Harvard Archives.

[DSB] *Dictionary of Scientific Biography*, Charles C. Gillispie (ed.). Scribner's, 1970- .

[Dup] A. Hunter Dupree, *Science in the Federal Government*, Harvard University Press, 1957. (Reprinted by Johns Hopkins University Press, 1986.)

[EB] *Encyclopaedia Britannica*, 1971 edition.

[GB] Garrett Birkhoff, *Selected Papers in Algebra and Topology*, J. Oliveira and G.-C. Rota, eds., Birkhäuser, 1987.

[GDB] George David Birkhoff, *Collected Papers*, 3 vols. Amer. Math. Soc., 1950.

[HH] Hugh Hawkins, *Between Harvard and America. The Educational Leadership of Charles W. Eliot*, Oxford University Press, 1972.

[JLC] Julian Lowell Coolidge, "Three Hundred Years of Mathematics at Harvard", *Amer. Math. Monthly* **50** (1943), 347-56.

[LAM] "Some leaders in American mathematics: 1891–1941", by Garrett Birkhoff. This is [Tar, pp. 29-78].

[LIn] Edward Weeks, *The Lowells and their Institute*, Little Brown, 1963.

[Love] James Lee Love, "The Lawrence Scientific School...1847-1906", Burlington, N.C. 1944. Copy available in Harvard Archives.

[Low] Abbott Lawrence Lowell, *At War with Academic Traditions in America*, Harvard University Press, 1934.

[May] Kenneth O. May, *The Mathematical Association of America: the First Fifty Years*, Math. Assn. of America, 1972.

[Mor] Samuel E. Morison (ed.), *The Development of Harvard University, 1869-1929*, Harvard University Press, 1930.

[NAS] Biographical Memoirs of the National Academy of Sciences, vols. 1- . U.S. National Academy of Sciences,

[Pei] "Benjamin Peirce", four "Reminiscences", followed by a Biographical Sketch by R.C. Archibald. *Amer. Math. Monthly* **32** (1925), 1-30.

[Pit] A.E. Pitcher, *A History of the Second Fifty Years of the American Mathematical Society*, 1939-1988, Amer. Math. Soc., 1988.

[Put] "The William Lowell Putnam Competition", by G. Birkhoff and L. E. Bush, *Amer. Math. Monthly* **72** (1965), 469–83.

[Qui] Josiah Quincy, *History of Harvard University*, 2 vols., Cambridge, 1840.

[S-G] David Eugene Smith and Jekuthiel Ginsburg, *A History of Mathematics in America before* 1900, Math. Assn. America, 1934. Arno Press reprint, 1980.

[SoF] *The Society of Fellows*, by George C. Homans and Orville T. Bailey, Harvard University Press, 1948. Revised (Crane Brinton, ed.), 1959.

[Tar] Dalton Tarwater (ed.), *The Bicentennial Tribute to American Mathematics*: 1776–1976, Math. Assn. America, 1977.

[TCH] Samuel E. Morison, *Three Centuries of Harvard*: 1636–1936, Harvard University Press, 1936.

[Tex] *Men and Institutions in America*, Dalton Tarwater (ed.), Graduate Studies, Texas Tech. University, vol. 13 (1976).

[Van] H. S. Vandiver, "Some of my recollections of George D. Birkhoff", *J. Math. Anal. Appl.* **7** (1963), 272–83.

[Whi] Paul A. Schilpp (ed.), *The Philosophy of Alfred North Whitehead*, Northwestern University Press, 1941.

[Yeo] Henry A. Yeomans, *Abbott Lawrence Lowell*, Harvard University Press, 1948.

THE SCIENTIFIC WORK OF MAXIME BÔCHER.

BY PROFESSOR GEORGE D. BIRKHOFF.

WITH the recent death of Professor Maxime Bôcher at only fifty-one years of age American mathematics has suffered a heavy loss. Our task in the following pages is to review and appreciate his notable mathematical work.*

His researches cluster about Laplace's equation $\Delta u = 0$, which is the very heart of modern analysis. Here one stands in natural contact with mathematical physics, the theory of linear differential equations both total and partial, the theory of functions of a complex variable, and thus directly or indirectly with a great part of mathematics.

His interest in the field of potential theory began in undergraduate days at Harvard University through courses given by Professors Byerly and B. O. Peirce. There is still on file at the Harvard library an undergraduate honor thesis entitled "A thesis on three systems of parabolic coördinates," written by him in 1888. Under the circumstances it was inevitable that he should use formal methods in dealing with his topic, but a purpose to penetrate further is found in the concluding sentences. No better opportunity for fulfilling such a purpose could have been granted than was given by his graduate work under Felix Klein at Göttingen (1888–1891).

In the lectures on Lamé's functions which Klein delivered in the winter of 1889–1890 his point of departure was the cyclidic coördinate system of Darboux. This sytem of coördinates was known to be so general as to include nearly all of the many types of coördinates useful in potential theory, and Wangerin had shown (1875–1876) how solutions of Laplace's equation existed in the form of triple products, each factor being a function of one of the three cyclidic coördinates. After presenting this earlier work Klein extended his "oscillation theorem" for the case of elliptic coördinates (1881) to the more general cyclidic coördinates. By this means he was able to attack the problem of setting up a potential function taking on given values over the surface of a solid bounded by

---

* An account of his life and service by Professor Osgood will appear in a later number of the BULLETIN.

Reprinted with permission from the *Bulletin of the American Mathematical Society*, Volume 25, pp. 197–215.

Maxime Bôcher

six or fewer confocal cyclides. This function was given by a series of the triple "Lamé's" products discovered by Wangerin.

Klein also aimed to get at the various forms of series and integrals previously employed in potential theory as actual limiting cases, and thus to bring out the underlying unity in an extensive field of mathematics.

The task which Bôcher undertook was to carry through the program sketched by Klein. He did this admirably in his first mathematical paper "Ueber die Reihenentwickelungen der Potentialtheorie," which appeared in 1891 and which served both as a prize essay and as his doctor's dissertation at Göttingen.* But the space available was so brief that he was only able to outline results without giving their proofs.

One must look to his book with the same title,† published three years later, for an adequate treatment of the subject. Here is also to be found original work not outlined in his dissertation. It was characteristic that he did not call attention explicitly to the new advances although these formed his most important scientific work in the years 1891–1894. We turn now to a consideration of this book, which thus contains nearly all that he did before 1895.

Besides giving the classification of all types of confocal cyclides in the real domain and of the corresponding Lamé's products, as sketched by Klein, Bôcher determined to what extent the theorem of oscillation holds in the degenerate cases and found an interesting variety of possibilities.

The difficulties presented by these degenerate cases are decidedly greater than those of the general case when the singular points e_i ($i = 1, 2, 3, 4, 5$) of the Lamé's linear differential equation are regular with exponents 0, 1/2. A very simple degenerate case is that arising when two such points coincide in a single point and one of the two intervals (m_1, m_2), (n_1, n_2) under consideration ends at this point. By an extension of Klein's geometric method, he proved that the theorem of oscillation fails to hold even here.

More specifically, the facts are as follows. In the general case the oscillation theorem states that for any choice of integers m, n ($m, n \geq 0$) there is a unique choice of the two ac-

---

* This paper appears as (2) in the chronological list of papers given at the end of the present article. Hereafter footnote references to papers will be made by number.

† (15).

cessory parameters in the differential equation, yielding solutions u_1, u_2 such that u_1 vanishes at m_1 and m_2, and m times for $m_1 < x < m_2$, while u_2 vanishes at n_1 and n_2, and n times for $n_1 < x < n_2$. If now, for instance, m_1 lies at the double singular point $e_1 = e_2$, while $m_1 < m_2 < e_3 < e_4 < n_1 < n_2 < e_5$, there exist such solutions u_1, u_2 only if $n > r_m$ where r_m is an integer increasing indefinitely with m. But, to compensate for this deficiency of solutions of the boundary value problem, Bôcher found it necessary to introduce solutions u_{1k}, u_{2k} dependent on n and a continuous real parameter k such that u_1 vanishes at m_2 and infinitely often for $m_1 < x < m_2$ although remaining finite, while u_2 vanishes at n_1 and n_2, and n times for $n_1 < x < n_2$.

The corresponding expansion in Lamé's products presents a remarkable form under these circumstances, for it is made up of a series and an integral component. In another case this type of expansion takes the form of an integral augmented by a finite number of complementary terms, as he had pointed out in an important paper "On some applications of Bessel's functions with pure imaginary index,"[*] published in 1892 in the *Annals of Mathematics*.

Although dealing satisfactorily with the oscillation theorem in the case specified above and other similar cases, Bôcher did not discuss adequately the case in which three or more singular points unite to form an irregular singular point.[†] Indeed it appears that he fell into an error of reasoning as follows. If the irregular point be taken at $t = +\infty$ the Lamé's equation has the form

$$\frac{d^2 y}{dt^2} = \varphi y,$$

where in the case under consideration φ has a limit $\varphi_0 \neq 0$ as t becomes infinite. The lemma which Bôcher then sought to prove[‡] was that there always exists a solution y *finite* for $t \geq T$ and not identically zero. His proof for the case $\varphi_0 > 0$ is essentially correct. Here he interpreted the equation above as the equation of motion of a particle distant y from a point O of its line of motion and repelled from it with a force φy.

---

[*] (7). In passing, attention may also be called to a slightly earlier article (3) on Bessel's functions.
[†] See (15), p. 179.
[‡] See (15), p. 177.

The gist of the argument employed is that one can find an initial velocity of projection toward O just sufficient to carry it into that point as a limiting position. This part of the lemma constitutes a very simple and interesting theorem concerning a special type of irregular point. In the case $\varphi_0 < 0$, however, using a similar dynamical interpretation, he argued* "we have infinitely many oscillations as we approach $t = +\infty$, and since the attractive force is not infinitely weak, the amplitudes of the oscillations remain finite." This argument appears insufficient although the lemma as stated for Lamé's equation is probably correct.† To satisfactorily complete the discussion it would seem to be necessary to call in the explicit analytic theory of the irregular singular point, since the corresponding theory of the regular singular point is required in the simpler cases.‡

In his book Bôcher considered the boundary problem under *periodic* conditions, when the interval between two adjacent singular points is taken an even number of times and is regarded as closed; this case arises, for example, when the solid in the potential problem is a complete ellipsoid. Here the function φ in the linear differential equation above written is an even doubly periodic function with real period. By the aid of these properties of φ he reduced the new boundary problem to one of the ordinary type.

Likewise in treating the roots of Lamé's polynomials he made a distinct advance by extending the dynamical method of Stieltjes from the real axis to the complex plane. Thus he was able to prove that the roots of these polynomials lie within the triangle whose vertices are the three finite singular points of the corresponding Lamé's equation.

Finally we may note that at the end of his book he obtained all Lamé's products satisfying the equation $\Delta u + k^2 u = 0$.

The determinative effect of the dissertation and book upon the direction of Bôcher's later researches was very great. In the first place he had used sphere geometry and the algebra of elementary divisors as essential tools in analysis; his resulting interest in the fundamental parts of geometry and algebra never subsided, and some of his research lies in these fields.

---

* (15), p. 178. The translation is not literal.

† In this connection see (7), p. 150, footnotes.

‡ Since the above was written Professor Osgood has disposed of the question at issue by elementary means. See his note in this number of the BULLETIN.

But, more important still, he was brought into contact with open mathematical questions. The most vital of these questions from the purely mathematical point of view was doubtless the very difficult analytical question of convergence and representation presented by the series of Lamé's products. This was the outstanding problem which Klein emphasized,* but to which Bôcher seems never to have given particular attention. Another more practical direction of effort was afforded by the task of giving rigorous and accessible form to the work of Sturm and Klein on the real solutions of ordinary linear differential equations and then going on further in this overlooked but attractive field of research. It was primarily to this task that he now turned.

In 1897 he published an article in the BULLETIN† showing the immediate usefulness of Sturm's theorems for fixing the distribution of the roots of Bessel's functions with real index. A year later in the same place he presented the fundamentals of Sturm's work in simplified rigorous form, and gave the first analytic proof of Klein's theorem of oscillation.‡

Reference should also be made to his article on the boundary problems of ordinary differential equations which appeared in the German mathematical encyclopedia in 1900. This article together with his address on "Boundary problems in one dimension" before the Fifth International Congress of Mathematicians in 1912 give an excellent account of this field to the latter date.

Bôcher wrote a considerable number of other papers in this same field.§ Perhaps the most important of these are the three to which we will refer first and which appeared in the beginning volumes of the *Transactions*.

His paper "Application of a method of d'Alembert to the proof of Sturm's theorems of comparison" (1900) contained an elegant proof of what Bôcher had called the theorems of comparison. His method was entirely different from Sturm's, being based on the Riccati's resolvent equation, and was very simple.

In the second of these papers "On certain pairs of trans-

---

* See the concluding pages of his 1889–1890 lectures on Lamé's functions and his preface to Bôcher's book (15).
† (26). See (38) also.
‡ (30), (31), (35).
§ (32), (38), (42), (46), (48), (49), (50), (55), (65), (81), (85), (92), (93), (100).

cendental functions whose roots separate each other"* (1901) his starting point was the linear differential equation

$$y'' + py' + qy = 0,$$

and a pair of linear forms in y, y',

$$\Phi = \varphi_2 y' - \varphi_1 y, \quad \Psi = \psi_2 y' - \psi_1 y.$$

These latter satisfy a "homogeneous Riccati's equation"

$$(\varphi_1 \psi_2 - \varphi_2 \psi_1)(\Phi' \Psi - \Phi \Psi') + A\Phi^2 + B\Phi\Psi + C\Psi^2 = 0,$$

and Bôcher considered the relation of the roots of Φ, Ψ.

He notes first that Φ, Ψ cannot vanish together unless $\varphi_1 \psi_2 - \varphi_2 \psi_1 = 0$, for otherwise $y = y' = 0$. In order that Φ, Ψ cannot vanish together it is thus sufficient to assume $\varphi_1 \psi_2 - \varphi_2 \psi_1 \neq 0$. Also if $\Phi = 0$, then $\Phi' \neq 0$ if $C \neq 0$, by the above equation. A like remark holds for Ψ. Hence the roots of Φ, Ψ are simple if $A \neq 0$, $C \neq 0$.

Under these hypotheses between any pair of adjacent roots of Φ there must be a root of Ψ. For if Ψ has no such root the homogeneous Riccati's equation at these roots shows that Φ' has one and the same sign at both roots, which is impossible. Likewise between any pair of adjacent roots of Ψ there must be a root of Φ.

Hence *the roots of Φ, Ψ separate each other if*

$$\varphi_1 \psi_2 - \varphi_2 \psi_1 \neq 0, \; A \neq 0, \; C \neq 0.$$

This is the third theorem of the paper. The sixth theorem gives similar conditions sufficient to ensure cyclical separation of the roots of three linear forms.

Here Bôcher not only achieved greater generality and simplicity than Sturm but, as I wish to point out, he has reached a maximum of generality.

For, let y_1, y_2 be any pair of linearly independent solutions yielding the values Φ_1, Φ_2 and Ψ_1, Ψ_2 of Φ and Ψ. Then

$$\Phi = c_1 \Phi_1 + c_2 \Phi_2, \quad \Psi = c_1 \Psi_1 + c_2 \Psi_2$$

are the general values of Φ, Ψ. If Φ_1, Φ_2 are regarded as the homogeneous coordinates of a point P in the projective line, Φ vanishes if P coincides with $E \equiv (-c_2, c_1)$; similarly Ψ vanishes if $Q = (\Psi_1, \Psi_2)$ coincides with the same point E.

---

* See also (100).

Clearly the roots of Φ, Ψ will only be distinct for all values of c_1, c_2 if $\varphi_1\psi_2 - \varphi_2\psi_1 \neq 0$. Moreover, if these roots are to separate each other for all values of c_1, c_2, the points P, Q must pass *any point* E in alternation. This is only possible if P, Q never reverse their direction of motion; in other words the Wronskians of Φ_1, Φ_2 and of Ψ_1, Ψ_2 must be of invariant signs. Taking into account the fact that $y_1 y_2' - y_1' y_2$ is not zero, this gives precisely the conditions $A \neq 0$, $C \neq 0$.

This same geometric interpretation shows a similar generality in the other theorems.

Of like completeness is the third paper "On the real solutions of systems of two homogeneous linear differential equations of the first order" (1902), where he treated analogous questions and also derived comparison theorems.

It was a matter of primary interest with him to vary proofs of known theorems as well as to discover new theorems. An illustration in point is afforded by his treatment of the elementary separation theorem for the roots of linearly independent solutions y_1, y_2 of an ordinary linear differential equation of the second order.

Here he first gave a very brief proof* based on the function y_1/y_2: if y_1 vanishes at a and b but not for $a < x < b$, while y_2 is not zero for $a \leq x \leq b$, then the derivative of y_1/y_2 is of one sign for $a < x < b$ since $y_1 y_2' - y_1' y_2 \neq 0$. This is impossible. By this argument and a like argument based on y_2/y_1 it follows that the roots of y_1, y_2 separate each other. In the same place† he isolates a geometric proof implicitly given by Klein depending on the fact that if y_1, y_2 be taken as homogeneous coördinates of a point in the projective line then $y_1 y_2' - y_1' y_2 \neq 0$ is the condition that this point moves continually in one sense. Later he gave a second analytic proof based on the function

$$\frac{y_1'}{y_1} - \frac{y_2'}{y_2}, \ddagger$$

and also a second geometric proof* based on the vector $y_1 + \sqrt{-1}\, y_2$ in the complex plane which will rotate continually in one sense if $y_1 y_2' - y_2 y_1' \neq 0$.

---

* (26), p. 210.
† Footnote, p. 210.
‡ (48).
§ (99), pp. 46–47.

It was not easy for him to believe that the methods of Sturm were inadequate to deal with any particular boundary problem in one dimension. The problem for periodic conditions, which had been formulated by him in his encyclopedia article, was first successfully attacked by Mason in 1903–1904 by means of the calculus of variations. In a very interesting note published in 1905,* Bôcher showed that the principal result fell out immediately by the methods of Sturm, and that these methods were applicable under much more general conditions. Likewise in his address before the Fifth International Congress of Mathematicians alluded to above he noted that the equation

$$\frac{d}{dx}\left(k\frac{du}{dx} \right) + (\lambda g - l)u = 0, \qquad l < 0,$$

(λ a parameter) comes directly under the case treated by Sturm after division by $|\lambda|$ even if g changes sign. This simple remark disposed of the necessity of treating this case separately, as had been done earlier.

Bôcher was interested in all phases of the theory of ordinary linear differential equations with real independent variable. Having seen the gap in the theory of the regular singular point for real independent variable when the coefficients are not analytic, he proved that theorems analogous to those given by Fuchs in the complex domain are true.† It was necessary here to replace the power series treatment by a variation of the method of successive approximation which has been seen later to afford a new approach to the theory of the regular singular point in the complex domain.

He also did some work in the field of fundamental existence theorems for linear differential equations.‡ He showed that it is sufficient to impose the condition of integrability (joined with other conditions) upon the coefficients in place of Peano's condition of continuity,§ and thus advanced beyond Peano. Bôcher seems also to have been the first to prove that the solutions of a linear differential system are continuous functionals of the coefficients.‖

---

* (65).
† (37), (40), (41).
‡ (32), (37), (56).
§ (56), p. 311.
‖ (56), p. 315; (55), p. 208.

In 1901 he published a paper on "Green's functions in space of one dimension," in which he pointed out that the Green's function for the equation of Laplace in one dimension $y'' = 0$, exhibited by Burkhardt in 1894, might be extended to the general nth order ordinary linear differential equation with fairly general boundary conditions. These extended Green's functions have turned out to be of great importance. Later he returned to the subject of Green's functions with the most general linear boundary conditions and set up these functions for linear difference equations.* Also he extended the notion of adjoint boundary conditions to very general cases.†

We have now referred briefly to the most important of his researches on ordinary linear differential equations with *real* independent variable. In this domain his best work is perhaps to be found. Directly springing from this field were his researches on linear dependence of functions of a single real variable‡—an important topic which he was the first to isolate sufficiently from the field of linear differential equations.

His paper on "The roots of polynomials which satisfy certain linear differential equations of the second order"§ lies in the field of ordinary linear differential equations with a *complex* variable. Here he generalizes further the extension of the method of Stieltjes which he had employed in dealing with Lamé's polynomials.

The series arising in mathematical physics had been Bôcher's point of departure. Indeed it is the existence of these series which constitutes the main importance of the boundary value problems of linear differential equations. Nevertheless he gave special attention only to Fourier's series which he took up in an expository article in the *Annals of Mathematics* for 1906.‖ Here he called attention to the remarkable phenomenon exhibited by a Fourier's series near a point of discontinuity, previously noted by Gibbs and called "Gibbs's phenomenon" by Bôcher who gave the first adequate treatment of it.¶

His contributions to the theory of the harmonic function in two dimensions are elegant and distinctly important.

---

* (81).
† (85).
‡ (43), (45), (47), (51), (97).
§ (29).
‖ (67). See also (89).
¶ Reference may also be made here to the short note on infinite series (60).

The first of these occurs incidentally in his paper "Gauss's third proof of the fundamental theorem of algebra."[*] It consists in a proof of the average value theorem by means of Gauss's theorem for the circle, which in polar coordinates r, φ is

$$\int_0^{2\pi} \frac{\partial u}{\partial r} d\varphi = 0.$$

Integrating with respect to r from 0 to a and reversing the order of integration, we get

$$\int_0^{2\pi} (u(a, \varphi) - u(0, \varphi))d\varphi = 0,$$

whence the average value theorem follows at once. This very neat proof was probably suggested by the artifice used by Gauss in his third proof of the fundamental theorem of algebra.

The "Note on Poisson's integral" (1898) gives a more natural interpretation of Poisson's integral than had been stated before. By the average value theorem a harmonic function is the average of its values on any circle with its center at the given point. He generalized this theorem in the spirit of the geometry of inversion and thus reached a visual interpretation of Poisson's integral which may be formulated as follows: The value of a harmonic function at any point within a circle is the average of its values as read by an observer at the point who turns with uniform angular velocity, if the rays of light to his eye take the form of circular arcs orthogonal to the given circle.

According to Riemann's program, the theory of harmonic functions requires a development independent of the theory of functions of a complex variable. In 1905 Bôcher demonstrated[†] that a harmonic function could not become infinite at a point unless it was of the form $C \log r + v$, where C is a constant, r is the distance from a variable point to the given point and v is harmonic at that point. This theorem corresponds to the fundamental theorem in functions of a complex variable which states that if $f(z)$ becomes infinite at the isolated singular point $z = a$, then $f(z)$ is of the form $(z - a)^{-r}g(z)$ where r is a positive integer and $g(z)$ is analytic and not zero

[*] (17), p. 206.
[†] (59).

at $z = a$. He demonstrated further that a similar theorem holds for large classes of linear partial differential equations.

Another extremely interesting paper "On harmonic functions in two dimensions" appeared in 1906. Here he defines u to be harmonic if it is single valued and continuous with continuous first partial derivatives and satisfies Gauss's theorem for every circle. If u possessed continuous second partial derivatives also it would then follow at once by Green's theorem that u is harmonic in the customary sense. But it is the merit of Bôcher's paper to have proved that u is harmonic in the ordinary sense without further assumptions. On the basis of the definition made, the average value theorem is first deduced as outlined above. Also if s', n' are the new variables s, n after an inversion (taking circles into circles) we have

$$0 = \int \frac{\partial u}{\partial n} ds = \int \frac{\partial u}{\partial n'} ds'$$

along corresponding circles, since $ds'/dn' = ds/dn$ (the inversion being conformal). Thus u is "harmonic" in the transformed plane also, so that the definition is invariant under inversion. Hence Poisson's integral formula, which comes from the average value theorem by inversion, also holds, and u is harmonic in the ordinary sense.

He also determined the precise region of convergence of the real power series in x, y for any harmonic function $u(x, y)$.[*]

In connection with his papers on harmonic functions in two dimensions it is natural to call to mind his early paper "On the differential equation $\Delta u + k^2 u = 0$" (1893), which is taken in two dimensions. The "u-functions" so defined give a generalization of harmonic functions which he treated by means of the fact that $u(x, y)e^{kz}$ satisfies Laplace's equation in three dimensions. A similar method had been employed earlier by Klein.

Practically none of Bôcher's work lies directly in the field of functions of a complex variable.[†]

We have still to consider his contributions in the fields of algebra and geometry. In the early paper on the fundamental theorem of algebra cited above he made clear how, by taking for granted a few theorems in functions of a complex variable,

---

[*] (74).
[†] See (78), however.

an immediate proof could be given; and then he went on to show that by elimination of these theorems, the proof could be given a second more fundamental form and finally a third form due to Gauss and involving only distinctly elementary theorems. In a second paper* he simplified Gauss's proof very considerably by replacing Gauss's auxiliary function zf'/f by $1/f$. Here $f = 0$ is the given equation.

Here and elsewhere he succeeded in simplifying an apparently definitive proof. This kind of work was congenial to Bôcher, who believed that mathematics was capable of almost indefinite simplification, and that such simplification was of the highest consequence.

In the paper with the title "A problem in statics and its relation to certain algebraic invariants" (1904) he employed a dynamical method similar to his extension of the method of Stieltjes in order to develop an interpretation of the roots of covariants as the positions of equilibrium of particles in the complex plane. Thus if f_1, f_2 are polynomials of the same degree in the homogeneous variables x_1, x_2, the vanishing of their Jacobian determines the points of equilibrium in the field of force under the inverse first power law due to particles of "mass" 1 at the roots of f_1 and of "mass" $-$ 1 at the roots of f_2 in the x_1/x_2 plane.

We shall not refer to his geometrical papers† save to mention the one entitled "Einige Sätze über projective Spiegelung" (1893) in which he proves that conics in different planes may be projectively reflected into each other through a pair of lines in four ways, and also that the general collineation of space may be represented as the product of a rigid motion and a projective reflection through a pair of lines.

Besides this original research he undertook various more or less didactic articles with characteristic unselfishness.‡ However, just as in the article on Fourier's series, matter of an original cast is nearly always present.

The same may be said of his books,§ even of the most elementary. We have already considered his book on the series of potential theory. Of the others, the most significant are his Algebra, where a satisfactory exposition of the elementary

---

* (18).
† (6), (8), (12), (13), (53).
‡ (14), (20), (24), (39), (66), (67), (70), (73), (83), (92).
§ (15), (71), (77), (94), (95), (99).

divisor theory is given, his Cambridge tract on integral equations,* and his Paris 1913–14 lectures "Leçons sur les Méthodes de Sturm." In the last is given the first complete discussion of the convergence of the series used in the method of successive approximations. This furnishes another good instance of Bôcher's power to seize on important theorems which have been missed although near at hand. In concluding this brief survey it is worth while noting that a few of his papers are fairly popular in character.†

In a recent one of these, "Mathématiques et mathématiciens Français" (1914), while speaking of the characteristics of American creative work in all fields (page 9), Bôcher says "Ce qu'il y a de plus caractéristique dans la meilleure production intellectuelle américaine, c'est la finesse et le contrôle voulu des moyens et des effets. La faute la plus commune dans ce que nous avons fait de mieux, ce n'est pas l'excès de force, mais plutôt son défaut" and later (page 10) "Ce que je viens de dire se rapporte aussi bien aux mathématiques qu'à toute autre branche de la production intellectuelle en Amérique." There can be no doubt that this characterization is applicable to his own mathematical production. His papers excel in simplicity and elegance, and nearly all of them treat subjects of great importance to marked advantage. The *usefulness* of his papers is exceptional,‡

In amount and quality his production exceeds that of any American mathematician of earlier date in the field of pure mathematics.

Because of this fact and the weight he has added to our mathematical traditions in other ways, Maxime Bôcher will ever remain a memorable personality in American mathematics.

List of Bôcher's Writings.‖

1888.

(1) The meteorological labors of Dove, Redfield and Espy. *American Meteorological Journal*, vol. 5, No. 1, pp. 1–13, May.

---

* In connection with this, attention should be called to a short note on integral equations listed as (84) below.

† (1), (9), (11), (82), (90), (91). His first paper "On the meteorological labors of Dove, Redfield and Espy" was a youthful essay written about the time of his graduation from Harvard University.

‡ This is brought out clearly in Professor Osgood's Lehrbuch der Funktionentheorie, vol. 1.

‖ Substantially as compiled by him.

1891.

(2) Über die Reihenentwickelungen der Potentialtheorie. Gekrönte Preisschrift und Dissertation. Göttingen, Kästner. 4 + 66 pp.

1892.

(3) On Bessel's functions of the second kind. *Annals of Mathematics,* vol. 6, No. 4, pp. 85–90, Jan.
(4) Pockels on the differential equation $\Delta u + k^2 u = 0$ [Review]. *Annals of Mathematics,* vol. 6, No. 4, pp. 90–92, Jan.
(5) Geometry not mathematics [Letter to editor]. *Nation,* vol. 54, No. 1390, p. 131, Feb.
(6) On a nine-point conic. *Annals of Mathematics,* vol. 6, No. 5, p. 132, March.
(7) On some applications of Bessel's functions with pure imaginary index. *Annals of Mathematics,* vol. 6, No. 6, pp. 137–160, May.
(8) Note on the nine-point conic. *Annals of Mathematics,* vol. 6, No. 7, p. 178, June.
(9) Collineation as a mode of motion. *Bulletin of the New York Mathematical Society,* vol. 1, No. 10, pp. 225–231, July.

1893.

(10) On the differential equation $\Delta u + k^2 u = 0$. *American Journal of Mathematics,* vol. 15, No. 1, pp. 78–83, Jan.
(11) A bit of mathematical history. *Bulletin of the New York Mathematical Society,* vol. 2, No. 5, pp. 107–109, Feb.
(12) Some propositions concerning the geometric representation of imaginaries. *Annals of Mathematics,* vol. 7, No. 3, pp. 70–72, March.
(13) Einige Sätze über projective Spiegelung. *Mathematische Annalen,* vol. 43, No. 4, pp. 598–600.
(14) Chapter IX, Historical Summary, pp. 267–275. An Elementary Treatise on Fourier's Series and Spherical, Cylindrical and Ellipsoidal Harmonics. By W. E. Byerly. Boston, Ginn.

1894.

(15) Über die Reihenentwickelungen der Potentialtheorie. Mit einem Vorwort von Felix Klein. Leipzig, Teubner, 8 + 258 pp.

1895.

(16) Hayward's Vector Algebra [Review]. *Bulletin of the American Mathematical Society,* ser. 2, vol. 1, No. 5, pp. 111–115, Feb.
(17) Gauss's third proof of the fundamental theorem of algebra. *Bulletin of the American Mathematical Society,* ser. 2, vol. 1, No. 8, pp. 205–209, May.
(18) Simplification of Gauss's third proof that every algebraic equation has a root. *American Journal of Mathematics,* vol. 17, No. 3, pp. 266–268, July.
(19) General equation of the second degree [Set of formulas on a card]. Harvard University Press.

1896.

(20) On Cauchy's theorem concerning complex integrals. *Bulletin of the American Mathematical Society,* ser. 2, vol. 2, No. 5, pp. 146–149, Feb.
(21) Bessel's functions [Review]. *Bulletin of the American Mathematical Society,* ser. 2, vol. 2, No. 8, pp. 255–265, May.

(22) Linear differential equations and their applications. [Report by T. S. Fiske of a lecture at the Buffalo Colloquium]. *Bulletin of the American Mathematical Society*, ser. 2, vol. 3, No. 2, pp. 52-55, Nov.

(23) Heffter's Linear Differential Equations [Review]. *Bulletin of the American Mathematical Society*, ser. 2, vol. 3, No. 2, pp. 86-92, Nov.

(24) Regular points of linear differential equations of the second order. Cambridge, Harvard University Press, 23 pp.

1897.

(25) Schlesinger's Linear Differential Equations [Review]. *Bulletin of the American Mathematical Society*, ser. 2, vol. 3, No. 4, pp. 146-153, Jan.

(26) On certain methods of Sturm and their application to the roots of Bessel's functions. *Bulletin of the American Mathematical Society*, ser. 2, vol. 3, No. 6, pp. 205-213, March.

(27) Review of Bailey and Woods: Plane and Solid Analytic Geometry. *Bulletin of the American Mathematical Society*, ser. 2, vol. 3, No. 9, pp. 351-352, June.

1898.

(28) Examples of the construction of Riemann's surfaces for the inverse of rational functions by the method of conformal representation. By C. L. Bouton with an introduction by Maxime Bôcher. *Annals of Mathematics*, vol. 12, No. 1, pp. 1-26, Feb.

(29) The roots of polynomials which satisfy certain linear differential equations of the second order. *Bulletin of the American Mathematical Society*, ser. 2, vol. 4, No. 6, pp. 256-258, March.

(30) The theorems of oscillation of Sturm and Klein (first paper). *Bulletin of the American Mathematical Society*, ser. 2, vol. 4, No. 7, pp. 295-313, April.

(31) The theorems of oscillation of Sturm and Klein (second paper). *Bulletin of the American Mathematical Society*, ser. 2, vol. 4, No. 8, pp. 365-376, May.

(32) Note on some points in the theory of linear differential equations. *Annals of Mathematics*, vol. 12, No. 2, pp. 45-53, May.

(33) Note on Poisson's integral, *Bulletin of the American Mathematical Society*, ser. 2, vol. 4, No. 9, pp. 424-426, June.

(34) Niewenglowski's Geometry [Review]. *Bulletin of the American Mathematical Society*, ser. 2, vol. 4, No. 9, pp. 448-452, June.

(35) The theorems of oscillation of Sturm and Klein (third paper). *Bulletin of the American Mathematical Society*, ser. 2, vol. 5, No. 1, pp. 22-43, Oct.

1899.

(36) Burkhardt's Theory of Functions [Review]. *Bulletin of the American Mathematical Society*, ser. 2, vol. 5, No. 4, pp. 181-185, Jan.

(37) On singular points of linear differential equations with real coefficients. *Bulletin of the American Mathematical Society*, ser. 2, vol. 5, No. 6, pp. 275-281, March.

(38) An elementary proof that Bessel's functions of the zeroth order have an infinite number of real roots. *Bulletin of the American Mathematical Society*, ser. 2, vol. 5, No. 8, pp. 385-388, May.

(39) Examples in the theory of functions. *Annals of Mathematics*, ser. 2, vol. 1, No. 1, pp. 37-40, Oct.

1900.

(40) On regular singular points of linear differential equations of the second order whose coefficients are not necessarily analytic. *Transactions of the American Mathematical Society*, vol. 1, No. 1, pp. 40-52, Jan.; also No. 4, p. 507, Oct.

(41) Some theorems concerning linear differential equations of the second order. *Bulletin of the American Mathematical Society*, ser. 2, vol. 6, No. 7, pp. 279–280, April.

(42) Application of a method of d'Alembert to the proof of Sturm's theorems of comparison. *Transactions of the American Mathematical Society*, vol. 1, No. 4, pp. 414–420, Oct.

(43) On linear dependence of functions of one variable. *Bulletin of the American Mathematical Society*, ser. 2, vol. 7, No. 3, pp. 120–121, Dec.

(44) Randwertaufgaben bei gewöhnlichen Differentialgleichungen. Encyklopädie der mathematischen Wissenschaften, II A 7a, pp. 437–463, Leipzig, Teubner.

1901.

(45) The theory of linear dependence. *Annals of Mathematics*, ser. 2, vol. 2, No. 2, pp. 81–96, Jan.

(46) Green's functions in space of one dimension. *Bulletin of the American Mathematical Society*, ser. 2, vol. 7, No. 7, pp. 297–299, April.

(47) Certain cases in which the vanishing of the Wronskian is a sufficient condition for linear dependence. *Transactions of the American Mathematical Society*, vol. 2, No. 2, pp. 139–149, April.

(48) An elementary proof of a theorem of Sturm. *Transactions of the American Mathematical Society*, vol. 2, No. 2, pp. 150–151, April.

(49) Non-oscillatory linear differential equations of the second order. *Bulletin of the American Mathematical Society*, ser. 2, vol. 7, No. 8, pp. 333–340, May.

(50) On certain pairs of transcendental functions whose roots separate each other. *Transactions of the American Mathematical Society*, vol. 2, No. 4, pp. 428–436, Oct.

(51) On Wronskians of functions of a real variable. *Bulletin of the American Mathematical Society*, ser. 2, vol. 8, No. 2, pp. 53–63, Nov.

(52) Picard's Traité d'Analyse [Review]. *Bulletin of the American Mathematical Society*, ser. 2, vol. 8, No. 3, pp. 124–128, Dec.

1902.

(53) Some applications of the method of abridged notation. *Annals of Mathematics*, ser. 2, vol. 3, No. 2, pp. 45–54, Jan.

(54) Review of Schlesinger: Einführung in die Theorie der Differentialgleichungen mit einer unabhängigen Variabeln. *Bulletin of the American Mathematical Society*, ser. 2, vol. 8, No. 4, pp. 168–169, Jan.

(55) On the real solutions of two homogeneous linear differential equations of the first order. *Transactions of the American Mathematical Society*, vol. 3, No. 2, pp. 196–215, April.

(56) On systems of linear differential equations of the first order, *American Journal of Mathematics*, vol. 24, No. 4, pp. 311–318, Oct.

(57) Review of Gauss' Wissenschaftliches Tagebuch. *Bulletin of the American Mathematical Society*, ser. 2, vol. 9, No. 2, pp. 125–126, Nov.

1903.

(58) The Elements of Plane Analytic Geometry. By George R. Briggs. Revised and enlarged by Maxime Bôcher. New York, Wiley, 4 + 191 p.

(59) Singular points of functions which satisfy partial differential equations of the elliptic type. *Bulletin of the American Mathematical Society*, ser. 2, vol. 9, No. 9, pp. 455–465, June.

(60) On the uniformity of the convergence of certain absolutely convergent series. *Annals of Mathematics*, ser. 2, vol. 4, No. 4, pp. 159–160, July.

1904.

(61) Contribution to Sprechsaal für die Encyklopädie der Mathematischen Wissenschaften. *Archiv der Mathematik und Physik*, vol. 7, No. 3, p. 181, Feb.

(62) The fundamental conceptions and methods of mathematics. Address delivered before the Department of Mathematics of the International Congress of Arts and Science, St. Louis, Sept. 20, 1904. *Bulletin of the American Mathematical Society*, ser. 2, vol. 11, No. 3, pp. 115–135, Dec. Also in Congress of Arts and Science Universal Exposition, St. Louis, 1904, vol. 1. Boston, Houghton and Mifflin, 1905, pp. 456–473.

(63) A problem in statics and its relation to certain algebraic invariants. *Proceedings of the American Academy of Arts and Sciences*, vol. 40, No. 11, pp. 469–484, Dec.

1905.

(64) Linear differential equations with discontinuous coefficients. *Annals of Mathematics*, ser. 2, vol. 6, No. 3, pp. 97–111 (49–63), April.

(65) Sur les équations différentielles linéaires du second ordre à solution périodique. *Comptes Rendus de l'Académie des Sciences*, vol. 140, No. 14, pp. 928–931, April.

(66) A problem in analytic geometry with a moral. *Annals of Mathematics*, ser. 2, vol. 7, No. 1, pp. 44–48, Oct.

1906.

(67) Introduction to the theory of Fourier's series. *Annals of Mathematics*, vol. 7, No. 2, and No. 3, pp. 81–152, Jan. and April.

(68) On harmonic functions in two dimensions. *Proceedings of the American Academy of Arts and Sciences*, vol. 41, No. 26, pp. 577–583, March.

(69) Review of Picard: Sur le Développement de l'Analyse, etc. *Science*, n. s., vol. 23, No. 598, p. 912, June.

(70) Another proof of the theorem concerning artificial singularities. *Annals of Mathematics*, ser. 2, vol. 7, No. 4, pp. 163–164, July.

1907.

(71) Introduction to Higher Algebra. By Maxime Bôcher. Prepared for publication with the cooperation of E. P. R. Duval. New York, Macmillan, 11 + 321 pp.*

1908.

(72) Review of Bromwich: Quadratic Forms and their Classification by Means of Invariant Factors. *Bulletin of the American Mathematical Society*, ser. 2, vol. 14, No. 4, pp. 194–195, Jan.

(73) On the small forced vibrations of systems with one degree of freedom. *Annals of Mathematics*, ser. 2, vol. 10, No. 1, pp. 1–8, Oct.

1909.

(74) On the regions of convergence of power-series which represent two-dimensional harmonic functions. *Transactions of the American Mathematical Society*, vol. 10, No. 2, pp. 271–278, April.

---

* A German translation appeared in 1909: Einführung in die höhere Algebra. Deutsch von Hans Beck. Mit einem Geleitwort von Eduard Study. Leipzig, Teubner, 12 + 348 pp.

(75) Review of Runge: Analytische Geometrie der Ebene. *Bulletin of the American Mathematical Society*, ser. 2, vol. 16, No. 1, pp. 30–33, Oct.

(76) Review of d'Adhémar: Exercices et Leçons d'Analyse. *Bulletin of the American Mathematical Society*, ser. 2, vol. 16, No. 2, pp. 87–88, Nov.

(77) An introduction to the study of integral equations. Cambridge Tracts in Mathematics and Mathematical Physics, No. 10, Cambridge, England, University Press, 72 pp.*

1910.

(78) On semi-analytic functions of two variables. *Annals of Mathematics*, ser. 2, vol. 12, No. 1, pp. 18–26, Oct.

(79) Kowalewski's Determinants [Review]. *Bulletin of the American Mathematical Society*, ser. 2, vol. 18, No. 3, pp. 120–140, Dec.

1911.

(80) The published and unpublished work of Charles Sturm on algebraic and differential equations. Presidential address delivered before the American Mathematical Society, April 28, 1911. *Bulletin of the American Mathematical Society*, ser. 2, vol. 18, No. 1, pp. 1–18, Oct.

(81) Boundary problems and Green's functions for linear differential and difference equations. *Annals of Mathematics*, ser. 2, vol. 13, No. 2, pp. 71–88, Dec.

(82) Graduate work in mathematics in universities and in other institutions of like grade in the United States. General report. *United States Bureau of Education Bulletin*, No. 6, pp. 7–20. Also in *Bulletin of the American Mathematical Society*, ser. 2, vol. 18, No. 3, pp. 122–137, Dec.

1912.

(83) On linear equations with an infinite number of variables. By Maxime Bôcher and Louis Brand. *Annals of Mathematics*, ser. 2, vol. 13, No. 4, pp. 167–186, June.

(84) A simple proof of a fundamental theorem in the theory of integral equations. *Annals of Mathematics*, ser. 2, vol. 14, No. 2, pp. 84–85, Dec.

1913.

(85) Applications and generalizations of the conception of adjoint systems, *Transactions of the American Mathematical Society*, vol. 14, No. 4, pp. 403–420, Oct.

(86) Doctorates conferred by American universities [Letter to the editor], *Science*, n. s., vol. 38, No. 981, p. 546, Oct.

(87) Boundary problems in one dimension [A lecture delivered Aug. 27, 1912]. Proceedings of the Fifth International Congress of Mathematicians, Cambridge, England, University Press, vol. 1, pp. 163–195.

1914.

(88) The infinite regions of various geometries. *Bulletin of the American Mathematical Society*, ser. 2, vol. 20, No. 4, pp. 185–200, Jan.†

(89) On Gibbs's phenomenon. *Journal für die reine und angewandte Mathematik*, vol. 144, No. 1, pp. 41–47, Jan.

---

* A second edition appeared in 1914.

† See vol. 22, (1915) No. 1, p. 40, Oct.

(90) Mathématiques et mathématiciens français. *Rerue Internationale de l'Enseignement*, vol. 67, No. 1, pp. 20–31, Jan.

(91) Charles Sturm et les mathématiques modernes. *Revue du Mois*, vol. 17, No. 97, pp. 88–104, Jan.

(92) On a small variation which renders a linear differential system incompatible. *Bulletin of the American Mathematical Society*, ser. 2, vol. 21, No. 1, pp. 1–6, Oct.

(93) The smallest characteristic number in a certain exceptional case. *Bulletin of the American Mathematical Society*, ser. 2, vol. 21, No. 1, pp. 66–99, Oct.

1915.

(94) Trigonometry with the theory and use of logarithms. By Maxime Bôcher and H. D. Gaylord, New York, Holt, 9 + 142 pp.

(95) Plane analytic geometry with introductory chapters on the differential calculus. New York, Holt, 13 + 235 pp.

1916.

(96) Review of Gibb: A Course in Interpolation etc. and Carse and Shearer: A Course in Fourier's Analysis etc. *Bulletin of the American Mathematical Society*, ser. 2, vol. 22, No. 7, pp. 359–361, April.

(97) On the Wronskian test for linear dependence, *Annals of Mathematics*, ser. 2, vol. 17, No. 4, pp. 167–168, June.

(98) Syllabus of a Brief Course in Solid Analytic Geometry. Lancaster, New Era Press, 10 p.

(99) Leçons sur les méthodes de Sturm dans la théorie des équations différentielles linéaires et leurs développements modernes. Professées à la Sorbonne en 1913–14. Recueillies et rédigées par G. Julia. Paris, Gauthier-Villars, 6 + 118 pp.

1917.

(100) Note supplementary to the paper "On certain pairs of transcendental functions whose roots separate each other." *Transactions of the American Mathematical Society*, vol. 18, No. 4, pp. 519–521, Oct.

1918

(101) Concerning direction cosines and Hesse's normal form. *American Mathematical Monthly*, vol. 25, No. 7, pp. 308–310, Sept.

Joseph L. Walsh (1895–1973) was educated at Harvard, receiving a bachelor's degree in 1916 and a Ph.D. in 1920. His thesis adviser was G. D. Birkhoff. He was on the Harvard faculty from 1921 until his retirement in 1966, when he moved to a special chair at the University of Maryland. He did basic research in complex approximation theory, conformal mapping, harmonic functions, and orthogonal expansions. The Walsh functions, a complete orthonormal extension of the Rademacher functions, became important in digital communication. Walsh was President of the AMS and a member of the National Academy of Sciences. His biographical sketch of Osgood is published here for the first time by permission of the Harvard University Archives.

William Fogg Osgood [1]

J. L. WALSH

William Fogg Osgood (March 10, 1864–July 22, 1943) was born in Boston, Massachusetts, the son of William and Mary Rogers (Gannett) Osgood. He prepared for college at the Boston Latin School, entered Harvard in 1882, and was graduated with the A.B. degree in 1886, second in his class of 286 members. He remained at Harvard for one year of graduate work in mathematics, received the degree of A.M. in 1887, and then went to Germany to continue his mathematical studies. During Osgood's study at Harvard, the great Benjamin Peirce (1809–1880), who had towered like a giant over the entire United States, was no longer there. James Mills Peirce (1834–1906), son of Benjamin, was in the Mathematics Department, and served also later (1890–1895) as Dean of the Graduate School and (1895–1898) as Dean of the Faculty of Arts and Sciences. William Elwood Byerly was also a member of the Department (1876–1913), and is remembered for his excellent teaching and his texts on the Calculus and on Fourier's Series and Spherical Harmonics. Benjamin Osgood Peirce (1854–1914) was a mathematical physicist, noted for his table of integrals and his book on Newtonian Potential Theory. Osgood was influenced by all three of those named — they were later his colleagues in the department — and also by Frank Nelson Cole.

---

[1] Reproduced with permission of Harvard University Archives, from the papers of Joseph L. Walsh.

William Fogg Osgood

Cole graduated from Harvard with the Class of 1882, studied in Leipzig from 1882 to 1885, where he attended lectures on the theory of functions by Felix Klein, and then returned to Harvard for two years, where he too lectured on the theory of functions, following Klein's exposition.

Felix Klein left Leipzig for Göttingen in 1886, and Osgood went to Göttingen in 1887 to study with him, Klein (Ph.D., Göttingen, 1871) had become famous at an early age, especially because of his Erlanger Program, in which he proposed to study and classify geometries (Euclidean, hyperbolic, projective, descriptive, etc.) according to the groups of transformations under which they remain invariant; thus Euclidean geometry is invariant under the group of rigid motions. The group idea was a central unifying concept that dominated research in geometry for many decades. Klein was also interested in the theory of functions, following the great Göttingen tradition, especially in automorphic functions. Later he took a leading part in organizing the *Enzyklopädie der Mathematischen Wissenchaften*, the object of which was to summarize in one collection all mathematical research up to 1900. Klein also had an abiding interest in elementary mathematics, on the teaching of which he exerted great influence both in Germany and elsewhere.

The mathematical atmosphere in Europe in 1887 was one of great activity. It included a clash of ideals, the use of intuition and arguments borrowed from physical sciences, as represented by Bernhard Riemann (1826–1865) and his school, versus the ideal of strict rigorous proof as represented by Karl Weierstrass (1815–1897), then active in Berlin. Osgood throughout his mathematical career chose the best from the two schools, using intuition in its proper place to suggest results and their proofs, but relying ultimately on rigorous logical demonstrations. The influence of Klein on "the arithmetizing of mathematics" remained with Osgood during the whole of his later life.

Osgood did not receive his Ph.D. from Göttingen. He went to Erlangen for the year 1889–1890, where he wrote a thesis, "Zur Theorie der zum algebraischen Gebilde $y^m = R(x)$ gehörigen Ablesschen Functionen." He received the degree there in 1890 and shortly after married Theresa Ruprecht of Göttingen, and then returned to Harvard.

Osgood's thesis was a study of Abelian integrals of the first, second, and third kinds, based on previous work by Klein and Max Noether. He expresses in the thesis his gratitude to Max Noether for aid. He seldom mentioned the thesis in later life; on the one occasion that he mentioned it to me he tossed it off with "Oh, they wrote it for me." Nevertheless, it was part of the theory of functions, to which he devoted so much of his later life.

In 1890 Osgood returned to the Harvard Department of Mathematics, and remained for his long period of devotion to the science and to Harvard. At about this time a large number of Americans were returning from graduate work in Germany with the ambition to raise the scientific level of mathematics in this country. There was no spirit of research at Harvard then, except what Osgood himself brought, but a year later Maxime Bôcher (A.B., Harvard, 1888; Ph.D., Göttingen, 1891) joined him there, also a student greatly influenced by Felix Klein, and a man of mathematical background and ideals similar to those of Osgood. They were very close friends both personally and in scientific work until Bôcher's death in 1918.

Osgood's scientific articles are impressive as to their high quality. In 1897 he published a deep investigation into the subject of uniform convergence of sequences of real continuous functions, a topic then as always of considerable importance. He found it necessary to correct some erroneous results on the part of du Bois Reymond, and established the important theorem that a bounded sequence of continuous functions on a finite interval, convergent there to a continuous function, can be integrated term by term. Shortly thereafter, A. Schoenflies was commissioned by the Deutsche Mathematiker-Vereinigung to write a report on the subject of Point Set Theory. Schoenflies wrote to Osgood, a much younger and less illustrious man, that he did not consider Osgood's results correct. The letter replied in the spirit that he was surprised at Schoenflies' remarkable procedure, to judge a paper without reading it. When Schoenflies' report appeared (1900), it devoted a number of pages to an exposition of Osgood's paper. Osgood's result, incidentally, as extended to non-continuous but measurable functions, became a model for Lebesgue in his new theory of integration (1907).

In 1898 Osgood published an important paper on the solutions of the differential equation $y' = f(x, y)$ satisfying the prescribed initial conditions $y(a) = b$. Until then it had been hypothesised that $f(x, y)$ should satisfy a Lipschitz condition in y: $|f(x, y_1) - f(x, y_1)| \leq M|y_1 - y_2|$, from which it follows that a unique solution exists. Osgood showed that if $f(x, y)$ is merely continuous there exists at least one solution, and indeed a maximal solution and a minimal solution, which bracket any other solution. He also gave a new sufficient condition for uniqueness.

In 1900 Osgood established, by methods due to H. Poincaré, the Riemann mapping theorem, namely that an arbitrary simply connected region of the plane with at least two boundary points, can be mapped uniformly and conformally onto the interior of a circle. This is a theorem of great importance, stated by Riemann and long conjectured to be true, but without a satisfactory proof. Some of the greatest European mathematicians (e.g., H. Poincaré, H. A. Schwarz) had previously attempted to find a proof but without success. This theorem remains as Osgood's outstanding single result.

Klein had invited Osgood to collaborate in the writing of the *Enzyklopädie*, and in 1901 appeared Osgood's article "Allgemeine Theorie der analytischen Funktionen a) einer und b) mehrerer komplexen Grössen." This was a deep, scholarly, historical report on the fundamental processes and results of mathematical analysis, giving not merely the facts but including numerous and detailed references to the mathematical literature. The writing of it gave Osgood an unparalleled familiarity with the literature of the field.

In 1901 and 1902 Osgood published on sufficient conditions in the Calculus of Variations, conditions which are still important and known by his name. He published in 1903 an example of a Jordan curve with positive area, then a new phenomenon. In 1913 he published with E. H. Taylor a proof of the one-to-oneness and continuity on the boundary of the function mapping a Jordan region onto the interior of a circle; this fact had been conjectured from physical considerations by Osgood in his *Enzyklopädie* article, but without demonstration. The proof was by use of potential theory, and a simultaneous proof by functional-theoretic methods was given by C. Carathéodory.

In 1922 Osgood published a paper on the motion of the gyroscope, in which he showed that intrinsic equations for the motion introduce simplifications and make the entire theory more intelligible.

From time to time Osgood devoted himself to the study of several complex variables; this topic is included in his *Enzyklopädie* article. He published a number of papers, gave a colloquium to the American Mathematical Society (1914) on the subject, and presented the first systematic treatment in his *Funktionentheorie*. He handled there such topics as implicit function theorems, factorization, singular points of analytic transformations, algebraic functions and their integrals, uniformization in the small and in the large.

It will be noted that Osgood always did his research on problems that were both intrinsically important and classical in origin — "problems with a pedigree," as he used to say. He once quoted to me with approval a German professor's reply to a student who had presented to him an original question together with the solution, which was by no means trivial: "Ich bestreite Ihnen das Recht, ein beliebiges Problem zu stellen und aufzulösen."

Osgood loved to teach, at all levels. His exposition was not always thoroughly transparent, but was accurate, rigorous, and stimulating, invariably with emphasis on classical problems and results. This may have been due in some measure to his great familiarity with the literature through writing the *Enzyklopädie* article. He also told me on one occasion that his own preference as a field of research was real variables rather than complex, but that circumstances had constrained him to deal with the latter; this may also have been a reference to the *Enzyklopädie*.

Osgood's great work of exposition and pedagogy was his *Funktionentheorie*, first published in 1907 and of which four later editions were published. Its purpose was to present systematically and thoroughly the fundamental methods and results of analysis, with applications to the theory of functions of a real and of a complex variable. It was more systematic and more rigorous that the French traités d'analyse, also far more rigorous than, say, Forsyth's theory of functions. It was a moment to the care, orderliness, rigor, and didactic skill of its author. When G. Pólya visited Harvard for the first time, I asked him whom he wanted most to meet. He replied "Osgood, the man from whom I learned function theory" — even though he knew Osgood only from his book. Osgood generously gives Bôcher part of the credit for the *Funktionentheorie*, for the two men discussed with each other many of the topics contained in it. The book became an absolutely standard work wherever higher mathematics was studied.

Osgood had previously (1897) written a pamphlet on Infinite Series, in which he set forth much of the theory of series needed in the Calculus, and his text on the Calculus dates from 1907. This too was written in a careful exact style, that showed on every page that the author knew profoundly the material he was presenting and its background both historically and logically. It showed too that Osgood knew the higher developments of mathematics and how to prepare the student for them. The depth of Osgood's interest in the teaching of the calculus is indicated also by his choice of that topic for his address as retiring president of the American Mathematical Society in 1907.

Osgood wrote other texts for undergraduates, in 1921 an Analytic Geometry with W. C. Graustein, which again was scholarly and rigorous, and in 1921 a revision of his *Calculus*, now called *Introduction to the Calculus*. In 1925 he published his *Advanced Calculus*, a masterly treatment of a subject that he had long taught and that had long fascinated him. He published a text on Mechanics in 1937, the outgrowth of a course he had frequently given, and containing a number of novel problems from his own experience.

After Osgood's retirement from Harvard in 1933 he spent two years (1934–1936) teaching at the National University of Peking. Two books in English of his lectures there were prepared by his students and published there in 1936: *Functions of Real Variables* and *Functions of a Complex Variable*. Both books borrowed largely from the *Funktionentheorie*.

Osgood did not direct the Ph.D. theses of many students; the theses he did direct were those of C. W. Mcg. Blake, L. D. Ames, E. H. Taylor, and (with C. L. Bouton) G. R. Clements. I asked him in 1917 to direct my own thesis, hopefully on some subject connected with the expansion of analytic functions, such as Borel's method of summation. He threw up his hands, "I know nothing about it."

Osgood's influence throughout the world was very great, through the soundness and depth of his *Funktionentheorie*, through the results of his own research, and through his stimulating yet painstaking teaching of both undergraduates and graduate students. He was intentionally raising the scientific level of mathematics in America and elsewhere, and had a great part in this process by his productive work, scholarly textbooks, and excellent classroom teaching.

Osgood's favorite recreations were touring in his motor car, and smoking cigars. For the latter, he smoked until little of the cigar was left, then inserted the small blade of a penknife in the stub so as to have a convenient way to continue.

Osgood was a kindly man, somewhat reserved and formal to outsiders, but warm and tender to those who knew him. He had three children by Mrs. Teresa Ruprecht Osgood: William Ruprecht, Freida Bertha (Mrs. Walter Sitz, now deceased), Rudolph Ruprecht. His years of retirement were happy ones. He married Mrs. Celeste Phelpes Morse in 1932, and died in 1943. He was buried in Forest Hills Cemetery, Boston.

Harold L. Dorwart received an A. B. from Washington and Jefferson College in 1924. He then joined the Yale department as a teaching assistant and graduate student. After teaching at Williams College in 1928–1930, he returned to Yale to complete a Ph.D. in 1931 under the guidance of Oystein Ore. He then taught at Williams, at Washington and Jefferson, and at Trinity College (Hartford), where he served as chairman of the mathematics department and as dean of the college, retiring in 1968. He is the author of numerous articles on irreducibility criteria, the Tarry–Escott problem, configurations and other geometric topics, and a book, The Geometry of Incidence.

Mathematics and Yale in the Nineteen Twenties

HAROLD L. DORWART

Mathematics today occupies a position of considerable importance and respect in all American colleges and universities, and the prolific work of American mathematicians is highly regarded all over the world. However, as recently as fifty to seventy-five years ago, the situation was somewhat different. An article by E. J. McShane in the February 1976 issue of the *American Mathematical Monthly* contains the following sentence: "At the beginning of this century the set of mathematicians in the United States (hardly to be thought of as a mathematical community) consisted almost exclusively of a few professors in colleges and universities, a very few of whom tried to get some research done in the time that their teaching did not fill." Contrast this with three sentences from a 1984 National Research Council report that describes the present situation: "The mathematical sciences research community in the United States [now] has over 10,000 members. About 9,000 of them are faculty members in educational institutions and have research as their primary or secondary activity. They are part of the larger group of 14,000 doctoral mathematical scientists for whom teaching or research is the primary/secondary activity."

The decade from the early 1920s to the early 1930s appears to be a transitional period from what is now probably called classical mathematics to the so-called modern mathematics that we know today. I say "so-called" because "modern" is always a relative term and much of our present-day mathematics had its roots and early development before 1920. As a result of the reforms and reorganization of 1918–1919 (thoroughly described and documented in George Wilson Pierson's two-volume history of Yale and in Brooks Mather Kelley's more recent one-volume history), undergraduate teaching was given great importance in this period. Hence there is good reason to consider the decade of the twenties as an interesting one for mathematics at Yale. But before considering this period in detail, a brief historical introduction will be given. This will be drawn in part from an article by P. F. Smith [4].

Arithmetic and geometry were a part of the curriculum at the founding of Yale College; algebra soon followed; and in 1743 fluxions (calculus) was offered to juniors. The first professorship at Yale was established in 1755 and was for sacred theology. The second, in 1770, was for mathematics, natural philosophy, and astronomy. (This was thirty years before the professorship in ancient languages was established.) At Cambridge University, the chairs in theology were also the oldest and those in mathematics and astronomy were established in the seventeenth and eighteenth centuries. The importance of astronomy in those days was due to its application to ocean navigation in a mercantile Atlantic civilization. The Reverend Jeremiah Day of the class of 1795 held the chair of mathematics and natural philosophy from 1803 to 1820 and continued to lecture on the latter subject after he became the fifth president of Yale in 1817. The title professor of mathematics appears for the first time in the catalogue of 1841–1842, and the incumbent of the chair was Anthony D. Stanley of the class of 1830. Stanley died in 1854 and instruction in mathematics was then given by a tutor, H. A. Newton, class of 1850, "whose mathematical talents were so unusual that he was elected to the chair in 1855 at the age of twenty-five." He held this professorship until his death in 1896.

In 1861, Yale became the first university in America to award a Ph.D. degree. The first Ph.D. in mathematics was awarded in 1862 to John Hunter Worrall (title of dissertation unknown). Charles Greene Rockwood followed in 1866 with a mathematics dissertation entitled "The daily motion of a brick tower caused by solar heat." In 1863, Josiah Willard Gibbs (for whom the Gibbs phenomenon in Fourier analysis and the American Mathematical Society Gibbs Lectures are named) received a Ph.D. in physics with a dissertation called "The form of the teeth of wheels in spur gearing." Among the other early Ph.D.s in mathematics at Yale were Andrew Wheeler Phillips (1877, "On three-bar motion"), later professor of mathematics and dean of the graduate school at Yale; and Eliakim Hastings Moore (1885, "Extensions of certain theorms of Clifford and Cayley in the geometry of n dimensions"), a major figure who went on to an outstanding career at the University of Chicago. (Parshall [3] gives an account of E. H. Moore's influence on American mathematics.) The first woman to obtain a Ph.D. in mathematics from

Yale was Charlotte Cynthia Barnum (B.A. Vassar College; Yale Ph.D. 1895, "Functions having linear or surfaces of discontinuity").

Actually, the early degrees at Yale were classified retroactively after the university was reorganized by departments in 1919. Writing in 1927, Wilbur Cross [1] explains as follows:

> Some difficulty has been encountered in the distribution of doctorates according to the Departments of Study as now constituted. In the early days there was only the Department of Philosophy and the Arts, which covered all subjects in which instruction was given. It was then the custom of a student to submit to a Committee the course of study which he desired to pursue. The program for which he received approval might be of a rather heterogeneous character provided he had in mind a specific question for investigation. Thus the late Josiah Willard Gibbs, who obtained his degree in 1863, enrolled as a student in philology and mathematics and took as the subject of his dissertation a problem in mechanics such as would now lie within the field of physics. Subsequently were organized several general departments, some of which in the last twenty-five years have been divided as an inevitable result of the progress of knowledge. Thus there often arises doubt concerning the proper assignment of recipients of the Ph.D. degree. In these cases the guiding principle has been the subject of the dissertation as presented for the degree.

During the nineteenth century most of the activity in advanced mathematics took place in England and on the Continent, particularly in Germany and in France, and that was where Americans went for their training. Newton became the first Yale mathematician to do this, studying in Paris for a year before assuming his duties as professor at Yale. He later "made important researches on the origin of meteoric showers ...."

James Pierpont (1869–1941), whose doctorate was obtained at the University of Vienna, was appointed lecturer at Yale in 1894 and was made a professor four years later. He taught courses that were called modern mathematical analysis at that time. In 1896, Percey F. Smith (1867–1956) — after studying at the Universities of Göttingen, Berlin, and Paris from 1894 to 1896 — as an assistant professor at Yale offered courses in advanced geometry. In 1907, Ernest W. Brown (1867–1938), a native of Hull, England, whose undergraduate and graduate studies were at Cambridge University and whose chief work had been in mathematical astronomy, joined the Yale mathematics department. Thus even in those far-off days Yale was probably in the forefront of advanced mathematical education in the United States.

David Eugene Smith in his 1906 *History of Modern Mathematics* states: "... a remarkable change is at present passing over the mathematical work done in the universities and colleges of this country. Courses that a short time ago were offered in only a few of our leading universities are now not uncommon in institutions of college rank. They are often given by men who have taken advanced degrees in mathematics at Göttingen, Berlin, Paris, or other leading universities abroad, and they are awakening a great interest in the modern field. A recent investigation in 1903 showed that 67 students in ten American institutions were taking courses in the theory of functions, 11 in the theory of elliptic functions, 94 in projective geometry, 26 in the theory of invariants, 45 in the theory of groups, and 46 in the modern advanced theory of equations, courses which only a few years ago were rarely given in this country.

Yale was probably one of the institutions described by Smith.

The usual tests for the professional esteem in which a mathematics professor is held by his peers have been his publication record, and whether or not he was a top-level officer of the American Mathematical Society and/or a Colloquium or Josiah Willard Gibbs Lecturer, both of which carry considerable prestige. In addition to his numerous published papers, Professor Pierpont's books on complex and real variables were highly regarded in the early years of this century. He was the Colloquium Lecturer in 1896 (the first year in which this series of lectures was given) and was the Gibbs Lecturer in 1925 (the third one in this series). E. W. Brown's chief work was on the theory of motion of the moon, and many decades of his mathematical calculations "compelled the revision of astronomical tables of the entire solar system." He was President of the AMS in 1915–1916, Colloquium Lecturer in 1901, and Gibbs Lecturer in 1927. In 1937 he was awarded the Watson medal of the National Academy of Science for distinguished contributions to astronomical science. Professor Smith's publications were mainly in the field of college and university textbooks and several of these were widely used during his lifetime. He was editor of the *Transactions of the American Mathematical Society* from 1917 to 1920.

These three men were still active when I came to Yale as a graduate student and teaching assistant in 1924 and I had at least one course with each of them. All were interesting in appearance but in quite different ways. Professor Pierpont was a large man with full beard and fairly long hair — both almost white — at a time when beards and long hair were rarely seen. He seldom wore a hat; always carried a book bag; and frequently wore a cape as an outergarment. He was fond of going to the movies in the afternoon — apparently regardless of what kind of picture was being shown — and was an eye-catching sight as he walked at a quick pace across the New Haven Green with hair flying and book bag over his shoulder.

The only course I had with Professor Pierpont was complex variables, in which he used his own text (written many years earlier). He kept a copy locked in a drawer of the classroom desk and each class period would remove the book and ask some member of the class what sections had been assigned for the day. After glancing over these he would frequently say, "I think there's a better way to prove that theorem." Usually he was right, but sometimes he would flounder around for most of the period to no avail. Even in these situations we were able to see how a creative mind worked, and that there was frequently much trial and error in mathematics.

Professor Pierpont was normally pleasant and good-natured but sometimes he became very agitated. I remember once when I inadvertently aroused his ire. He had asked a question for which I volunteered an answer. Immediately he demanded the basis for my statement, and when I replied "Granville's *Calculus*," the storm broke. I had not been warned that W. A. Granville, Yale Ph.B. and Ph.D. 1897, had refused to take Professor Pierpont's advice when he wrote what became one of the most popular and best selling of the early calculus textbooks, but one that did not exhibit much rigor. Fortunately we soon took up the subject of conformal representation and I made a series of drawings (which I enjoyed doing) that greatly pleased my instructor.

As a result of his years abroad, Professor Pierpont always bemoaned the fact that American professors were not better paid and held in higher regard by the general public. In fact he frequently said to graduate students that he couldn't understand why anyone would want to prepare for teaching in an American college. Four years after he retired in 1934 and moved to San Mateo, California, I sent him a reprint of a paper I had published. In his reply, after thanking me and saying that hearing from me brought back the "happy days at Yale," he characteristically went on to say, "You see, I have gone as far West as the Pacific Ocean lets me. Thousands of bums, down and outers, et al., have been stopped by the same means. But we are all looking for 'the more abundant life' and other administrative favors."

Percey F. Smith , Yale Ph.B. 1888, Ph.D. 1891, spend his entire academic career at Yale (except for the two years of study abroad that has already been mentioned). He was instructor, 1888–1894, assistant professor, 1895–1900, and professor, 1900–1936. Moreover, he was chairman most (if not all) of the years he held professorial rank. Tall, always dressed in a three-piece suit, and with an authoritative mien, he guided the department with a firm hand. The effect that he had on me can best be explained by describing what happened when I presented myself to him for my language exams. He pulled a bound volume of a foreign periodical from his bookshelf, opened it, and told me to read. Fortunately, the language was French — in which I was reasonably proficient — and the article concerned a geometric topic with which I was familiar, so I had no difficulty. He then opened the book to other sections, stopping me each time I appeared to be able to give a reasonable translation. Finally he closed the book and said, "Well, you passed." I immediately said, "But I wanted to take the German exam also." Even after almost sixty years have gone by I can still see the look of utter amazement on his face as he replied, "Just what language do you think you have been reading for the past five minutes?"

In the early 1920s Professor Smith had a serious medical problem which resulted in the amputation of one leg in 1925. But — except for a short period of recuperation — this did not slow him down. He could still drive his car, and he managed his crutches with considerable ease. His office was on the ground floor of Old Sheff and he taught in a basement classroom in nearby North Sheff which had a ground-level parking lot in the rear. I can remember having one course with him "Geometrical Transformations and Continuous Groups" (Lie Theory) in that room where I was the only member of the class. Lectures were conducted just as formally as if the room were filled with students.

Ernest W. Brown was a genial Englishman who continually smoked cigarettes. He was an avid reader of detective stories but complained that he had difficulty in finding ones he had not already read or where, after reading the first dozen or so pages, he couldn't guess who committed the crime. About 1925 or 1926 there was a total eclipse (I remember going to the top of East Rock early one cold winter morning with a piece of smoked glass to observe it) and Professor Brown received considerable publicity because his work on the New Lunar Tables (published by the Yale University Press) enabled prediction of the time of the eclipse to be made with great accuracy.

Additional appointments to the department between 1906 and 1916 under the presidency of Arthur Twining Hadley (Smith and Pierpont had been appointed when the second Timothy Dwight was president) and the universities where each man pursued his main mathematical studies were: W. R. Longley (Chicago), E. J. Miles (Chicago), J. I. Tracey (Johns Hopkins), J. K. Whittemore (Harvard), and W. A. Wilson (Yale). Professor Longley's specialties were periodic orbits and differential equations (when a department of astronomy was set up at Yale, Professor Longley was listed in that department as well as in the department of mathematics). From 1926 to 1937 Longley was an associate editor of the *Bulletin of the American Mathematical Society*, and he ended his career as chairman of the mathematics department from 1945 to 1949. Miles had concentrated on calculus of variations, Tracey on projective rational curves, Whittemore on differential geometry, Wilson on real variables (a large portion of his Ph.D. thesis was included in one of Professor Pierpont's books on analysis), and they each taught graduate courses and directed dissertations in these areas. These five and the three mentioned earlier constituted the tenured faculty in mathematics in 1924. The following vignettes are a part of a sixty-year-old memory of these men.

W. R. Longley — a pleasant, dignified professor — conducted his class in a somewhat formal manner. In the one course I took with him, near the beginning he assigned to each member of the class a specific topic that we were to digest on our own and then make a presentation to the class near the end of the term. I do not remember this procedure being followed in any other course that I took at Yale.

E. J. Miles was an enthusiastic and well-liked professor. He once told me that he had always hoped to write an elementary, popular, understandable book on calculus that he would call *Change*. Each year, Miles taught an advanced calculus course that was required for first-year graduate students and that could be elected by qualified upperclassmen — a course that was particularly popular with engineering students. The year I took the course, in order to accommodate all who wanted to register, Miles scheduled the course from 7 to 8 a.m. Quite naturally during that early hour some students who had been up late the night before became sleepy. The instructor conducted his class by pacing back and forth in front of the blackboard with a piece of chalk in his right hand and with a yardstick held firmly in his left hand. Whenever he spotted a drowsy student, that individual received a vigorous poke in the stomach accompanied by a thunderous "Do you see?" — uttered as if it were one word "J'see?" This procedure — unusual in a college class — was highly effective.

J. I. Tracey was also a greatly admired and approachable professor. I remember that he daily rode a bicycle from his home to his office. Students who might otherwise be lonely at Thanksgiving were frequently invited to share dinner with his family. Although Tracey's field of specialization was generally considered in the late twenties to be a largely "worked out" and even a "dead end" area, so popular was he with graduate students that three of the four Ph.D. dissertations of 1927 and 1928 were directed by him. The fourth was directed by Miles.

J. K. Whittemore — a true gentleman scholar whose lectures on differential geometry were polished gems — frequently used material from somewhat obscure French sources. A lecture usually concluded with several "Exercises" for the class members, some of which were routine but a number of which were minor research projects. Unfortunately, his approach and methods were what E. T. Bell has described as the "classical differential geometry which was that of the majority of professionals in the 1920s," at a time when the tensor analysis of Ricci and Levi–Civita was just becoming known. I believe that Professor Whittemore later spent a year at Princeton and adapted to the new orientation in his field.

W. A. Wilson conducted lectures in real variables in an extremely expeditious manner. His classroom had blackboards on three sides and he would start at the back of one side of the room — talking and writing rapidly. He would continue across the front of the room and along the other side, and almost always would fill the last bit of blackboard just as the bell rang for the end of the period, when he would then make his exit from the room. Wilson's students of course were copying notes furiously and rarely had time to digest what was being said. I have often wished I could have taken this course in a later period when "handouts" of notes might have been available, and when a real classroom discussion could have been held.

These men formed a mature group with Smith and Brown each 57, Pierpont 55, and the youngest probably at least 40. There were also usually six to eight young instructors and teaching assistants in the department who each stayed for several years and then moved elsewhere. In 1924, Professor Smith [4] wrote that the program of studies for the highest degrees in mathematics is "calculated to develop the student's power and interest in research either in analysis, geometry, or applied mathematics." Note the absence of the two areas — topology and modern abstract algebra — that were soon to assume what one might almost call overriding importance in mathematics.

Although the word topology was not used during the 1920s, the forerunner of this subject — analysis situs — was beginning to attract some attention. Actually, H. M. Gehman, fresh from a year as an NRC fellow at the University of Texas where he worked with R. L. Moore, who was one of the pioneers in the field, was at Yale from 1926 to 1929 and gave a course in the subject in at least one of these years. (To my later regret, I thought it would be a passing fad and did not take the course!) By 1930–1931, Professor Wilson was offering two courses entitled Functions of Real Variables and Analysis Situs I and II.

Although the axiomatic method had had a certain amount of earlier usage in various parts of mathematics such as geometry, it was the appearance of van der Waerden's textbook on modern algebra in 1930 that first placed algebra on an equal footing with analysis and started the great rush in American universities to that field and to the development of abstract postulational mathematics. In 1927 Øystein Ore (1899–1968), a Norwegian who had studied at Göttingen and the Sorbonne and had received his Ph.D. in 1924 from Oslo University, came to Yale as an assistant professor. He at once offered a course in theory of algebraic numbers, a topic that may be thought of as a prerequisite for modern algebra, which was a field where Ore had been an active participant on the Continent. Three years later Ore was made Sterling Professor and also given the new title of director of graduate studies in mathematics. Professor Ore was the Colloquium Lecturer in 1941. From 1936 to 1945 he was chairman and recruited many fine mathematicians to the department, beginning with Marshall Stone in 1931 (who later went to

Harvard and then to Chicago, was President of the AMS in 1943–1944, Colloquium Lecturer in 1939, and Gibbs Lecturer in 1956), and Einar Hille in 1933 (President of the AMS in 1947–1948, and Colloquium Lecturer in 1944). These were followed by mathematicians of the caliber of G. A. Hedlund (Colloquium Lecturer in 1949), and Nathan Jacobson (President of the AMS 1971–1972, and Colloquium Lecturer in 1955).

Now that — after a historical introduction — we have considered some of the graduate mathematics courses offered at Yale in the mid and late twenties and the men who taught these courses, it is time to consider the changes in undergraduate courses brought about as a result of the 1918–1919 reforms and reorganization, and the effects of these on both faculty and students. Three major objectives of the reforms and reorganization were:

(1) more effective undergraduate teaching;

(2) a common freshman year for Ac (the academic department of Yale College) and Sheff (the Sheffield Scientific School) with a specially designated freshman faculty (a 1984 National Institute of Education report on college teaching urges colleges to "reallocate faculty" so that the "finest instructors" are assigned to freshmen); and

(3) faculty members to be organized by departments (i.e., history, physics, etc.) rather than by schools (Ac, Sheff) as they had been previously.

Designing a common mathematics course for future Ac and Sheff students that would be interesting, that would accommodate the varying degrees of preparation of entering students, and that would be a good introductory course for those continuing their mathematics, was a real challenge to the department. The average number of freshmen admissions in the 1920–1929 decade was approximately 850 and — while mathematics was not a required subject — almost half of each class did enroll in the freshman course. At most American colleges in the first two decades of the twentieth century, the freshman course in mathematics consisted of college algebra and trigonometry with heavy emphasis on the computational aspects (Horner's method for irrational roots of equations, solutions of plane and spherical triangles, etc.). This was good training for those who would later use logarithmic tables extensively but not very inspiring for the others. The sophomore course was usually a four-hour semester course in plane and solid analytic geometry, followed by a four-hour course in differential calculus. The junior year course was devoted to integral calculus and related topics. This is the program I followed as an undergraduate.

The freshman course designed by the Yale department would now be considered to be of a standard type, but in the early twenties it was a daring break with tradition. It integrated (no pun intended) the two branches of calculus together with the requisite amount of analytic geometry introduced when needed. Professors Longley and Wilson wrote the textbook (available in published form shortly before 1924) and the course was a joy to teach. The material was new to almost all students and — since no degree of rigor was demanded — most of them found the course relatively easy, but they were still challenged by some difficult problems. And there were always the Barge prize exams for the very best students.

For freshmen with poor or insufficient preparation, 2 two-hour semester courses (noncredit as I recall it) were provided. The first one was standard trigonometry and the second one (very interesting to me at least) was called solid geometry and mensuration and used a booklet written by Professor Longley where formulas for the volume of a cone, pyramid, sphere, etc., were found by a summation method (essentially that of integral calculus). I was permitted to teach this course entirely on my own in 1924–1925 but find no record of it after that year.

In his 1976 *Yale: A short history*, Pierson mentions "an odd consequence" of the fact that in the period we are considering almost all of Yale's scholar-teachers taught undergraduates. "The able Yale seniors went elsewhere for their graduate and professional training; they felt they already knew Yale's great men in their own fields." However, at that time it was probably also true that Yale was not the best place for training in modern mathematics. Two outstanding Yale seniors of this period who went elsewhere were Hassler Whitney (Yale Ph.B. 1928, Harvard Ph.D. 1932) and Saunders Mac Lane (Yale Ph.B. 1930, Chicago M.A. 1931, Göttingen D.Phil. 1934). One who returned was Marshall Hall, Jr. (Yale A.B. 1932, Yale Ph.D. 1936), but he studied at Cambridge University in 1932–1933 and would have stayed there for his Ph.D. had funds been available [2].

Aside from graduate studies and teaching, what was it like to live in New Haven in the twenties? Although the early 1920s was a fairly prosperous period, all graduate students appeared to be poor. My 1924–1925 letter of appointment (signed by Robert Maynard Hutchins as secretary of the university) stated that my salary would be "$500 per year, together with tuition in the graduate school." It went up $100 in each of the two following years and finally in 1927–1928 I was promoted to the rank of instructor with the almost unbelievable salary of $1,000. Actually once you learned how, and learned to exercise a certain degree of restraint, you could live on very little at that time. Room rent was low, good and inexpensive meals could be obtained at the Commons cafeteria, many lectures and concerts were free, the athletic department gave graduate students a break on football seats, and by standing in line in the alley next to the Shubert Theater you could buy second balcony seats for Broadway tryouts. Certainly none of us had cars and in fact not even bicycles (although we sometimes had to step lively to keep out of the way of undergraduates on motorcycles). But there was excellent and cheap public transportation. My memory is that it was still all trolley cars in 1924 but there may have been a few bus lines later on. Walking and hiking had always been my favorite outdoor activity since boy scout days, and even if you had only a couple of hours to spare you could take a trolley ride to East or West Rock parks and then enjoy the many trails that they contained. If a half day or more was available, say on a weekend or during a vacation, an expedition to Sleeping Giant State Park or to Lighthouse Point could be taken. The city of New Haven, with its many ethnic neighborhoods to be explored, was always a source of interest and I never felt any fear when wandering about during the day or early evening.

Perhaps the best way to end this rambling discussion and these reminiscences will be with another quotation from Professor Pierson: "As the decade of the twenties recedes, it will more and more come to be regarded as one of Yale's notable periods, an era of great vigor and achievement. For all their turbulence, these were the years of self-appraisal, self-strengthening, and striking out along new lines." I am happy that I had the privilege of being there at that time.

REFERENCES

[1] Wilbur L. Cross, "Doctors of Philosophy," *Bulletin of Yale University*, Nov. 1, 1927.

[2] Marshall Hall, Jr., "Mathematical biography," *A Century of Mathematics in America, Part I* (Amer. Math. Soc., Providence, R.I., 1988), pp. 367–373.

[3] Karen H. Parshall, "Eliakim Hastings Moore and the founding of a mathematical community in America, 1892–1902," *Annals of Science* **41** (1984), pp. 313–333. Reprinted in this volume, pp. 155–176.

[4] Percey F. Smith, "The Department of Mathematics," *The Yale Alumni Weekly*, October 10, 1924, pp. 95–96. Reprinted in this volume, pp. 121–126.

The Scientific Style
of Josiah Willard Gibbs

Martin J. Klein

The scientific writings of Josiah Willard Gibbs quickly acquired the well-deserved reputation for difficulty that they continue to enjoy in the scientific community. About fifteen years after their initial publication Wilhelm Ostwald translated Gibbs's papers on thermodynamics into German and collected them in one volume. In the preface to this book Ostwald warned his readers that they were embarking on a study that would "demand extraordinary attentiveness and devotion." He pointed out that Gibbs had chosen his mode of exposition, "abstract and often hard to understand," in order to achieve "the greatest possible generality in his investigation and the greatest possible precision in his expression."[1] As a result these papers were full of treasures that had yet to be unearthed. In the same year that Ostwald's translation appeared, Lord Rayleigh wrote to Gibbs urging him to expand his papers into a treatise on thermodynamics, so as to make his ideas more accessible. Rayleigh found Gibbs's original exposition "too condensed and too difficult for most, I might say all, readers."[2] (In assessing this remark one must remember that Rayleigh himself could be so terse as to be almost cryptic.) Even Einstein, who once referred to Gibbs's book on

Some of the material in this chapter has already been published in my paper "The Early Papers of J. Willard Gibbs: A Transformation of Thermodynamics," in *Human Implications of Scientific Advance. Proceedings of the XVth International Congress of the History of Science. Edinburgh 10-15 August 1977.*, ed. E. G. Forbes (Edinburgh: Edinburgh University Press, pp. 330-341. Some of the research reported here was done under a grant from the National Science Foundation.

statistical mechanics as "a masterpiece," qualified his praise by adding, "although it is hard to read and the main points have to be read between the lines."[3]

The abstract, general, and concise form in which Gibbs set forth his work made it difficult for scientists to master his methods and to survey his results. The same qualities in Gibbs's writing offer difficulties to the historian who approaches these papers as historical documents. His task is to enliven those records of scientific activity in the past, "to capture the processes in the course of which those records were produced and became what they are."[4] The historian wants to find out things like the questions Gibbs was answering when he formulated his theoretical systems, the choices available to Gibbs and his contemporaries in dealing with these questions, the various contexts within which he worked. The finished form of Gibbs's papers leaves very few clues for pursuing such historical studies. And yet their form is the completely appropriate expression of Gibbs's way of doing science, of his remarkable combination of the physicist's drive to understand the natural world with the mathematician's concern for logical structures.

These papers also show that "the most abstract and most algebraic scientific work can nevertheless reflect its author's temperament like a faithful mirror," as Pierre Duhem remarked in his essay on Gibbs.[5] It was Henry A. Bumstead, Gibbs's former student, who described his teacher as having been "of a retiring disposition,"[6] and Duhem seized on that phrase, recognizing that it characterized Gibbs's scientific style, and even what one might call his scientific personality.[7] That "retiring disposition" is so faithfully expressed in Gibbs's writings that it is often hard to realize just how much new insight lies behind even his way of posing the scientific issues he discussed, much less his way of resolving them.

In this paper I want to explore and illustrate some of those features of Gibbs's science that give it the individual character that is so distinctive. I shall concentrate on his first publications, the papers in which he introduced himself to the scientific public in 1873. These two works on geometrical methods in thermodynamics had but few readers at the time of their appearance, and their importance to the development of that subject has rarely been recognized by scientists or historians of science. Nevertheless they do exhibit the same characteristics as their author's longer and more famous writings; like them the first papers bear the unmistakable signs of the lion's claw.

When Josiah Willard Gibbs was appointed Professor of Mathematical Physics at Yale in 1871, he had already spent almost all of his thirty-two years in New Haven, Connecticut.[8] He would rarely leave it again except for summer holidays in the mountains. Gibbs's father, also Josiah Willard Gibbs, was the first Yale graduate in a family that had already sent four generations of its sons to Harvard College. The elder Gibbs was a distinguished philologist, Professor of Sacred Literature at Yale. He was known as "a genuine scholar," and despite the difference between their fields some of his intellectual traits resemble those of his son. "Mr. Gibbs loved system, and was never satisfied until he had cast his material into the proper form. His essays on special topics are marked by the nicest logical arrangement."[9] The younger Gibbs graduated from Yale College in 1858 having won a string of prizes and scholarships for excellence in Latin and especially mathematics. He continued his studies at Yale and was one of the first few scholars to be granted a Ph.D. by an American university. Yale had begun to award this degree in 1861, and Gibbs received his in 1863 in the field of engineering.

The dissertation Gibbs wrote bears the title, "On the Form of the Teeth of Wheels in Spur Gearing," not exactly what one might have expected from what we know about his later activities.[10] But as Gibbs pointed out in his first paragraph, "the subject reduces to one of plane geometry," and the thesis is really an exercise in that field. It was not published until 1947, at which time its editor wrote that in reading the thesis, one "feels that he is gradually reaching the summit of an intellectual structure that is firmly founded and well joined." The editor's subsequent comments on the style of this work could apply with only minor changes to much that Gibbs would write in later years. "If [the reader] has a natural friendliness for the niceties of geometrical reasoning, he will be rewarded with a sense of satisfaction akin to that felt upon completing, say, a book of Euclid; if he is not so endowed, he had perhaps better not trouble himself with the austerities of style and extreme economy (one might almost say parsimony) in the use of words that characterize the entire work."[11]

After he received his doctorate Gibbs was appointed a Tutor at Yale and spent the next three years teaching Latin and natural philosophy (physics) to undergraduates. During this period Gibbs continued to develop his engineering interests, and in 1866 he obtained a patent on his design of an improved brake for railway cars.[12] That same year he presented a paper to the Connecticut Academy of Arts and Sciences, of which he had been a member

J. W. Gibbs

(Yale University Archives, Manuscripts and Archives, Yale University Library)

since 1858, on "The Proper Magnitude of the Units of Length, and of other quantities used in Mechanics."[13] This unpublished paper includes a remarkably clear discussion of the dual roles played by the concept of mass in the structure of mechanics—inertial mass and gravitational mass—and of the confusion introduced by writers who would define mass as quantity of matter. "Yet it is evident," Gibbs wrote, "that when the matter of the bodies compared is different in kind, we cannot strictly speaking say that the quantity of matter of one is equal, greater, or less than that of the other. All that we have a right to say, except when the matter is the same in kind, is that the gravity is proportioned to the inertia. To say *that*, is to express a great law of nature,—a law by the way of that class which we learn by experience and not by a priori reasoning. It might have been otherwise, but its truth is abundantly attested by experience. But to say, that the intensities of these two properties are both proportioned to the quantity of matter, is to bring in an element of which we know *nothing*."[14]

In August 1866 Gibbs left New Haven for three years of study in Europe, his only extended absence from his native city. He spent a year each at the universities of Paris, Berlin, and Heidelberg, attending lectures on mathematics and physics and reading widely in both fields. While our information on his activities during this period is rather scanty, one thing is certain.[15] Gibbs did not work as a research student with any of the great scientists whose lectures he attended or whose papers he studied. Nor is there any indication in the notebooks he kept while in Europe that he had yet begun any research of his own or even decided what line he would try to follow in his work. Gibbs had apparently decided to use his time at the three scientific centers to broaden and deepen the somewhat limited education in mathematics and physics he had acquired at Yale, and to inform himself about the current concerns of those who were actively working in these areas. He would then be prepared to choose the subjects of his own researches after he returned to New Haven.

Gibbs's future was still uncertain when he came back in the summer of 1869. Of the next two years we know only that he taught French for at least one term at Yale's Sheffield Scientific School, and that he probably worked out his modification of Watt's governor for the steam engine at this time.[16] Gibbs was evidently able to manage financially on what he had inherited from his father, especially since he continued to live in the family home with his unmarried sister Anna and with his sister Julia and Julia's husband, Addison Van Name, Gibbs's college classmate who had become the

Librarian of Yale. This fortunate state of financial independence was certainly known within the academic community of New Haven when Gibbs was appointed on July 13, 1871 to a newly created position as Professor of Mathematical Physics. It explains to some extent the chilling phrase "without salary" that forms part of the official record of that appointment in the minutes of the Yale Corporation meeting of that date.[17] The new chair was not yet endowed, so there was no money available for paying its incumbent. In any case his teaching duties would be light, since the appointment was in the small graduate department; Gibbs actually taught only one or two students a year during his first decade or so in the new professorship. (Not until 1880, when Gibbs was on the point of accepting an attractive offer from the exciting, new, research-oriented Johns Hopkins University, did Yale pay him a regular salary.)

Gibbs's appointment to the chair of mathematical physics preceded his first published research by two years. Although this now seems like an inversion of the normal order of events, it was not so extraordinary at the time. Benjamin Silliman had been invited to become Yale's first professor of both chemistry and natural history while he was reading the law in New Haven, and before he had reached his twenty-second birthday.[18] Silliman knew essentially nothing of either science but was persuaded by Yale's President Timothy Dwight, who insisted that "the study will be full of interest and gratification, and the presentation which you will be able to make of it to the college classes and the public will afford much instruction and delight."[19] It was only after his appointment in 1802 that Silliman embarked on the systematic study of the sciences he was to profess. Half a century after Silliman's appointment another graduate of Yale was made professor of mathematics "at the early age of twenty-five," only five years after his graduation from the college. This was Hubert Anson Newton, who was then given a year's leave of absence to study modern mathematics in Paris.[20] Newton, only nine years older than Gibbs, was one of his teachers, and must have been one of those who strongly supported Gibbs's appointment to the faculty.

Evidently the thirty-two-year-old Gibbs, with his brilliant college record, his doctoral degree, his demonstrated abilities as an engineer, and his three years of postdoctoral study in Europe, had far more impressive qualifications for a professorship than most of his colleagues had had when they were appointed. Yale had every reason to express its confidence in Gibbs, who had, after all, been a member of this small academic community since his birth.

Gibbs began his first paper, "Graphical Methods in the Thermo-dynamics of Fluids,"[21] by remarking that although geometrical representations of thermodynamic concepts were in "general use" and had done "good service," they had not yet been developed with the "variety and generality of which they are capable." Such representations had been restricted to diagrams whose rectilinear coordinates denote volume and pressure, and he proposed to discuss a range of alternatives, "preferable . . . in many cases in respect of distinctness or of convenience." This beginning suggested that the paper would be primarily didactic, and was likely to be rather removed from the current concerns of scientists already actively involved in thermodynamic research. What followed seemed to bear out this suggestion, as Gibbs went on to list the quantities relevant to his discussion and to write down the relationships among them.

The quantities appropriate for describing the body in any given state were its volume v, pressure p, absolute temperature t, energy ϵ, and entropy η. In addition to these functions of the body's state, there were the work done W, and the heat received H, by the body in passing from one state to another. These quantities were related by the equations

$$d\epsilon = dH - dW, \tag{1}$$

$$dW = pdv, \tag{2}$$

$$dH = td\eta. \tag{3}$$

The first and third equations express the definitions of the state functions energy and entropy whose existence is required by the first and second laws of thermodynamics, respectively, while equation (2) is just the expression for the mechanical work done by an expanding fluid.[22] Gibbs then eliminated the work and heat to obtain the equation

$$d\epsilon = td\eta - pdv, \tag{4}$$

which he referred to as the differential form of "the fundamental thermodynamic equation of the fluid," the equation expressing the energy as a function of entropy and volume.

Gibbs used hardly any more words than I have just used in stating these matters, as though he too were simply reminding his readers of familiar, widely known truths, and were writing them down only to establish the notation of his paper. That is just what one might expect as the starting point of a first scientific paper, one that promised to be only a modest didactic exercise. Could there have been any doubt or disagreement that this was the proper starting point for any treatment of thermodynamics?

To answer this question one must look not at Gibbs's paper but rather at the general state of thermodynamics in 1873, when it was written. Almost a quarter of a century had gone by since Rudolf Clausius had set the subject on its proper, dual foundation, and William Thomson had endorsed and developed the idea that there are *two* basic laws of thermodynamics.[23] Clausius, especially, had explored the second law in a series of memoirs, searching for "the real nature of the theorem."[24] He was convinced that that "real nature" had been found with the help of his analysis of transformations, first for cyclic processes in 1854 and then for the general case in 1862.[25] It was not until 1865 that Clausius invented the word entropy as a suitable name for what he had been calling "the transformational content of the body."[26] The new word made it possible to state the second law in the brief but portentous form: "The entropy of the universe tends toward a maximum," but Clausius did not view entropy as the basic concept for understanding that law. He preferred to express the physical meaning of the second law in terms of the concept of disgregation, another word that he coined, a concept that never became part of the accepted structure of thermodynamics.[27] Clausius restricted his use of entropy to its convenient role as a summarizing concept; in the memoir of 1865 where it was introduced, he derived the experimentally useful consequences of the two laws without using the entropy function, or even the internal energy function.

Ten years later, when Clausius reworked his articles on thermodynamics into a treatise that would make a convenient textbook, he had not changed his mind about the status of both entropy and energy.[28] Although he showed how the two laws guarantee the existence of these two state functions, Clausius eliminated them from his working equations as soon as possible. In contrast to Gibbs, Clausius kept the original thermodynamic concepts, work and heat, at the center of his thinking, although he did devote a chapter to showing how the energy and entropy of a system could be determined from experimentally measurable quantities.

Since Clausius gave entropy such a secondary position in his writings, it is not surprising that his contemporaries paid little or no attention to the concept. Thomson had his own methods which bypassed the need for introducing the entropy function, as did Carl Neumann.[29] W. J. M. Rankine had independently introduced the same concept in 1854,[30] calling it the thermodynamic function, and using it in his book, *A Manual of the Steam Engine and Other Prime Movers* in 1859.[31] Although this must have been a popular text, since it was already in its sixth edition in 1873, Rankine's thermodynamic function was not used by many of his readers. As James Clerk Maxwell once put it, Rankine's exposition of fundamental concepts often "strained [the reader's] powers of deglutition"; and as for his statement of the second law, "its actual meaning is inscrutable."[32]

Clausius's word, entropy, did enter the thermodynamic literature in English, but only by an act of misappropriation. Peter Guthrie Tait liked Clausius's "excellent word," but preferred to use it in his *Sketch of Thermodynamics* in 1868 as the name of quite another concept. "It would only confuse the student," he wrote, "if we were to endeavor to invent another term for our purpose."[33] And so, acting like Lewis Carroll's Humpty Dumpty, Tait chose to make entropy mean available energy, a usage unfortunately followed by his friend Maxwell in the first edition of his *Theory of Heat* a few years later.[34]

If we now return to the opening pages of Gibbs's first paper, we see that his statements were by no means the conventional wisdom of scientists in 1873. Few, if any, of those working in thermodynamics would have chosen the approach that Gibbs set forth as though it were obvious. From the beginning he emphasized the properties of material systems rather than "the motive power of heat." As a result the state functions, energy and entropy, necessarily took precedence over those quantities that depend on the process carried out by the system—the work and heat it exchanges with its surroundings. No wonder then that Gibbs used the term fundamental equation for the relation of the system's energy to its entropy and volume, since "from it . . . may be derived all the thermodynamic properties of the fluid."[35] For Gibbs thermodynamics was the theory of the properties of matter at equilibrium, rather than the mechanical theory of heat that Clausius and Thomson had seen. But as it was expressed by a man "of a retiring disposition," this important change in viewpoint could easily be overlooked by his readers.

When Gibbs wrote of the "general use and good service" given by the geometrical representation of thermodynamic propositions, he was thinking of something more than mere illustrations. In the four decades since Clapeyron had introduced the indicator diagram (or pressure-volume diagram) into his exposition of Carnot's ideas, that

diagram had been developed into a valuable technique.[36] Rankine
had shown how this geometrical method could be used for "the
solution of new questions especially those relating to the action of
heat in all classes of engines," and for presenting "in a systematic
form, those theoretical principles which are applicable to all methods
of transforming heat to motive power by means of the changes of
volume of an elastic substance."[37] To see how far Rankine could go
in this fashion, one should examine his geometrical derivation of the
equation for the difference between the heat capacities of a fluid
at constant pressure and at constant volume. Rankine used these
methods extensively in his *Manual of the Steam Engine,* as did the
authors of other texts, such as Gustav Zeuner.[38]

Gibbs set out to free the geometrical approach from the limitations imposed by the particular choice of volume and pressure as
basic variables. He wanted to find "a general graphical method which
can exhibit at once all the thermodynamic properties of a fluid
concerned in reversible processes, and serve alike for the demonstration of general theorems and the numerical solution of particular
problems."[39] To this end he considered the general properties of any
diagram in which the states of the fluid were mapped continuously
on the points of a plane. The thermodynamic properties of the fluid
would then be expressed in the geometrical properties of the several
families of curves connecting states of equal volume, of equal temperature, of equal entropy, and so forth.

Since the equations relating work to pressure and volume, and
heat to temperature and entropy (equations (2) and (3) above), are
of exactly the same form, the temperature-entropy diagram must
share many of the useful features of the pressure-volume diagram. As
Gibbs pointed out, it has the additional advantage that the universal
nature of Carnot's ideal cycle appears directly, since this cycle is
always represented by a rectangle in the temperature-entropy diagram, regardless of the properties of the working substance. Gibbs
saw that the real advantage of the temperature-entropy diagram is
that it "makes the second law of thermodynamics so prominent, and
gives it so clear and elementary an expression." He meant that
although there is no formal difference between representing the work
done in a process as the area under its curve in the pressure-volume
diagram, and representing the heat exchanged in the process as the
area under its curve in the temperature-entropy diagram, the former
representation was only an expression of the definition of mechanical work. The latter, however, was "nothing more nor less than a
geometrical expression of the second law of thermodynamics in its

application to fluids, in a form exceedingly convenient for use, and from which the analytical expression of the same law can, if desired, be at once obtained."[40]

The potential value of a diagram so closely analogous to the pressure-volume diagram was seen by others at about the same time, but in more limited inquiries. In December 1872 the Royal Academy of Sciences at Brussels published a paper on the second law of thermodynamics by the civil engineer Th. Belpaire,[41] a paper described by one of the Academy's referees as "the first truly original work on this subject written in Belgium."[42] Belpaire's "new demonstration of the second law" was more original than it was cogent, unfortunately, but he did nevertheless introduce a diagram like Gibbs's and use it effectively. Several years later J. Macfarlane Gray, Chief Examiner of Marine Engineers for the Board of Trade in London, independently began to use such a diagram in his own analyses of engines.[43] When Gray presented the diagram and described its uses before the Institution of Naval Architects in 1880, he was told of Gibbs's work—one wonders by whom—which he proceeded to obtain and read. Gray reported that "Professor Gibbs's paper was a very high-class production," despite its "revelling in mathematics," and that he would limit his own claim to asserting that he had done "what others had only said could be done," namely, applied the diagram "for practical use by engineers."[44]

Gibbs did not consider the temperature-entropy diagram worth an extended discussion, and devoted only three of his thirty-two pages to it. The diagram "whose substantial advantages over any other method" made it most interesting and worth discussing in more detail, was that in which the coordinates of the point denoting the state were the volume and the entropy of the body.[45] The very nature of the fundamental thermodynamic equation, which expresses energy in terms of entropy and volume, would suggest the importance of an entropy-volume diagram. Such a diagram has a variable scale factor. In other words, the ratio of the work done in a small cyclic process to the area enclosed by the cycle representing that process in the diagram varies from one part of the volume-entropy plane to another. Both the pressure-volume and temperature-entropy diagrams have constant scale factors. Although a variable scale factor might offer difficulties, or at least some awkwardness, for engineering purposes, it was a definite advantage in studying the properties of matter at equilibrium.

Gibbs showed this advantage in his analysis of the states in which vapor, liquid, and solid coexist at a definite, unique set of values of the pressure and temperature. The scale factor, which is $(\partial p/\partial \eta)$ in this diagram, vanishes for such states. The region of coexistence of the three phases is represented by the interior of a triangle in the entropy-volume diagram, and, as Gibbs remarked, the information conveyed this way "can be but imperfectly represented" in any other diagram.

Some features of the thermodynamic diagrams, that is to say some aspects of the families of curves representing thermodynamic properties, are independent of the choice of coordinates. Gibbs carefully examined these invariant features, and especially the order of the curves of different kinds (such as the isobars, isotherms, and isentropics) as they cross at any point, and the geometrical nature of these intersections, which could involve contacts of higher order.

In closing his paper, Gibbs pointed out that what he had done was to start from the analytical expression of the laws of thermodynamics, taken as known, and "to show how the same relations may be expressed geometrically." The process could have been reversed. "It would, however, be easy, starting from the first and second laws of thermodynamics as usually enunciated, to arrive at the same results without the aid of analytical formulae,—to arrive, for example at the conception of energy, of entropy, of absolute temperature, in the construction of the diagram without the analytical definitions of these quantities, and to obtain the various properties of the diagram without the analytical expression of the thermodynamic properties which they involve." And Gibbs also emphasized the essential point, "that when the diagram is only used to demonstrate or illustrate general theorems, it is not necessary . . . to assume any particular method of forming the diagram; it is enough to suppose the different states of the body to be represented continuously by points upon a sheet."[46]

Gibbs's "natural friendliness for the niceties of geometrical reasoning," already demonstrated in his thesis, is very much in evidence in his work on thermodynamics. This surely was a major factor in the enthusiastic response his work evoked from Maxwell, who preferred to argue geometrically rather than analytically whenever possible, and who derived the four thermodynamic relations that bear his name by completely geometrical reasoning.[47]

Gibbs did not tell his readers what had drawn his attention to thermodynamics as the subject for his first professorial research. He had not attended lectures in this field during his years of study in Europe, and there was no stimulus for his work coming from his colleagues at Yale. What made this untried, but mature and independent thinker, as he at once showed himself to be, select this particular set of problems to begin with? It is true that in 1872 the pages of the *Philosophical Magazine,* probably the most widely read British journal of physics, carried a number of articles on thermodynamics. They were full of lively, pointed, and even angry words on the subject as Tait and Clausius debated the history, which to them meant the priorities, of the discovery of the second law.[48] Gibbs could hardly have avoided noticing this dispute, in which the names of Thomson and Maxwell were mentioned, and it might have prompted him to do some reading. But controversy repelled rather than attracted Gibbs, and he did not enter the debate in progress.

Another possible source for Gibbs's interest in thermodynamics is Maxwell's *Theory of Heat,* which appeared in London in 1871 and in a New York edition the following year.[49] The book was widely read, going through a number of editions within a few years. In 1873 J. D. van der Waals wrote of it as "the little book which is surely to be found in the hands of every physicist."[50] Although it appeared in a series described by its publisher as "text-books of science adapted for the use of artisans and of students in public and science schools,"[51] Maxwell did not keep his writing at an elementary level. Tait even thought that some of it was probably "more difficult to follow than any other of his writings,"[52] which was saying a great deal. In any event Maxwell did not hesitate to include discussions of whatever interested him in the latest developments of thermodynamics as his book went into its successive editions. One such development described in Maxwell's first edition was Thomas Andrews's recent discovery of the continuity of the two fluid states of matter, liquid and gas.[53] Whether Gibbs learned of Andrews's work from Maxwell's book or came across the paper Andrews presented to the Royal Society of London by reading the *Philosophical Transactions* for 1869, he certainly knew about Andrews's discovery of the critical point when he wrote his first paper. There is a footnote referring to Andrews at the place where Gibbs discusses the possible second order contact of the isobar and the isotherm corresponding to a particular state of the fluid: "An example of this is doubtless to be found at the critical point of a fluid."

Andrews's Royal Society paper reported the results of a decade of experimental work. It was the high point of his scientific career and Andrews was well aware of it, writing to his wife: "I really begin to think that Dame Nature has at last been kind to me, and rewarded me with a discovery of a higher order than I ever expected to make."[54] His careful measurements of the isotherms of carbon dioxide established the existence of a critical temperature: if the gas were compressed isothermally below this temperature it would eventually begin to liquefy and become a two-phase system with a visible surface of separation between gas and liquid. Further compression would lead to complete transformation of gas into liquid, the liquid then being almost incompressible. In sharp contrast to this behavior, an isothermal compression of the gas at a temperature above the critical one would never lead to two phases, although the density would eventually take on values appropriate to the liquid state. Above the critical temperature there was no distinction between liquid and gas. It was always possible to pass from a state clearly liquid to one that was equally clearly gas without ever going through the discontinuous two-phase region. These remarkable properties were by no means peculiar to carbon dioxide. They are "generally true of all bodies which can be obtained as gases and liquids," as Andrews confirmed by studies on some half dozen substances.[55] He did not theorize at all about the implications of his discovery, attempting neither a kinetic-molecular explanation nor a thermodynamic analysis.

Andrews's discovery—a new, surprising, and general feature of the behavior of matter, as yet quite unexplained—would have been just the sort of thing to attract Gibbs's attention as that promising new professor of mathematical physics sought for a suitable subject on which to work. As he advised one of his students many years later, "one good use to which anybody might put a superior training in pure mathematics was to the study of the problems set us by nature."[56]

By the time Gibbs wrote his second paper, which appeared only a few months after the first, his physical interests were much more apparent.[57] Although this paper, "A Method of Geometrical Representation of the Thermodynamic Properties of Substances by Means of Surfaces," might seem to be a mere extension of the geometrical methods from two dimensions to three, one does not have to read far to see that Gibbs is doing something quite different. The emphasis is no longer on methods as such but rather on the phenomena to be explained. His problem was to characterize the behavior of matter at equilibrium, to determine the nature of the equilibrium state of a body that can be solid, liquid, or gas, or some combination of these, according to the circumstances. Gibbs goes directly to a single, fundamental three-dimensional representation in which the three rectangular coordinates are the energy, entropy, and volume of the body, and the equilibrium states constitute a surface whose properties he proceeded to explore.

There was only one precedent for using a three-dimensional representation of equilibrium states, and that a very recent one. James Thomson, William's older brother and former collaborator in studies on heat, had introduced the thermodynamic surface in pressure-volume-temperature space to assist his thinking about Andrews's results.[58] Thomson was Andrews's colleague at Queen's College, Belfast, and had been thinking about Andrews's work and trying to interpret it since 1862, though he did not publish his ideas until 1871. Gibbs was aware of Thomson's publication and cited it at the start of his own paper. (It is possible, of course, that it was Thomson's work that had set Gibbs thinking about thermodynamics.) Thomson's "chief object" in his paper was to argue that Andrews had not gone far enough in his claim that the liquid and gas phases are continuously related. Below the critical temperature Andrews's isotherms included a straight line segment parallel to the axis of volume. This represented the states in which gas and liquid could co-exist at a fixed pressure, in proportions varying from all gas to all liquid. This pressure of the saturated vapor depends only on the temperature, for a given substance. At both ends of the two-phase region the slope of the isothermal curve changes abruptly, producing what Thomson called a "practical breach of continuity." He proposed to smooth this out by introducing a "theoretical continuity" that would require the isotherm to include "conditions of pressure, temperature, and volume in unstable equilibrium."[59] The isothermal curves that Thomson sketched freehand look much like the theoretical isotherms derived on quite different grounds by van der Waals in his dissertation two years later.

The crucial question about the isothermal curves that neither Thomson nor van der Waals could answer was: where must the horizontal line segment be drawn? In other words, what is the condition that determines the pressure at which gas and liquid can co-exist in equilibrium at a specified temperature? (This temperature must be below the critical temperature.) The problem of finding the condition for equilibrium between two phases had a relatively long history,[60] but not even Maxwell (who included a discussion of Thomson's work in his "elementary textbook") had been able to solve it.

What was missing from all these attempts was nothing less than the second law of thermodynamics. Gibbs, who started from the thermodynamic laws, settled the question in his usual brief and elegant manner. The fundamental equation—relating energy, entropy, and volume for a homogeneous phase—corresponds to what he called the primitive surface. It includes all equilibrium states, regardless of their stability. When the system consists of several homogeneous parts, its states form the derived surface. This is constructed by recognizing that "the volume, entropy, and energy of the whole body are equal to the sums of the volumes, entropies, and energies respectively of the parts, while the pressure and temperature of the whole are the same as those of each of the parts." In a two-phase system the point representing the compound state must then lie on the straight line joining the two pure (that is, single-phase) state points which are themselves on the primitive surface. The pressure and temperature are the same at all points on this line. But the direction of the tangent plane at any point of the primitive surface is determined by the pressure and temperature, since we have from equation (4),

$$p = -\left(\frac{\partial \epsilon}{\partial v}\right)_\eta \qquad , \qquad (5)$$

and also

$$t = \left(\frac{\partial \epsilon}{\partial \eta}\right)_v \qquad . \qquad (6)$$

Since the line joining the two points on the primitive surface that represent the two phases in equilibrium must lie in the tangent planes at both points, and since those planes are parallel, they must be the same plane. This condition, that there be a common tangent plane for the points representing two phases in equilibrium, is easily expressed analytically in the form

$$\epsilon_2 - \epsilon_1 = t(\eta_2 - \eta_1) - p(v_2 - v_1), \qquad (7)$$

where the subscripts 1 and 2 refer to the two phases and where p, t are the common values of pressure and temperature.

Gibbs referred to this second paper almost twenty years later in a letter to Wilhelm Ostwald. "It contains, I believe, the first solution of a problem of considerable importance, viz: the additional condition (besides equality of temperature and pressure) which is necessary in order that two states of a substance shall be in equilibrium in contact with each other. The matter seems simple enough now, yet it appears to have given considerable difficulty to physicists. . . . I suppose that Maxwell referred especially to this question when he said . . . that by means of this model, problems which had long resisted the efforts of himself and others could be solved at once."[61] Maxwell expressed his appreciation of Gibbs's thermodynamic surface by constructing a model of it for water and sending a cast of it to Gibbs. He also included a fourteen-page discussion of the Gibbs surface in the 1875 edition of his textbook, giving more details of its properties than Gibbs had. In that same year Maxwell developed an alternative form of the equilibrium condition—the Maxwell construction—which states that the horizontal, two-phase portion of the isotherm must cut off equal areas above and below in the van der Waals-Thomson loop. The proof involved a direct application of the second law carried out with what may be called genuinely Maxwellian ingenuity. Clausius independently arrived at Maxwell's result five years later by a slightly different argument.[62] He had apparently missed Maxwell's discussion of the Gibbs surface, and there is no sign that he ever read any of the reprints Gibbs regularly sent to him.

Gibbs arrived at a new, profound understanding of the critical point by analyzing the conditions for the stability of states of thermodynamic equilibrium. He showed that, for a system surrounded by a medium of constant pressure P and constant temperature T, the quantity $(\epsilon - T\eta + Pv)$ must be stationary in an equilibrium state, and must be a minimum if that equilibrium is to be stable. The possible instabilities are of two kinds. The first, "absolute instability," corresponds to states like those of a supercooled gas, stable against small variations but not against a radical split into two phases. The other, "essential instability," corresponds to states like those in the inner part of a van der Waals loop. The critical point is the common limit of both regions of instability; it is itself stable against both types of change. Gibbs's analysis of the critical point led to a series of explicit conditions that must be fulfilled, most of which had not been pointed out before.

Once again Gibbs's "retiring disposition" meant that the transformation of thermodynamics he had accomplished in these early papers was not properly appreciated. In 1902 Paul Saurel was quite justified in remarking: "It does not seem to be generally known that Gibbs, in his memoir on the energy surface, has given in outline a very elegant theory of the critical state of a one-component system and of the continuity of the liquid and gaseous states."[63] And in 1979, over a century after their publication, Arthur Wightman wrote: "For those who like their physics stated in simple general mathematical terms, the version of thermodynamics offered by

Gibbs's first two papers can scarcely be improved. Nevertheless, apart from its impact on Maxwell, it had very little influence on late nineteenth century textbooks. The notion of 'fundamental equation' and the simple expression it gives for the laws of thermodynamics . . . only became available with the publication of 'neo-Gibbsian' textbooks and monographs in the mid-twentieth century."[64]

When Gibbs accepted the Rumford Medal awarded to him by the American Academy of Arts and Sciences at Boston, he wrote: "One of the principal objects of theoretical research in any department of knowledge is to find the point of view from which the subject appears in its greatest simplicity."[65] His efforts to achieve that point of view were central to all his scientific activity, and he never presented his work to the public until he was satisfied with the logical structure he had constructed. But as a consequence, his readers are "deprived of the advantage of seeing his great structures in process of building . . . and of being in such ways encouraged to make for themselves attempts similar in character, however small their scale."[66]

Notes

1. J. W. Gibbs, *Thermodynamische Studien,* trans. W. Ostwald (Leipzig: W. Engelmann, 1892), p. v.

2. Lord Rayleigh to J. W. Gibbs, June 5, 1892.

3. A. Einstein to M. Besso, June 23, 1918, in A. Einstein, M. Besso, *Correspondance 1903-1955,* ed. P. Speziali (Paris: Hermann, 1972), p. 126.

4. E. Panofsky, *Meaning in the Visual Arts* (Garden City, N.Y.: Doubleday & Company, 1955), p. 24.

5. P. Duhem, *Josiah Willard Gibbs à propos de la publication de ses mémoires scientifiques* (Paris: A. Hermann, 1908), p. 6.

6. H. A. Bumstead, "Josiah Willard Gibbs," in *The Scientific Papers of J. Willard Gibbs,* eds. H. A. Bumstead and R. G. Van Name, 2 vols. (New York: Longmans, Green, and Co., 1906; reprinted 1961), I, p. xxiii. This collection will be noted as *Sci. Pap.* and page references to Gibbs's papers will refer to this edition.

7. Duhem, *Josiah Willard Gibbs,* p. 10.

8. Full information on Gibbs's life will be found in L. P. Wheeler, *Josiah Willard Gibbs. The History of a Great Mind* 2nd ed. (New Haven: Yale University Press, 1952).

9. See Wheeler, *Josiah Willard Gibbs,* p. 9. The quotations are descriptions of the elder Gibbs.

10. Gibbs's thesis is printed in *The Early Work of Willard Gibbs in Applied Mechanics,* eds. L. P. Wheeler, E. O. Waters, and S. W. Dudley (New York: H. Schuman, 1947), pp. 7-39.

11. E. O. Waters, "Commentary upon the Gibbs Monograph, *On the Form of the Teeth of Wheels in Spur Gearing,*" in Wheeler et al., eds., *The Early Work of Willard Gibbs,* p. 43.

12. S. W. Dudley, "An Improved Railway Car Brake and Notes on Other Mechanical Inventions," in Wheeler et al., eds., *The Early Work of Willard Gibbs,* pp. 51-61.

13. Printed in Wheeler, *Josiah Willard Gibbs,* Appendix II, pp. 207-218.

14. Ibid., p. 209.

15. See ibid., pp. 40-45 for a summary of Gibbs's European notebooks.

16. Ibid., pp. 54-55, 259. For Gibbs's governor see also L. P. Wheeler, "The Gibbs Governor for Steam Engines," in Wheeler et al., eds., *The Early Work of Willard Gibbs,* pp. 63-78.

17. Wheeler, *Josiah Willard Gibbs,* p. 57. For the later history of Gibbs's salary see pp. 59-60, 90-93.

18. L. G. Wilson, "Benjamin Silliman: A Biographical Sketch," in *Benjamin Silliman and His Circle: Studies on the Influence of Benjamin Silliman on Science in America,* ed. L. G. Wilson (New York: Science History Publications, 1979), pp. 1-10.

19. Quoted by J. C. Greene in ibid., p. 12.

20. J. W. Gibbs, "Hubert Anson Newton," *Sci. Pap.* II, p. 269.

21. J. W. Gibbs, "Graphical Methods in the Thermodynamics of Fluids," *Transactions of the Connecticut Academy* 2 (1873), pp. 309-342. Reprinted in *Sci. Pap.* I, pp. 1-32.

22. I have used Gibbs's notation except for adding the bar on d to denote an inexact differential as đ. This notation is due to C. Neumann, *Vorlesungen über die mechanische Theorie der Wärme* (Leipzig: B. G. Teubner, 1875). See p. ix.

23. See J. W. Gibbs, "Rudolf Julius Emanuel Clausius," *Sci. Pap.* II, pp. 261-267. Also see D. S. L. Cardwell, *From Watt to Clausius* (Ithaca, N.Y.: Cornell University Press, 1971).

24. R. Clausius, "Ueber eine veränderte Form des zweiten Hauptsatzes der mechanischen Wärmetheorie," *Ann. Phys.* 169 (1854), pp. 481-506.

25. R. Clausius, "Ueber die Anwendung des Satzes von der Aequivalenz der Verwandlungen auf die innere Arbeit," *Ann. Phys.* 192 (1862), pp. 73-112.

26. R. Clausius, "Ueber verschiedene für die Anwendung bequeme Formen der Hauptgleichungen der mechanischen Wärmetheorie," *Ann. Phys.* 201 (1865), pp. 353-400.

27. For an analysis of disgregation see M. J. Klein, "Gibbs on Clausius," *Historical Studies in the Physical Sciences* 1 (1969), pp. 127-149, and also K. Hutchison, "Der Ursprung der Entropiefunktion bei Rankine und Clausius," *Annals of Science* 30 (1973), pp. 341-364.

28. R. Clausius, *Die mechanische Wärmetheorie,* Vol. I (Braunschweig: F. Vieweg & Sohn, 1876).

29. See note 22.

30. W. J. M. Rankine, "On the Geometrical Representation of the Expansive Action of Heat, and the Theory of Thermodynamic Engines," *Philosophical Transactions of the Royal Society* 144 (1854), pp. 115-176. Reprinted in W. J. Macquorn Rankine, *Miscellaneous Scientific Papers,* ed. W. J. Millar (London: C. Griffin and Co., 1881), pp. 339-409. See also Hutchison, "Der Ursprung der Entropiefunktion."

31. W. J. M. Rankine, *A Manual of the Steam Engine and Other Prime Movers* (London: C. Griffin and Co., 1859). I have used the Sixth Edition, 1873.

32. J. C. Maxwell, "Tait's *Thermodynamics,*" *Nature* 17 (1878), p. 257. Reprinted in *The Scientific Papers of James Clerk Maxwell,* ed. W. D. Niven, 2 vols. (Cambridge: At the University Press, 1890; reprinted 2 vols. in 1 New York, 1965), II, pp. 663-664.

33. P. G. Tait, *Sketch of Thermodynamics* (Edinburgh: Edmundston and Douglas, 1868), p. 100.

34. J. C. Maxwell, *Theory of Heat* (London: Longmans, Green and Co., 1871), p. 186. Reading Gibbs's first paper persuaded Maxwell to correct the error he had "imbibed" from Tait. See my paper cited in note 27. Tait's misusage of "entropy" can also be found in G. Krebs, *Einleitung in die mechanische Wärmetheorie* (Leipzig: B. G. Teubner, 1874), pp. 216-218.

35. An approach similar to that used by Gibbs, deriving "all the properties of the body

that one considers in thermodynamics" from a single characteristic function, is to be found in two notes by François Massieu, "Sur les fonctions caractéristiques des divers fluides," *Comptes Rendus* 69 (1869), pp. 858-862, 1057-1061. Massieu does not carry the discussion very far in these notes, and there is no reason to think that Gibbs knew them in 1873. He does refer to them in 1875 in his work on heterogeneous equilibrium. See *Sci. Pap.* I, p. 86 fn.

36. É. Clapeyron, "Mémoire sur la puissance motrice de la chaleur," *Journal de l'École Polytechnique* 14 (1834), pp. 153-190. The earlier history of the indicator diagram is discussed in D. S. L. Cardwell, note 23, pp. 80-83, 220-221. Cardwell remarks that the diagram had been a closely guarded trade secret of the firm of Boulton and Watt, and that John Farey, the English engineer, learned about it only in 1826 in Russia, presumably from one of Boulton and Watt's engineers working there. Since Clapeyron was also in Russia during the late 1820s he may have acquired his knowledge of the indicator diagram in the same way.

37. W. J. M. Rankine, *Papers*, note 30, pp. 359-360.

38. G. Zeuner, *Grundzüge der mechanischen Wärmetheorie* (Leipzig: Arthur Felix, 2nd ed., 1866).

39. J. W. Gibbs, *Sci. Pap.* I, p. 1.

40. J. W. Gibbs, *Sci. Pap.* I, p. 11.

41. T. Belpaire, "Note sur le second principe de la thermodynamique," *Bulletins de l'Académie Royale des Sciences, des Lettres, et des Beaux-Arts de Belgique,* 34 (1872), pp. 509-526. It seems safe to conclude that Gibbs had not read this paper even though the *Bulletins* were received by the Connecticut Academy on an exchange basis: the pages of Belpaire's paper in the Connecticut Academy's volume of the *Bulletins* for 1872, now in the Yale University library, were still uncut in the early summer of 1977.

42. See the "Rapport" on Belpaire's paper by F. Folie, in the same journal, pp. 448-451.

43. J. M. Gray, "The Rationalization of Regnault's Experiments on Steam," *Proceedings of the Institution of Mechanical Engineers* (1889), pp. 399-450. Discussion of this paper, see pp. 451-468. See in particular pp. 412-414.

44. Ibid., pp. 463-464.

45. J. W. Gibbs, *Sci. Pap.* I, pp. 20-28.

46. J. W. Gibbs, *Sci. Pap.* I, p. 32.

47. J. C. Maxwell, *Theory of Heat,* 5th ed. (London: Longmans, Green, and Co., 1877), pp. 165-171. For his geometrical preferences see, for example, L. Campbell and W. Garnett, *The Life of James Clerk Maxwell* (London: Macmillan and Co., 1882; reprinted in New York: Johnson Reprint Corporation, 1969), p. 175.

48. For a discussion and references see my paper in note 27. Also see C. G. Knott, *Life and Scientific Work of Peter Guthrie Tait* (Cambridge: At the University Press, 1911), pp. 208-226.

49. See note 34.

50. J. D. van der Waals, *Over de continuiteit van den gas-en vloeistoftoestand* (Leiden: A. W. Sijthoff, 1873), p. 81.

51. Advertisement of the publisher (Longmans, Green, and Co.) for the series, printed at the end of the text in some editions.

52. P. G. Tait, "Clerk-Maxwell's Scientific Work," *Nature* 21 (1880), pp. 317-321.

53. T. Andrews, "On the Continuity of the Gaseous and Liquid States of Matter," *Phil. Trans. Roy. Soc.* 159 (1869), pp. 575-590. Reprinted in *The Scientific Papers of Thomas Andrews* (London: Macmillan and Co., 1889), pp. 296-317. Further references are to this edition.

54. Andrews, *Scientific Papers,* p. xxxi.

55. Ibid., p. 315.

56. In June 1902 Gibbs advised his former student, Edwin B. Wilson, who was leaving for a year's study in Paris, to take some work in applied mathematics. "He ventured the opinion that one good use to which anybody might put a superior training in pure mathematics was to the study of the problems set us by nature." E. B. Wilson, "Reminiscences of Gibbs by a Student and Colleague," *Scientific Monthly* 32 (1931), pp. 210-227. Quotation from p. 221.

57. J. W. Gibbs, "A Method of Geometrical Representation of the Thermodynamic Properties of Substances by Means of Surfaces," *Transactions of the Connecticut Academy* 2 (1873), pp. 382-404. Reprinted in *Sci. Pap.* I, pp. 33-54.

58. James Thomson, "Considerations on the Abrupt Change at Boiling or Condensing in Reference to the Continuity of the Fluid State of Matter," *Proc. Roy. Soc.* 20 (1871), pp. 1-8. Reprinted in James Thomson, *Collected Papers in Physics and Engineering* (Cambridge: At the University Press, 1912), pp. 278-286. References will be to this reprinting. Related papers and unpublished notes by Thomson are to be found at pp. 276-277 and pp. 286-333. Although Gibbs does not refer to them, he may well have read Thomson's papers to the British Association for the Advancement of Science in 1871 and 1872 (pp. 286-291, 297-307) in which the triple point is first named and discussed.

59. See Thomson, *Collected Papers*, p. 279.

60. See van der Waals, *Over de continuiteit*, pp. 120-121.

61. J. W. Gibbs to W. Ostwald, March 27, 1891. Printed in *Aus dem wissenschaftlichen Briefwechsel Wilhelm Ostwalds*, ed. H.-G. Körber (Berlin: Akademie-Verlag, 1961) I, pp. 97-98.

62. R. Clausius, "On the Behavior of Carbonic Acid in Relation to Pressure, Volume, and Temperature," *Philosophical Magazine* 9 (1880), pp. 393-408. See particularly pp. 405-407.

63. P. Saurel, "On the Critical State of a One-Component System," *Journal of Physical Chemistry* 6 (1902), pp. 474-491.

64. A. S. Wightman, "Introduction: Convexity and the Notion of Equilibrium State in Thermodynamics and Statistical Mechanics," in R. B. Israel, *Convexity in the Theory of Lattice Gases* (Princeton: Princeton University Press, 1979), pp. ix-lxxxv. Quote from p. xiii.

65. Quoted in Wheeler, *Josiah Willard Gibbs*, pp. 88-89.

66. H. A. Bumstead in J. W. Gibbs, *Sci. Pap.* I, p. xxiv.

The Department of Mathematics [1]

PROFESSOR PERCEY F. SMITH, '88 S.

A century ago the Faculty of Yale College consisted of nine professors and six tutors, with an attendance of about two hundred and fifty students. Mathematics was a prescribed subject for Freshmen and Sophomores, also for Juniors in part, arithmetic, algebra, Euclid, conic sections, spherical geometry, trigonometry, navigation and surveying making a sequence of subjects extending through two and a half years. Fluxions, as differential and integral calculus was then called, was an optional study in the second half of Junior year. Instruction in mathematics was in charge of a Professor of Mathematics and Natural Philosophy. All exact and natural science as then taught did not presumably surpass the intellectual powers of one individual. Reverend Jeremiah Day, of the Class of 1795, held the chair of Mathematics and Natural Philosophy from 1803–1820, and continued to lecture on the latter subject after his elevation to the Presidency of the University in 1817. It is an interesting fact that a scholar other than a theologian should have been chosen for this office in those early days, and this scholar a mathematician. The title, Professor of Mathematics, appears for the first time in the catalog of 1841–1842, the incumbent of the chair being Anthony D. Stanley, Class of 1830. There appears, however, to have been no change in the curriculum in the matter of mathematical subjects until 1854. In that year the professorship of mathematics fell vacant owing to the death of Professor Stanley, and instruction in the subject was in charge of a tutor, H. A. Newton of the Class of 1850. Newton's mathematical talents were so unusual that he was elected to the chair in 1855 at the age of twenty-five, and assumed his duties after a year of study in Paris. A significant announcement appeared in the catalog of 1854-1855, probably to be traced to the influence and enthusiasm of the young tutor Newton: "Students desirous of pursuing the higher branches of mathematics are allowed to choose Analytic Geometry in place of the regular

---

[1] Reprinted with permission from the October 1924 issue of the *Yale Alumni Magazine* ; copyright by the Yale Alumni Publications, Inc.

The Departmental Staff

Standing, left to right: Messrs. Foster, Miles, Elliott, Betz, Tracey, Mikesh, Hill, Vatnsdal. Seated: Messrs. Whittemore, Patton, Longley, Moore, Smith, Shook, Brown, Schwartz, Wilson. Professor Pierpont does not appear in the picture.

(Yale University Archives, Manuscripts and Archives, Yale University Library)

mathematics (Navigation) in the third term of Sophomore year, and Differential and Integral Calculus during the first two terms of Junior year, in place of Greek or Latin." It is interesting also to note that analytic geometry and calculus were prescribed studies in the School of Engineering established in 1852 as a section of the graduate Department of Philosophy and the Arts. Further advanced subjects were offered by Professor Newton in 1862–1863, under the title "Pure and Mixed Mathematics" as courses leading to the degree of Doctor of Philosophy, doubtless the first occasion in the United States when such courses were announced for graduate students.

The Yale Corporation was the first to establish this degree in America (in 1860). The creation of the degree acted immediately as a stimulus to research. One of the first graduate students to complete advanced studies was J. Willard Gibbs, B.A. 1858, who qualified for the degree Ph.D. in 1863. At this point in our narrative it is appropriate to note that Professor Gibbs, following his appointment as Professor of Mathematical Physics, offered for many years one or more courses in mathematics for graduate students. Those who were fortunate enough to be of the small group of listeners to his presentation of multiple algebras, or his exposition of the system of vector analysis which he created, will not easily forget the painstaking care with which he endeavored to smooth out the difficulties of these subjects. It is indeed a significant fact that graduate instruction in mathematics should have been in the earliest years in charge of two outstanding scholars of the calibre of Newton and Gibbs.

Space does not permit the writer to set down in detail an account of the services to the Department of other former members. Passing mention is however due Professor Andrew W. Phillips, for many years Dean of the Graduate School. His native kindliness and tact were of inestimable benefit in laying the foundations for this strong arm of the University.

This brief sketch will serve as historical background for the account of the present organization, aims and personnel which follows.

Organization and Aims of the Instruction

When the Freshman year was established in 1919 as a separate School of the University, the place of mathematics in the curriculum for that year became a question of some difficulty. Two objects in planning the course were sought. First, it was necessary to base the work upon such preparation as could be depended upon at entrance, and to present subject matter which the Freshman electing mathematics would find interesting and inspiring, either as the completion of his mathematical studies, or as a stimulus for more advanced work. Second, as the first year of a well-defined two-year course for Freshmen planning to enter the Sheffield Scientific School, careful consideration had to be given to coordination with a second year of required mathematics. Much time and thought were given to working out the details, and it is believed that a successful course has now been developed, embracing topics in analytic geometry and differential and integral calculus, constituting a course of three hours a week throughout the year. A longer course of five exercises per week is given in the first term to Freshmen entering without trigonometry, and in the second term to those who have not passed solid geometry. Instruction in mathematics in the Freshman Year is carried along with the utmost thoroughness and attention to detail. This year twenty-six divisions of Freshmen are taking the subject, and sixteen members of the Department take part in the instruction. Frequent conferences of this group are held under the chairmanship of Professor W. R. Longley for discussion of progress and future plans. It is a pleasant duty at this point to set down a deserved tribute to the splendid spirit and enthusiasm shown by all individuals of this group. As an indication of the reaction to this devoted work of the staff, it may be pointed out that seventy-five Freshmen have contested annually in the examination for the Barge mathematical prizes, unmistakable testimony to the interest aroused in the study.

In the group courses of the Sheffield Scientific School, mathematics is prescribed in Sophomore year for all students in engineering and applied science. Without dwelling on variations caused by pressure of time due to the crowding in of technical courses, it may suffice to state that the subject matter in mathematics is carefully chosen from the differential and integral calculus with a view to its importance and usefulness to the student in his professional training. Careful selection of this material has resulted in sifting down the content to topics closely related to those met by the student in other courses. Juniors and Seniors may elect courses in advanced calculus, mechanics, theory of statistics, or some other one of the group of advanced mathematical studies regularly offered by the Department.

Frequent changes in the course of study in Yale College in recent years have had an influence on the number of under-graduates electing mathematics in Sophomore year and later. There is, however, satisfaction in recording that a remarkable revival of interest in the course offered in calculus by Professor Pierpont has been manifested in the last three years. The number of Sophomores electing this course has increased from fifteen to forty-five. It appears that the Freshman course may stimulate interest in the science, apart from the success of the instructor named in conducting the course.

Reference has already been made to advanced courses offered by the Department to undergraduates. These are directly introductory to the curriculum of the Graduate School. There is little that need be said of the program of studies laid down for candidates for the higher degrees, save that it forms a well-outlined plan calculated to develop the student's power and interest in research either in analysis, geometry, or applied mathematics. For teachers, not candidates for a degree, the Department offers courses primarily of pedagogical interest, carried out in close coordination with the Department of Education. One further item may be dwelt upon as evidence of the intention to offer as broad a training as is consistent with the standards of the Graduate School. This concerns the subject of mathematical statistics. Recognizing the importance of this modern science in its relations to insurance, finance, biology, and other natural sciences, in addition to an elementary course offered to undergraduates an advanced course was established some years ago which met with instant favor. At present the Department needs on its staff a member who shall specialize along this line. The importance of offering thorough courses in statistical theory can hardly be over-emphasized.

Graduate instruction at Yale in mathematics began in the '60s with Gibbs and Newton. In common with other American universities, the great developments of the science by continental mathematicians in the years following 1860 found tardy recognition at Yale. Courses in modern mathematical analysis may fairly be said to have begun at Yale with the appointment of Professor Pierpont as Lecturer in 1894. These were supplemented by courses in modern geometry by Professor Smith in 1896. The tradition established at Yale by the contributions of Loomis and Newton to astronomical science and meteorology was maintained by the appointment in 1907 of Professor E. W. Brown, now Sterling Professor of Mathematics. Professor Brown's contributions to the lunar theory, and in particular his work on the New Lunar Tables published by the Yale University Press, give him an unquestioned place among the outstanding scholars in the field of mathematical astronomy.

In the reorganization of the University in 1919 it appeared that the suggestions of the Alumni Committee on Plan for University Development laid great stress on effective and inspiring classroom instruction. With the emphasis so placed, it was inevitable that for a time more attention in the organization of the staff should be paid to this phase of the work than was really warranted. Research and productive scholarship on the part of younger instructors were relegated to a subsidiary place. But it will be granted that the inspiring teacher must be full of enthusiasm for his subject. If his teaching is not to sink to the level of mere coaching, he must maintain his interest in his science and show this by reasonable productive scholarship. With the resumption of normal conditions in graduate study, it is again possible to make appointments to the grade of instructor of promising young scholars, who have already completed the requirements for the doctorate, and who show unmistakable gifts as teachers.

THE DEPARTMENTAL STAFF

There are now sixteen members of the teaching staff: four professors, one associate professor, three assistant professors, four instructors, and four assistants. It is interesting to note that these men are recruited from many institutions, and to bring out this fact the name of the university or universities at which each individual pursued his main mathematical studies is set down: E. W. Brown (Cambridge University), H. L. Dorwart (Yale), W. W. Elliott (Cornell), M. C. Foster (Yale), L. S. Hill (Columbia and Chicago), W. R. Longley (Chicago), J. S. Mikesh (Harvard and Minneapolis), E. J. Miles (Chicago), T. W. Moore (Yale), James Pierpont (Vienna), P. D. Schwartz (Yale), C. A. Shook (Johns Hopkins), P. F. Smith (Yale), J. I. Tracey (Johns Hopkins), J. K. Whittemore (Harvard), W. A. Wilson (Yale).

The connection with the University of several members of the Department extends over a term of years. These periods for those of professorial rank are as follows: Brown (17 years), Longley (18), Miles (13), Pierpont (30), Smith (36), Tracey (12), Whittemore (8), Wilson (15). Of these, Smith and Wilson pursued their undergraduate studies at Yale. All other members of the Department have joined the staff after graduation or later in their careers.

In planning the routine work of instruction it is considered important that the classroom teaching should be reasonable in amount, so that each instructor may have time for other activities and other lines of University work. A large amount of effort is demanded from the staff by assignments to various committees. Correction of tests and daily notebook work adds to the hours of routine labor. Teaching is at best an exacting profession, and when possible the schedule made out for any instructor is planned so that there may be variety in his work.

In the instruction for Freshmen and Sophomores the daily work is carried on by assignments from textbooks. These have been prepared by members of the staff, thus carrying on a Yale tradition dating back to President Day, whose mathematical textbooks were used by generations of Yale men. The educational influence of a well-written textbook is far-reaching, and the care and success with which work of this kind has been done by members of the staff now and in the past have made a distinct contribution to the service of Yale University to education in the English-speaking world.

Saunders Mac Lane studied at Yale, Chicago, and at the University of Göttingen, where he received his doctorate in 1934 under the supervision of Paul Bernays and Hermann Weyl. After early positions at Harvard, he returned to the University of Chicago in 1947. His research has ranged through algebra, logic, algebraic topology, and category theory. Among his books are Homology, Categories for the Working Mathematician, *and (with Garrett Birkhoff) the influential text* A Survey of Modern Algebra. *His numerous honors include a Chauvenet Prize and a Distinguished Service Award from the MAA, and a Steele Prize from the AMS. He served as president of the MAA in 1951–1953, as president of the AMS in 1973–1974, and as vice-president of the National Academy of Sciences in 1973–1981.*

Mathematics at the University of Chicago
A Brief History

SAUNDERS MAC LANE

Since its inception, the University of Chicago has had an active department of mathematics. Indeed, in two periods, 1892–1909 and 1947–1959, this department has been perhaps the dominant one in the United States. This essay will sketch the development of the department.

THE FIRST MOORE DEPARTMENT

The University of Chicago opened its doors in 1892. William Rainey Harper, the first president of the university, immediately started out to emphasize active graduate study, bringing in a number of university presidents to be his department heads. For acting head of mathematics, he found Eliakim H. Moore (1862–1932), a lively young mathematician (Ph.D. Yale 1885, student at Berlin, Germany 1885–1886), then an associate professor at Northwestern. Moore came in 1892 and immediately found and appointed two excellent German mathematicians, then at Clark University: Oskar Bolza (1857–1936), a student of Weierstrass in the calculus of variations, and Heinrich Maschke (1853–1908), a geometer — both had been students at Berlin and at Göttingen. They constituted the core of the first department at

E. H. Moore

(University of Chicago Archives)

Chicago. G. A. Bliss [1] (who studied at Chicago then) has written of them: "Moore was brilliant and aggressive in his scholarship, Bolza rapid and thorough, and Maschke more brilliant, sagacious and without doubt one of the most delightful lecturers on geometry of all times". This team almost immediately made Chicago the leading department of mathematics in the United States.

In the period 1892–1910, Chicago awarded 39 doctorates in mathematics (far surpassing the next institutions: Cornell, Harvard, and Johns Hopkins). Even more striking is the quality of the first doctorates. The first (1896) was Leonard Eugene Dickson, who subsequently did decisive research on algebra [5] and in number theory [4]. He and five others in this group of 39 Ph.D.s subsequently held the office of president of the American Mathematical Society (AMS). These five (with their subsequent institutions) were, with year of Ph.D.:

1900	Gilbert Ames Bliss	University of Chicago
1903	Oswald Veblen	Princeton
1905	Robert Lee Moore	Texas
1907	George David Birkhoff	Harvard
1910	Theophil Henry Hildebrandt	University of Michigan

All except Bliss were doctoral students of E. H. Moore although it appears that Veblen actually directed most of R. L. Moore's thesis work! In the next generation, these mathematicians were probably the dominant figures at their institutions (Dickson was also dominant at Chicago). Note that this includes the three institutions generally regarded as the leading ones in mathematics from 1910–1940: Chicago, Harvard, and Princeton. At Texas, R. L. Moore was a great individualist, while Hildebrandt, as a long-time chairman at Michigan, set the style for a major state university. Birkhoff was interested in differential equations and dynamics. In 1912, Henri Poincaré, the leading French mathematician, had formulated and left unproven his "last geometric problem". Birkhoff provided the proof in 1913 and was in consequence soon regarded as the leading American mathematician. Veblen also had a major role in the development of topology and mathematical logic at Princeton University and later (from 1932) administered the mathematical group at the Institute for Advanced Study.

Clearly these results indicate that remarkable and aggressive advanced mathematical education took place at Chicago. G. A. Bliss [1] writes, "Those of us who were students in those early years remember well the intensely alert interest of these three men (Bolza, Maschke, and Moore) in the papers which they themselves and others read before the club ... Mathematics ... came first in the minds of these leaders". In the files of the mathematical club, which met biweekly, I have also found a rapturous account of a visit by Dickson, who in 1897 returned for a brief visit after his year of study

in Paris and Leipzig, to report on the current mathematical developments
in Europe. Some echoes of this sense of excitement were still present when
I was a graduate student (1930–1931) in Chicago. I took a seminar on the
"Hellinger integral", conducted by E. H. Moore, with assistance from R.
W. Barnard. Moore, realizing that I knew little about Hellinger's integral,
asked me to present E. Zermelo's [13] famous second proof that the axiom
of choice implies that every set can be well-ordered. I gave what I thought
was a very clear lecture, but after Barnard and the two Chinese students had
left, Moore took me aside and spent an hour explaining to me what was really involved and what I should have said in my lecture. It was a thrilling
experience — one which reflects in brief the excitement of the early days at
Chicago.

What sorts of mathematics were studied then?

The archives at the University of Chicago library contain lecture notes
taken by Benjamin L. Remick (1894–1900) on the following subjects:

By Bólza, with academic quarter indicated:

Functions of a Complex Variable	Autumn, 1894
Notes on Quaternions	Autumn, 1894
Hyperelliptic Functions	Autumn, 1897
Invariants I and II	Winter, 1897, 1898
Theory of Abstract Groups	Summer, 1899

By Maschke:

Higher Plane Curves	Autumn, 1894
Analytical Mechanics	Spring, 1895
Algebraic Surfaces	Spring, 1895
Weierstrass on Elliptic Functions	Winter, 1895
Higher Plane Curves	Winter, 1897
Algebraic Surfaces	Spring, 1897
Linear Differential Equations	Spring, 1897

By E. H. Moore:

Projective Geometry	Autumn, 1896
Theory of Numbers	Autumn, 1897
General Arithmetic I and II	Winter, Spring, 1898

Also Harris Hancock (later at the University of Cincinnati) lectured on the
calculus of variations (Spring, 1895), while George A. Miller (later at the University of Illinois) lectured on permutation groups (Summer, 1898). There
was a closely related astronomy department, which emphasized mathematical astronomy. F. R. Moulton lectured there on general astronomy (Autumn,

Óskar Bólza

(University of Chicago Archives)

Heinrich Maschke

(University of Chicago Archives)

1896), and Kurt Laves on "the three body problem" (Spring, 1897). In 1902, the full list of graduate courses given reads as follows:

Autumn:	Theory of Equations	Bolza
	Projective Geometry	Moore
	Modern Geometry	Maschke
	Theory of Functions	Bolza
	Finite Groups	Dickson
Winter:	Theory of Equations II	Bolza
	History of Mathematics	Epsteen
	Higher Plane Curves	Maschke
	Theory of Functions II	Bolza
	Linear Substitution Groups	Maschke
Spring:	History of Mathematics	Epsteen
	Teaching Laboratory	Moore
	Vector Analysis	Lunn
	Linear Differential Equations	Maschke

These courses cover most of the topics of mathematics then of current interest. In the mathematics club, Moore spoke of finite fields (presumably his proof classifying all such), on Peano's space-filling curve, and on his elegant system of generators for the symmetry group S_n. Current concerns in finite group theory started then: J. W. A. Young spoke "On Hölder's enumeration of all simple groups of orders at most 200." In 1904, J. H. M. Wedderburn, a visitor from Scotland, proved his famous theorem that every finite division ring must be a commutative field. Algebra was there with a vengeance.

The original faculty at Chicago included some junior members: Jacob William Albert Young (Ph.D. from Clark University, presumably a student of Bolza); he retired as associate professor in 1926 and Harris Hancock (1892–1900). Then several of Chicago's own Ph.D.s were appointed to the faculty:

Herbert Ellsworth Slaught (1861–1937); Ph.D. 1898. From assistant professor (1894) to professor (1913–1931). Slaught was primarily concerned with mathematical education and with assisting students. He was (with lively support from E. H. Moore) one of the principal founders of the Mathematical Association of America in 1916.

Leonard Eugene Dickson (1874–1950); Ph.D., 1896. From assistant professor (1900) to professor (1910–1939). His massive and scholarly *History of the Theory of Numbers* was a landmark [4], while his monograph *Algebras and Their Arithmetics* was translated into German and had a major influence on the German school of abstract algebra [5]. He was a powerful and assertive mathematician who directed at least 64 doctoral theses. It is rumored that he consciously had two classes of doctoral students: the regular ones and the really

promising ones (such as C. C. MacDuffee, 1921, who later went to
Wisconsin; C. G. Latimer 1924, to Kentucky; Burton W. Jones 1928,
to Cornell and Colorado; A. A. Albert 1928, to Chicago; Gordon Pall,
1929, to IIT; Alexander Oppenheim 1930, to Singapore; Arnold E.
Ross 1931, to Notre Dame and Ohio State; R. D. James, 1928, to
Berkeley and British Columbia; and Ralph Hull 1932, to Purdue).
One can contemplate with amazement the wide influence exerted by
Dickson. I can also recall his course (1930) in number theory, taught
from a book of that title which he had written with sparse precision:
He expected his students (Hull, James, Mac Lane, et al.) to have
understood every argument and every shift in notation.

Arthur Constant Lunn (1877–1949); Ph.D. 1904, rose from associate
professor in applied mathematics (1902) to professor (1923–1942).
He had accumulated a massive knowledge of all of classical mathematical physics and lectured on this in enthusiastic but rambling
ways that did not end with the formal end of the class hour. It
was rumored among the students that a 1926 paper of his containing
some of the new ideas of quantum mechanics had been rejected by
some uncomprehending editor. Whether or not this was true, Professor Lunn was discouraged but very knowledgeable when I listened
to him in 1930–1931.

2. THE SECOND MOORE DEPARTMENT

In the period 1908–1910, the verve and dynamism of the original Chicago
department appears to have been gradually lost. Professor Maschke died in
1908; in 1910 Professor Bolza returned to Germany (Freiburg in Baden) but
kept up an essentially nominal "nonresident" professorship at Chicago.
He was still alert when I visited him in Freiburg in 1933. E. H. Moore
developed his interest in postulational generality to a form of "general analysis" which tracked properties of integral equations in terms of functions "on
a general range". The first form of his general analysis was presented in his
colloquium lectures of the AMS at Yale University in 1906, and published (in
considerably altered form) in 1910. At that time, it was very much in order
to find the ideas underlying the existence theorems for solutions of integral
equations — and indeed this objective led David Hilbert to his study of what
are now called Hilbert spaces. Moore's formulation of these ideas did not
succeed, in part because of his delay in publishing. He continued to work on
it for 20 or more years, developing a second form of general analysis which
was written in a logistic notation derived from Peano — a notation which was
precise but hard to read, and which did not include formal logical rules of

inference. In 1930–1931, his general analysis was presented in a six-quarter sequence of courses at Chicago; there were not many students. This version has been written up by Moore's student and associate R. W. Barnard; it is a monument to a timely but failed initiative [11].

This then is the sense in which the initial department at Chicago (Moore, Bolza, and Maschke) came to its effective end in 1907–1910. A new team appeared: Dickson, as already noted, plus Bliss and Wilczynski:

> Gilbert A. Bliss (1876–1951), Ph.D., 1900, Chicago, wrote his thesis with Bolza. After teaching at Minnesota, Chicago, Missouri, and Princeton, he became an associate professor at Chicago in 1908, Professor in 1913, and chairman in 1927 until his retirement in 1941. His interests covered many fields of analysis — algebraic functions, implicit function theorems, and the theory of exterior ballistics. He was an enthusiast for the calculus of variations.

> Ernst Julius Wilczynski (1876–1932) received his Ph.D. at Berlin in 1897. After teaching at the University of California (1898–1907) and at the University of Illinois (1907–1910), he became an associate professor at Chicago in 1910, and full professor (1914–1926). He published voluminously and enthusiastically, especially in his favorite subject of projective differential geometry, where the local properties of curves and surfaces were analyzed in terms of canonical power series expansions. Clearly, Wilczynski was appointed at Chicago as a successor to the previous geometer, Maschke.

There were also two more junior appointments in this period:

> Mayme Irwin Logsdon (1881–1967) received her Ph.D. at Chicago in 1921, with a thesis on equivalence of pairs of hermitian forms, directed by Dickson. She was an instructor at Chicago from 1921 and rose to be an associate professor (1930–1946). It is my own observation that one of her duties was that of advising and helping many women who were graduate students at Chicago; moreover, she taught a survey course required of all undergraduates. After her retirement from Chicago, she taught for many years at the University of Miami, in Coral Gables, Florida.

> Ernest P. Lane (1886–1969) received his Ph.D. at Chicago in 1918 with a thesis in geometry, and returned to Chicago as assistant professor in 1923. He was a meticulous man, and an enthusiast for projective differential geometry.

Thus, in the period 1910–1927, the team at Chicago, headed by E. H. Moore, consisted primarily of Bliss, Dickson, and Wilczynski. There were

Left to right: Dickson, Wilczynski, Lunn, MacMillan, Moore, Slaught and Moulton

(Courtesy of Saunders Mac Lane)

many Ph.D.s in this period — 115 of them. Some of the more memorable, classified by subject, were the following:

E. H. Moore directed theses in general analysis:

1910   Anna Pell (later Anna Pell-Wheeler), subsequently a professor at Bryn Mawr College. In 1927 she delivered colloquium lectures for the AMS on the "Theory of quadratic forms in infinitely many variables and applications."

1912   E. W. Chittenden, who became a leader in point-set topology at the University of Iowa.

1916   W. L. Hart, later of Minnesota, a prolific author of textbooks.

1924   Mark H. Ingraham, later chairman and dean at Wisconsin.

1926   H. L. Smith, later a leader at Louisiana State University; he worked on the Moore–Smith limits well known in topology.

1926   R. W. Barnard, later Moore's amanuensis at Chicago.

G. A. Bliss directed theses in analysis:

1914   W. V. Lovitt, who taught at Colorado College and wrote a book on integral equations.

1924   L. M. Graves, subsequently professor at Chicago.

L. E. Dickson directed theses in algebra; for example:

1921   M. I. Logsdon
1921   C. C. MacDuffee
1924   C. G. Latimer

C. J. Wilczynski directed theses in geometry:

1915   Archibald Henderson, who became influential at North Carolina.

1918   E. P. Lane, later professor at Chicago.

1921   Edwin R. Carus, who later founded the Carus monographs (MAA).

1925   V. G. Grove, subsequently chairman at Michigan State.

Applied Mathematics: 1913, E. J. Moulton, to Northwestern.

Clearly, Moore was still a dominant influence. However, none of the Ph.D.s of this period achieved the profundity in mathematical research of the best five earlier Ph.D.s. Many, however, did rise to influential positions in important universities, as indicated below:

Bryn Mawr: Anna Pell-Wheller

Cornell: Burton W. Jones (Ph.D., 1928)

Louisiana State University: H. L. Smith

Michigan State: V. G. Grove

Northwestern: E. J. Moulton

University of Iowa: E. W. Chittenden

Colorado College: W. V. Lovitt

Wisconsin: M. H. Ingraham, C. C. MacDuffee

In 1924, E. H. Moore reported proudly that the department had by then trained 116 Ph.D.s, plus 15 more in mathematical astronomy, and that 52 of this total of 131 were already full professors at their respective institutions. In 1928 (according to the lists in the *Bulletin of the American Mathematical Society*), 45 Ph.D.s were granted in mathematics in the United States, of whom 12 (according to the Bulletin) or 14 (according to department records) were at Chicago. The nearest competing institutions were Minnesota (four Ph.D.s) and Cornell and Johns Hopkins, with three Ph.D.s each.

My conclusion is this: Chicago had become in part a Ph.D. mill in mathematics.

3. THE BLISS DEPARTMENT

In 1927 G. A. Bliss became chairman at Chicago, while E. H. Moore continued as head — by then largely a titular formality. This ushered in a new period which lasted all during Bliss' terms as chairman (1927–1941). At about this time there were a number of new appointments to the faculty:

> E. P. Lane was promoted to an associate professorship in 1927.
> R. W. Barnard was appointed assistant professor in 1926.
> L. M. Graves was appointed assistant professor in 1926.
> Walter Bartkey (Ph.D. Chicago 1926) became an assistant professor of applied mathematics and statistics in 1927; he was subsequently dean of the division of physical sciences at Chicago (1945–1955).
> Ralph G. Sanger (1905–1960), Ph.D. Chicago 1931, became instructor in 1930 and assistant professor in 1940–1946; he later moved to Kansas State University.

I note explicitly that every one of these appointees had received his Ph.D. at Chicago.

In the period 1927–1941, Bliss (who retired in 1941) and Dickson (who retired in 1939) were the dominant figures in the department. In the eleven year period from 1927–1937, there were 117 Ph.D.s awarded. Of these theses, Bliss, with the occasional cooperation of Graves, directed 35, of which 34 were in the calculus of variations. (In the prior 21-year period 1906–1926, there had been 17 theses devoted to this calculus.) Dickson directed 32 theses. In this period, Dickson's interests shifted from his earlier enthusiasms for quadratic forms and division algebras to an extensive (and somewhat numerical) study of aspects of Waring's problem: for each exponent n, find a number k such that every (or every sufficiently large) integer is a sum of at most k n-th powers. At that time, the then newer methods of analytic number theory could prove the "sufficiently large" part; Dickson was concerned with a corresponding explicit bound and with the calculation of what happened below that bound. The topics of Dickson's 32 theses projects were distributed as follows: thirteen on quadratic forms, twelve on Waring's problem, six on division algebras, and five on general topics in number theory. There were four additional algebraic theses directed by A. A. Albert. In geometry, Lane directed 20 theses and Logsdon two. Moore and Barnard together directed six theses. There were seven in aspects of applied mathematics and ten others on assorted topics. All told, this period represents an intense concentration on the calculus of variations and on number theory.

In this period there were some outstanding results. Two Chicago Ph.D.s went on to become president of the AMS: A. Adrian Albert (Ph.D. 1928, with Dickson) and E. J. McShane (Ph.D. 1930, with Bliss). In the calculus of variations, I note four: W. L. Duren, Jr. (Ph.D. 1929), who soon played an important leadership role at Tulane and later at Virginia, M. R. Hestenes (Ph.D. 1932) was later influential at UCLA, Alston S. Housholder (Ph.D. 1937) who shifted his interests and became a leader in numerical analysis at Oak Ridge, while Herman Goldstine (Ph.D. 1936) was associated with von Neumann in the development of the stored program computer. As already mentioned, some half dozen of the Dickson Ph.D.s did effective work in number theory. Mina Rees (a Dickson Ph.D., 1931) subsequently was the first program officer for mathematics at the Office of Naval Research. Her leadership there set the style for the subsequent mathematics program at the NSF; subsequently, Dr. Rees became founding president of the Graduate School and University Center of the newly established City University of New York. In 1983, she was awarded the Public Welfare Medal of the National Academy of Sciences.

In functional analysis Leon Alaoglu (Ph.D. 1938, with L. M. Graves) became famous for his theorem that the closed unit ball in the dual space of a Banach space is compact in the weak-star topology. After teaching at Harvard and Purdue and doing more research, he became a senior scientist at the Lockheed Aircraft Corporation. Malcolm Smiley took his Ph.D. in the calculus of variations in 1937, but then switched to active research in algebra. Ivan Niven, a Ph.D. of Dickson's in 1938, studied then with Hans Rademacher at the University of Pennsylvania. After teaching at Illinois and Purdue, he went to the University of Oregon and did decisive research on uniform distribution of sequences modulo m. Frederick Valentine (Ph.D. 1937 in the calculus of variations) was subsequently at UCLA where he published an important book on convex sets.

To summarize: In this period the department at Chicago trained a few outstanding research mathematicians and a number of effective members of this community — plus produced a large number of essentially routine theses. Was this because there was an undue concentration on a few special fields, or because the presence of so many graduate students meant that the faculty was forced into finding routine topics? In some cases they may have failed to appreciate students' potential. I do not know. I do clearly recall my own experience as a graduate student at Chicago (1930–1931). Since the calculus of variations was evidently a major issue there, I signed up for Professor Bliss' two-quarter course in this subject. Sometime well into the first quarter I had trouble putting the (to my mind necessary) ε's and δ's into his rather sketchy proof of the properties of fields of extremals. So I ventured to ask Professor Bliss how to do this. At once he produced all the needed epsilons, with great skill — but he also made it very clear to me that I did not need to concern myself with such details; graduate students were expected to get chiefly an overall impression of the shape of the subject. Some years later, I had occasion to study Bliss' book on algebraic functions; I observed then that this book correctly reproduced the suitable German sources but did not press on to get a real understanding of why things worked out and what the Riemann–Roch theorem really meant.

There were lighter moments. Professor Bliss liked to kid his students. One day in his lectures on the calculus of variations, he recounted his own earlier experiences in Paris. After he sat down in the large lecture amphitheatre there, an impressive and formally dressed man entered and went to the front. Bliss thought it was the professor himself, but no, it was just his assistant who cleaned the blackboards and set the lights. When the professor finally arrived, all the students stood up. At this point in his story, Bliss observed that American students do not pay proper respect to their professors. So the class agreed on suitable steps; I was the only member owning a tuxedo. The next day, arrayed in that tuxedo, I knocked on the door for Professor Bliss to report that his class awaited him. When he came in they all rose in his honor.

Of the six students of Moore and Barnard during this period, only Y. K. Wong (Ph.D. 1931) continued substantial activity. With Moore, he had studied matrices and their reciprocals; in his later research (at the University of North Carolina) he was concerned with the use of Minkowski–Leontief matrices in economics (Wong, [12]).

There remains the fascinating question: In the early days, Moore had been dynamic and remarkably effective in training graduate students. What changed? As I have already noted, he was still an alert critic when I knew him in 1930, and he had continued to work diligently on his form of general analysis. But he did not publish. According to an obituary by Bliss, he published only two substantial research papers after 1915, both in 1922, and one of them with H. L. Smith on the important concept of the Moore–Smith limit. At the start, Moore had been in lively contact with many current developments in mathematics. I conjecture that he had gradually lost that contact, in part because of a heavy preoccupation with his own ideas in general anlaysis, and in part because he may have depended on the exchange of ideas with his contemporaries Bolza and Maschke, while the newer and younger appointments at Chicago did not provide an effective such exchange.

4. Appointments by Bliss

In this period (1927–1941), there were a number of other appointments to the faculty, as follows:

A. Adrian Albert (1905–1972), Ph.D. Chicago 1928; assistant professor (1931) to professor (1941).

M. R. Hestenes (1906–), Ph.D. Chicago 1932; assistant professor (1937), associate professor (1942–1947) later influential in numerical analysis and combinatorics at UCLA.

W. T. Reid (1907–1977), Ph.D. Texas 1927; instructor, Chicago (1931), associate professor (1942–1944); later a professor at Northwestern, Iowa, and Oklahoma.

With these appointments, note the emphasis on the two fields of algebra and the calculus of variations — and on Ph.D.s from Chicago. (Reid came from Texas, but had spent the years 1929–1931 as a postdoctoral fellow at Chicago.)

Later on, the department made real attempts to appoint mathematicians not from Chicago and in new fields; two of them, as follows, did not last:

Saunders Mac Lane (Ph.D. Göttingen 1934); instructor, Chicago (1937–1938), then to Harvard. I believe that my appointment in 1937 at Chicago was due to the intervention of President Hutchins. At any rate, I had met Hutchins in 1929, and he had personally arranged to get me a graduate fellowship at Chicago for 1930–.

Norman Earl Steenrod (1910–1971), Ph.D. Princeton 1936; assistant professor Chicago (1939–1942); then to the University of Michigan as assistant professor (1942), in 1945 to Princeton.

Otto F. G. Schilling (1911–1973), Ph.D. Marburg 1934; instructor Chicago, (1939) to professor (1958); in 1961 to Purdue.

There are, to be sure, rumors of appointments which were *not* made. Thus, the famous German analyst and number theorist Carl Ludwig Siegel left Göttingen in the spring of 1940 and escaped via Norway to the United States. It then became clear that he needed a suitable position in this country; rumor has it that G. A. Bliss knew this but did not act on this possibility; soon Siegel became a professor at the Institute for Advanced Study in Princeton.

The appointment of Steenrod, who soon became a noted topologist, may well have been stimulated by the use of topology and the related theory of critical points (Marston Morse) in the calculus of variations. Up until this point the appointment policy at Chicago seems to have been based on what I might call the "inheritance principle": If X has been an outstanding professor in field F, appoint as his successor the best person in F, if possible the best student of X. Let me re-examine the appointments at Chicago in this light.

Bolza was outstanding in analysis and had written an authoritative book on the calculus of variations. Shortly after he left in 1908, his best student, G. A. Bliss, was appointed. Subsequently, three students of Bliss were appointed: Graves, Sanger, and Hestenes, as well as W. T. Reid from Texas. There resulted a great concentration on such topics as variants of the problem of Bolza in the calculus of variations, but the school at Chicago missed out on the major development of the subject in the early 1930s, as represented by the work of Marston Morse on the calculus of variations in the large. Chicago was of course aware of this work, but did nothing much about it. Specifically, in the spring of 1931, Bliss conducted a seminar on this topic, and assigned Mac Lane to report on Betti numbers and their meaning. Mac Lane thereupon studied the (then unique) text by Veblen, and reported on the Betti numbers but not on their meaning (which he did not really understand).

In geometry, the death of that notable geometer Maschke in 1908 was soon followed by the appointment of another geometer, Wilczynski, in 1910, and then, upon his retirement in 1926, by the promotion in 1927 of his best student E. P. Lane. The special emphasis on the subfield of projective differential geometry (as in Lane's subsequent book) gradually lost its importance, both in Chicago and in Shanghai (where the senior professor Buchin Su worked in this field). In 1939, George Whitehead, one of the graduate students, asked Professor Lane for a thesis topic in projective differential geometry. Instead of giving him a topic, Lane gave Whitehead the good advice to work in the newer field of topology with Steenrod; Whitehead later (at MIT) became a leader in this field.

In general analysis, E. H. Moore had considerable influence on Lawrence Graves; then in 1928, Moore's student Barnard was appointed to the faculty. However, Moore did not work out the possible connection between his general analysis and the study (at other centers) of Hilbert spaces and of functional analysis. Moore was a great enthusiast for infinite matrices, postulational methods, and Peano. In early work, Peano had the axioms for a (two-dimensional, real) vector space. I never learned about these axioms from Moore — and had to learn them in 1932 from Herman Weyl in Göttingen (who had clearly formulated them in his 1917 book on relativity). This is another small piece of evidence that Moore had lost contact.

In mathematical astronomy and applied mathematics, Kurt Laves (first appointed about 1894) was the first faculty member. Then various Chicago Ph.D.s were appointed in astronomy or in mathematics. The most outstanding was perhaps F. R. Moulton; others were W. D. Macmillan, A. C. Lunn, and then Walter Bartky in 1926. Perhaps because of his activity in military research during WW II, Bartky's interests shifted to administrative matters; he became a dean and finally a vice president for research at the university from 1956 to 1958. In this sense, the line of inheritance in applied mathematics died out, not to be renewed until the appointment in 1963 of two former students (at Chicago) of S. Chandrasekhar.

E. H. Moore provided the initial impetus in algebra, group theory, and number theory; the appointment of his first Ph.D. student Dickson in 1900 was a strong step. Then in 1931, Dickson's best student, Adrian Albert, was appointed. In 1945, Albert's recommendation brought the appointment of Irving Kaplansky (Ph.D. from Harvard; Mac Lane's first student). Albert kept the interest in algebra generally and in group theory in particular alive, and in 1961 organized a "special year" on group theory at Chicago. It was during this year that Walter Feit (M.S. Chicago, 1951) and John Thompson (Ph.D. Chicago, 1959) worked out their "odd order" paper with the remarkable proof that every finite simple group is either cyclic or of even order. This was a vital step toward the subsequent classification of all finite simple groups.

Thus, in algebra the inheritance theory of appointments worked splendidly, while in other fields, as noted, it was not successful in the long run.

5. THE LANE DEPARTMENT

When G. A. Bliss retired in 1941, E. P. Lane became chairman of the department of mathematics. He made several attempts to revive and strengthen the department, but the times were not propitious, largely because of the onset of WW II. When President Robert M. Hutchins (with considerable administrative courage) brought the Manhattan project on atomic energy to Chicago, it was soon housed in the department's treasured building, Eckhart Hall, and the mathematicians were moved out to one of the towers of Harper Library. There were no new appointments till the postwar appointment of Kaplansky (1945). There were 21 Ph.D.s (1941–1946), including Whitehead (1941), the algebraists R. D. Schafer (1942) and Daniel Zelinsky (1946), and in the calculus of variations the very young mathematician J. Ernest Wilkins (1942) who later did notable research in applied mathematics.

PH.D.s TO WOMEN AT CHICAGO

In the 39 years 1908–1946, the department awarded 51 Ph.D.s to women out of a total of 270 Ph.D.s in mathematics in that period. It is likely that more Ph.D.s were awarded to women at Chicago than at any other American university in this period. Chicago had been coeducational from the start, but 1908 was the year when the first Ph.D. was awarded to a woman — Mary Emily Sinclair, who subsequently became professor and chairman at Oberlin College. Also, 1946 is the year when Marshall Stone, as a new chairman, came to drastically change the direction of the department; this determines the period 1908–1946 which I chose for this list.

Thanks to Marlene Tuttle of the alumni relations office of the University of Chicago I have been able to collect definitive information on almost all of these 51 women mathematicians. In particular, I could locate the college or university where they subsequently taught mathematics. After classifying these institutions as women's colleges, coeducational colleges (e.g., Oberlin), universities, or research universities, I get the following table based on one chosen institution for each Ph.D.; in a few cases there was a change of institution:

Subsequent Academic Employment of Women Ph.D.s

Ph.D.	Date: 1908–1931	1932–1946
Women's college	8	12
College	7	6
University	5	5
Research university	6	1
Total	26	24

In one case (in the second period) I was unable to locate any subsequent teaching employment; I believe that the individual was married and did not take up teaching. But note that of the 51 listed, 50 did engage in teaching, most of them at just one institution and for a considerable period. The Ph.D.s from Chicago provided an effective source of faculty — especially at women's colleges. Note that teaching loads at such colleges were then quite heavy.

I have ventured to classify seven of the institutions as "research universities", although that term was not then in use. The word appeared only later as a label for those universities which seek to acquire substantial research funds from the government. At any rate the seven research universities listed above were (in chronological order of the degrees) Wisconsin, Berkeley, Minnesota, Chicago, Illinois, Northwestern, and Illinois again, the last in 1932. It will make my classification clear if I list the five universities (1932–1946) as Kent State, University of Utah, University of Oklahoma, University of Alabama, and Temple University. All told, this tabulation indicates clearly that in all this period very few of the women went or were sent to major research universities. (In 1916, the University of California at Berkeley did not then have its present standing in mathematics.)

In reporting this situation, I deliberately said "were sent," because in those days positions for new doctorates in mathematics were managed by what is now called the "old-boy network". At present this is a term of opprobrium; at that time it referred to a placement system for a small number of graduates that in fact worked much more efficiently than the present system, which inevitably is applied to much larger numbers and involves massive employment interviews at the January AMS meetings, plus pious declarations of equal opportunity in advertisements which (especially today) make it clear that opportunity beckons at X university only if your research lies in a field already X-represented.

The old-boy network functioned as follows: all the active mathematicians such as Veblen at Princeton, Bliss at Chicago, or Birkhoff at Harvard (plus many others, such as Hildebrandt at Michigan) had pretty shrewd ideas as to the level of mathematical activity at many schools, and they also had quite detailed (but perhaps mistaken) knowledge of the qualities of their own current products. So when they heard that Oberlin College or the women's college of North Erehwon or the University of W had a vacancy, they knew which of their graduates would be an appropriate candidate there, and they acted accordingly. (Of course, the candidate's professor was also an actor in this network.) The system did make mistakes. For example, in 1957, Michigan sent to Chicago letters of recommendation for a new Ph.D., one Steven Smale. The letters were not especially enthusiastic. At that time, the department had few vacancies for an instructor, so Smale was appointed at Chicago in the college mathematics staff, then separate from the department and intended primarily for undergraduate teaching. This goes to show that there can be misjudgements about research potential.

At any rate, the table above makes it clear that Chicago did not normally send its women Ph.D.s to universities anxious to acquire research hot-shots.

I also tabulated published research papers in mathematics (1931–1960) for the 25 women Ph.D.s in the second period. In ten cases, I found 14 publications all told, in most cases the publication of the thesis, but I note that one woman had three publications. I have not tabulated research publications for men Ph.D.s from the period (1932–1946); some were prolific, while others hardly published.

On the evidence, I summarize thus: In this period, women were encouraged to study for the Ph.D. degree at Chicago, and there was a role model on the staff to help and support them (Mayme I. Logsdon). But these women students were not really expected to do any substantial research after graduation; the doctorate was it, and in many cases the thesis topic was chosen to suit. This last sentence agrees with my own recollection of the situation and atmosphere at Chicago during my graduate study there in 1930–1931. I might add that for some of the men students there was the same low level of research expectations — but not for all.

For completeness, I add that in the following period (1946–1960) at Chicago there were exactly four Ph.D.s granted to women; among those, one to Mary Weiss (Ph.D. with Zygmund 1957) who made an impressive research career. I note that the women's liberation movement was yet to come, and that there apparently were very few women graduate students present. My own course records (in basic graduate courses in algebra and topology in this period) show 38 women out of 267 students all told in my courses — about 14 percent. It appears that women students began graduate work, but that few went on to the Ph.D.

6. THE STONE DEPARTMENT

Robert Maynard Hutchins, president of the University of Chicago (1929–1951), had brought the Manhattan project to the university during WW II, and with it many notable scientists including Enrico Fermi, James Franck, and Harold Urey. As the war drew to a close, he and his advisors decided to try to hold these men and their associates at Chicago. For this purpose, he established two research institutes, now known as the Fermi and the Franck Institutes. He and his advisors realized that there should be a much-needed strengthening of the department of mathematics. With the advice of John von Neumann (who had been associated with the Manhattan project), they approached Marshall H. Stone, then a professor at Harvard, suggesting (after some talk of a deanship) that he come to Chicago as chairman of mathematics. Stone had thought deeply about the conditions which would support a great department of mathematics at a level well above that then present at Harvard or Princeton. After receiving suitable assurances from President Hutchins, Stone came to Chicago in 1946. He thereupon brought together what was in effect a whole new department. In each such new case, I specify the dates of their activity on the Chicago faculty:

As professors:

> André Weil (1947–1958), a notable (and contentious) French mathematician, one of the leading members of the Bourbaki group. He had just published his fundamental book on the foundations of algebraic geometry, containing his proof of the Riemann hypothesis for function fields.

> Antoni Zygmund (1947–1980), a Polish analyst, interested in Fourier Analysis and harmonic analysis. Zygmund had been in this country at Mount Holyoke and then at the University of Pennsylvania.

> Saunders Mac Lane (1947–1982), from Harvard; he was at that time active in studying the cohomology of groups and the related cohomology of Eilenberg–Mac Lane spaces in topology.

> Shiing–Shen Chern (1949–1959), an outstanding Chinese mathematician with interests in differential geometry and topology (for example, his characteristic classes).

As assistant professors, Stone brought to Chicago:

> Paul R. Halmos (1946–1961), working in measure theory and Hilbert space. He had just published his elegant exposition "Finite Dimensional Vector Spaces", with a presentation influenced by his contacts with John von Neumann.

> Irving E. Segal (1948–1960), an enthusiast for rings of operators in Hilbert space and their application to quantum mechanics.

> Edwin H. Spanier (1948–1959), a young and knowledgeable algebraic topologist from Princeton; he had recently finished his Ph.D. under the direction of Norman Steenrod.

Of the previous department, Albert and Kaplansky were immediately enthusiastic members of this new team which then read:

Professors: Albert, Chern, Mac Lane, Stone, Weil, and Zygmund.

Assistant Professors: Halmos, Kaplansky, Segal, and Spanier.

Of the other previous members, Hestenes soon left for UCLA and J. L. Kelley, who had briefly been an assistant professor, left for Berkeley. Professors Barnard, Graves, Lane, and Schilling stayed on (in most cases until retirement); they cooperated but were not really full members of the new dispensation.

The new group covered quite a variety of fields. There were exciting graduate courses, and some clashes of opinion (for example, between Weil and Segal). Weil, in continuing the tradition of Hadamard's seminar in Paris, taught a course called "Mathematics 400" in which the students were required to report on a paper of current research interest *not* in their own field; a few students were discouraged by his severe criticisms but many others were encouraged to broaden their interests. Under Stone's encouragement, a whole new graduate program was laid out, with three-quarter sequences in algebra, analysis, and geometry (see Mac Lane [8]).

This was the immediate postwar period, when many ex-soldiers could take up advanced study under the G. I. Bill. Thus, there were many lively students at Chicago; in the period 1948–1960 there were 114 Ph.D.s granted by the department. Among the recipients were a number of subsequently active people, including:

1950: A. P. Calderón, R. V. Kadison, and I. M. Singer

1951: Murray Gerstenhaber, E. A. Michael, and Alex Rosenberg

1952: Arlen Brown and I. B. Fleischer

1953: Katsumi Nomizu

1954: Louis Auslander and Bert Kostant

1955:   Errett Bishop, Edward Nelson, Eli Stein, and Harold Widom

1956:   R. E. Block, W. A. Howard, Anil Nerode, and Guido Weiss

1957:   B. Abrahamson, Donald Ornstein, Ray Kunze, and Mary Weiss

1958:   Paul Cohen, Moe Hirsch, and E. L. Lima

1959:   Hyman Bass, John G. Thompson, and Joseph Wolf

1960:   Steve Chase, A. L. Liulevicius, and R. H. Szczarba.

The qualities of this group of graduates, in my view, match the qualities of
the best graduates of the first group of the Moore department. For example,
by 1988, eight of those listed just above had been elected to membership
in the section of mathematics of the National Academy of Sciences; by that
date in the whole country, about 30 had been elected from those with Ph.D.s
from these years 1948–1960; Princeton, with six, contributed the next largest
contingent.

In this period at Chicago, there was a ferment of ideas, stimulated by the
newly assembled faculty and reflected in the development of the remarkable
group of students who came to Chicago to study. Reports of this excitement
came to other universities; often students came after hearing such reports
(I can name several such cases). This serves to emphasize the observation
that a great department develops in some part because of the presence of
outstanding students there. (This is true also of Göttingen in 1930–1933 and
Harvard in 1934–1948 in my own experience.)

By 1952, Marshall Stone had grown weary of the continued struggle with
the administration for new resources; Mac Lane succeeded him as chairman
(1952–1958). The department continued in similar activity until about 1959,
when it suddenly came apart. In 1958, Weil left to go to the Institute for
Advanced Study, Chern and Spanier left to go to Berkeley in 1959, Segal left
for MIT in 1960, and Halmos left for Michigan in 1961. Those departures
essentially brought to a close the Stone Age. The department was soon rebuilt
under A. A. Albert (chairman, 1958–1962, and dean, 1962–1974) and Irving
Kaplansky (chairman 1962–1967). (This later period will not be described in
this essay.) But there had been just this one period 1945–1960 when Chicago,
in its new style, was without doubt the leading department of mathematics
in the country.

7. Why the Change?

One may wonder why the Stone Age came to such an abrupt end. In some
part, this may just be the inevitability of changes in human situations; people grow older and shift their interests. Nationally, Sputnik in 1957 stimulated much more extended government support for mathematics in the period
1958–1960; one result was that there soon were more mathematics departments of major standing — for example, Berkeley and MIT. There are also

explanations "internal" to the University of Chicago. After 1950, Marshall Stone traveled frequently, and clearly the loss of his presence and leadership made a difference. Mac Lane may have made mistakes as chairman; Albert (unpublished) and Halmos (published) evidently thought so. A major observation is this: In the period 1949–1957, except for temporary instructors, there were no new appointments to the faculty; there was one appointment in 1958. This suggests that there was not a sufficient inflow of ideas.

The top administration of the university had changed. Robert Maynard Hutchins resigned as president in 1951; the new president or "chancellor" (1951–1960) was Lawrence A. Kimpton. It seems clear that the trustees instructed Kimpton to pay attention to the neighborhood and to achieve a balanced budget for the university. This he did, but there were intellectual costs. For example, about 1954 André Weil noted an important paper of a young man, Felix E. Browder, on partial differential equations; Browder came to visit and gave a talk. The department proposed his appointment as assistant professor, but the administration declined to act: they had observed that Browder's father, Earl Browder, had been head of the communist party in the United States. In fact, Felix had been born in Moscow. A decision on appointments on such shaky grounds would never have happened while Hutchins was president (and indeed Browder was subsequently appointed to the faculty). There are other examples of the intellectual ineptitude of the Kimpton administration. Perhaps universities cannot maintain great departments without outstanding academic leadership at the top — leadership which was subsequently restored at Chicago.

8. Requirements for Good Departments

On the basis of this and other examples, it is tempting to speculate: What does it take to make a great department of mathematics?

(1) Outstanding faculty, preferably younger; in particular, including some not on tenure.

(2) Numerous lively students, helping to prod the faculty.

(3) Exciting fields of study, preferably some new thrusts, and certainly several different fields — perhaps even a clash of interests between fields.

(4) Several instructors (e.g., postdoctorals or temporary instructors), again bringing in new ideas.

(5) Active contacts between people, e.g., colloquiums, mathematics clubs, seminars, and (important) meetings at tea.

(6) Understanding support by the university administration.

(7) An active sense of common purpose.

These conditions seem to me to have been met in the examples of great departments which I personally know: Göttingen (1930–1933), Harvard (1930–1960), Princeton, and Chicago (1897–1908, Moore; 1947–1960, Stone). When two or more of these conditions fail, a department can lose momentum. When they are present, real advance is possible.

9. THE BLISS DEPARTMENT REVIEWED

Was this department just a "diploma mill," as asserted above, or are there other aspects? This will now be reconsidered in the light of Bill Duren's recollections [6] and the autobiographical notes of Bliss [3] himself, all as cited below. I have also profited from a considerable discussion with Herman Goldstine, who served as a research assistant to Bliss in the mid thirties, when Bliss was preparing his book on the calculus of variations.

In the period 1920–1935 there were many graduate students at Chicago, and hence quite large advanced classes; this is very different from the present case when at Chicago there may be 15 or 20 advanced (post master's) courses offered in a given quarter, with no requirement for a minimum number of attendees. It is reported that in the twenties the department of education at Chicago arranged for special trains from Texas to bring the students for the summer quarter. In some departments, it became the custom for teachers elsewhere to come to Chicago summer after summer so as to finally arrive at a Ph.D., and indeed this happened then in some cases in mathematics. It would be unthinkable now, only in part because the summer quarter has shrunk. Mathematics Ph.D.s from Chicago were stationed in influential positions at universities throughout the Midwest and the South, and of course they sent their best undergraduate students to Chicago for graduate study. The activity of the department must be judged in the light of this massive input of students. According to Goldstine, Bliss felt that there was in the United States a great need for well-trained teachers of mathematics, and that Chicago was ideally placed to fill that need. In his autobiographical note, Bliss says that the merit of a department of mathematics should not be rated by an index such as the average number of research papers per Ph.D.; at Chicago there were just too many students to expect them all to do research. He implies that what really matters is the research done by a few outstanding students, while in the faculty itself what matters most is the research done by a few outstanding professors, such as Dickson (whom he names). All this took place long before the present widespread conviction that every department member is expected to do research to get promotions and government grants. In department meetings, Bliss often depended for advice on H. S. Everett, whose formal position was that of extension professor, and who was not interested in research. Everett was indeed effective in helpfully correcting student's papers in correspondence courses, and this activity did indeed bring students to Chicago — for example, I. M. Singer, post WW II.

Bliss said: "The real purpose of graduate work in mathematics, or in any other subject, is to train the student to recognize what men call the truth and to give him what is usually his first experience in working out the truth in some specific field".

If graduate work at Chicago in this period is judged on this basis, it must be accounted a rousing success — as for example with the Ph.D.s to women noted above.

The autobiographical note [3] by Bliss also exhibits the development of his interest in the calculus of variations. After studying mathematical astronomy with F. R. Moulton, he switched to mathematics and Bolza, and soon came across a copy of the 1879 lecture notes by Weierstrass on the calculus of variations. They were fascinating, as might well be, because it was there that rigorous proof was finally brought to fruition in this centuries-old subject; the dissemination of such Weierstrass notes had a wide effect. It may be that this initial enthusiasm was the leading principle of all of his career — there he found additional problems in a more general setting in which the truth could be teased out, and which students could handle. All these truths were brought together in his treatise [2], published at the end of his life, which can be viewed as a systematic extension of the Weierstrass method to all the variants of the "problem of Bolza". Moreover, the ideas there were then ready to hand, so that when Pontrjagin and others much later saw that the calculus of variations was adapted to the study of optimal control, Bliss's student Hestenes brought it all together in his 1966 book [1].

In a recent issue of the *Mathematical Intelligencer*, I have argued that many mathematicians today may specialize so narrowly on their first research field that they miss important connections. This may not be new.

As noted above, Bliss became chairman in 1927; I have argued that there might then have been more widely spread appointments to the faculty, with less emphasis on inheritance (and indeed, Sanger may have been regarded as the successor to Slaught). But in 1927 there may have been a different objective: a new mathematics building. Up until that time, the department had been housed on the upper floors of Ryerson, the physics lab. Bliss laid the plans for constructing Eckhart Hall next door as a building for mathematics, with a fine common room, central offices for mathematics and mathematical astronomy, ample faculty offices, and even space for graduate students. (In 1930, as a beginning graduate student, I occupied a fourth floor office which 40 years later served as the office for a full professor.) Eckhart Hall may well have set a pattern for mathematics buildings; at any rate it is reported that Oswald Veblen in Princeton kept track of Eckhart as he planned for the construction of Fine Hall for the Princeton Mathematics department.

These important things said, I return to my harsh judgment that by 1930 the department at Chicago had ceased to be really first class. This conclusion is not so much based on the various items of evidence assembled above, but on my own direct experience.

In the fall of 1929, as a senior at Yale, I chanced to meet Robert Maynard Hutchins, recently law dean at Yale and newly president at Chicago. He knew of my academic interests; finding that I intended to study mathematics, he told me that Chicago had an outstanding department, and that I should come there. Some weeks later, he wrote me to offer me a fellowship in the (then handsome) amount of $1,000. I accepted.

When I came in the fall of 1930, I attended Moore's seminar, as above, and signed up for courses with the leading members of the department:

Dickson's course on number theory presented a good treatment of the representation of integers by quadratic forms, but there was no indication of the connections of this with algebraic number fields, a subject with which I had a passing acquaintance. Dickson's own current interests were in the computations for the Waring problem, but when Landau came to give a visiting lecture, I could see that the center of interest was with the new ideas of analytic number theory (Hardy–Littlewood, Vinogradoff, and Landau). I learned something about approximations on major and minor arcs of the unit circle, but that was not a Chicago subject.

Lane lectured on projective differential geometry. I had never studied differential geometry, nor did Lane teach it; this left a serious gap in my background, not even adequately filled when in 1933 Herman Weyl's warning before my oral exam led me to bone up on the first and second quadratic forms of a surface. Despite this lack of background, I took Lane's course. I soon noticed an older student up in the front row with an open notebook in which he made only occasional careful entries; eventually, I learned that he had taken the course once before, and was now bringing his notes up to date with the latest refinements. At the time I was deeply offended by this display of pedantry.

The calculus of variations with Bliss (two quarters) taught me all about the brachistrochrone (I did not care) and about fields of extremals (I did), but I did not really learn anything about the connections with geometrical optics (I found this out in Göttingen) or about the connections with Hamiltonian mechanics, which I had to tease out later on my own. Bliss knew that there was Morse theory, but it was not taught at Chicago.

When I signed up for a course in the philosophy department with Mortimer Adler, Bliss disapproved.

Barnard supervised my M.A. thesis, which was an unsuccessful attempt to discover universal algebra. Barnard was then much taken up with Moore's use of functions on a general range, meaning functions $X \to F$, where X is an arbitrary set and F is a field — reals, complexes, or quaternions. Goldstine and I both think that Moore's emphasis on this "general" idea may have blinded him to other axiomatic approaches to functional analysis; I did not learn about Banach spaces until 1934 at Harvard.

Moore himself was in poor health.

At that time, President Hutchins was beginning to press for his new college devoted to general education; Bliss and other senior faculty members strongly opposed his ideas. This did not help the department.

My conclusions were not clearly formulated at that time, but they really came to this: The department of mathematics at Chicago in 1930–1931 was no longer outstanding in attention to current research. With Moore ill, there was no one on the faculty under whose direction I would have liked to write a Ph.D. thesis. I did not say this but simply put it that I had the the wanderlust and that I wanted to study logic — so I took off for Göttingen. There I did indeed find a great mathematics department. I can still recall the excitement at the start of each new semester with many new courses at hand: Lie groups (Herglotz) or group representations (Weyl) or Dirichlet series (Landau) or PDE (Lewy) or representation of algebras (Noether) or logic (Bernays). And there were many lively fellow students (many more, and on a level hardly present at Chicago): Gerhard Gentzen, Fritz John, Hans Schwerdtfeger, Kurt Schütte, Peter Sherk, Oswald Teichmüller, and Ernst Witt, for examples [10].

The conclusion seems to be that there are times when certain developments achieve a vibrancy and excitement with ample contacts with current departments which serve to stimulate faculty and students alike. May this analysis perhaps help to encourage more such cases.

Acknowledgments

Though the opinions voiced in this article are my own, I have been much helped by incisive comments from Bill Duren, Herman Goldstine, E. J. Mc Shane, Ivan Niven, Mina Rees, Arnold Ross, Alice Turner Schafer, and George Whitehead. All of them had direct experience with the mathematics department at Chicago.

References

[1] Bliss, G. A., Eliakim Hastings Moore, *Bull. Amer. Math. Soc.*, **39** (1933), 831–838; and *The Scientific Work of Eliakim Hastings Moore*, **40** (1934), 501–514.

[2] _____, *Lectures on the Calculus of Variations*, The University of Chicago Press, Chicago, Ill., 1946, ix+296 pp.

[3] ____, Autobiographical Notes, *Amer. Math. Monthly*, **59** (1952), 595–606.

[4] Dickson, L. E., *History of the Theory of Numbers*, Carnegie Institute of Washington, Washington, D.C., vol. 1 (1919), 485 pp., vol. 2 (1920), 803 pp., vol. 3 (1923), 312 pp.

[5] ____, *Algebras and their Arithmetics*, The University of Chicago Press, Chicago, Ill., 1923, 241 pp.

[6] Duren, W. L., Jr., Graduate Student at Chicago in the Twenties, *Amer. Math. Monthly*, **83** (1967), 243–248.

[7] Hestenes, M. R., *Calculus of Variations and Optimal Control Theory*, Wiley, New York–London–Sydney, 1966.

[8] Mac Lane, S., Of Course and Courses, *Amer. Math. Monthly*, **61** (1954), 151–157.

[9] Mac Lane S., Gilbert Ames Bliss, *Biographical Memoirs, Yearbook of the American Philosophical Society from* 1951, 288–291.

[10] Mac Lane, S., Mathematics at the University of Göttingen, 1931–1933, in *Emmy Noether, a tribute to her life and work*, J. Brewer and M. Smith (editors), Marcel Dekker, New York, NY 1982.

[11] Moore, E. H. (with the cooperation of R. W. Barnard), *General Analysis*, Part I, 231 pp. (1935), Part II, 255 pp. (1939), The American Philosophical Society, Philadelphia, PA.

[12] Wong, Y. K., Quasi-inverses Associated with Minkowski–Leontief Matrices, *Econometrica*, **22** (1954), 350–355.

[13] Zermelo, E., Neuer Beweis für die Möglichkeit linearer Wohlordnung, *Math. Ann.*, **65** (1908), 107–128.

Eliakim Hastings Moore and the Founding of a Mathematical Community in America, 1892–1902

KAREN HUNGER PARSHALL

Department of Mathematical Sciences, Sweet Briar College,
Sweet Briar, Virginia 24595, U.S.A.

Received 30 March 1984

Summary

In 1892, Eliakim Hastings Moore accepted the task of building a mathematics department at the University of Chicago. Working in close conjuction with the other original department members, Oskar Bolza and Heinrich Maschke, Moore established a stimulating mathematical environment not only at the University of Chicago, but also in the Midwest region and in the United States in general. In 1893, he helped organize an international congress of mathematicians. He followed this in 1896 with the organization of the Midwest Section of the New York City-based American Mathematical Society. He became the first editor-in-chief of the Society's *Transactions* in 1899, and rose to the presidency of the Society in 1901.

Contents

1. Introduction

The final decade of the nineteenth century and the first decade of the twentieth marked an incredible period of growth in American mathematics. During this time advanced mathematics, including mathematics at the research level, firmly took hold in many American universities. No longer did college mathematics mean merely arithmetic, trigonometry, the rudiments of algebra and geometry, and a smattering of calculus. No longer were American students essentially forced to travel to the great universities of Europe if they wished to study the modern advances in mathematics seriously. Whereas in the 1870s they would have been limited to working under Benjamin Peirce (1809–80) at Harvard, Hubert A. Newton (1830–96) at Yale,[1] or James Joseph Sylvester (1814–97) at Johns Hopkins, by the 1890s a dozen or more American universities could boast able research mathematicians. Furthermore, by 1910 several of these schools had native-son professors who enjoyed, or would soon enjoy, international reputations, a situation which was unprecedented in the history of American

---

[1] Hubert Anson Newton became a tutor at Yale in 1853 after having earned his undergraduate degree there in 1850. At the age of twenty-five he was appointed full professor in mathematics, a position he held until his death in 1896. Although his primary scientific interest lay in the studies of meteorites and comets, he was the mathematical mentor both of Josiah Willard Gibbs (1839–1903) and E. H. Moore. For a review of Newton's life and work, see J. Willard Gibbs, "Hubert Anson Newton", *The American Journal of Science*, 4th series, 3 (1897), 359–78.

Reprinted with permission from the *Annals of Science*, Volume **41**, 1984, pp. 313–333.

mathematics up to this time. At the University of Chicago, Eliakim Hastings Moore (1862–1932) mastered and extended the foundational work in geometry of David Hilbert (1862–1943), in addition to making important contributions to group theory, while his student, Leonard Eugene Dickson (1874–1954), pursued group theory and the study of linear associative algebras. At Harvard, William Osgood (1864–1943) worked in complex variables as well as in other branches of analysis, while Maxime Bôcher (1867–1918) advanced the theory of integral equations. At Princeton, another of E. H. Moore's students, Oswald Veblen (1880–1960), did seminal work in geometry, while perhaps Moore's most famous student, George D. Birkhoff (1884–1944), concentrated primarily on dynamics and mathematical physics before moving on to Harvard in 1912. With this list as evidence, it comes as no real surprise that many consider E. H. Moore to have been the prime driving force which finally turned the United States from a mathematical wasteland into a leader in the field.[2]

Moore's expertise in parenting new mathematicians, although important to the subsequent development of mathematics in America, proved to be only one of his lesser talents. First and foremost, he was a mathematical activist bursting with new ideas and full of the vigour needed to implement them successfully. During the decade from 1892–1902 his energy produced not only a major centre of mathematics at the newly-formed University of Chicago, but also a firm and growing national commitment to serious mathematical activity. In this study we shall follow the young Moore to his appointment as professor at Chicago in 1892 and through his term as president of the American Mathematical Society in 1902. In tracing these ten years of his life, we shall see American mathematics develop from early puberty to young adulthood.

2.   The Mathematics Faculty at the University of Chicago, 1892

E. H. Moore was born on 26 January, 1862, in Marietta, Ohio, a small town on the Ohio–West Virginia border.[3] The son of of a scholarly Methodist minister and the grandson of a well-to-do United States Congressman, the young Eliakim had perhaps more opportunities than most children growing up in rural Ohio during those immediately *postbellum* days. While his grandfather was in Congress, for example, Moore spent a summer in Washington employed as a congressional messenger. More important for our story though, he worked one summer as a research assistant to Ormond Stone (1847–1933), then the head of the Cincinnati Observatory. This

[2] See, for example, Judith V. Grabiner, "Mathematics in America: The First Hundred Years," and Garrett Birkhoff, "Some Leaders in American Mathematics: 1891–1941", both in *The Bicentennial Tribute to American Mathematics*, edited by Dalton Tarwater (n.p.: Mathematical Association of America, 1977), pp. 9–24 and 25–78, respectively. The semicentennial addresses of Eric Temple Bell and George D. Birkhoff also reflect this opinion. See Eric T. Bell, "Fifty Years of Algebra in America", and George D. Birkhoff, "Fifty Years of American Mathematics", both in *Semicentennial Addresses of the American Mathematical Society*, 2 vols, edited by Raymond C. Archibald (New York: American Mathematical Society, 1938), ii, 1–34 and 270–315 respectively.

Many of E. H. Moore's correspondents also hold this view. Moore's papers are housed in the Department of Special Collections at the University of Chicago in nineteen boxes. The first four boxes contain his correspondence, but unfortunately the number of letters dating from the period of our study are exceedingly few and unilluminating. The bulk of the preserved correspondence dates from 1918 to 1932.

[3] The majority of the biographical details presented here come from Gilbert A. Bliss's article "Eliakim Hastings Moore", *Bulletin of the American Mathematical Society*, 2nd series, 39 (1933), 831–8. Bliss followed this up with a careful look at Moore's mathematical contributions in "The Scientific Work of Eliakim Hastings Moore", *Bulletin of the American Mathematical Society*, 2nd series, 40 (1934), 501–14. For other biographical sketches of Moore, see Archibald, ed. (footnote 2) i, 144–50; and Leonard E. Dickson, "Eliakim Hastings Moore", *Science*, new series, 77(1933), 79–80. Articles on Moore and the majority of the other people discussed in this paper also appear in *Dictionary of Scientific Biography*, edited by C. C. Gillispie, 16 vols (New York:Charles Scribner's Sons, 1970–1980).

association, in addition to providing Moore with a firsthand look at scientific research, gave him his first real taste of mathematics. As a researcher Stone leaned much more toward the mathematical than toward the observational side of astronomy. Also, as a teacher, 'he had keen insight in the choosing of able students, and, once they came under his influence, the ability of turning their interests permanently to scientific careers'.[4] Whether Stone was a primary or only a secondary influence on the later career of the teenaged E. H. Moore, it is impossible to say, but Moore did proceed to Yale where he came under the influence of another mathematician–astronomer, Hubert A. Newton.

As an undergraduate at Yale, Moore excelled in astronomy, English, and Latin, in addition to mathematics, taking prizes in all of these subjects at one time or another during his academic career. In 1883 he received the A.B. degree as valedictorian of his class. Two years later he earned a Ph.D. in mathematics under Newton for his thesis entitled 'Extensions of Certain Theorems of Clifford and Cayley in the Geometry of n Dimensions'.[5] Realizing that Moore had advanced as far as an American education at the time allowed, Newton encouraged his mathematical son to continue his studies in Germany. There, as Newton knew from personal experience,[6] he could immerse himself in the mainstream of mathematics rather than remaining in the American backwater.

With money lent to him by his advisor, Moore travelled to Germany in the summer of 1885. His first stop was Göttingen, where he studied German as well as mathematics before moving on to Berlin for the winter semester. At that time Karl Weierstrass (1815–97) and Leopold Kronecker (1823–91) dominated the Berlin mathematical scene, and Kronecker's abstract thinking dominated Moore's Berlin experience. As with so many other young American mathematicians who studied abroad, the ideas which Moore encountered in Germany dominated his mathematical thinking for the remainder of his career.

His *Wanderjahr* completed, Moore returned to the United States to begin his professional life. His first position took him to the Academy at Northwestern University in 1886, a rather remote outpost of American academe. After serving there as an instructor for one year, he moved back East to Yale where he assumed the temporary post of tutor. Finally, in 1886 he obtained his first permanent university post at Northwestern, and two years later he was promoted to associate professor. With his livelihood thus secured, Moore could have happily remained at Northwestern for the rest of his career, but another opportunity, the chance to forge his own Department of Mathematics, came his way.

---

[4] Charles P. Oliver, "Ormond Stone", *Popular Astronomy*, 41 (1933), 295–8 (p. 296).

[5] E. H. Moore, "Extensions of Certain Theorems of Clifford and Cayley in the Geometry of *n* Dimensions", *Transactions of the Connecticut Academy of Arts and Sciences*, 7 (1885), 9–26. J. Willard Gibbs also published his dissertation as well as much of his subsequent work in this journal.

[6] When Newton was appointed to the professorship in mathematics in 1855, he was immediately granted a year's leave of absence so that he could better prepare himself for the job. He spent his year in Paris listening to Michel Chasles's (1793–1880) lectures on geometry at the Sorbonne. These lectures left a permanent impression on Newton even though the majority of his subsequent work focused on astronomical problems. See Gibbs (footnote 1) p. 360 and p. 373. In fact, this undying interest manifested itself almost thirty years later in the geometrical topic which Newton and Moore worked out for the latter's dissertation research. In urging Moore to study in Europe, Newton most likely wanted to assure his student of an equally lasting and rewarding experience.

Whether because of the original and promising research which Moore had done after earning his degree or because of 'his agressive genius as a rising young scholar'[7], William Rainey Harper (1856–1906), the president-elect of the new University of Chicago, stole Moore from Northwestern with an offer of a full professorship and an acting chairmanship of the department of mathematics.[8] This selection proved fortunate not only for the University of Chicago, but also for American mathematics as a whole. The academic and administrative superstructure which Harper built, guided by his own attitudes and beliefs and from Rockefeller's money, served as the foundation upon which Moore built not only a department but also an American mathematical community.

On 13 May, 1889, after several years of pressure from education-minded Baptists, John D. Rockefeller (1839–1937) agreed to put up $600,000 toward the endowment of a college to be located in Chicago, provided an additional $400,000 could be raised by 1 June, 1891. In September 1890, Rockefeller supplemented his already generous contribution by pledging another $1,000,000 'for the support of theological and graduate studies'.[9] In keeping with his wishes, $800,000 of this sum went toward graduate work. Thus, from the very beginning a precedent was set at Chicago for the liberal promotion of graduate education and research. Rockefeller, however, had no real interest in personally supervising the running and policy-making of the university. Since his attitude was essentially one of *laissez-faire*, someone had to take charge and lay both the philosophical and the physical foundations of the school. Harper energetically and enthusiastically accepted these tasks.

A true scholar himself, Harper deeply believed in both inspired teaching and first-rate research. He felt that a strong University of Chicago depended on this twofold foundation. As a consequence he sought to draw scholars who excelled in both domains

---

[7] Bliss, "E. H. Moore" (footnote 3) p. 833. In 1892 seven years after receiving his Ph.D., Moore had only six publications to his credit, and one of those was a research announcement. These six papers, however, ranged widely in subject matter from geometry to group theory to the theory of elliptic functions. Although breadth and depth do not always go hand in hand, E. H. Moore made important contributions to these and other fields during the period from 1892 to 1902.

Moore's interest in geometry dated to the time of his dissertation, 1885. At first his interest centred on so-called algebraic geometry, geometry studied by algebraic means in the spirit of Felix Klein's *Erlanger Programm* of 1872. Moore's early work concentrated on algebraic properties of surfaces and of curves on these surfaces. In 1899 with the publication of David Hilbert's *Grundlagen der Geometrie*, however, Moore became totally involved in a very different aspect of geometry, its foundations. Among his contributions here was his proof that two of Hilbert's axioms were redundant.

As a natural outgrowth of his approach to geometry prior to 1899, Moore also studied the theory of groups. In 1893 he proved the important fact that every finite field was a Galois field, that is, every finite field had a primitive element over a field with p elements, p finite. Later while under Moore's influence, Joseph H. M. Wedderburn proved his famous theorem which stated that Galois fields were in fact exactly the finite division algebras.

Finally, the theory of functions, which first captured Moore's attention in 1890, became his all-consuming passion in later life, for he utilized his interest in both foundations and analysis to approach the theory of functions with the proper degree of abstraction. Rather than studying individual functions, the basic tenet of his 'general analysis' was to study classes of functions and to move from one such class to another by means of functional transformations. Modern functional analysis shares this same basic tenet.

For further information on Moore's mathematics, see Bliss, "The Scientific Work" (footnote 3) pp. 501–14.

[8] Harper's raid on other colleges and universities earned him and the University of Chicago quite a bit of notoriety in 1892. The most striking exodus occurred at Clark University, where almost half of the faculty left their positions to go to Chicago in the wake of disagreements with the administration. As we shall soon see, one of the three original members of the mathematics department, Oskar Bolza, was one such fugitive. For a detailed description of Harper's efforts to assemble a faculty, see Richard J. Storr, *A History of the University of Chicago: Harper's University The Beginnings* (Chicago:, 1966), pp. 65–85.

[9] Ibid., p. 47.

to the new university, and he did everything within his power to plan for and to provide a congenial and stimulating environment for them. As part of his plan for making the University a haven for the research-oriented members of the faculty, Harper put forth, among others, the following two suggestions in his decennial report:

4. There should be established Research Professorships, the occupants of which might lecture or not according to the best interests of the work in which they are engaged. This is practically the character of the Professorships in the [Yerkes] Observatory. There should be chairs in other Departments, perhaps a chair in every Department, to which there might be made a permanent appointment, or which might be occupied for a longer or shorter period by the various members of the Department capable of doing research work...

6. Arrangements should be made to encourage a larger number of men to devote six months of the twelve to research and investigation, their lecture work and teaching being confined to the other six months. This plan has already been adopted in several individual cases. It is very desirable to place the advantages of this arrangement at the command of the others. With the privilege thus secured of living a year abroad and a year at home, the highest results may be achieved.[10]

With policies such as these, Chicago could not fail to attract the best people in the various fields.[11]

Harper also made a special commitment to the pure sciences among which he included mathematics. When pressure was applied early on to establish an engineering school at Chicago, Harper did not take money away from the infant departments of science to finance the venture. Although he supported the idea and agreed that such an addition would benefit the University, he refused to sacrifice the progress that had already been made in the sciences. He tried to circumvent the problem of capital in a variety of ways. In 1896 he sought to annex the Chicago Manual Training School, thereby bringing to the University existing facilities and a substantial endowment. During the first few years of the new century, he aimed to join forces with the Armour Institute of Technology. Both of these attempts failed,[12] yet Harper's original conviction did not change. In defence of his unwillingness to divert funds from the sciences he wrote:

First, it seemed upon the whole wise to devote the entire energy of the institution in scientific lines to departments of pure science, with the purpose of establishing these upon a strong foundation. This work being finished, there would be ample opportunity for the other work, and the other work would be all the stronger when it came, because of the earlier and more stable foundation of pure science. Second, it was also thought wise not to lay too much emphasis on the practical

---

[10] *The President's Report: Administration—The Decennial Publications*, 1st series, vol. I (Chicago: University of Chicago Press, 1903), p. xxv.

[11] That Chicago did attract good people, at least in mathematics, is reflected in Maxime Bôcher, David R. Curtiss, Percey F. Smith, and Edward B. Van Vleck, "Graduate Work in Universities and in Other Institutions of Like Grade in the United States", *Bulletin of the American Mathematical Society*, (1911), 122–37. As they put it, '... the opening of the University of Chicago in 1892 may almost be said to mark an epoch in the development of graduate [mathematical] instruction in the West and Middle West, for, though that university had from the start an undergraduate department, it stood out, through the character of its faculty and the emphasis laid on research work, as a strong exponent of the graduate idea'. (p. 125).

[12] Storr (footnote 8), pp. 134–5.

side of education at the start ... The greater danger was that pure science might be left without provision.[13]

In Harper's view the University's strength depended as much on its excellence in pure science as on its excellence in the humanities. If these areas were firmly established and well-represented, the rest would follow suit.

To insure that his commitments became reality, Harper drew up a meticulous blueprint for the organization of the University. In this master plan, he made provisions for the appointments of so-called Head Professors, or department chairmen, to whom he entrusted the task of building the various departments. Rather than bringing in a 'few younger instructors and allowing the work to grow more gradually under the domination of a single spirit',[14] Harper's aim was to 'bring together the largest possible number of men who had already shown their strength in their several departments, each one of whom, representing a different training and a different set of ideas, would contribute much to the ultimate constitution of the University'.[15] He wanted to create a lively atmosphere where many new and fresh ideas met, clashed, and ultimately meshed into the best of all possible university situations. In this way the president defined the overall philosophy of the institution, and the Head Professors set the working guidelines which best suited their respective departments. The University was thus tailored to the constituent parts. The fit was optimal—not too loose, not too tight.

Consistent with his policy of providing the superstructure on which to build, Harper compiled a list of duties and powers of the Head Professor.

> He was to supervise the entire work of his department in general, prepare all entrance examination papers and approve all course examinations prepared by other instructors, arrange course offerings from quarter to quarter, examine all theses offered in the department, determine the textbooks to be used, edit any appropriate papers or journals, conduct a club or seminar, consult with the librarian about needed books and periodicals, consult with the President on appointments of instructors, and countersign the course certificates in the department.[16]

Most certainly, E. H. Moore, Harper's choice for acting Head Professor of the Department of Mathematics, fulfilled these duties very well.[17]

In the autumn of 1892, the University of Chicago opened its doors for business, and E. H. Moore and his colleagues set about the job of organizing their department. During its first year of operation, the University counted three mathematicians among its ranks: Moore as professor, Oskar Bolza (1857–1942) as associate professor, and Heinrich Maschke (1853–1908) as assistant professor. We have already seen how painlessly Harper snared Moore. He had a much more difficult time securing the rest of the department. Even when he had finally succeeded, there was no guarantee that the men he had assembled would fit together to form a unified whole.

---

[13] *The President's Report* (footnote 10), p. xviii.

[14] Ibid.

[15] Ibid.

[16] Storr (footnote 8), p. 62.

[17] Because of his relative youth and inexperience, Harper made Moore acting department chairman in case he proved unsuited for the job. With a long list of successes already behind him and many more to come, Moore became permanent head in 1896.

After hiring Moore, Harper, in his quest for talent, approached Oskar Bolza with an offer to leave his position as associate[18] at Clark University for an associate professorship in his new Mathematics Department at Chicago. Ostensibly this should have been the perfect opportunity for one who had come to America to find permanency and who had yet to find a permanent job. Furthermore, Harper was giving him the chance to help mould the department right from the beginning.

Bolza first came to the United States in 1888 with his parents' blessings and financial support. Thirty-five years old by this time, Bolza had had a difficult academic career in Germany.[19] In 1875 he went to Berlin to study at the university. However, in accordance with his father's wishes he was simultaneously enrolled in the vocational academy there, a most unusual situation for a German student at this time. His father's plan was to secure a classical university education for his son while preparing him to take over the family-owned factory. After two semesters of this divided loyalty, Bolza left the technical school and decided to study physics. By 1878, however, he realized that he would never make an experimental physicist, and so he opted for pure mathematics. After two years of study and not much progress toward his Ph.D., Bolza turned to preparation for the State qualifying examination for *Gymnasium* teachers. Unfortunately, it took him three years to pass the test. After an additional year of practice teaching, Bolza finally had a profession in 1884, almost ten years after entering the university.

Thoughts of a Ph.D. in mathematics still with him, he worked on his research while teaching in Freiburg, and Klein accepted his results as a dissertation in 1886. As a result of his work, Bolza left Freiburg for Göttingen for the 1886–87 academic year, where he and his younger friend Maschke participated in a private mathematical seminar with Klein. The experience of working in such close conjuction with the intense and insightful Klein shattered Bolza's self-confidence. He doubted his ability to gain a university position (his age also being against him), and his post in the *Gymnasium* proved too draining on him physically.

Following the lead of one of his friends, Bolza left Germany for America in 1888. There, for the winter term of 1888–89, he secured a readership on Klein's recommendation at Johns Hopkins, which he followed up with a position at Clark the following academic year. Indeed, the outlook seemed brighter in America, but a permanent position still eluded him. Moreover, halfway through his second year at Clark, his father died in Germany after a long illness. In the summer of 1891 he returned home to Freiburg to comfort his mother and to coax her into staying with him in Worcester for the following year. His initial plan was to remain at Clark for one more year and then to return with his mother to Germany. Harper's arrival in Worcester and his subsequent offer at first only raised doubts in Bolza's mind about his decision to leave America. Nevertheless, he refused the call to Chicago and proceeded with his travel plans. Just ten days before he was to leave, another plea inveigling him into coming for at least one year arrived from Harper. Bolza's colleague, Henry Taber (1860–1936), urged him to accept the offer and finally convinced him to go to Chicago and talk to Harper in

[18] Not to be confused with the rank of associate professor, 'associate' or 'tutor' was one of the various grades of temporary, as opposed to permanent faculty.

[19] The following biographical account of Bolza's life has been condensed from his very readable and highly interesting autobiography, *Aus Meinem Leben* (Munich: Verlag Ernst Reinhardt, 1936). See particularly pp. 18–25 for his account of his early experiences in America. G. A. Bliss used this same source in his article, "Oskar Bolza: In Memoriam", *Bulletin of the American Mathematical Society*, 50 (1944), 478–89, especially pp. 478–80.

person. This one side trip gained not only Bolza's acceptance, but also a position for his longtime friend, Heinrich Maschke.

Bolza and Maschke had first met in 1875 while both were students at Berlin.[20] Maschke received his Ph.D under Felix Klein at Göttingen in 1880, six years before Bolza finished, and immediately took a teaching position at the *Luisenstädtische Gymnasium* in Berlin. Like Bolza he felt the physical burden of the teaching load at the *Gymnasium* and longed for a university position. However, the competition for such positions was intense, and it was difficult to change from the station of *Gymnasium* teacher to that of university professor.[21] At his friend's urging, he finally came to the United States in 1891. Before leaving, he retrained himself as an electrician in order to insure his livelihood. As Bolza later explained, 'he did not wish to rely entirely upon the doubtful chances of finding a university position'.[22] This proved to be a wise move for he quickly found a job as an electrician at the Weston Electric Instrument Company in Newark. Although this met Maschke's financial needs, it did not satisfy the intellectual needs of a man who yearned for the chance to prove himself as a mathematician. Harper's solicitation of Bolza provided him with just such a chance.

During the course of his negotiations with Harper in Chicago, Bolza agreed to take the job but only on the condition that Harper also hire Maschke. At first this seemed impossible due to a lack of funds, but at the last moment a new gift came through.[23] Harper offered Maschke an assistant professorship, and the staff was complete. All that remained was for these three men to combine their efforts and form a real, working department.

In the beginning both Maschke and Bolza had their doubts about the success of their new association. As Bolza so candidly expressed it:

> Of the greatest importance now was the question of how the relations in the mathematics department would turn out. As already mentioned, E. H. Moore, formerly an associate professor at Northwestern University in Evanston, Ill., had been called to head the Mathematics Department. According to the '*Führerprinzip*' which reigned at the University, Moore answered only to the President, whose complete trust he possessed, and otherwise was theoretically the absolute ruler of the department. He was almost five years younger than I, even more than eight years younger than Maschke and was at that time little known. In addition to that, Maschke and I were foreigners who for many years had been close friends and who had lived in the absolute freedom of the German university. All of these were factors which could have risked the inner peace of the department.
>
> But on the one hand, Maschke and I recognized early on Moore's towering ability.... And on the other hand, Moore was extremely tactful, considerate, and unselfish and consulted us as equals on all important questions which concerned the department. Thus, it was not hard for us to accept him willingly as a leader whom we gladly and cheerfully served, even more so since he also manifested

---

[20] This information on Maschke's life may be found in Oskar Bolza, "Heinrich Maschke: His Life Work", *Bulletin of the American Mathematical Society*, 15 (1908), 85–95 (p. 86).

[21] For a clear picture of the educational hierarchy in Germany at this time, see Lewis Pyenson, *Neohumanism and the Persistence of Pure Mathematics in Wilhelmian Germany* (Philadelphia: American Philosophical Society, 1983), especially chapter 3, pp. 16–26.

[22] Ibid.

[23] Bolza (footnote 19), p. 25.

extraordinary organizational abilities and a forward-charging, selfless enthusiasm in the interest of the department.[24]

By all accounts, then, the members of this mathematical triumvirate perfectly complemented each other as mathematicians, as teachers, and as administrators. In the words of one who knew and studied under them all, 'Moore was brilliant and aggressive in his scholarship, Bolza rapid and thorough, and Maschke more deliberate but sagacious'.[25] With these character traits pooled into one corporate entity, it is little wonder that good, carefully thought out ideas ensued and that they were efficiently and effectively implemented.

3. E. H. Moore's organizational years 1892–1902

The first major undertaking of this young department involved the World's Columbian Exposition held in Chicago in 1893 to commemorate the four-hundredth anniversary of the discovery of America. It resulted in Chicago becoming 'overnight *the* leading centre of American [mathematical] research, with Moore as its inspired chief'.[26]

The organizers of the Exposition conceived, in addition to the displays, amusements, and cultural activities associated with a world's fair, of a series of congresses which would represent the current intellectual activities of the world and bring together the leading experts in many diverse fields.[27] The success of such an endeavour would insure that the 1893 Columbian Exposition in Chicago surpassed the landmark Paris Exposition of 1889 in at least one way since the idea of holding congresses had been implemented in Paris on only a limited scale. Furthermore, by emphasizing the intellectual aspects of world culture, the organizers would somewhat play down the nationalistic and materialistic sides of the world fair.[28] These high ideals served the mathematics department at the University of Chicago well. Of the 214 local organizing committees charged by the World Congress Organization to contact the world's leaders in the various fields,[29] the Chicago mathematics department of Moore, Bolza, and Maschke together with Henry S. White (1861–1943) of Northwestern formed the committee on mathematics. Thus, the fledgling department at Chicago was almost instantly thrust into the limelight in its unofficial role as host to the mathematicians of the world.

The congress proved quite successful, especially as an impetus to Moore and his colleagues at Chicago. In all, forty-five people representing nineteen states as well as Austria, Germany, and Italy actually attended the meeting, and mathematicians of France, Russia, and Switzerland were well-represented by their submitted papers.[30]

---

[24] Ibid, p. 26. The translation is mine.

[25] Bliss, "E. H. Moore" (footnote 3) p. 833.

[26] Garrett Birkhoff (footnote 2), p. 31. Birkhoff's emphasis.

[27] Reid Badger, *The Great American Fair: The World's Columbian Exposition and American Culture* (Chicago, 1979), pp. 77–8. I would like to thank Professor David Rowe for this reference and for his helpful remarks concerning Felix Klein's participation in the mathematical congress and in the Evanston colloquium.

[28] Ibid., p. 78.

[29] Ibid.

[30] Of the forty-five people actually in attendance, forty-one were American, one was Austrian, two were German, and one was Italian. Thus, the American contingent to this 'international' congress definitely outnumbered the participants from all other countries combined by a margin of ten to one. Of the thirty-nine papers read, however, only thirteen were delivered by Americans. Of the remaining papers sixteen were German, three Italian, one Swiss, three French, two Austrian, and one Russian. Most notable among these foreign contributors were Hermite, Hilbert, Minkowski, Netto, Max Noether, Study, and Weber. Although populated primarily by Americans, the meeting truly was international in content.

Furthermore, as part of its contingent to the fair, the Prussian Ministry of Culture sent Felix Klein to the congress as its official representative.[31] He brought with him recent articles by over a dozen of his countrymen in an effort to disseminate German mathematics more fully and in order to pave the way for a meaningful exchange of ideas between mathematicians working in Germany and those working elsewhere in the world. Of all the foreign nations actively participating in some aspect of the fair, only the Prussian government made such a gesture to the mathematical congress. Considering the fact that two of the members of the local committee, Bolza and Maschke, were native Germans who had obtained their Ph.D.'s under Klein, and that the other two members, Moore and White,[32] had studied in Germany, this gesture served to underscore the enormous debt that American mathematics owed to Germany. The American participation at all levels at the congress proved, however, that mathematics was finally established in this country.

At the beginning of September, immediately after the close of the congress, Moore and his colleagues continued their mathematical activity by attending a two-week-long colloquium given by Felix Klein at Northwestern University.[33] Klein had been trying to work out plans for such a series of talks in the United States for some time. Thus, when Moore, Bolza, Maschke, and White approached him about the congress, he offered to give the follow-up lectures free of charge. This colloquium served a dual purpose: first, it allowed the twenty-four people in attendance to hear more from Klein and to talk to him one-on-one, and second, it brought home the point to Moore and White that regular and frequent colloquia involving the mathematics departments at Chicago and Northwestern should definitely be held in the future.[34]

After Klein returned to Germany, the two departments did set up a schedule of meetings to discuss mathematics and to present papers. By 1896, however, the idea of organizing something bigger had occurred to Moore and his friends. In December of that year, Moore took the initiative and mailed a circular to professors and students of mathematics who lived and worked within a reasonable train ride of Chicago. As its title proclaimed, the letter represented 'a call to a conference in Chicago' on 31 December, 1896, and 1 January, 1897, in Ryerson Laboratory at the University of Chicago. At the meeting, Moore proposed to discuss the feasibility of organizing a Chicago section of what had since become the American Mathematical Society.

In April 1894, with the papers read at the international congress amassed and organized, E. H. Moore approached the New York Mathematical Society for money toward the publication of the proceedings, thereby performing the Head Professor's prescribed task of editing 'any appropriate papers or journals'. The Society in conjunction with the publisher, Macmillan and Company, worked out an equitable

---

[31] Eduard Study (1862–1930), professor extraordinarius at the University of Marburg in Germany, also accompanied Klein.

[32] White also obtained his Ph.D. under Klein at Göttingen.

[33] These lectures were given in English and ranged widely over the areas of geometry and algebra. Meticulous notes, virtually transcripts, were taken during each of the twelve lectures by Alexander Ziwet (1853–1928) who was then an assistant professor at the University of Michigan. After Klein's careful revisions, Ziwet saw the lectures through the press. See Felix Klein, *The Evanston Colloquium: Lectures on Mathematics* (New York, 1894). This volume, although not affiliated with the New York (as of 1 July 1894, American) Mathematical Society, served as the impetus and the archetype for the series of AMS sponsored colloquia which began in Buffalo, New York in 1896.

[34] Archibald, ed. (footnote 2), i, 74–5.

agreement for financing the venture, and the volume appeared later that year.[35] As Raymond C. Archibald (1875–1955) explained, 'this major publication enterprise, transcending local considerations and sentiment quickened the desire of the Society for a name indicative of its national or continental character'.[36] On 1 July, 1894, largely as a result of E. H. Moore's initiative and insight, this desire became reality, and the *American* Mathematical Society met for the first time.

It did not take Moore and his Midwestern colleagues long to realize, however, that a national name did not necessarily imply a national organization. After 1894 as before, the monthly meetings of the Society took place in New York City, which effectively prevented all but the Northeastern mathematicians from attending and participating regularly. Moore's idea of a formally sanctioned Chicago section, then, would serve to involve the Midwesterners officially in the Society's activities. Such a group would provide necessary mathematical interaction for those located in the Midwest who could not pack up and go to New York each year, much less each month. Moore stressed the supposed national character of the Society in his letter and strongly hinted at its obligations toward the non-Eastern members. As he put it:

> Our Society represents the organized mathematical interests of this country. Its function is to promote those interests in all possible ways.
>
> Do we not need most of all frequent meetings? Those who have attended the summer meetings know the keen stimulus and inspiration resulting from personal contact—inside and outside the stated meeting—with colleagues from other institutions. The regular monthly meetings of the Society afford similar opportunities to those who live in the vicinity of New York.
>
> By the organization of *sections* of the Society can similar advantages be secured for other parts of the country? Shall, for instance, a Chicago section be organized?[37]

In short, the creation of a section of the Society of Chicago would have served to shift some of the mathematical power and activity from New York to Chicago. Under the shrewd leadership of Moore and his associates, then, the Department of Mathematics at the University of Chicago could have achieved preeminence. This was precisely what happened.

The Chicago section, which was formally sanctioned in the By-laws of the American Mathematical Society in 1897, had its first official meeting on 24 April, 1897. At that time Moore was elected chairman of the group, a post he held until 1902, and Thomas F. Holgate (1859–1945) of Northwestern was elected secretary. The energy of these two men, as well as of the other Midwestern mathematicians, reflected itself in the statistics based on the section's first three years. From April, 1897, to April, 1900, a total of between eleven and twenty-one actual members of the Society attended each meeting in addition to numerous non-members, and a total of 106 papers were delivered by forty-one different speakers.[38] Furthermore, the research presented was for the most part of a

[35] Such a volume was in fact published. See *Mathematical Papers Read at the International Mathematical Congress Held in Connection with the World's Columbian Exposition: Chicago 1893* edited by Eliakim H. Moore, Oskar Bolza, Heinrich Maschke, and Henry S. White, (New York, 1896). The facts and figures which follow are based on information in the introduciton of the book, pp. vii–xvi.

[36] Archibald, ed. (footnote 2), I, 7.

[37] Ibid, p. 75.

[38] Ibid, p. 77. A statistical study of the first twenty-five years of the Chicago Section may be found in Arnold Dresden, "A Report on the Scientific Work of the Chicago Section, 1897–1922, "*Bulletin of the American Mathematical Society*, 28 (1922), 303–7.

very high quality. Those in attendance were privy not only to the latest work of Moore, Bolza, and Maschke but also to the thesis research of such rising stars as Dickson, Veblen, George Birkhoff, and Robert L. Moore (1882–1974) by the end of the first decade of the twentieth century.[39]

This fundamental research developed in an environment characterized not only by the activity of the Chicago section of the Society, but also by Moore's continued enthusiasm and industriousness. After the formation of the local group and after its settling in period, Moore had the opportunity to devote his thoughts to other projects which would benefit his department, the Society, and American mathematics as a whole. Once again, in keeping with the list of duties of the Head Professor, Moore became involved in the movement to found a new mathematics journal which would provide a much needed forum for American mathematicians.

It may be surprising that mathematicians at this time felt the need for another journal considering that there were already three major American journals dedicated solely to the publication of mathematical research: the *American Journal of Mathematics* founded by James Joseph Sylvester and William E. Story (1850–1930) at Johns Hopkins in 1878; the *Annals of Mathematics* begun by Ormond Stone at the University of Virginia in 1884; the *Bulletin of the American Mathematical Society* first issued in 1891 under the direction of Thomas S. Fiske (1865–1944).

In the eyes of Moore and others, each of these journals had shortcomings. The avowedly *American Journal*, for example, published an inordinate number of papers by foreign mathematicians, thereby depriving American contributors of precious space. During its first ten years of operation, fully one-third of the ninety papers which appeared came from foreign sources.[40] The editorship, first under J. J. Sylvester (until 1883) and then under Simon Newcomb (1835–1909), distinctly favoured and valued European over American mathematics, or so it was believed. Whereas the *American Journal* published advanced research from foreign sectors, the *Annals* printed American 'papers of intermediate difficulty and more popular character'.[41] This was perfectly in keeping with Stone's original plan, however. He felt the need for a journal that catered more to the mathematical enthusiast than to the high-powered mathematician. Finally, the *Bulletin* proclaimed its mission as an 'Historical and Critical Review of Mathematical Science' right on its title page, so full-length research articles were not appropriate there either. The United States sorely needed a journal under editors who

---

[39] Joseph H. W. Wedderburn (1882–1948) also first presented his beautiful theorem on finite division algebras before the Chicago Section in April of 1905 while he was a visiting fellow at the University of Chicago. For a discussion of this theorem and the environment in which it was proved, see Karen Hunger Parshall, 'In Pursuit of the Finite Division Algebra Theorem and Beyond, *Archives internationales d'Histoire des Sciences*, 33 (1983), 274–99.

[40] Thomas S. Fiske, "Mathematical Progress in America, "*Bulletin of the American Mathematical Society*, 11 (1905), 238–46 (p. 239). Furthermore, another third of the articles came from people associated in one way or another with Johns Hopkins.

[41] J. J. Lick, "Ormond Stone—1847–1933", *Bulletin of the American Mathematical Society*, 39 (1933), 318–9 (p. 318). In the first twelve volumes, which appeared while the journal was still housed at the University of Virginia, only three foreigners contributed (if we exclude a letter from Cayley which was published there but which he did not actually submit), making the *Annals* decidedly American. The popular nature of the journal was characterized by a section of problems and solutions in each issue, and the articles which came out during the first few years of operation were definitely at an intermediate level. This gradually changed after 1890, however, when people like Bôcher, Moore, Dickson, Osgood, and Maschke began sending articles there. In fact, Dickson's Ph.D. thesis appeared in volume twelve. See L. E. Dickson, "The Analytic Representations of Substitutions on a Power of a Prime Number of Letters with a Discussion of the Linear Group", *Annals of Mathematics*, 12 (1897–1898), 65–120, 161–83.

recognized and emphasized the merits of Americans working at an advanced level in the field of mathematics.

In the spring of 1898, Thomas S. Fiske of Columbia University, the secretary of the Americam Mathematical Society, proposed to the governing council that the Society approach Newcomb with the suggestion of making the *American Journal* a joint venture between Johns Hopkins and the Society. A committee consisting of Fiske, Moore, Newcomb, Maxime Bôcher of Harvard, and James Pierpont (1886–1938) of Yale was appointed to draw up a proposal to submit to Johns Hopkins. With Newcomb on the drafting committee (he was also president of the Society), acceptance of the proposal by the authorities at Hopkins should have been immediate. Apparently though, Newcomb's strongest allegiance lay with his journal and not with his Society. In the last stages of the bargaining, he and the University refused to accept the committee's first and third recommendations, namely,

> (1) That the *American Journal of Mathematics* shall bear upon its title page the inscription Founded by the Johns Hopkins University. Published under the auspices of the Johns Hopkins University and the American Mathematical Society ... (3) That the *Journal* shall have a board of seven editors, of whom two shall be selected by The Johns Hopkins University, and five by the Council of the American Mathematical Society ...[42]

They wanted no mention of the Society on the title page and complete control over the editorial board.

This sort of bickering did not appeal to Moore and to many of the other members. Therefore, when asked later by Newcomb and Fiske if he would consider serving on the editorial board of the *Journal*, Moore refused. So many other people approached for the position also refused that the plan was abandoned altogether. With this option no longer viable, the problem became one of financing a totally new journal without further jeopardizing relations with Johns Hopkins University and the editorial board of the *American Journal*. William F. Osgood recalled the mixed feelings at the lively New York meeting which followed the collapse of the *American Journal* plans in this way:

> A number of the younger men spoke informally and all were agreed on the desirability of a journal, the one difficulty being the financial one. But a few of the older men had been contributors to the *American Journal*. They had followed with enthusiasm the development of the Johns Hopkins and it was natural that they should feel a certain loyalty to its publications. These views had not, however, been expressed at the meeting. Few of the younger men knew that they existed, and little would they have cared when the interests of mathematics were at stake. Young men are impetuous, and when they are sure they are right, proceed directly to reach their ends. Not till later do they learn the importance of listening to a minority. which is wrong, but sincere and of winning it over if possible, without sacrificing their main objects.[43]

This difference of opinion and divided loyalty dissolved completely away when Maxime Bôcher hit upon the idea of referring to the new journal not as a 'journal' but as the 'transactions' of the Society. Certainly a publication under that name would present no direct competition to any other extant or future mathematical journal. Thus, the

---

[42] Archibald, ed. (footnote 2) I, 56. A detailed account of the negotiations with Newcomb appears on pp. 56–8.

[43] Ibid, p. 58.

Society founded its *Transactions* on 25 February, 1899, and Moore, Fiske, and Ernest W. Brown (1866–1938) of Yale were appointed editors with Moore as editor-in-chief.[44]

Brown vividly conveyed the sense of adventure and uncertainty associated with the early years of the journal as well as Moore's matchless contributions when he wrote:

> I think that none of us knew much if anything about running a mathematical journal—I certainly didn't. But we were young and could still learn. I like to think of the immense amount of trouble we all took—and especially Moore—to get the best information, the best printing, the best editing and the best papers before the first number appeared [January, 1900]. And the work did not stop there. We wrestled with our younger contributors to try to get them to put their ideas into good form. The refereeing was a very serious business.... Most of it in those days was, I believe, done by Moore himself though he sought outside assistance whenever possible.[45]

Right from the start this editorial staff maintained only the highest standards of style and mathematical quality. In the first issue, for example, there were papers by Moore, Bolza, and Maschke, as well as work by Bôcher, Osgood, Dickson, and the foreign mathematicians Paul Gordan (1837–1912) and Eduoard Goursat (1858–1936).[46] Although contributions from abroad were encouraged, they were not included unless they had previously been presented at one of the Society's meetings. In this way the editors remained faithful to the transactional nature of their publication and avoided all possible accusations of an anti-American bias.[47] They also invited European mathematicians to take an interest in the American Mathematical Society and its proceedings. The willingness of such prominent Europeans to accept this invitation signalled a coming of age of American mathematics, and once again we find E. H. Moore at the forefront of such developments.

With virtually complete control over the Society's official standards for publishable research, Moore wielded great power over his fellow mathematicians, forcing them to strive for the very best intellectually. He also enjoyed growing influence within the political sphere of American mathematics. When the first number of the *Transactions* appeared in 1900, Moore had already completed a two-year tenure as vice-president of the Society. Apparently his political acumen was as sharp as his mathematical ability, for in December of 1900 he was elected president of the A.M.S. His colleagues had awarded him their highest honour.[48]

[44] Ibid, pp. 58–9.

[45] Ibid, p. 60.

[46] This list became only more impressive as the Moore years (1900–1908) went by, including names like Hilbert, Eisenhart, Hadamard, Fréchet, Veblen, Wedderburn, Poincaré, George D. Birkhoff, and R. L. Moore.

[47] Archibald, ed. (footnote 2), I, 59. Whereas one third of the *American Journal's* papers during its first ten years came from abroad, under Moore each volume of the *Transactions* was only 12·5 per cent European on the average. Foreign contributors numbered 4 in 27, 4 in 21, 2 in 28, 4 in 29, 4 in 28, 4 in 29, 2 in 31, 3 in 28 in the first eight volumes, respectively.

[48] Interestingly enough, in 1903 Moore was ranked the number one mathematician in the United States by a group of ten mathematicians selected by James McKeen Cattell for *American Men of Science*. The rankings, which remained secret until they were published thirty years later in the fifth edition of this directory, were: 1) Moore, 2) George W. Hill of the Nautical Almanac Office, 3) William F. Osgood of Harvard, 4) Maxime Bôcher also of Harvard, 5) Oskar Bolza of Chicago and Simon Newcomb of Johns Hopkins, 6) Frank Morley also of Hopkins, 7) Ernest W. Brown of Haverford, later of Yale, 8) Henry S. White of Northwestern, 9) L. E. Dickson of Chicago, and George A. Miller of Stanford in 1903, later of the University of Illinois. Since Cattell gave no indication as to the criteria used in selecting the ranking committee, this list is of somewhat questionable value. See *American Men of Science: A Biographical Dictionary*, 5th ed edited by James McKeen Cattell and Jaques Cattell (New York, 1933), pp. 1261–78 (p. 1269).

Moore's two-year term as president marked several interesting changes in the Society. The sixth president,[49] he was the first Midwesterner. His efforts over the previous eight years had finally won national respectability for his region of the country, and his single-mindedness of purpose had resulted in a clear shift of power. Moore was also the first pure mathematician to accede to the presidency. Although not as widely renowned as the astronomer–presidents, George W. Hill (1838–1914) and Simon Newcomb, Moore had gained international recognition as a mathematician, receiving an honourary degree from the University of Göttingen in 1899 for his achievements in pure research. Finally, Moore was by far the youngest man to have served as the Society's president, assuming that position at the age of thirty-eight. John McClintock (1840–1916) and Robert S. Woodward (1849–1924), the next youngest of the six, were both fifty at the time of their inaugurations. Thus, the American mathematical community learned that not only could its leader hail from the Midwest, but he could also be a younger man at the peak of his creative career.

Compared with the preceding eight years, Moore's tenure as president of the A.M.S. was a relatively calm period. By the time he came to power, the Society boasted a steadily increasing membership, financial solvency, and two important periodicals. As we have seen, Moore had already done much in an unofficial capacity to assure this success. Not content to glide through on past achievements, however, Moore adopted a very important cause to promote during his presidency, namely, the advancement of mathematics education in America.

As a university level educator, Moore became painfully aware of the vast differences in mathematical preparation among the students entering the university at the turn of the twentieth century. He also recognized that college or university training in mathematics varied greatly from institution to institution. As a solution to these problems of non-uniformity, Moore advocated drawing up national mathematics requirements for admission into colleges and technological schools as well as setting standards for degrees in mathematics nationwide. Since he viewed the American Mathematical Society as the logical vehicle for instituting, supporting, and maintaining such standards, he set up various committees during his term of office to study and make recommendations on this issue.

As a teacher of mathematics, E. H. Moore was also deeply concerned about the pedagogy of mathematics. A follower of his Chicago colleague John Dewey (1859–1952), Moore renounced the standard lecture method of teaching mathematics in favour of a laboratory method in the Deweyan sense. For Dewey, learning must always be motivated be experience. In presenting new subject matter to a child, for example, 'the lack of any organic connection with what the child has already seen and felt and loved makes the material purely formal and symbolic... any fact, whether of arithmetic, or geography, or grammar, which is not led up to and into out of something which has previously occupied a significant position in the child's life, for its own sake, is forced into [the] position'[50] of being a 'base or mere symbol',[51] that is, one which has no real meaning to the child. Thus, in order to teach elementary mathematical concepts, the instructor must fundamentally relate the new ideas to the child by means

---

[49] In chronological order, the other presidents were John H. VanAmringe (1835–1915), John McClintock (1840–1916), George W. Hill (1838–1914), Simon Newcomb (1835–1909), and Robert S. Woodward (1849–1924). For biographical sketches of these men, see Archibald, ed. (footnote 2) I, 110–44.

[50] John Dewey, *The Middle Works, 1899–1924*, edited by Jo Ann Boydston, vol. 2, (*1902–1903: The Child and the Curriculum* (Carbondale: Southern Illinois University Press, 1976), pp. 286–7.

[51] Ibid.

of some hands-on type of experience or, more broadly, through so-called manual training. For instance, a conception of number might come through measurement of baking powder for a cake and a conception of number applied to chemistry might come from baking a cake with too much or too little of this essential ingredient. At the Laboratory School affiliated with Dewey's Department of Pedagogy at the University of Chicago, the teachers implemented precisely this experiential philosophy of education in teaching children at the elementary level. Dewey himself admitted, however, that '...the first person who succeeds in working out the real correlation of mathematics with science and advanced form of manual training, will have done more to simplify the problems of *secondary* education than any other one thing that I can think of.'[52]

E. H. Moore attempted to meet this challenge with his adaptation of the laboratory method to secondary and university level instruction in mathematics and physics. Moore sought to relate mathematical and physical concepts to the student's realm of experience by actively engaging the student in experimentation of an intrinsically interesting nature. Thus, 'in the physics laboratory it is undesirable to introduce experiments which teach the use of calipers or of the vernier or of the slide rule. Instead of such uninteresting experiments of limited purpose, the students should be directed to extremely interesting problems which involve the use of these instruments, and thus be led to learn the instruments as a matter of course, and not as a matter of difficulty. Just so the smaller elements of mathematical routine can be made to attach themselves to laboratory problems, arousing and retaining the interest of the students.'[53] As for the more difficult problem of presenting theorems and their proofs, experience must again prevail. The instructor must first convince the student of a theorem's truth at an intuitive level whether by means of experimentation or computation or graphic depiction. Then, 'in most cases, much of the proof should be secured by the research work of the students themselves.'[54]

Moore pushed for the widespread adoption of the laboratory method in teaching elementary, secondary, and university level mathematics.[55] As vice-president and then president of the American Mathematical Society, Moore saw this body as his primary vehicle for instituting this change. American mathematics education needed help according to Moore, and he wanted to guarantee its reform from the bottom up. He underscored his commitment to these ideals by reserving his final official forum as the Society's president, his retiring address, for a final plea for change.[56]

---

[52] John Dewey, *Lectures in the Philosophy of Education: 1899*, edited by Reginald D. Archimbault (New York: Random House, 1966), p. 295. My emphasis.

[53] E. H. Moore, "On the Foundations of Mathematics", *Science*, new series, 17 (1903), 401–16 (p. 412). See also John Dewey, *The Early Works, 1882–1898*, edited by Jo Ann Boydston, vol. 5, (*1895–1898*): "A Pedagogical Experiment" (Carbondale: Southern Illinois University Press, 1972), pp. 244–6 (p. 246).

[54] E. H. Moore, (footnote 53) pp. 412–13. This also serves as one of the fundamental ideas in Robert L. Moore's celebrated method of teaching mathematics. See the end of Section 4 below.

[55] As implemented by E. H. Moore and his colleagues, the laboratory method emphasized the "...fundamentals and their graphical interpretations. The courses were so-called laboratory courses, meeting two hours each day, and requiring no outside work from the students. It might be added parenthetically that, as with many such new plans, the amount of work required of the instructor was exceedingly great." (Bliss, footnote 3, p. 834).

[56] See note 53 above.

After stepping down from his national post, he set an example by implementing these ideas in his own department at the University of Chicago. Along with Bolza and Maschke, Moore built up a major teaching institution at Chicago which turned out some of the foremost American mathematicians of the first half of the twentieth century. In order to understand fully Moore's role in the founding of an American mathematical community, we must finally turn to the mathematical children he fathered, the second generation, and to the intellectual environment in which they matured.

4. The Mathematics Department at Chicago and Moore's early students

As we have already said, President Harper as well as Moore and his colleagues tried to stimulate a lively intellectual atmosphere at the University. In the Department of Mathematics, part of this effort was reflected in the vast number of journals which Moore, in his capacity as chairman, had ordered through the library. He clearly realized the importance of keeping abreast of the literature. Primarily through subscriptions to the greatest possible number of journals would his faculty and students have the opportunity to study trends in research, to learn of new, open problems, and to remain on top of the current developments in their respective areas.

In 1903, Harper's decennial report listed the journals received by the mathematics library. They numbered an incredible thirty-eight.[57] A mathematician looking for a reference could consult five American journals, three from England, nine from France, ten in German, one from Holland, nine from Italy, one in Portuguese, and one from Sweden. Fifteen years earlier Florian Cajori (1859–1930) surveyed the major colleges and universities and inquired as to the number of mathematics journals they took.[58] Of the 168 which responded, only Johns Hopkins, the U.S. Naval Academy, and Columbia claimed to subscribe to almost all mathematics journals currently in print (perhaps twenty-one or so in 1888). The vast majority of the rest (117) took none at all. Harvard, Yale, and Princeton were conspicuously absent from the list of schools which answered Cajori's survey, and of the 117 institutions which took no journals virtually all were small liberal arts colleges. Nonetheless, these figures suggest that by the standards of 1903 the University of Chicago possessed vast mathematical resources in relation to most other schools.

It was also extremely fortunate to have the faculty resource of Moore, Bolza, and Maschke. Following President Harper's wishes, these three men founded the Mathematical Club in 1892 as soon as the University opened.[59] They conceived of it as a workshop where research papers in preliminary versions were presented, ripped apart, and put together again. Gilbert A. Bliss (1876–1961) described the atmosphere at the club meetings well when he wrote:

> Those of us who were students in those early years remember well the tensely alert interest of these three men in the papers which they themselves and others read before the Club. They were enthusiasts devoted to the study of mathematics, and aggressively acquainted with the activities of the mathematicians in a wide variety

---

[57] *The President's Report*, pp. 247–63. These pages contain an alphabetical listing of all of the periodicals to which the University of Chicago library subscribed. The mathematics journals may be gleaned from this list.

[58] Florian Cajori, *The Teaching and History of Mathematics in the United States* (Washington, D. C.: Government Printing Office, 1890), pp. 296–302. The numbers which follow are based on Cajori's data.

[59] Department of Mathematics, "The Junior Mathematical Club of the University of Chicago", with an Historical Sketch by Herbert E. Slaught, Chicago, 1906–41, p. 1. (Typewritten.)

of domains. The speaker before the Club knew well that the excellence of his paper would be fully appreciated, but also that its weaknesses would be discovered and thoroughly discussed. Mathematics, as accurate as our powers of logic permit us to make it, came first in the minds of these leaders in the youthful department at Chicago, but it was accompanied by a friendship for others having serious mathematical interests which many who experienced their encouragement will never forget.[60]

The experience of presenting a paper or an idea before such a dynamic and inquiring group of minds must have been rewarding indeed, not to mention frustrating. After one of these sessions of questioning and probing, though, results emerged more polished, papers more succinct, ideas more concrete. The renown of the second generation of standard-bearers which emerged from this environment testified to the success of Moore's efforts in building both his department and a national mathematical community.

In all Moore supervised the dissertations of thirty Ph.D. candidates during his forty years at the University of Chicago, and although the names of many of his later students are not well-known, a list of his students from 1896 to 1907 contains some of the brightest stars of twentieth century mathematics.[61] The algebraist Leonard E. Dickson, the geometer Oswald Veblen, the mathematical physicist George D. Birkoff, and the topologist Robert L. Moore[62] each grew up on E. H. Moore's brand of mathematical thinking and matured into independent-minded mathematicians who made seminal contributions to their respective fields. Together these four mathematicians published thirty books and over six hundred papers in addition to directing the research of almost two hundred Ph.D.'s[63]. Furthermore, each of them edited major journals, served as president of the A.M.S., and won election to the National Academy of Sciences. Like their mathematical father, they also built or maintained exceedingly strong departments at their respective institutions.

Dickson, who received the University of Chicago's first doctorate in mathematics in 1896, became an assistant professor there in 1900 and remained at Chicago until his retirement in 1939.[64] During this time he sustained the research momentum of the original triumvirate after Maschke's death in 1908 precipitated Bolza's return to Germany two years later. He filled this large gap with his prodigious output of papers, books, and students[65] and engendered an entire school of ring theorists which eventually scattered all over the United States.

Veblen came along a bit later, receiving his Ph.D. in 1903. After staying on at Chicago for two more years as an associate, he joined the ranks of Woodrow Wilson's

---

[60] Ibid.(footnote 3), p. 833.

[61] For a complete list of Moore's students, see Bliss (footnote 3), p. 834.

[62] Although R. L. Moore's name was listed in the reference cited in note 52, it did not appear in Archibald's sketch of Moore's life. See Archibald, editor (footnote 2) I, 144–50 (p. 146). As R. L. Wilder noted in his obituary of R. L. Moore, the records needed to clear up this matter are incomplete. See R. L. Wilder, "Robert Lee Moore, 1882–1974", *Bulletin of the American Mathematical Society*, 82 (1976), 417–27 (p. 419).

[63] These numbers breakdown in the following way. Of the papers, Dickson had over 280, Veblen had 69, Birkhoff had 203, and R. L. Moore had 68. As for the books, Dickson published a phenomenal eighteen, Velben six, Birkhoff five, and R. L. Moore one. Finally Dickson directed 64 students, Veblen 14, Birkhoff 45, and R. L. Moore 50.

[64] For further biographical information on Dickson, see Archibald, ed. (footnote 2) I, 183–94, and A. A. Albert, "Leonard Eugene Dickson: 1874–1954", *Bulletin of the American Mathematical Society*, 61 (1955), 331–45.

[65] See note 63 above.

(1856–1924) preceptors[66] at Princeton. A full professor by 1910, Veblen left Princeton University in 1932 to become professor at the newly-created Institute for Advanced Study. As one of the early organizers, he put together the Institute's world famous School of Mathematics which boasted such luminaries as James W. Alexander (1888–1971), Albert Einstein (1879–1955), John von Neumann (1903–57), Hermann Weyl (1885–1955), and Veblen himself on its initial faculty. A haven of research for research's sake, the Institute, as Veblen conceived of it, was a breeding ground for new results where younger but gifted scholars could go as temporary members to work alone, together, or with the members of the permanent faculty in a perfectly conducive research environment.[67]

Like Veblen, Birkhoff also taught as one of Princeton's 'preceptor guys,'[68] but in 1912 he accepted an assistant professorship at his undergraduate alma mater, Harvard. Although Birkhoff wrote his dissertation under Moore and looked to him for advice in his earlier years, his research was far afield of Moore's interests. Inspired by the writings of Henri Poincaré (1854–1912), Birkhoff ardently pursued questions involving dynamical systems such as the three body problem and made good advances in these and other areas.[69] At Harvard his abilities were recognized by the large number of students[70] who worked under his direction and by the administration which appointed him Dean of the Faculty of Arts and Sciences in 1936. Birkhoff's international influence was reflected in the many awards and honours he received from universities and societies around the world. Perhaps the greatest testimony to this respect was his appointment to the presidency of the International Congress of Mathematicians which was to have been held in Cambridge, Massachesetts in 1940.[71]

Less well known worldwide, Robert L. Moore secured national renown both for his research in point-set topology and for his famous 'Moore method' of teaching. A true Texan, 'he was proud, steadfast, and ever ready to defend his ideas, but appreciative of (often delighted with) an opponent who openly opposed him'.[72] His uncompromising dedication to his teaching and to his method resulted in an extremely large school of University-of-Texas-bred topologists all of whom bore the R. L. Moore stamp. The method, which bore a distinct resemblance to E. H. Moore's laboratory method of teaching mathematics, involved discovery. Students, rather than being spoonfed theorems and their proofs, were skillfully led to 'discover' the mathematics for themselves.[73] In this way R. L. Moore believed that the student's ability to produce

---

[66] The preceptorial system was the brainchild of Woodrow Wilson who was President of Princeton from 1902 to 1910. It amounted to a sort of small group system where a young and energetic faculty member read on a given subject with a small number of undergraduates in order to stimulate intellectual activity. There were no lectures, necessarily, just discussions among thoughtful people. For a more detailed description of the system as conceived by Wilson, see Woodrow Wilson, "The Preceptorial System at Princeton", *Educational Review*, 39, (1910) 385–90.

[67] For further biographical information on Veblen, see Archibald, ed. (footnote 2) I, 206–11, and Saunders Mac Lane, "Oswald Veblen: 1880–1960," *Biographical Memoirs of the National Academy of Sciences*, 37 (1964), 325–41.

[68] H. S. Taylor, "Joseph Henry Maclagen Wedderburn: 1882–1948, "*Obituary Notices of Fellows of the Royal Society*, 6 (1948–49), 619–25 (p. 620). Wedderburn was another of the first preceptors probably owing to Veblen's recommendation.

[69] For more on George Birkhoff's life, see Archibald, ed. (footnote 2) I, 212–18, and Marston Morse, "George David Birkhoff and His Mathematical Work", *Bulletin of the American Mathematical Society*, 52 (1946), 357–91. Morse also goes into great detail on the various aspects of Birkhoff's work.

[70] See note 63 above.

[71] This Congress, which was postponed due to the war in Europe, was eventually held in Cambridge in 1950. Since G. D. Birkhoff had died in the interim, the presidency went to Oswald Veblen. See Mac Lane (footnote 67) p. 334.

[72] Wilder (footnote 62) p. 418.

totally original results later on was enhanced. As one of his more famous students admitted though, 'it was a unique method employed by a unique man in a unique situation'.[74] Like the laboratory method applied to mathematics, few could implement this 'Moore method' with the success of the original innovator.[75]

The accomplishments outlined above reflect the successes and hard work not of E. H. Moore but of his gifted students. Nevertheless, these four careers suggest an impressive genealogical tree. Rooted in the European and especially in the German mathematical tradition, this tree has Moore as its trunk and his four distinguished students as its main branches. At its crown hundreds of other branches, many strong and some weak, represent the subsequent mathematical generations. A flourishing heritage, it symbolizes the firmly-grounded mathematical community which has grown during the twentieth century in America due in large part to the energy and enthusiasm of E. H. Moore.

This account of Moore's career from 1892 when he assumed his professorship at the University of Chicago to roughly 1902 when he completed his term as president of the American Mathematical Society details not only the most productive period of his professional life but also the most formative years of American mathematics. Moore came along at a time when mathematics in the United States was struggling to leave its infancy and childhood behind and to develop into a mature and productive adult.

5. Conclusion

In the 1870s and 1880s, mathematics had been born on the North American continent through the efforts of men like J. Willard Gibbs of Yale, Benjamin Peirce of Harvard, and James J. Sylvester of Johns Hopkins, but rapid growth during that period had been an impossibility. Since the American educational system had only begun to offer more advanced training in mathematics, the level of sophistication of the students had not reached sufficient heights.

By the turn of the twentieth century, however, this situation had begun to change largely as a result of the efforts of Eliakim Hastings Moore. Based at the University of Chicago, Moore was both a dynamic administrator and a first-rate mathematician. His administrative duties began in 1892 when William Rainey Harper chose him as chairman of the mathematics department of the new University of Chicago. In this post Moore worked diligently for mathematics at a local, national, and international level. From his organization of the international mathematics congress at the World's Columbian Exposition to his founding of the Chicago Section of the American Mathematical Society to his presidency of this same body, Moore's activities bore the clear stamp of his well-thought-out ideas and persistence.

---

[73] Compare E. H. Moore (footnote 53) pp. 412–13.

[74] Wilder (footnote 62) p. 418.

[75] E. H. Moore had one other student of importance during the period from 1896 to 1907 who deserves mention. Although never particularly productive as a research mathematician, Herbert E. Slaught (1861–1937) carried on Moore's pedagogical and organizational traditions. Slaught became an editor of the *American Mathematical Monthly* in 1907, and sought to have this publication taken over by the A.M.S. Because of its more popular and less research-oriented nature, the Society refused. Believing firmly in the need for an organization devoted to teachers of mathematics and college students, Slaught organized the Mathematical Association of America in 1915 and became its president in 1919. The *Monthly* became the official publication of this new association. For more on Slaught and his contributions to mathematics teaching in America, see W. D. Cairns, "Herbert Ellsworth Slaught—Editor and Organizer", *American Mathematical Monthly*, 45 (1938), 1–4, and Gilbert A. Bliss, "Herbert Ellsworth Slaught—Teacher and Friend", *American Mathematical Monthly*, 45 (1938), 5–10.

As a professor of mathematics, his deep concern for his subject and for his students always characterized his career. His students Leonard E. Dickson, Oswald Veblen, George D. Birkhoff, and Robert L. Moore, nurtured on Moore's brand of abstract thinking, also went on to make seminal contributions to mathematics and to its organization. E. H. Moore limited neither his interests nor his efforts to his doctoral students, however. He also pushed throughout his early career for the pedagogical reform of mathematics at especially the secondary and college levels. L. E. Dickson's closing words in his obituary of Moore summed up the man's achievements quite succinctly: 'Moore's work easily places him among the world's great mathematicians. In America, his various accomplishments made him the leader. But he was a leader who was universally loved, and this was because he was at the same time a prince of a man'.[76]

[76] Dickson (footnote 3) p. 80.

GRADUATE STUDENT AT CHICAGO IN THE TWENTIES

W. L. DUREN, JR.*

As an undergraduate at Tulane in New Orleans, 1922-'26, I was programmed to go to the University of Chicago and study celestial mechanics with F. R. Moulton. My teacher, H. E. Buchanan, had been a student of Moulton. That was an example of the great strength of the University of Chicago. Its PhD graduates made up a large part of the faculties of universities throughout the Mississippi Valley, Midwest and Southwest. So they sent their good students back to Chicago for graduate work. I went there first in the summer of 1926 and came to stay in 1928. In the interim I studied Moulton's *Celestial Mechanics* and some of his papers in orbit theory. I met Moulton at a sectional meeting of the MAA where he was the invited speaker. He was a man of great charm and energy and was most encouraging to me. But by the time I got to Chicago in 1928 Moulton had resigned. I was told that he felt it was an ethical requirement, since he and his wife were getting a divorce. On the advice of T. F. Cope, another former student of Buchanan, who was working with Bliss, I turned to Bliss as an advisor in the calculus of variations.

It was a down cycle for mathematics at Chicago. All the great schools have their downs as well as ups, partly because great men retire, partly because their lines of investigation dry up. At Chicago at that time a young student could see the holdovers of the great period, 1892–1920, in Eliakim Hastings Moore, officially retired, Leonard E. Dickson, rounding out his work in algebra, Gilbert A. Bliss, busy with administration and planning for the projected Eckhart Hall. Also there was Herbert E. Slaught, teacher and doer, one of the original organizers of the Mathematical Association of America and its MONTHLY, even if he played only a supporting role in mathematics itself. He had an extrovert, friendly personality that reached out and got hold of you, whether he was organizing a department social or the Mathematical Association of America. He was the teacher of teachers and key figure in Chicago's hold on education in the midwest and south. Every graduate department needs a man like Slaught if it is fortunate enough to find one. He was being succeeded by Ralph G. Sanger, a student of Bliss, an outstanding undergraduate teacher, though not the organizer Slaught was.

The University of Chicago was founded in 1892 with substantial financial support from John D. Rockefeller. William Rainey Harper, the first president, had bold educational ideas, one of which was that the United States was ready for a primarily graduate university, not just a college with graduate school attached. Harper brought E. H. Moore from Yale to establish his department of mathematics. Moore's graduate teaching was done in a research laboratory setting. That is, students read and presented papers from journals, usually German, and tried to develop new theorems based on them. The general subject of these seminars was a pre-Banach form of geometric analysis that Moore called "general analysis." It was itself not altogether successful. But even if general analysis did not succeed, Moore's seminars on it generated a surprising number of new results in general topology, among them the Moore theorem on iterated limits and Moore-Smith convergence. Moore's seminars also produced some outstanding mathematicians. His earlier students had included G. D. Birkhoff, Oswald Veblen, T. H. Hildebrant and R. L. Moore, who took off in different mathematical directions. R. L. Moore developed the teaching method into an intensive research training regimen of his own, which was very successful in producing research mathematicians at the University of Texas.

I studied general analysis with other members of the faculty including R. W. Barnard, whom

---

* With the help of Antoinette Killen Huston, who earned her way as graduate student by serving as secretary to Mr. Bliss. She was Adrian Albert's first student, receiving her PhD degree in 1934.

Reprinted with permission from *American Mathematical Monthly*, Volume **83**, No. 4, April 1976, pp. 243–247.

Moore had designated as his successor and whose notes record the second form of the theory, [Am. Philosophical Soc., *Memoirs*, v. 1, Philadelphia, 1935]. Instead of taking the general analysis courses, my old friend E. J. McShane, from New Orleans, worked in Moore's small seminar in the foundations of mathematics. Although he was officially a student of Bliss, I think he was in a sense Moore's last student.

Moore himself was meticulous in manners and dress. He would stop you in the hall, gently remove a pen from an outside pocket and suggest that you keep it in the inside pocket of your jacket. Nobody thought of not wearing a jacket. But Moore was less gentle if you used your left hand as an eraser, and he displayed towering anger at intellectual dishonesty. To understand him and his times one must read his retiring address as President of the Society [*Science*, March 1903]. In those days the Society accepted responsibility for teaching mathematics and Moore's address was largely devoted to the organization of teaching, the curriculum, and the ideas of some of the great teachers of the time, Boltzman, Klein, Poincaré, and, in this country, J. W. A. Young and John Dewey, whose ideas Moore supported by proposing a mathematics laboratory. This address was adopted as a sort of charter by the National Council of Teachers of Mathematics and republished in its first Yearbook (1925). By the time I got to Chicago the Association had been formed to relieve the Society of concern for college education, and NCTM to relieve it of responsibility for the school curriculum and training teachers. In the top universities only research brought prestige, even if a few, like Slaught, upheld the importance of teaching.

L. E. Dickson's students tended to identify themselves strongly as number theorists or algebraists. I felt this particularly in Adrian A. Albert, Gordon Pall and Arnold Ross. All his life Albert strongly identified himself, first as an algebraist, later with mathematics as an institution and certainly with the University of Chicago. I remember him as an advanced graduate student walking into Dickson's class in number theory that he was visiting, smiling and self confident. He knew where he was going. Dickson was teaching from the galley sheets of his new *Introduction to the Theory of Numbers* [University of Chicago Press, 1929] with its novel emphasis on the representation of integers by quadratic forms. I think he requested Albert to sit in for his comments on this aspect. He was tremendously proud of Albert. I remember A^3 too with his beautiful young wife, Frieda, at the perennial department bridge parties. He had superb mental powers; he could read a page at a glance. One could see even then that as heir apparent to Dickson he would do his own mathematics rather than a continuation of Dickson's, however much he admired Dickson.

In the conventional sense Dickson was not much of a teacher. I think his students learned from him by emulating him as a research mathematician more than being taught by him. Moreover, he took them to the frontier of research, for the subject matter of his courses was usually new mathematics in the making. As Antoinette Huston said, "He made you want to be with him intellectually. When you are young, reaching for the stars, that is what it is all about." He was good to his students, kept his promises to them and backed them up. Yet he could be a terror. He would sometimes fly into a rage at the department bridge games, which he appeared to take seriously. And he was relentless when he smelled blood in the oral examination of some hapless, cringing victim. He was an indefatigable worker and in public a great showman, with the flair of a rough and ready Texan. An enduring bit in the legend is his blurt: "Thank God that number theory is unsullied by any application." He liked to repeat it himself as well as his account of his and his wife's honeymoon, which he said was a success, except that he got only two papers written.

The theme of beauty for its own sake was expressed more surprisingly by another Texan who worked in mechanics and potential theory, W. D. MacMillan. According to the story he had come to Chicago as a mature man, without a college education, to sell his cattle. Having sold them, he went to

Chicago's Yerkes Observatory to see the Texas stars through the telescope. He was so fascinated that he stayed on to get his degrees in rapid succession, all *summa cum laude.* Then he remained as a member of the faculty. One day in his course on potential theory he wrote some important partial differential equations on the board with obvious pleasure, drawing the partial derivative signs with a flourish. Standing back to admire these equations, he said: "That is just beautiful. People who ask, 'What's it good for?', they make me tired! Like when you show a man the Grand Canyon for the first time and you stand there as you do, saying nothing for a while." And we could see that old Mac was really looking at the Grand Canyon. "Then he turns to you and asks, 'What's it good for?' What would you do? Why, you would kick him off the cliff!" And old Mac kicked a chair halfway across the room. He was a prodigy, a good lecturer, an absolutely fascinating personality with a twinkling wit. Some of his work was outstanding, yet he had few doctoral students.

Celestial mechanics was being carried on by the young Walter Bartky, who was, I think, Moulton's last student. But celestial mechanics had gone into a barren period and Bartky with his superb talents turned to other applications of differential equations, to statistics and to administration.

Lawrence M. Graves was the principal hope of the department for carrying on the calculus of variations, which he did in the spirit of functional analysis. He was my favorite professor because he knew a lot of mathematics, knew it well, and in an unassuming way was glad to share it with you. Although he taught Moore's general analysis, he pointed out the difficulties in it to me. His own brand of functional analysis was more oriented towards the use of the Fréchet differential in Banach space.

Research in geometry at Chicago was a continuation of Wylczinski's projective differential geometry. There was no topology, though we heard that Veblen's students studied something called *analysis situs* at Princeton. I knew so little about the subject that years later when I wanted to prepare for Morse theory I spent months studying Kuratowski's point set topology before it dawned on me that what I wanted was algebraic topology. E. P. Lane and his students carried on the study of projective differential geometry using rather crude analytical methods, that is, expansions in which one neglected higher order terms. We who were not Lane's students tended to look on it with disdain as being non-rigorous. But the structure of the theory was beautiful, I thought. Lane was honest about the shortcomings of the methods, though he did not know how to overcome them.

Lane was a very fine man. I had come to Chicago in 1926 to run the high hurdles in the National Intercollegiate Track and Field Meet at Soldiers Field. I placed in the finals and some members of the U. S. Olympic Committee urged me to keep working for the 1928 Olympics. So I worked on the Stagg Field track until an accident set off a series of leg infections. I was very sick in Billings Hospital in the days before antibiotics and it was Lane who came to the hospital to see me and make sure that I got the best available care. The only way I was ever able to express my thanks to him was to do a similar service to some of my own students in later years. I guess that is the only way we ever thank our teachers.

Bliss was an outstanding master of the lecture-discussion. He could come into a class in calculus of variations obviously unprepared, because of the demands of his chairmanship, and still deliver an elegant lecture, drawing the students into each deduction or calculation, as he looked at us quizzically and waited for us to tell him what to write. His students learned their calculus of variations very thoroughly. Yet we did not work together, except in so far as we presented class assignments. Each research student reported to Bliss by appointment. The subject itself had come to be too narrowly defined as the study of local, interior minimum points for certain prescribed functionals given by integrals of a special form. Generalization came only at the cost of excessive notational and analytic complications. It was like defining the ordinary calculus to consist exclusively of the chapter on maxima and minima. A sure sign of the decadence of the subject was Bliss's project to produce a history of it, like Dickson's *History of the Theory of Numbers*. The history reached publication only in the form of certain theses imbedded in *Contributions to the Calculus of Variations*, 4 vols, 1930–1944, University of Chicago Press.

It is perhaps surprising that this narrowly prescribed regimen turned out men who did important work in entirely different areas as, for example, A. S. Householder did in biomathematics and numerical analysis, and Herman Goldstine did in computer theory. Among all of us Magnus Hestenes has been most faithful to the spirit of Bliss's teaching in carrying on research in the calculus of variations. Yet when Pontryagin's Optimal control papers revived interest in the subject many years later, students of Bliss were easily able to get into it. Optimal control theory really contained relatively little that was correct and not in the calculus of variations. In fact, optimal control was anticipated by the thesis of Carl H. Denbow, *loc. cit.*

Quantum mechanics was breaking wide open in the twenties. Bliss himself got into it with his students by studying Max Born's elegant canonical variable treatment of the Bohr theory. While that was going on, Sommerfeld's *Wellenmechanische Ergänzungsband* to his *Atombau und Spektrallinien* [Vieweg, Braunschweig, 1929] came out. It was the first connected treatment of the new wave mechanics formulation of quantum mechanics due to de Broglie and Schrödinger. We dropped everything to study wave mechanics. Bliss was a remarkably knowledgeable mathematical physicist and quite expert in the boundary value problems of partial differential equations. That was not so remarkable in a mathematician of his generation. The narrowing of the definition of a mathematician and withdrawal into abstract specializations was just beginning. In fact Bliss had been chief of mathematical ballistics for the U.S. Government in World War I, and later was commissioned to do a mathematical study of proportionate representation for purposes of reassigning Congressional districts. Bliss did not follow up his move into quantum mechanics but returned to the classical calculus of variations.

There were always more students in summers with all the teachers who came. Visiting professors like Warren Weaver, E. T. Bell, C. C. Mac Duffee and Dunham Jackson came to teach. And there was the memorable visit of G. H. Hardy which was supposed to provide a uniting of Hardy's analytic approach to Waring's theorem with Dickson's algebraic approach. Even with this infusion of talent, the offerings of the department were rather narrow. Besides having no topology as such, more surprisingly, there was little in complex function theory. And I do not recall being in a seminar, either a research or journal seminar. Essentially all teaching was done in lectures. Yet the only one of the abler students who I remember taking the initiative to go elsewhere was Saunders MacLane, when he did not find at Chicago what he was looking for.

I once asked Edwin B. Wilson, a famed universalist among mathematicians, how he came to switch from analysis to statistics at Yale. With a humorous twinkle he said: "An immutable law of academia is that the course must go on, no matter if all of the substance and spirit has gone out of it with the passing of the original teacher. So when (Josiah Willard) Gibbs retired, his courses had to go on. And the department said: 'Wilson, you are it'." A graduate student at Chicago in the late twenties could see this immutable academic law in effect. In each line of study of the, then passing, old Chicago department, a younger Chicago PhD had been designated to carry on the work. If, in one's immaturity, this was not apparent, the point was made loud and clear in a blast from Dickson during a colloquium with graduate students present. Dickson charged the chairman with permitting the department to slide into second rate status. It was true that the spirit of original investigation had given way to diligent exposition in some of these fields. In some cases the fields themselves had gone sterile.

It was the lot of Bliss to preside over this ebb cycle of the department. He did an impressive best possible with what he had, with high mathematical standards, firmly, kindly and quietly. Most of the difficulties he had inherited. Bliss was able to appoint some outstanding young men but, if he had asked for the massive financial outlay to bring in established leading mathematicians to make a new start like the original one under President Harper, the support would not have been forthcoming, even with a mathematician, Max Mason, as president and certainly not with the young Robert M. Hutchins, bent primarily on establishing his new college. It took the Manhattan Project, the first nuclear pile under the Stagg Field bleachers and Enrico Fermi to convince Hutchins of the importance of physical science and mathematics and to throw massive resources into the reorganization of the department near the end of World War II. Such reorganizations are necessary from time to time in every graduate department. They can be effective only when the time is right. It is the mark of a great university to recognize the necessity to break the immutable law of academia, and the opportunity, and to do it when the time is right. However, there were deep hurts, symbolized by Bliss's refusal ever again to set foot in Eckhart Hall to his death. But this is really getting ahead of my story.

It was no ebb cycle for the University of Chicago as a whole in the twenties. There was intellectual excitement in many places in the university. I attended the physics colloquia where the great innovators of the day came to talk. With Mr. Bliss's grudging consent, I took Arthur Compton's course in X-rays. He already had the Nobel Prize for his work on the phenomena of X-rays colliding with electrons. Yet he seemed so naïvely simple minded to me, far less expert and mentally profound than other physicists in the department. Somewhere in here Einstein came for a brief visit. He permitted himself to be escorted by the physics graduate students for a tour of their experiments. To one he offered a suggestion. The brash young man explained immediately why it could not work. Einstein shook his head sadly. "My ideas are never good," he said.

Michelson, another Noble Prizeman, was around, though retired. So was the great geologist, Chamberlin, with his planetesimal hypothesis in cosmology. In biology and biochemistry the great breakthroughs on the chemical nature of the steroid hormones and their effects on growth and development were excitingly unfolding. Young Sewall Wright was attracting students to his mathematical genetics. Economics promised a real breakthrough, though as it turned out, it was slow in coming. Linguistics was burgeoning. Anthropology and archeology were still actively following up the results of digs in Egypt, Turkey and Mesopotamia. The great debates over the truth of theories of relativity and quantum mechanics were raging. What was later to be planet Pluto had been observed as "Planet X" but heated arguments persisted on what it really was. On Sundays the University Chapel produced a succession of the leading Christian and Jewish spokesmen of the day. The textbook, *The Nature of the World and of Man*, H. H. Newman ed., University of Chicago Press, 1926, by illustrious Chicago faculty members was the best survey of physical and biological knowledge for college students that I have ever seen, though now dated, of course.

And outside the university the dangerous and ugly city of Chicago nevertheless had its charms, cultural and otherwise, that could take up all the time (and money) of a country boy. One could hear Mary Garden or Rosa Raisa at the Chicago Opera by getting a job as usher or super, or attend a fiesta in honor of the patron saint of some Halstead Street community that maintained its identity with the home village in the old country. One could drink wine at Alexander's clandestine speakeasy. For recall that it was Prohibition and the height of the bootlegging days of Al Capone and rival gangs. The famous Valentine Day massacre was just one of the lurid stories in the Chicago Tribune. We students formed an informal protective association to promulgate rules to optimize safety for oneself and date. One old boy from Georgia, a graduate student in history, was so impressed by our admonition never to approach a car asking him to get in, that, when a police car challenged him with order to stop, he just took off in a blaze of speed. Caught later, out of breath, his one phone call brought some of us to police court to testify to his character. The officer who had made the arrest moved to dismiss the charges on the condition that "the defendant appear at Soldiers Field next Saturday and run for our company in the policemen's track meet." But it was grim business. Police, armed with machine guns, in such a car once arrested me on suspicion of rape on the Midway (not guilty!). Other students were mugged, raped, robbed and even killed.

Like today it was a time of inflation and most of us were poor. I had a full fellowship of $410, of which $210 had to be returned in tuition for three quarters. A dormitory room cost $135 out of what was left. We could get cheap meals at the Commons, and on Sundays one could go to the Merit Cafeteria and splurge on a plate-sized slab of roast beef. It cost 28¢ but it was worth it. We all looked forward to a teaching job, I think. Those jobs required 15 hours of teaching for about $2700. Soon the depression hit and, if we were lucky, we kept our jobs with salary cut to $2400. Some beginning salaries for Chicago PhD's were as low as $1800 in the early thirties.

Before closing these recollections I must write something about women as graduate students in those times, not long after the victory of women's suffrage. Only years later did I learn that it was considered unladylike to study mathematics. Many of the graduate students in mathematics were women. In fact there were 26 women PhD's in mathematics at Chicago between 1920 and 1935. I shall mention only a few by name. Mayme I. Logsdon (1921) was in the faculty of the department. Mina Rees (1931) was already showing the kind of ability that led her to a distinguished administrative career at Hunter College and CUNY. She did more than any other person to gain federal support for mathematics through her position as chief, Mathematics Branch ONR, when the National Science Foundation was established. Others included Abba Newton (1933), chairman at Vassar, and Frances Baker (1934) also of Vassar, Julia Wells Bower (1933), chairman at Connecticut College, Marie Litzinger (1934), chairman, Mt. Holyoke, Lois Griffiths (1927) Northwestern, Beatrice Hagen (1930) Penn State, and Gweneth Humphreys (1935) Randolph Macon. Graduate students married graduate students, though of necessity only after the man had his degree. In the department Virginia Haun married E. J. McShane. Emily Chandler, student of Dickson, married Henry Pixley and continued her publishing and teaching career at the University of Detroit. Antoinette Killen married Ralph Huston. They both later taught at Rensselaer Polytech. Aline Huke married a non-Chicago mathematician, Orrin Frink, and continued her teaching at Penn State. Jewel Hughes Bushey was in the department of Hunter College. These, and a number of others, were able to continue their professional work in spite of family obligations. Even intermarriage between departments was permitted! My wife to be, Mary Hardesty, was in zoology. We got our PhD degrees in the same commencement.

Looking back on those days, I wonder if the current women's liberation has even yet succeeded in pushing the professional status of women to the level already reached in the twenties. Maybe this time women can hold their gains in universities.

DEPARTMENT OF APPLIED MATHEMATICS AND COMPUTER SCIENCE, UNIVERSITY OF VIRGINIA, CHARLOTTES-
VILLE, VA 22901.

Reminiscences of Mathematics at Chicago [1]

MARSHALL H. STONE

In 1946 I moved to the University of Chicago. An important reason for this move was the opportunity to participate in the rehabilitation of a mathematics department that had once had a brilliant role in American mathematics but had suffered a decline, accelerated by World War II. During the war the activity of the department fell to a low level and its ranks were depleted by retirements and resignations. The administration may have welcomed some of these changes, because they removed persons who had opposed some of its policies. Be that as it may, the University resolved at the close of the war to rebuild the department.

The decision may have been influenced by the plans to create new institutes of physics, metallurgy and biology on foundations laid by the University's role in the Manhattan Project. President Hutchins had seized the opportunity of retaining many of the atomic scientists brought to Chicago by this project, and had succeeded in making a series of brilliant appointments in physics, chemistry, and related fields. Something similar clearly needed to be done when the University started filling the vacancies that had accumulated in mathematics. Professors Dickson, Bliss, and Logsdon had all retired fairly recently, and Professors W. T. Reid and Sanger had resigned to take positions elsewhere. The five vacancies that had resulted offered a splendid challenge to anyone mindful of Chicago's great contribution in the past and desirous of ensuring its continuation in the future.

When the University of Chicago was founded under the presidency of William Rainey Harper at the end of the 19th century, mathematics was encouraged and vigorously supported. Under the leadership of Eliakim Hastings Moore, Bolza and Maschke it quickly became a brilliant center of mathematical study and research. Among its early students were such mathematicians as Leonard Dickson, Oswald Veblen, George Birkhoff, and R. L. Moore, destined to future positions of leadership in research and teaching. Some of

Marshall H. Stone

these students remained at Chicago as members of the faculty. Messrs. Dickson, Bliss, Lane, Reid, and Magnus Hestenes were among them.

Algebra, functional analysis, calculus of variations and projective differential geometry were fields in which Chicago obtained special distinction. With the passage of time, retirements and new appointments had brought a much increased emphasis on the calculus of variations and a certain tendency to inbreeding. When such outstanding mathematicians as E. H. Moore or Wilczynski, a brilliant pioneer in projective differential geometry, retired from the department, replacements of comparable ability were not found. Thus in 1945 the situation was ripe for a revival.

A second, and perhaps even more important, reason for the move to Chicago was my conviction that the time was also ripe for a fundamental revision of graduate and undergraduate mathematical education.

The invitation to Chicago confronted me with a very difficult question: "Could the elaboration of a modernized curriculum be carried out more successfully at Harvard or at Chicago?"

When President Hutchins invited me to visit the University in the summer of 1945, it was with the purpose of interviewing me as a possible candidate for the deanship of the Division of Physical Sciences. After two or three days of conferences with department heads, I was called to Mr. Hutchins' residence, where he announced that he would offer me not the deanship but a distinguished service professorship in the Department of Mathematics.

The negotiations over this offer occupied nearly a year, during which I sought the answer to the question with which it confronted me. It soon became clear that the situation at Harvard was not ripe for the kind of change to which I hoped to dedicate my energies in the decade following the war. However, it was by no means clear that circumstances would be any more propitious at Chicago than they seemed to be at Harvard. In consulting some of my friends and colleagues, I was advised by the more astute among them to come to a clear understanding with the Chicago administration concerning its intentions.

There are those who believe that I went to Chicago to execute plans that the administration there already had in mind. Nothing could be farther from the truth. In fact, my negotiations were directed towards developing detailed plans for reviving the Chicago Department of Mathematics and obtaining some kind of commitment from the administration to implement them. Some of the best advice given me confirmed my own instinct that I should not join the University of Chicago unless I were made chairman of the department and thus given some measure of authority over its development. Earlier experiences had taught me that administrative promises of wholehearted interest in academic improvements were too often untrustworthy. I therefore asked the University of Chicago to commit itself to the development program that was under discussion, at least to the extent of offering me the chairmanship.

This created a problem for the University, as the department had to be consulted about the matter, and responded by voting unanimously that Professor Lane should be retained in the office. As I was unwilling to move merely on the basis of a promise to appoint me to the chairmanship at some later time, the administration was brought around to arranging the appointment, and I to accept it. Mr. Lane, a very fine gentleman in every sense of the word, never showed any resentment. Neither of us ever referred to the matter, and he served as an active and very loyal member of the department until he retired several years later. I was very grateful to him for the grace and selflessness he displayed in circumstances that might have justified a quite different attitude.

Even though the University made no specific detailed commitments to establish the program I had proposed during these year-long negotiations, I was ready to accept the chairmanship as an earnest of forthcoming support. I felt confident that with some show of firmness on my part the program could be established. In this optimistic spirit I decided to go to Chicago, despite the very generous terms on which Harvard wished to retain me.

Regardless of what many seem to believe, rebuilding the Chicago Department of Mathematics was an up-hill fight all the way. The University was not about to implement the plans I had proposed in our negotiations without resisting and raising objections at every step. The department's loyalty to Mr. Lane had the fortunate consequence for me that I felt released from any formal obligation to submit my recommendations to the department for approval. Although I consulted my colleagues on occasion, I became an autocrat in making my recommendations. I like to think that I am not by nature an autocrat, and that the later years of my chairmanship provided evidence of this belief. At the beginning, however, I took a strong line in what I was doing in order to make the department a truly great one.

The first recommendation sent up to the administration was to offer an appointment to Hassler Whitney. The suggestion was promptly rejected by Mr. Hutchins' second in command. It took some time to persuade the administration to reverse this action and to make an offer to Professor Whitney. When the offer was made, he declined it, and remained at Harvard for a short time before moving to the Institute for Advanced Study.

The next offer I had in mind was one to Andre Weil. He was a somewhat controversial personality, and I found a good deal of hesitation, if not reluctance, on the part of the administration to accept my recommendation.

In fact, while the recommendation eventually received favorable treatment in principle, the administration made its offer with a substantial reduction in the salary that had been proposed; and I was forced to advise Professor Weil, who was then in Brazil, that the offer was not acceptable. When he declined the offer, I was in a position to take the matter up at the highest level. Though I had to go to an 8 a.m. appointment suffering from a fairly high fever, in order to discuss the appointment with Mr. Hutchins I was rewarded by his willngness to renew the offer on the terms I had originally proposed. Professor Weil's acceptance of the improved offer was an important event in the history of the University of Chicago and in the history of American mathematics.

My conversation with Mr. Hutchins brought me an unexpected bonus. At its conclusion he turned to me and asked, "When shall we invite Mr. Mac Lane?" I was happy to be able to reply, "Mr. Hutchins, I have been discussing the possibility with Saunders and believe that he would give favorable consideration to a good offer whenever you are ready to make it." That offer was made soon afterwards and was accepted.

Hand-to-mouth budgeting

One explanation doubtless was to be found in the University's hand-to-mouth practices in budgeting. This would appear to have been the reason why one evening I was given indirect assurances from Mr. Hutchins that S. S. Chern would be offered a professorship, only to be informed by Vice-President Harrison the next morning that the offer would not be made. Such casual, not to say arbitrary, treatment of a crucial recommendation naturally evoked a strong protest. In the presence of the dean of the Division of Physical Sciences I told Mr. Harrison that if the appointment were not made, I would not be a candidate for reappointment as chairman when my three-year term expired. Some of my colleagues who were informed of the situation called on the dean a few hours later to associate themselves with this protest. Happily, the protest was successful, the offer was made to Professor Chern, and he accepted it. This was the stormiest incident in a stormy period. Fortunately the period was a fairly short one, and at the roughest times Mr. Hutchins always backed me unreservedly.

As soon as the department had been brought up to strength by this series of new appointments, we could turn our attention to a thorough study of the curriculum and the requirements for higher degrees in mathematics. The group that was about to undertake the task of redesigning the department's work was magnificently equipped for what it had to do. It included, in alphabetical order, Adrian Albert, R. W. Barnard, Lawrence Graves, Paul Halmos, Magnus Hestenes, Irving Kapansky, J. L. Kelley, E. P. Lane, Saunders Mac Lane, Otto Schilling, Irving Segal, M. H. Stone, Andre Weil, and Antoni Zygmund. Among them were great mathematicians, and great teachers, and leading specialists in almost every branch of pure mathematics. Some were new to the University, others familiar with its history and traditions. We were all resolved to make Chicago the leading center in mathematical research and education it had always aspired to be. We had to bring great patience and open minds to the time-consuming discussion that ranged from general principles to detailed mathematical questions. The presence of a separate and quite independent College mathematics staff did not relieve us of the obligation to establish a new undergraduate curriculum beside the new graduate program.

Two aims on which we came to early agreement were to make course requirements more flexible and to limit examinations and other required tasks to those having some educational value.

This streamlined program of studies, the unusual distinction of the mathematics faculty, and a rich offering of courses and seminars have attracted many very promising young mathematicians to the University of Chicago ever since the late '40s. The successful coordination of these factors was reinforced by the concentration of all departmental activities in Eckhart Hall with its offices (for faculty and graduate students), classrooms, and library. As most members of the department lived near the University and generally spent their days in Eckhart, close contact between faculty and students was easily established and maintained. (This had been foreseen and planned for by Professor G. A. Bliss when he counseled the architect engaged to build Eckhart Hall.) It was one of the reasons why the mathematical life at Chicago became so spontaneous and intense. By helping create conditions so favorable for such mathematical activity, Professor Bliss earned the eternal gratitude of his University and his department. Anyone who reads the roster of Chicago doctorates since the later '40s cannot but be impressed by the prominence and influence many of them have enjoyed in American — indeed in world — mathematics. It is probably fair to credit the Chicago program with an important role in stimulating and guiding the development of these mathematicians during a crucial phase of their careers. If this is done, the program must be considered as a highly effective one.

As I have described it, the Chicago program made one conspicuous omission — it provided no place for applied mathematics. During my correspondence of '45–'46 with the Chicago administration I had insisted that applied mathematics should be a concern of the department, and I had outlined plans for expanding the department by adding four positions for professors of applied subjects. I had also hoped that it would be possible to bring about closer cooperation than had existed in the past between the Departments of Mathematics and Physics.

Circumstances were unfavorable. The University felt little pressure to increase its offerings in applied mathematics. It had no engineering school, and rather recently had even rejected a bequest that would have endowed one. Several of its scientific departments offered courses in the applications of mathematics to specific fields such as biology, chemistry and meteorology.

The Department of Physics and the Fermi Institute had already worked out an entirely new program in physics and were in no mood to modify it in the light of subsequent changes that might take place in the Mathematics Department.

However, many students of physics elected advanced mathematics courses of potential interest for them — for example, those dealing with Hilbert space or operator theory, subjects prominently represented among the specialties cultivated in the Mathematics Department.

On the other hand, there was pressure for the creation of a Department of Statistics, exerted particularly by the economists of the Cowles Foundation. A committee was appointed to make recommendations to the administration for the future of statistics with Professor Allen Wallis, Professor Tjalling Koopmans, and myself as members. Its report led to the creation of a Committee on Statistics, Mr. Hutchins being firmly opposed to the proliferation of departments.

The committee enjoyed powers of appointment and eventually of recommendation for higher degrees. It was housed in Eckhart and developed informal ties with the Department of Mathematics.

At a somewhat later time a similar committee was set up to bring the instruction in applied mathematics into focus by coordinating the courses offered in several different departments and eventually recommending higher degrees.

Long before that, however, the Department of Mathematics had sounded out the dean of the division, a physicist, about the possibility of a joint appointment for Freeman Dyson, a young English physicist then visiting the United States on a research grant. We had invited him to Chicago for lectures on some brilliant work in number theory, that had marked him as a mathematician of unusual talent. We were impressed by his lectures and realized that he was well qualified to establish a much needed link between the two departments.

However, Dean Zachariasen quickly stifled our initiative with a simple question, "Who is Dyson?"*

By 1952 I realized that it was time for the Department of Mathematics to be led by someone whose moves the administration had not learned to predict. It was also time for the department to increase its material support by entering into research contracts with the government.

Fortunately there were several colleagues who were more than qualified to take over. The two most conspicuous were Saunders Mac Lane and Adrian Albert. The choice fell first on Professor Mac Lane, who served for the next six years.

---

*He was soon to become a permanent member of the Institute of Advanced Study.

Under the strong leadership of these two gifted mathematicians and their younger successors the department experienced many changes, but flourished mightily and was able to maintain its acknowledged position at the top of American mathematics.

The Stone Age of Mathematics on the Midway [1]

FELIX E. BROWDER

After its foundation as a distinguished department by E. H. Moore in the 1890s, the single most decisive event in the history of the Department of Mathematics at the University was the assumption of its chairmanship by Marshall Harvey Stone in 1946. Stone arrived in Chicago from a professorship at Harvard as the newly appointed Andrew MacLeish distinguished service professor of mathematics as well as chairman of the department. Within a year or two, he had transformed a department of dwindling prestige and vitality once more into the strongest mathematics department in the U.S. (and at that point probably in the world). This remarkable transformation, which endowed the department with a continuing vitality during the trials of the following decades is unparalleled, to the writer's knowledge, in modern academic history for its speed and dramatic effect. This was no easy victory on the basis of great infusions of outside money for bringing in men and building research facilities. It was completely a triumph for Stone's sureness of judgment in men and his determination and strength of character in getting done what he knew had to be done.

Stone's account of the transformation which he formed and led is unparalleled for its candor and its objectivity (despite the strong flavor of Stone's personality) and for its remarkably open presentation of the process by which academic decisions are reached and leadership exerted. The central problem of academic life for institutions which aspire to excellence and to greatness is precisely the achievement of that excellence and that greatness.

Within every academic institution, policy leadership falls into two patterns. The most common pattern, which is the basis of the ongoing routine of the institution's existence, falls within the rational-bureaucratic mold (to use the classical terminology of Max Weber) in terms of rationalized general policies and procedures to be applied uniformly to an array of cases in the context of a balance of special interests and influences. The other pattern, which is less common, is that of charismatic leadership in which the individual judgment

---

and personal qualities of the administrator play a fundamental role in both the choice and nature of the policy decisions which are made, and in their acceptance by those who are affected by them. Stone's account gives us a picture of the most highly developed form of charismatic leadership, one which turned out to be enormously successful. What is most interesting about it is the question it raises about the role of charismatic leadership in the search for academic excellence.

To my knowledge, there is no case in which academic excellence in any reasonably high degree has been achieved and maintained without an infusion of charismatic leadership, either public or behind the scenes. Yet to an ever greater degree, it has become increasingly incompatible with the growing pressures and struggles of interests that tend to dominate the organized life of our universities.

When Marshall Harvey Stone arrived at the University in 1946 to play such a distinctive role, he was a relatively young man (43) and a mathematician of great distinction and great reputation. He had spent most of his academic life at Harvard, getting his Ph.D. degree there in the late '20s with the dominant personality of the Harvard department, George David Birkhoff who had himself been a student of E. H. Moore at Chicago. Stone had done fundamental work in a number of widely-known directions, in particular on the spectral theory of unbounded self-adjoint operators in Hilbert space and on the applications of the algebraic properties of Boolean algebras in the study of rings of continuous functions. He was an inner member of the country's mathematical establishment, having obtained a full professorship at Harvard as well as such honors as election to the National Academy of Sciences. He was profoundly involved in the growing trend toward putting mathematics research and education on an abstract or axiomatic foundation, and was sharply influenced by the efforts of the Bourbaki school in France in this direction, which achieved a major impact in the years after the end of World War II.

Most important of all, Stone was a man of forceful character and unquestioned integrity, with a strong insight into the mathematical quality of others.

Stone's fundamental achievement at Chicago was to bring together a faculty group of unprecedented quality. In the senior faculty he appointed four very diverse men with widely different personal styles and mathematical tastes. The most important of these was undoubtedly Andre Weil, the dominant figure of the Bourbaki group, who was, then and now, one of the decisive taste-makers of the mathematical world, as well as a brilliant research mathematician in his own work.

S. S. Chern, who was to be the central figure of differential geometry in the world, was brought from his haven at the Princeton Institute after his departure from China.

Antoni Zygmund, who became the central figure of the American school of classical Fourier analysis, which he was to build up single-handed, came from the University of Pennsylvania.

Saunders Mac Lane, who had been Stone's colleague and sympathizer in the abstract program as applied to algebra, came from a professorship at Harvard.

Together with Adrian Albert, who had been Dickson's prize student at Chicago and a longtime member of the Chicago department, these men formed the central group of the new Stone department at the University.

To do full justice to the kind of revolution that Stone brought about in Chicago mathematics, one needs to perform the unedifying task of acknowledging the decay of the department in the late '30s and early '40s. The great prestige and intellectual vitality that had been created under the long reign of E. H. Moore as chairman had not been maintained after his retirement from the chairmanship at the end of the '20s. His successors, G. A. Bliss and E. P. Lane, were not Moore's equals in either mathematical insight or standards. Especially under Bliss' regime, a strong tendency to inbreeding was in action, and as the great elder figures of the department died or retired, they were not replaced by younger mathematicians of equal caliber. Some of the most promising of those who came into the deparetment soon left. There was one principal exception: Adrian Albert. But despite his distinction as an algebraist in the Dickson tradition, Albert at that time had neither the influence nor the vision to bring about the kind of radical transformation that the department needed, and that Stone brought about.

The insights that Stone provides in his first-hand account of his great achievements and of how they were brought about provide us once more with a dramatic vindication of the decisive importance of the special qualities of significant individuals as the major agents of the development of academic institutions. In academic terms, Marshall Stone served as a great revolutionary and a great traditionalist. The revolution he made is the only kind which has a permanent significance – a revolution that founds or renovates an intense and vital tradition.

The Emergence of Princeton as a
World Center for
Mathematical Research, 1896-1939

1. Introduction

In 1896 the College of New Jersey changed its name to Princeton University, reflecting its ambitions for graduate education and research. At the time, Princeton, like other American universities, was primarily a teaching institution that made few significant contributions to mathematics. Just four decades later, by the mid-1930s, Princeton had become a world center for mathematical research and advanced education.[1] This paper reviews some social and institutional factors significant in this rapid rise to excellence.[2]

The decade of the 1930s was a critical period for American research mathematics generally, and for Princeton in particular. The charter of the Institute for Advanced Study in 1930 and the completion of a university mathematics building (Fine Hall) in 1931 frame the opening of the period in Princeton; the completion of separate quarters (Fuld Hall) for the institute mathematicians in 1939 and the entrance of the United States into World War II effectively close it. During this decade, Princeton had the unique atmosphere of an exclusive and highly productive mathematical club. This social environment changed after the war with the increase in university personnel and the move of the institute to separate quarters, and the uniqueness was challenged by the improvement of mathematical research and advanced education at other American institutions.

I appreciate the useful comments and corrections of Richard Askey, Saunders Mac Lane, and V. Frederick Rickey to the conference presentation and of Garrett Birkhoff and Albert W. Tucker to later drafts of the manuscript. The writing of this paper was stimulated by my participation in the Princeton Mathematical Community in the 1930s Oral History Project, which resulted from the efforts of Charles Gillespie, Frederik Nebeker, and especially Albert Tucker and from the financial support of the Alfred P. Sloan Foundation.

Reprinted with permission from *History and Philosophy of Modern Mathematics*, edited by William Aspray and Philip Kitcher, University of Minnesota Press, Minneapolis, 1988, pp. 346–366.

2. A Fine Start

Efforts to establish a research program in mathematics at Princeton University began in the first decade of the twentieth century at the hands of Henry Burchard Fine. Fine had completed an undergraduate major in classics at Princeton in 1880 and remained until 1884, first as a fellow in experimental physics and then as a tutor in mathematics. The latter position brought him into contact with George Bruce Halsted, a mathematics instructor fresh from his dissertation work under J. J. Sylvester at Johns Hopkins University.[3] Fine wrote of their relationship in a testimonial:

> I am glad of this opportunity of acknowledging my obligations to Dr. Halsted. Though all my early prejudices and previous training had been in favor of classical study, through his influence I was turned from the Classics to Mathematics, and under his instruction or direction almost all of my mathematical training had been acquired. (Eisenhart 1950, 31-32)

On Halsted's advice, Fine traveled to Leipzig in the spring of 1884 to study with Felix Klein. Halsted's ability to inspire proved greater than his ability to teach Fine mathematics, for Klein found Fine to know no German and little mathematics. Nevertheless, Fine was encouraged to attend lectures. He progressed quickly and was awarded the Ph.D. after only a year for his solution to a problem in algebraic geometry. In the summer of 1885, and again in 1891, Fine visited Berlin to study with Leopold Kronecker. Fine's first book and several of his papers are testimony to the profound influence of Kronecker (Fine 1891, 1892, 1914).

Fine returned to Princeton in the fall of 1885 as an assistant professor of mathematics with an admiration for the German system, which provided opportunities for young mathematicians to work closely with established researchers. He progressed steadily through the ranks. In 1889 he was promoted to professor and in 1898 to Dod Professor; by 1900, he was the senior member of the department. During Woodrow Wilson's tenure as university president, from 1903 to 1911, Fine's career and his influence on Princeton mathematics advanced most rapidly.[4] Fine was appointed chairman of mathematics (1904-28), dean of the faculty (1903-12), and dean of the science departments (1909-28). When Wilson resigned to run for governor of New Jersey in 1911, Fine served as acting president of the university until John Grier Hibben was appointed president.

Fine published several research papers in geometry and numerical analysis, but he was most prominent as a textbook writer (Fine 1905, 1927; Fine and Thompson 1909) and an institution builder. In the latter capacity, he was one of two professors of mathematics to support Thomas Fiske's 1888 plan to found the New York Mathematical Society, which became the American Mathematical Society (AMS) after an international meeting at the time of the Chicago World's Fair in 1893. In 1911 and 1912, Fine served as president of the society.

When Wilson was called to the Princeton presidency in 1903, his first priority was to match the quality of the educational program to the upgraded status of a university. At Wilson's instigation, the preceptorial system was introduced in 1905 to provide smaller classes and more personalized instruction. Fine was a strong proponent of the system, and he recognized the opportunity to strengthen the mathematics program through the new appointments the preceptorial system required.[5]

At the time of Wilson's appointment, the mathematics department numbered eight members—none more distinguished a researcher than Fine. Undergraduate teaching loads were heavy, salaries low, and opportunities for research limited. The department had an office in the library (East Pyne Hall), but most individual faculty members had to work at home.

Fine planned to build a strong research program in mathematics slowly through appointments to young mathematicians with research promise. In 1905 he appointed the young American mathematicians Gilbert Bliss, Luther Eisenhart, Oswald Veblen, and John Wesley Young to preceptorships. In a move that was uncharacteristic of American mathematical institutions, Fine also sought to hire English-speaking European mathematicians.[6] That same year he hired James Jeans to a professorship in applied mathematics and offered a senior position to Jeans's fellow Englishman, Arthur Eddington, who declined it in favor of a post at the Greenwich Observatory. When Bliss left for Chicago and Young for Illinois in 1908,[7] Fine replaced them with the promising American mathematician G. D. Birkhoff and the young Scottish algebraist J. H. M. Wedderburn. Birkhoff left for Harvard in 1912 and was replaced in 1913 by the Parisian Pierre Boutroux. Fine added the Swede Thomas Gronwall in 1913, Princeton-born and -educated James Alexander in 1915, another young Swede, Einar Hille, in 1922, and Paris-trained Solomon Lefschetz in 1924. Thus, between 1905 and 1925 many of the young mathematicians who were to become leaders in American mathematics were members of the Princeton faculty.[8] Princeton was beginning to collect mathematical talent that rivaled that of the established world centers: Göttingen, Berlin, Paris, Cambridge, Harvard, and Chicago.

3. The 1920s

Although individual members of the mathematics faculty carried on intensive research activities, Princeton remained principally a teaching institution in the 1920s. As at most American universities during those years, the Princeton faculty was saddled with heavy undergraduate teaching loads and had little money to improve facilities or research opportunities. The European mathematicians who came to Princeton recognized this clearly. As Einar Hille remembers his first year there, in 1922-23: "Princeton was somewhat of a disappointment. There were in power old undergraduate teachers Gillespie, McInnes, Thompson. I think that during my first term there I had two divisions of trigonometry with endless homework" (Hille 1962). Solomon Lefschetz confirmed this situation:

> When I came in 1924 there were only seven men there engaged in mathematical research.[9] These were Fine, Eisenhart, Veblen, Wedderburn, Alexander, Einar Hille and myself. In the beginning we had no quarters. Everyone worked at home. Two rooms in Palmer [Laboratory of Physics] had been assigned to us. One was used as a library, and the other for everything else! Only three members of the department had offices. Fine and Eisenhart [as administrators] had offices in Nassau Hall, and Veblen had an office in Palmer. (Bienen 1970, 18-19)

The situation began to change around 1924 when an effort was made to raise funds to support mathematical research. With the turnover in the preceptorial rank and the disinterest of Wedderburn and others in institutional matters, the responsibility for building the Princeton research program devolved to Fine, Eisenhart, and Veblen. The first step was taken by Veblen during his term in 1923-24 as president of the American Mathematical Society.[10] In an effort to improve American mathematics nationally, he arranged for mathematicians to be included in the National Research Council fellowship program already established for physicists and chemists. He also established an endowment fund for the AMS and raised funds to subsidize its publications.

Within Princeton, the move to improve the research environment was spearheaded by Fine with the assistance of Eisenhart and Veblen. As dean of the sciences, Fine assumed responsibility for helping Princeton President Hibben to raise and allocate funds for research in the sciences. In a fund-raising document of 1926, Fine outlined the "means to the full realization for the purposes of the Mathematics Department":

(1) Endowment for Research Professorships.
(2) Improvement and increase of personnel with schedules compatible with better teaching and more research.
(3) A departmental research fund to meet changing conditions.
(4) A visiting professorship which might well bear the name of Boutroux [in memorium].
(5) A group of offices and other rooms for mathematical work, both undergraduate and advanced.
(6) Continued financial support for the *Annals of Mathematics.*
(7) A number of graduate scholarships. (Fine 1926)

It is instructive to compare this list of objectives outlined by Fine to a plan for an Institute for Mathematical Research proposed by Veblen in the period of 1924-26 to both the National Research Council and the General Education Board of the Rockefeller Foundation. Veblen's plan not only amplifies on the reasoning behind Fine's list, but also illustrates the greater vision of Veblen—realized in the 1930s with the founding of the Institute for Advanced Study.

Veblen's argument began with the premise that "the surest way of promoting such research [in pure science] is to provide the opportunities for competent men to devote themselves to it" (letter to H. J. Thorkelson, 21 October 1925, Veblen Papers). According to the American system, Veblen noted, this is a "by-product of teaching. The consequence has been that although our country has produced a great many men of high abilities, very few of them have an output which corresponds to their native gifts." Playing to the desire of funding organizations to build strong American research institutions, Veblen added that "men of considerably less ability have been able to do greater things in the European environment . . . [because] their time and energy have been free for the prosecution of their research." Elaborating this argument elsewhere, Veblen noted that his American colleagues taught nine to fifteen hours a week as compared with three hours by a mathematician in the College de France;[11] and that the American mathematician's primary task was the teaching of elementary subjects to freshmen and sophomores. These subjects were taught in the lycées and Gymnasia of Europe, and university research mathematicians there could concentrate on the teaching of more advanced subjects (letter to Vernon Kellogg, 10 June 1924, Veblen Papers).

The simplest solution, to Veblen's mind, was to provide research positions for mathematicians in which teaching duties would be limited or not required. But how was this to be accomplished? Veblen rejected the idea of "distinguished service professorships," which he was skeptical in general "would be held by men of high distinction, but who often would have passed the most active stage of research" ("Institute for Mathematical Research at Princeton"; undated, unsigned proposal, Veblen Papers).

Instead, he proposed an institute consisting of "a balanced group of first rank productive mathematicians who have opportunities for mathematical research comparable with opportunities ordinarily given those who conduct research, and train research workers, in the laboratory sciences" (document labeled "C. Mathematical Research," probably prepared for the Princeton Scientific Fund proposal to the General Education Board; undated, unsigned, Veblen Papers).

The institute he envisioned would consist of four or five senior mathematicians and an equal number of junior colleagues. The senior men would devote themselves "entirely to research and to the guidance of the research of younger men," though all institute members should "be free to offer occasional courses for advanced students" (letter to Vernon Kellogg, 10 June 1924, Veblen Papers). Beyond salaries, professorial needs were not very great: a library, a few offices, lecture rooms, a few computing machines, and money for stenographers and (human) computers. The institute could operate successfully, he maintained, either in conjunction with a university or independently.

Veblen's preference for an institute over individual research professorships was based on his and Eisenhart's assessment of the success of Göttingen over other German universities as a mathematical research center.

> In those cases where a good scientific tradition has been established and has subsequently broken down, it will be found that the organization was such as to depend on a single leader. The break in the tradition came when the leader died. But if instead of having a single outstanding figure, you have a group of men of different ages who are working together so that the replacements which take place are gradual, then if you have made a good start, the conservative forces inherent in such a group tend to maintain it. A good illustration of this is to be found in the mathematical tradition of Göttingen. While there have often been men of the first magnitude at Göttingen, there has always been a large group gathered together which has maintained itself so well that the prestige of the Mathematical Institute at Göttingen is, if

possible, greater now than it ever has been. During the same period of time the other German universities, which have depended for their eminence on particular individuals, have had vicissitudes of all kinds. The preeminence of Göttingen is due to the laws of statistics and the power of tradition. (Letter to H. J. Thorkelson, 21 October 1925, Veblen Papers)

As a further incentive for funding, Veblen sought to dedicate the institute to applied mathematics, a discipline he regarded as underrepresented in the United States. He pointed out that through the work of Eisenhart, Veblen, and Tracy Thomas in topology and differential geometry, Princeton had already initiated "a very definite programme" in this direction. "This programme embraces studies in the geometry of paths and analysis situs which are becoming more and more clearly the foundations of dynamics and the quantum theory" ("Institute for Mathematical Research at Princeton"; undated, unsigned proposal, Veblen Papers).

For whatever reasons, Veblen's plan for an institute was not realized at this time. No record of the National Research Council's response has been found. His proposal to the General Education Board was included as part of Princeton's general campaign to raise money to support fundamental work in the sciences, a campaign that Fine directed for President Hibben. This proposal did call for support for, among projects in astronomy, physics, chemistry, and biology, a mathematical institute with focus on applied mathematics; and it specifically referred to Veblen's earlier contacts with the board ("Memorandum for Dr. Wickliffe Rose . . . for Endowment of Research in the Fundamental Sciences"; undated, Veblen Papers). Although this grant was made, the money was not used to form the institute Veblen desired. One reason may have been the board's concern about the long-term productivity of research mathematicians:

> (1) that one cannot be absolutely sure that a man who is appointed to a research position will continue for the rest of his life to do research of a high grade, and (2) that supposing your first appointments to be of the right quality, it is not certain that this quality will be maintained through the long future. (As repeated by Veblen in letter to H. J. Thorkelson, 21 October 1925, Veblen Papers)

Although the institute was not funded, the umbrella grant for fundamental scientific research was. The board awarded Princeton $1 million on the condition that it raise twice that amount. By 1928 the university had raised the $2 million through alumni gifts, and one-fifth of the total amount ($600,000) was made available to the mathematics department. It was used to buy library materials, support the journal *Annals of Mathematics*,[12] reduce teaching loads, and pay salaries of visiting mathematicians.[13]

Most of the other objectives on Fine's list were also met. Soon after Fine became involved with the Fund Campaign Committee in 1926, he approached Thomas Jones, an old friend and former Princeton classmate who had made a fortune through a Chicago law practice and his presidency of the Mineral Point Zinc Company. Jones endowed the Fine Professorship, the most distinguished chair in American mathematics at the time. Together with his niece Gwenthalyn, Jones also provided $500,000 to the research fund and endowed three chairs, including the Jones Chair in Mathematical Physics, which was first held by Hermann Weyl in 1928-29.

Princeton was able to provide good financial support for doctoral and postdoctoral mathematicians in the late 1920s and the 1930s. It attracted more National Research Council fellows than any other U.S. university.[14] British and French students were supported by the Commonwealth and Procter Fellowship programs and American graduate students by university funds.

Of the items on Fine's list, an endowment for research professorships, a departmental research fund, a visiting professorship, support for *Annals of Mathematics*, and graduate scholarships were all met. Only two items caused difficulty: increase in personnel and housing. Both needs were met in the early 1930s.

4. Fine Hall

In the late 1920s, the University of Chicago began construction of Eckart Hall for its mathematics department.[15] Veblen, a Chicago Ph.D. with continuing ties to his alma mater, kept closely informed about the new mathematics building.[16] He recognized its potential value in nurturing a mathematics community and "so he worked on Dean Fine to have this as a goal in connection with the Scientific Research Fund" (private communication from A. W. Tucker, 1985). Although Fine understood the need for adequate space (item 6 on his list), he resisted Veblen's exhortations because he knew that money was not available for similar buildings for the other sciences. Psychology, in particular, Fine regarded as having a greater space need than mathematics.

Curiously, Fine's accidental death in 1928 made possible the realization of Veblen's plan. Within a few weeks after Fine's death, the Jones family offered funds for the construction and maintenance of a mathematics building in memory of Dean Fine. Veblen and Wedderburn were given responsibility for designing the new building. Veblen took charge and designed a building that, as Jones said, "any mathematician would be loath to leave" (*Princeton University Alumni Weekly* 1931, 113). The building was constructed of red brick and limestone in the "college Gothic" style of the universities of Paris and Oxford that Veblen so admired.[17] Fine Hall was situated adjacent to the Palmer Laboratory of Physics, with a connecting corridor to enable the physicists easy access to the library and Common Room.

Veblen attended to every last detail in the design and finishing of the building. He worked closely with a high-quality decorating firm from New York on the furnishings and insisted on extensive sound testing of the classrooms. All design features were carefully chosen to promote a research environment and communal interaction. As Veblen observed:

> The modern American university is a complicated organism devoted to a variety of purposes among which creative scholarship is sometimes overlooked. Those universities which do recognize it as one of their purposes are beginning to feel the necessity of providing centers about which people of like intellectual interests can group themselves for mutual encouragement and support, and where the young recruit and the old campaigner can have those informal and easy contacts that are so important to each of them. (*Princeton University Alumni Weekly* 1931, 112)

The top floor housed the library for mathematics and physics. An open central court provided natural lighting and quiet space with carrels for each graduate student and postdoctoral visitor. Conversation rooms with blackboards were available in each corner. Eisenhart convinced the university administration to transfer there from the main library all books of research value to mathematicians and physicists, and all researchers had 24-hour access to the collection. Subscriptions of *Annals of Mathematics* were traded to other institutions in order to build up a complete collection of research journals, including all of the major foreign journals.

On the second floor the faculty members had "studies"—not offices—some of which were large rooms lavishly appointed with fireplaces, carved oak paneling, leather sofas, oriental rugs, concealed blackboards and coat closets, and leaded windows with mathematical designs. To promote continued close ties with physics, the mathematical physicists were also assigned studies there.[18]

The second floor also housed a Common Room and a Professor's Lounge, following a tradition Veblen had admired at Oxford. Princeton faculty proved to be less interested than their Oxford colleagues in having a place to retire from their students, and the rather formal Professor's Lounge was seldom used. However, at most any time of day or night one could find graduate students, faculty members, or visitors in the Common Room discussing mathematics, playing Kriegspiel, Go, or chess, or sleeping. The social and perhaps intellectual zenith was attained each weekday afternoon when the mathematics and physics communities would gather for tea. Here twenty or thirty would meet to socialize and discuss their craft.

The first floor housed additional studies, the chairman's (Eisenhart's) office, and various classrooms sized to accommodate seminars or lectures, all with ample blackboard space, good acoustics, and carefully determined dimensions. Completing the club atmosphere was a locker room with showers to facilitate short breaks by faculty (especially Alexander) on the nearby tennis or squash courts.

Fine Hall succeeded in promoting a community atmosphere for the mathematical researchers. Because of the Depression and the rule against marriage for graduate students, many of these poor, single, male students rented small furnished rooms in town and ate meals in restaurants. The Common Room was their main living space. The many foreign faculty members, students, and visitors also congregated in Fine Hall, which they found to be a place of congeniality.[19] Faculty members regularly used their studies, which contributed to the close ties between students and staff. Between 1933 and 1939, when Fuld Hall was completed for the institute, Fine Hall also accommodated the institute's faculty and many visitors.

Today, many departments have facilities similar to, if more modest than, Fine Hall. But this environment, unusual for its time, promoted a sense of community that was impossible to foster at many universities, like Harvard, where the faculty was scattered across the campus without a common meeting place.[20] Albert Tucker, a graduate student and later a faculty member at Princeton in the 1930s, describes the importance of Fine Hall to his career:[21]

It was the amenities of Fine Hall that certainly caused me to come back to Princeton in 1933, after the year that I had my National Research Council Fellowship. I could have had a second year as a Fellow, just for the asking; but if I had, I would be supposed to spend it somewhere else than Fine Hall because that was where I had gotten my doctor's degree. The experiences that I had had at Cambridge, England, but particularly at Harvard and Chicago, made me long for the comforts, the social atmosphere, the library convenience of Fine Hall. Of course there were other things, Lefschetz, Eisenhart, and so on. But at the time, [when] Marston Morse told me I was a fool not to take the instructorship at Harvard . . . it was the opportunity to be in the Fine Hall community [that mattered]. (Oral history, 11 April 1984; The Princeton Mathematical Community in the 1930s Oral History Project, Princeton University Archives)

Between the time of the Jones gift in 1928 and the completion of Fine Hall in the fall of 1931, mathematics buildings opened at Chicago, Paris, Göttingen, and Jena. Yet, as one of the earliest and certainly the most successful building for this purpose, Fine Hall served frequently as an architectural model.[22] For example, Dartmouth, Wisconsin, Arizona State, and Western Australia modeled their mathematics buildings after Fine Hall.

5. The Institute for Advanced Study

By the time the department moved into Fine Hall in 1931, the conditions were favorable for research. Of Fine's original list of objectives only item 2, improvement and increase of personnel with schedules compatible with better teaching and more research, was unfulfilled. This was soon to be met by the founding of the Institute for Advanced Study.

The story of the institute is well known and need not be told in detail here.[23] Money from the Bamberger and Fuld family fortunes was donated in 1930 to endow an institute for advanced research to be situated in the state of New Jersey or a contiguous area. Abraham Flexner, retired from the General Education Board and the original proponent of this advanced research institute, was appointed as director. Flexner chose Princeton as the site because he believed its rural environment was suitable to pure scholarly endeavor and because the university possessed a good research library. He decided to focus the institute's activities initially in a single area, and he chose mathematics as that area for three reasons:

(1) It was fundamental.
(2) It required the least investment in plant or books.

(3) It had become obvious to me [Flexner] that I could secure greater agreement upon personnel in the field of mathematics than in any other subject. (Flexner 1940, 235)

Between 1930 and 1932, Flexner toured Europe and North America discussing his plans with leading scholars. Veblen impressed Flexner with his counsel,[24] and, once mathematics was chosen in 1932 as the institute's first mission, Veblen was given the first faculty appointment. On an earlier trip to Europe, Flexner had discussed with Albert Einstein and Hermann Weyl the prospect of their joining the institute faculty. Veblen assumed responsibility for selecting the other original faculty members of the institute's School of Mathematics. James Alexander, John von Neumann, Einstein, and Weyl joined the institute in 1933 and Marston Morse in 1935.

Flexner and Veblen assembled an impressive international group of research mathematicians. Alexander was a distinguished topologist. Einstein was already world-renowned for his contributions to theoretical physics. Morse was an accomplished Harvard mathematician known for his research in "analysis in the large." Veblen was the senior American geometer. Von Neumann was a brilliant young mathematician who had already made major contributions to logic, quantum mechanics, and analysis. Weyl, who was regarded as having the widest range of mathematical knowledge since Poincaré, was Hilbert's successor in the mathematics chair at Göttingen. The institute had many visitors each year and a few research associates (known as "permanent members") like Kurt Gödel, but these six constituted the regular faculty of the School of Mathematics for about ten years.[25]

Although the research environment was undoubtedly an attraction, social factors also contributed to their decisions to accept appointments. Alexander, whose family wealth freed him from the need to work, was relieved at the release from teaching responsibility that the institute position offered. The Nazis drove Einstein, von Neumann, and Weyl from Europe: von Neumann lost his position at Berlin for being Jewish,[26] and Weyl feared for his Jewish wife and found the Nazi interference in the affairs of the mathematics institute at Göttingen intolerable. Morse had a better opportunity at the institute to forward his research program through the availability of funds for visiting postdoctoral researchers, and it must also have been some comfort to escape from the Harvard environment where his wife had recently divorced him to marry another of the Harvard mathematicians.

During the 1932-33 academic year, Veblen rented office space for the institute in a commercial building near the Princeton campus while retaining his research study in Fine Hall. President Hibben was sympathetic to the institute's goals and offered a five-year lease on space in Fine Hall, which had been built with expansion of the department in mind and therefore exceeded current space needs. The offer was gratefully accepted and the institute settled into Fine Hall the following year.

In the early years, it was uncertain what direction the institute might take. The charter provided for it to become an educational institution and grant doctoral degrees—in competition with the university. Others may have envisioned it as an "Ivory Tower" where distinguished mathematicians could pursue their research without distraction. Veblen steered a third course, however, emphasizing both postdoctoral education and research. The institute's School of Mathematics realized the plans for an Institute for Mathematical Research that Veblen had outlined in 1925, with the exception that applied mathematics was only the principal, not the exclusive, subject of research.

Veblen envisioned a place where promising young Ph.D.s and established research mathematicians could interact and pursue common mathematical research interests without interruption by undergraduate teaching or other routine faculty duties. To that end, he arranged for adequate funds for visitors, including funds for many young Ph.D.s to work as research assistants of the permanent institute faculty.

There was free interchange between institute and university personnel. Following Veblen's example, institute faculty members offered advanced seminars open to both institute and university faculty members, visitors, and students. For example, when Gödel presented his famous institute lectures on incompleteness in 1934, graduate students Stephen Kleene and J. Barkley Rosser prepared the course notes for distribution. Institute members were commonly employed as informal advisors of doctoral dissertations, and during this period, when both university and institute mathematicians had offices in Fine Hall, it was difficult to determine the official affiliation of visitors.

6. Growth of the Department

The founding of the institute could have seriously weakened the university's mathematics program. With Veblen, Alexander, and von Neumann accepting positions at the institute and Hille leaving for a professorship at Yale in 1933, representing over a third of the faculty and a greater percentage of those productive in research, the university's strength in mathematics was seriously threatened.[27] Both the institute and the university recognized the advantages of cooperation. Both gained from having a larger community of permanent and visiting mathematicians. The university retained the services of Alexander, Veblen, and von Neumann at least to the degree that they were available to consult, help direct theses, referee articles for *Annals of Mathematics*, and present advanced seminars.[28] The institute gained in return a good research library and excellent physical facilities.

Nonetheless, the department confronted a major rebuilding project. The able leadership of Eisenhart and Lefschetz carried the department through this tumultuous period. As department chairman and dean of the faculty, Eisenhart was in a position to facilitate close ties with the institute, which he supported as favorable to the university. When a new dean of the faculty was appointed, the position of dean of the graduate school was assigned to Eisenhart.[29] Although his greatest interest had always been in the undergraduate mathematics program, he devoted himself to his new duties and was able to use his position within the university to secure adequate resources for the graduate program in mathematics.

Lefschetz was appointed to the Fine Research Professorship vacated by Veblen. Although this position did not require teaching, Lefschetz followed Veblen's example and taught a research seminar every semester, which was well attended by the graduate students. Lefschetz was also departmental representative to the Committee on the Graduate School, which enabled him to further protect the interests of the graduate mathematics program. It also stimulated him to take a personal interest in the graduate students: he meticulously inspected graduate applications to select the ten to twenty percent that he believed could handle the demanding but unstructured graduate program. He went to great lengths to meet each graduate student and monitor his progress, and he was adept at identifying dissertation problems of appropriate content and difficulty. He directed many dissertations, including those of Hugh Dowker, William Flexner, Ralph Fox, Paul Smith, Norman Steenrod, Albert Tucker, John Tukey, Robert Walker, and Henry Wallman in the period 1925-39. While Eisenhart administered the department and the relations with the rest of the university, Lefschetz built up the research and graduate programs.

After considering senior appointments to outsiders, the decision was made to continue the Fine approach and replace the departing senior mathematicians with promising junior researchers.

It is the desire of this department to maintain its distinction by giving constant attention to the question of its personnel. Because we publish the *Annals of Mathematics*, we know rather well who the good men of the country are. Furthermore, because of the presence here of the Institute for Advanced Study, we have an additional means of keeping in contact with this situation.

Salomon Bochner, H. F. Bohnenblust, E. J. McShane, Albert Tucker, and Samuel Wilks were given junior appointments in 1933, as were Norman Steenrod, Walter Strodt, E. W. Titt, and C. B. Tompkins later in the decade.

The department continued to concentrate in a few select fields and did not try to provide uniform coverage of all areas of mathematics. Research in geometry and topology was carried out by Eisenhart, Lefschetz, Tracy Thomas, Morris Knebelman, Tucker, and Steenrod in the university and by Veblen and Alexander in the institute; mathematical physics by Eugene Wigner and H. P. Robertson in the university and by Weyl, von Neumann, and Einstein in the institute; mathematical logic by Church in the university and Gödel in the institute; and analysis by Bohnenblust, Bochner, and McShane in the university and by von Neumann and Weyl in the institute. A new area of research was initiated with the hiring of S. S. Wilks, one of the pioneers in mathematical statistics.

Measured by number or productivity of postgraduate mathematicians, the Princeton mathematical community excelled in the 1930s. Between 1930 and 1939 the university produced 39 mathematics Ph.D.s, including (with others already mentioned) Carl Allendoerfer, John Bardeen, J. L. Barnes, A. L. Foster, Wallace Givens, Robert Greenwood, Israel Halperin, Banesh Hoffmann, Nathan Jacobson, Malcolm Robertson, Ernst Snapper, Abraham Taub, Alan Turing, J. L. Vanderslice, J. H. C. Whitehead, and Shaun Wylie. Between 1923 and 1941, the years of the NRC Fellowship program in mathematics, 59 mathematicians visited Princeton, including (among others already mentioned) A. A. Albert, Gustav Hedlund, Derrick Lehmer, Neal McCoy, Deane Montgomery, Charles Morrey, Hassler Whitney, and Leo Zippin. One hundred eighty-nine mathematicians were visitors to the institute between 1933 and 1939, notably Reinhold Baer, Valentine Bargmann, Paul Bernays, Garrett Birkhoff, Eduard Cech, A. H. Clifford, P. A. M. Dirac, Witold Hurewicz, Deane Montgomery, Marshall Stone, Stanislaw Ulam, Andre Weil, and Oscar Zariski.

7. Conclusion

Institutional factors clearly helped to shape the development of the mathematics programs in Princeton in the 1930s. Careful planning by Fine, Eisenhart, and Veblen over the preceding quarter-century placed Princeton in a position to establish a world-class center of mathematics once funds started to become available in the late 1920s.

Funding for these purposes was fairly easily acquired. It may have been fortunate that the Fuld and Bamberger families were willing to endow the Institute in 1930, but it was no accident that Veblen was ready with a strong plan to build it in Princeton and to devote it to mathematical research. It is clear that the trend in the 1920s of the great foundations to support American scholarship benefited Princeton mathematics. However, it must be remembered that the bulk of the money for the university's program in mathematics came through alumni gifts. The ease at raising these matching funds is perhaps indicative of the general wealth of the United States in the late 1920s.

The existing strength of the department and a workable plan for the future were undoubtedly strong factors in attracting financial support. Several principles were consistently applied by Fine, Veblen, and others over the first forty years of the century to build excellence in the department and later in the institute. Foremost was the emphasis on research, as demonstrated in appointments and promotions, training offered to graduate students, teaching loads, and many other ways. This ran counter to the well-established tradition of American colleges as undergraduate teaching institutions. Second was the attempt to build a community of mathematical researchers so that the "old campaigner" and the "young recruit" could exchange ideas. Third was the concentration on a few areas (topology, differential geometry, mathematical physics, and logic), instead of attempting to provide uniform coverage across all of mathematics. Fourth was the adoption of an international perspective. More than any other American university in the period 1905-40, Princeton sought out students, visitors, and faculty from around the world. When the Nazi peril disrupted European mathematics, Veblen and Weyl led the way in placing émigré mathematicians in American institutions, including some in Princeton (Reingold 1981). Fifth was the decision to build up a research community through the cultivation of young mathematical talent. Although the institute took great advantage of the Nazi situation in attracting Einstein, von Neumann, and Weyl to its faculty, most of the Princeton staff was hired at the junior level and promoted from within.

Finally, great attention was given to environmental factors that would affect the research community. Prominent among these was Fine Hall. It is hard to overemphasize the importance Princeton mathematicians of the 1930s attached to their physical quarters. Soon universities throughout the world came to recognize the value of a place where their mathematicians could gather to discuss mathematics, with excellent support facilities. Another factor was the editing of professional journals (*Annals of Mathematics* and *Studies, Journal of Symbolic Logic*, and *Annals of Mathematical Statistics*) at Princeton. They provided the faculty and their students with an outlet for research and gave the faculty some control over the direction of American research. These journals also provided extensive contacts with the wider mathematical community and a vehicle for scouting new talent for appointments. Financial support for graduate students and visitors and for reduced teaching loads of staff also promoted the growth of a large community focused on mathematical research.

The success in Princeton is even more remarkable when it is considered that it occurred at the same time as the Great Depression and the growth of Nazism. General economic circumstances severely depressed academic salaries, limited funds for graduate and postdoctoral support, and restricted job placement for Princeton Ph.D.s and junior faculty. The political disruption of European academics resulted in an influx of European mathematicians into the United States, further straining the appointment and promotion of American-bred and -trained mathematicians.

Notes

1. Others were even more lavish in their praise of Princeton: R. C. Archibald wrote of Princeton as "the greatest center of mathematical activity in this country" (Archibald 1938, 169); the Danish mathematician Harald Bohr referred to Princeton as "the mathematical center of the world" when addressing an international scientific audience in 1936 (Chaplin 1958).

2. This paper about Princeton tells part of a larger story of the emergence of mathematics research in U.S. institutions in the period 1875-1940. The full story involves the rise of mathematics at Brown, Chicago, Clark, Johns Hopkins, Harvard, and Yale universities in the first half of this period and at Berkeley, MIT, Michigan, Stanford, and Wisconsin near the end. Harvard and Chicago, in particular, have many parallels with Princeton. Some information on this topic can be found in (Archibald 1938; Birkhoff 1977; Bocher et al. 1911; Lewis 1976; Reid 1976). Dr. Uta Merzbach of the Smithsonian Institution is preparing a history of American mathematics and mathematical institutions.

In this paper, I have focused on social and institutional issues. I am planning additional papers on the contributions in the 1930s of Princeton mathematicians to topology and logic.

3. Halsted was widely influential in the early development of American mathematics, e.g. inspiring Leonard Dickson and R. L. Moore as well as Fine (Birkhoff 1976; Lewis 1976).

4. Fine and Wilson were lifelong friends, having first become acquainted while working on the editorial board of the student newspaper, the *Princetonian*.

5. Harvard adopted the preceptorial system in 1910, using it to finance graduate students and train them to teach, rather than hiring additional junior faculty members.

6. Other distinguished European research mathematicians that came to the United States in its formative period include J. J. Sylvester at Johns Hopkins, R. Perrault at Johns Hopkins and Clark, and Oskar Bolza and Heinrich Maschke at Clark and Chicago. The number of foreign mathematicians willing to accept appointments in the United States was small. Many of the senior American mathematicians in this period were not distinguished researchers. Thus, Fine's appointment strategy appears sound.

7. Bliss became a leading figure in the Chicago department and in American mathematics generally. After two years at the University of Illinois and a year as mathematics department chairman at the University of Kansas, Young devoted many years to building up the mathematics program at Dartmouth.

8. Note the interesting, but perhaps coincidental, pattern of junior appointments at the university: 1905-10, American, trained elsewhere; 1910-25, mostly European; 1925-40, mostly American, several having been trained at Princeton. Princeton was not the first experience with U.S. institutions for some of these young European mathematicians. For example, Wedderburn first spent a year at Chicago and Hille a year at Harvard.

9. By the standards of the major European centers of the time, or of major American universities when Lefschetz made this comment in 1970, seven was not a large number of research mathematicians for an institution. But few other American universities, if any, had that large a number in 1924.

10. Veblen's interest in fund-raising dates from after World War I, perhaps stemming from his wartime administrative experience at Aberdeen Proving Grounds in Maryland.

11. At most American universities in the 1920s, mathematicians taught twelve or more hours per week. According to Garrett Birkhoff, it was considered a great coup at Harvard in 1928 when the weekly load for a mathematician was reduced to $4\frac{1}{2}$ hours of lectures, $1\frac{1}{2}$ hours of theoretical tutoring, and 3 hours of graduate student supervision (private correspondence, 4 October 1985).

12. Princeton had assumed editorial responsibility for *Annals of Mathematics* in 1911. Previously it was edited at Harvard, and before that at the University of Virginia.

13. Long-term visitors to the department in the late 1920s and 1930s included: Paul Alexandroff and Heinz Hopf (1927-28); G. H. Hardy (1928-29); Thornton Fry, John von Neumann, and Eugene Wigner (1929-30, the last two returning in subsequent years); J. H. Roberts and J. H. van Vleck (1937-38); and C. Chevalley (1939-40).

14. Veblen was one of three members of the NRC Fellowship selection committee for mathematics and was thus positioned to assist Princeton mathematics. This arrangement is characteristic of the organization of American mathematics in the 1920s and 1930s, where the power was concentrated in a small number of individuals, including G. D. Birkhoff of Harvard, G. A. Bliss of Chicago, R. G. D. Richardson of Brown, Veblen, and perhaps a few others.

15. Increased wealth in the United States in the latter 1920s redounded on the universities. Chicago's ability to construct Eckart Hall, Harvard's reduction of the teaching load, and Princeton's ease at matching the Rockefeller grant through alumni contributions all indicate this improved economic condition.

16. Veblen was at the University of Chicago from 1900 to 1905, receiving his Ph.D. in 1903 under the direction of E. H. Moore. Veblen's drive to build up American mathematics may have been stimulated by his experience at the University of Chicago, where Moore had built a strong program that produced many prominent research mathematicians, including L. E. Dickson, G. A. Bliss, G. D. Birkhoff, and Veblen, as well as supplying several major midwestern universities with their mathematics chairmen.

17. Veblen drew many ideas for Fine Hall from a visit to Oxford in 1928-29. As Savilian Professor, G. H. Hardy was expected to lecture occasionally on geometry. To avoid this responsibility, he exchanged positions that year with Veblen, whose principal mathematical interest was geometry.

18. Physics and mathematics had shared quarters earlier in Palmer Laboratory. Pro-

fessor Condon of physics was a supporter and advisor on the planning of Fine Hall; there are many references to Condon's role in the Oswald Veblen Papers.

19. Weyl reported to his former colleagues in Göttingen that German was spoken as much as English in the institute, then located in Fine Hall (Reid 1976, 157). Garrett Birkhoff reports that in the latter 1930s the official language of the institute was jokingly said to be "broken English" (private communication, 24 October 1985).

20. Harvard did have a common room, Room O, in Widener Library, but it did not have the facilities or receive the use that Fine Hall did.

21. Although Tucker may attach more significance to the environment of Fine Hall than others might care to, more than twenty mathematicians commented on the amenities of Fine Hall in the Princeton Mathematics Community of the 1930s Oral History Project.

22. The Conference Board of the Mathematical Sciences used Fine Hall as a model in a book on mathematical facilities (Frame 1963).

23. The most accessible account is in (Flexner 1940, chap. 27 and 28).

24. Flexner and Veblen held similar views about education and research. Both were enamored with the European university systems (see Flexner 1930), and both felt a need "to emphasize scholarship and the capacity for severe intellectual efforts" (Flexner 1927, 10). Both saw a need for an environment where the research faculty would be free from "routine duties . . . —from administrative burdens, from secondary instruction, from distracting tasks undertaken to piece out a livelihood" (Flexner 1930, 10-11). Both saw the research institute as a "specialized and advanced university laboratory" (Flexner 1930, 35). However, they disagreed on the addition of schools of study to the institute, support of European mathematicians, a new institute building off the university campus, administrative responsibilities for institute faculty, and other matters (see Reingold and Reingold 1982, chap. 13).

25. There is little firm evidence of who else was considered or offered an original appointment at the institute. An offer to G. D. Birkhoff was declined. Harvard countered the offer by making Birkhoff a Cabot Fellow. The address pages of Veblen's 1932 diary list five groupings of names that may have been candidates for institute positions or people whose advice Veblen sought about candidates during a 1932 tour of Europe and North America. The names (separated by semicolons the way Veblen grouped them) are as follows: Dirac, Artin, Lefschetz, Morse; Alexandroff, Wiener, Kolmogorov, von Neumann; Albert, R. Brauer, Gödel, Douglas; Bernays, Peterson, Kloosterman, Heyting, Chapin; Deuring, McShane, Whitney, Mahler.

26. Von Neumann was teaching half-time at Berlin and half-time at Princeton. After wavering for a considerable time, the university offered him a full-time position. He chose the appointment at the institute instead.

27. The 1933-34 directory of members of the mathematics department lists professors L. P. Eisenhart, W. Gillespie, S. Lefschetz, J. H. M. Wedderburn, and E. P. Wigner; associate professors H. P. Robertson and T. Y. Thomas (on leave); assistant professors H. F. Bohnenblust, A. Church, and M. S. Knebelman; and instructors E. G. McShane, J. Singer, A. W. Tucker, and S. S. Wilks. There were others listed as part-time instructors, research assistants, and advanced fellows.

28. Additional information about mathematics journals edited at Princeton in the 1930s and their contributions to the research community can be found in an oral history I conducted with Albert Tucker on 13 April 1984 (Princeton Mathematical Community in the 1930s, Oral History PMC 34).

29. In 1932-33, Eisenhart headed the Princeton administration as dean of the faculty— between the death of President Hibben and the appointment of President Dodds. Dodds appointed Eisenhart to replace the physicist Augustus Trowbridge as dean of the Graduate School and made Robert Root the new dean of the faculty.

Bibliography

The following bibliography includes only published and widely distributed printed materials. Biographies of the many distinguished Princeton mathematicians and physicists discussed herein are excluded, except those that have been quoted. *Bulletin of the American Mathematical Society*, *Dictionary of Scientific Biography*, *Proceedings of the National Academy of Sciences*, and *Princeton Alumni Weekly* have carried biographies of many of these mathematicians.

Two fertile archival sources for this research are the Oswald Veblen Papers and the Princeton Mathematical Community in the 1930s Oral History Project. The Veblen Papers are held by the Manuscripts Division of the Library of Congress, Washington, D.C. The Oral History Project records are held by the Seeley G. Mudd Manuscript Library of Princeton University. Copies are on deposit at the American Philosophical Society in Philadelphia and the Charles Babbage Institute, University of Minnesota, Minneapolis.

Archibald, R. C. 1938. *A Semicentennial History of the American Mathematical Society: 1888-1938.* Semicentennial Publications, vol. 1. Providence, R. I.: American Mathematical Society.

Bienen, Leigh Buchanan. 1970. Notes Found in a Klein Bottle. *Princeton Alumni Weekly* (April 21), pp. 17-20.

Birkhoff, Garrett. 1977. Some Leaders in American Mathematics: 1891-1941. In *The Bicentennial Tribute to American Mathematics 1776-1976.* ed. J. Dalton Tarwater. Providence, R. I.: American Mathematical Society.

Bocher, M., Curtiss, D. R., Smith, P. F., and Van Vleck, E. B. 1911. Graduate Work in Mathematics. *Bulletin of the American Mathematical Society* 18: 122-37.

Chaplin, Virginia. 1958. A History of Mathematics at Princeton. *Princeton Alumni Weekly* (May 9).

Eisenhart, Katherine. 1950. Brief Biographies of Princeton Mathematicians and Physicists. Unpublished, Princeton.

Eisenhart, L. P. 1931. The Progress of Science: Henry Burchard Fine and the Fine Memorial Hall. *Scientific Monthly* 33: 565-68.

Fine, H. B. 1891. *The Number System of Algebra Treated Theoretically and Historically.* Boston: Lench, Showall, and Sanborn.

——. 1892. Kronecker and His Arithmetical Theory of the Algebraic Equation. *Bulletin of the New York Mathematical Society* 1: 173-84.

——. 1905. *A College Algebra.* Boston: Ginn.

——. 1914. An Unpublished Theorem of Kronecker Respecting Numerical Equations. *Bulletin of the American Mathematical Society* 20: 339-58.

——. 1926. The Role of Mathematics. Limited distribution pamphlet (unsigned). The Princeton Fund Committee, Princeton University Press.

——. 1927. *Calculus.* New York: Macmillan.

Fine, H. B., and Thompson, Henry Dallas. 1909. *Coordinate Geometry.* New York: Macmillan.

Flexner, Abraham. 1927. *Do Americans Really Value Education?* The Inglis Lecture. Cambridge, Mass.: Harvard University Press.

——. 1930. *Universities: American, English, German.* New York: Oxford University Press.

——. 1940. *I Remember: The Autobiography of Abraham Flexner.* New York: Simon and Schuster.

Frame, J. Sutherland. 1963. *Buildings and Facilities for the Mathematical Sciences.* Washington, D.C.: Conference Board of the Mathematical Sciences.

Grabiner, Judith V. 1977. Mathematics in America: The First Hundred Years. In *The Bicentennial Tribute to American Mathematics 1776-1976*, ed. J. Dalton Tarwater. Providence, R.I.: American Mathematical Society. 9-25.

Hille, Einar. 1962. In Retrospect. Unpublished notes dated 16 May, Yale University Mathematics Colloquium.

Lewis, Albert. 1976. George Bruce Halsted and the Development of American Mathematics. In *Men and Institutions in American Mathematics*, ed. J. Dalton Tarwater, John T. White, and John D. Miller. Lubbock, Tex.: Texas Tech Press, Graduate Studies 13, pp. 123-30.

Montgomery, Deane. 1963. Oswald Veblen. *Bulletin of the American Mathematical Society* 69: 26-36.

Princeton University Alumni Weekly. 1931. A Memorial to a Scholar-Teacher. 30 October, pp. 111-13.

Reid, Constance. 1976. *Courant in Göttigen and New York*. New York: Springer-Verlag.

Reingold, Nathan. 1981. Refugee Mathematicians in the United States of America, 1933-1941: Reception and Reaction. *Annals of Science* 38: 313-38.

Reingold, Nathan, and Reingold, Ida H., eds. 1982. *Science in America: A Documentary History, 1900-1939*. Chicago: University of Chicago Press.

Richardson, R. G. D. 1936. The Ph.D. Degree and Mathematical Research. *American Mathematical Monthly* (April), pp. 199-204.

Stone, Marshall H. 1976. Men and Institutions in American Mathematics. In *Men and Institutions in American Mathematics*, ed. J. Dalton Tarwater, John T. White, and John D. Miller. Lubbock, Tex.: Texas Tech Press, Graduate Studies 13, pp. 7-30.

Veblen, Oswald. 1929. Henry Burchard Fine—In Memoriam. *Bulletin of the American Mathematical Society* 35: 726-30.

Topology and Logic at Princeton[1]

SAUNDERS MAC LANE

In the period 1920–1935 it was the general opinion that the three leading departments of mathematics in the United States were (in alphabetical order) Chicago, Harvard, and Princeton. Since I have never been at Princeton except for visits, this attempt to describe some aspects of Princeton's eminence will be formulated in terms of activities there in just two fields in which I have knowledge: algebraic topology and mathematical logic.

Princeton started as a men's college and did not become a university until about 1902. When Woodrow Wilson, in 1902, became president of the university, he decided to improve the teaching and research by appointing a number of young men as "preceptors". In mathematics this led to the appointment at Princeton of three young mathematicians who had recently earned the Ph.D. at the University of Chicago, George D. Birkhoff (at Princeton 1909–1911), Robert Lee Moore (at Princeton 1906–1908), and Oswald Veblen (preceptor, 1905–1910). (All three were later presidents of the American Mathematical Society.) Birkhoff and Moore subsequently left Princeton, but Veblen stayed there. (At the time, H. B. Fine (1858–1928) was perhaps the most influential senior member of the department; he was dean from 1903 to 1928.)

From Chicago, Veblen had developed an active interest in geometry; for example, he had written a splendid two-volume work on projective geometry (the first volume in collaboration with J. W. Young). In topology, the Jordan curve theorem was of considerable interest, in part because of its use in one version of the Cauchy integral theorem. However, none of the published proofs were complete and correct; in 1905 Veblen had provided the first rigorous proof of the Jordan curve theorem.

Poincaré in his study of the topology of manifolds had defined the Betti numbers and the torsion coefficients of a manifold by subdividing the manifold into cells (for example, into simplices) and then determining the Betti

---

[1] (This is a portion of an MAA/AMS Invited Lecture given at the Centennial Celebration of the AMS in Providence, August 11, 1988.)

217

numbers from the incidence coefficients of the cells. It seemed clear that the resulting numbers depended only on the manifold and so should be independent of the choice of the cellular decomposition, but this had not been proved. To get this result, it was hoped that one could prove the so-called "Hauptvermutung der Topologie" — roughly speaking, that any two subdivisions of the same manifold could be further refined so as to be combinatorially isomorphic, and hence so as to give the same Betti numbers. However, this conjecture turned out to be very difficult to prove. In 1915, J. W. Alexander (at Princeton, once a student of Veblen) gave a proof of the invariance of the Betti numbers and the torsion coefficients of a manifold — but without proving the Hauptvermutung (Alexander's insight enabled him to bypass this hard question). Instead, he took two simplicial subdivisions and subjected each to repeated barycentric subdivisions so that each could be deformed into the other by a suitable simplicial approximation. This proved the invariance; it was a major advance; it remained for many years the method of choice for the invariance proof. (At the time, it may not have been fully understood in Europe, as may be suggested in the cursory review in the *Jahrbuch über die Forschritte der Mathematik*, **45** (1915), pp. 728–729, written by the famous differential geometer Wilhelm Blaschke.)

In 1916, Veblen gave the Colloquium Lectures for the AMS, choosing as his subject analysis situs, that is, in the present terminology, algebraic topology. This book, published in the colloquium series in 1922 (second edition, 1931), was the first adequate presentation in book form of the Poincaré (and Brouwer) description of the combinatorial topology of a polyhedron. It gave a careful definition of the incidence matrices and their reduction to determine the connectivity numbers (i.e., the Betti numbers modulo 2) as well as the torsion coefficients and the Betti numbers themselves. At that time, group theory was not used here, so these numbers were not treated as the invariant of homology groups. Veblen's book also contained an exposition of the fundamental group (the Poincaré group) of a polyhedron.

The Jordan curve theorem stated that a simple closed curve divides the plane into two connected regions, and so relates the Betti numbers of the curve to those of its complement. There had been a search for higher-dimensional generalizations. It was J. W. Alexander who provided the decisive result (in 1923), that for a polyhedron Q in $\mathbf{R}^n$ one has $\beta^{n-1}(Q) = \beta^0(\mathbf{R}^n - Q) - 1$, $\beta^i(Q) = \beta^{n-i-1}(\mathbf{R}^n - Q)$, $i \neq n - 1$. This is known as the Alexander duality.

In the next year, Alexander discovered his notable "horned sphere" — which is described and pictured in the article by Hassler Whitney in the first volume of *A Century of Mathematics in America*.

Solomon Lefschetz, who had earned his Ph.D. degree at Clark University and had become a professor at the University of Kansas, had made penetrating studies of the use of Betti numbers in algebraic geometry. In 1923 he came to Princeton University. There in 1924 he discovered his fixed point index.

Algebraic topology was essentially a new (20th century) branch of mathematics, and this sequence of fundamental discoveries at Princeton made the University perhaps the world leader in this field: Veblen's book, Alexander's invariance proof, his duality and his horned sphere, plus the Lefschetz fixed point index. In 1926–1927, two young European topologists, Paul Alexandroff from Moscow and Heinz Hopf from Berlin, had Rockefeller fellowships for study abroad — and came to Princeton for the year. It was there that Hopf found his well-known extension of the Lefschetz index (when he later lectured on this in Göttingen, it was Emmy Noether who persuaded him that the proof could be more perspicuous if done with homology groups instead of Betti numbers).

The lively atmosphere in Princeton topology stimulated many graduate students. Thus Veblen's students included T. Y. Thomas (in differential geometry) and J. H. C. Whitehead, who subsequently, as Wayneflete Professor at Oxford, trained a whole generation of British topologists. Notable students of Lefschetz at about this time included P. A. Smith, A. W. Tucker, Norman E. Steenrod, Henry Wallman, and C. H. Dowker. In 1928, Alexander discovered his famous knot polynomial, while in 1930 Lefschetz published his influential (if somewhat obscure) book; since Veblen had preempted the classical title *Analysis Situs*, Lefschetz had to choose a new title — hence *Topology*.

This brilliant initial start of topology at Princeton has continued to this day.

Now I turn to logic at Princeton. In 1924, in his retiring address as President of the AMS, Oswald Veblen said:

> The conclusion seems inescapable: that formal logic has to be
> taken over by the mathematicians. The fact is that there does
> not exist an adequate logic at the present time, and unless the
> mathematicians create one, no one else is likely to do so.

Unlike many words in a retiring address, this statement led to results. Veblen himself had (at least) two Ph.D. students in logic; one was A. A. Bennet, subsequently a professor at Brown University, where he was influential in helping to form the Association for Symbolic Logic. Another Veblen student in logic was Alonzo Church who in 1927 wrote a thesis on "Alternatives to Zermelo's Assumption" (i.e., the axiom of choice). Church stayed on at Princeton, where he had a number of Ph.D. students who made notable contributions to logic: in 1934 J. Barkely Rosser and Stephen C. Kleene, in 1945 Leon Henkin, and in 1959 Dana Scott. Church himself discovered the λ-calculus (now extensively used in LISP and other computer science languages) and formulated his famous thesis, which asserts in effect that the intuitive notion "effectively computable" is captured by the equivalent formal notions "recursive = Turning computable = λ-definable".

Among the main subsequent developments in logic at Princeton, I note here only the early presence of Alan Turing and the coming of Kurt Gödel to the Institute for Advanced Study, where Oswald Veblen was by then a leading figure. It is evident that Veblen had made good on his call to have mathematicians take up the study of formal logic.

Algebraic topology and mathematical logic are two examples of the remarkable leadership at Princeton in mathematics; I will not attempt to summarize work there in analysis or in relativity theory, or in differential geometry, but I do note the extraordinary quality of Princeton algebraists: In succession, J. H. M. Wedderburn, Claude Chevalley, and Emil Artin.

In Princeton, the mathematics department was long housed in elegant offices in Fine Hall — when there was lively discussion of all sorts of mathematics (and of games) at tea. In some ways, the bubbly enthusiasm of Solomon Lefschetz set an atmosphere of active discussion. For example, in 1940 when he was writing his second book on topology, he sent drafts of one section up to Whitney and Mac Lane at Harvard. The drafts were incorrect, we wrote back saying so — and every day for the next seven or eight days we received a new message from Lefschetz, with a new proposed version. It is no wonder that the local ditty about Lefschetz ran as follows:

> Here's to Lefschetz, Solomon L
>
> Irrepressible as hell
>
> When he's at last beneath the sod
>
> He'll then begin to heckle God.

Veblen, on the other hand, was older and more restrained — he loved British tweed jackets. So

> The only mathematician of note
>
> who takes four buttons to button his coat.

Finally, Alexander was shy. It is reported that once, when an unwelcome visitor knocked on his office door on the first floor of Fine Hall, Alexander simply stepped out the window.

It was a remarkable group.

References

J. W. Alexander, A proof of the invariance of certain constants in Analysis Situs. *Trans. Amer. Math. Soc.* **16** (1915), 145–154. A proof and extension of the Jordan-Brouwer separation theorem. *Trans. Amer. Math. Soc.* **23** (1923), 333–349. Topological invariants of knots and links. *Trans. Amer. Math. Soc.* **30** (1928), 275–300.

W. Aspray, The emergence of Princeton as a world center of modern mathematics 1896–1939, 346–366, *History and Philosophy of Modern Mathematics*, W. Aspray and Ph. Kitcher (editors), University of Minnesota Press, Minneapolis, 1988.

Church, Alonzo, The calculi of λ-conversion. *Annals of Math. Studies*, No. 6, Princeton University Press, Princeton, 1945, 82 pp.

Hopf, Heinz, Zur Algebra des Abbildungen von Mannigfaltigkeiten. *J. Reine Angew. Math.* **163** (1930), 77–88.

Lefschetz, S., *Topology*. American Math. Soc. Colloquium Publications, vol. 12, New York, 1930, 410 pp.

Mac Lane, S., Oswald Veblen (Biographical Memoirs, vol. 37). Nat. Acad. Sci., 1964, 325–341. Topology becomes algebraic with Vietoris and Noether. *J. Pure Appl. Algebra* **39** (1986), 305–307.

Veblen, Oswald, Theory of plane curves in nonmetrical analysis situs. *Trans. Amer. Math. Soc.* **6** (1905), 83–90. *Projective Geometry*, 2 vols (the first with J. W. Young) Ginn and Company, Boston, 1910–1918. Analysis Situs. Amer. Math. Soc. Colloquium Publications, vol. 5, part 2, 1922, 150 pp; 2nd edition 1931, 199 pp.

Whitney, Hassler, Moscow 1935: Topology Moving Toward America, 97–117, *A Century of Mathematics in America*, Part I, edited by Peter Duren, et al., American Mathematical Society, Providence, 1988.

Gian-Carlo Rota was born in Italy, where he went to school through the ninth grade. He attended high school in Quito, Ecuador, and entered Princeton University as a freshman in 1950. Three years later, he graduated summa cum laude and went to Yale, where he received a Ph.D. in 1956 with a thesis in functional analysis under the direction of J. T. Schwartz. After a position at Harvard, in 1959 he moved to MIT, where he is now professor of mathematics and philosophy. He has made basic contributions to operator theory, ergodic theory, and combinatorics. The AMS recently honored him with a Steele Prize for his seminal work in algebraic combinatorics. He is a senior fellow of the Los Alamos National Laboratory and a member of the National Academy of Sciences.

Fine Hall in its golden age: Remembrances of Princeton in the early fifties*

GIAN-CARLO ROTA

Our faith in mathematics is not likely to wane if we openly acknowledge that the personalities of even the greatest mathematicians may be as flawed as those of anyone else. The greater a mathematician, the more important it is to bring out the contradictions in his or her personality. Psychologists of the future, if they should ever read such accounts, may better succeed in explaining what we, blinded by prejudice, would rather not face up to.

The biographer who frankly admits his bias is, in my opinion, more honest than the one who, appealing to objectivity, conceals his bias in the selection of facts to be told. Rather than attempting to be objective, I have chosen to transcribe as faithfully as I can the inextricable twine of fact, opinion and idealization that I have found in my memories of what happened thirty-five years ago. I

---

*The present article is a draft for a chapter of a book which the author is under contract to write for the Sloan science series.

hope thereby to have told the truth. Every sentence I have written should be prefixed by an "It is my opinion that...."

I apologize to those readers who may find themselves rudely deprived of the comforts of myth.

ALONZO CHURCH

It cannot be a complete coincidence that several outstanding logicians of the twentieth century found shelter in asylums at some time in their lives: Cantor, Zermelo, Gödel, Peano, and Post are some. Alonzo Church was one of the saner among them, though in some ways his behavior must be classified as strange, even by mathematicians' standards.

He looked like a cross between a panda and a large owl. He spoke slowly in complete paragraphs which seemed to have been read out of a book, evenly and slowly enunciated, as by a talking machine. When interrupted, he would pause for an uncomfortably long period to recover the thread of the argument. He never made casual remarks: they did not belong in the baggage of formal logic. For example, he would not say: "It is raining." Such a statement, taken in isolation, makes no sense. (Whether it is actually raining or not does not matter; what matters is consistency.) He would say instead: "I must postpone my departure for Nassau Street, inasmuch as it is raining, a fact which I can verify by looking out the window." (These were not his exact words.) Gilbert Ryle has criticized philosophers for testing their theories of language with examples which are never used in ordinary speech. Church's discourse was precisely one such example.

He had unusual working habits. He could be seen in a corridor in Fine Hall at any time of day or night, rather like the Phantom of the Opera. Once, on Christmas day, I decided to go to the Fine Hall library (which was always open) to look up something. I met Church on the stairs. He greeted me without surprise.

He owned a sizable collection of science-fiction novels, most of which looked well thumbed. Each volume was mysteriously marked either with a circle or with a cross. Corrections to wrong page numberings in the table of contents had been penciled into several volumes.

His one year course in mathematical logic was one of Princeton University's great offerings. It attracted as many as four students in 1951 (none of them were philosophy students, it must be added, to philosophy's discredit). Every lecture began with a ten-minute ceremony of erasing the blackboard until it was absolutely spotless. We tried to save him the effort by erasing the board before his arrival, but to no avail. The ritual could not be disposed of; often it required water, soap, and brush, and was followed by another ten minutes of total silence while the blackboard was drying. Perhaps he was preparing the lecture while erasing; I don't think so. His lectures hardly needed any preparation. They were a literal repetition of the typewritten text he had written over a period of twenty years, a copy of which was to be found upstairs in the Fine Hall library. (The manuscript's pages had yellowed with the years, and smelled foul. Church's definitive treatise was not published for another five years.) Occasionally, one of the sentences spoken in class would be at variance with the text upstairs, and he would warn us in advance of the discrepancy between oral and written presentation. For greater precision, everything he said (except some fascinating side excursions which he invariably prefixed by a sentence like: "I will now interrupt and make a meta-mathematical [sic] remark") was carefully written down on the blackboard, in large English-style handwriting, like that of a grade-school teacher, complete with punctuation and paragraphs. Occasionally, he carelessly skipped a letter in a word. At first we pointed out these oversights, but we quickly learned that they would create a slight panic, so we kept our mouths shut. Once he had to use a variant of a previously proved theorem, which differed only by a change of notation. After a moment of silence, he turned to the class and said: "I could simply say 'likewise', but I'd better prove it again."

It may be asked why anyone would bother to sit in a lecture which was the literal repetition of an available text. Such a question would betray an oversimplified view of what goes on in a classroom. What one really learns in class is what one does not know at the time one is learning. The person lecturing to us was logic incarnate. His pauses, hesitations, emphases, his betrayals of emotion (however rare), and sundry other nonverbal phenomena taught us a lot more logic than any written text could. We learned to think in unison with him as he spoke, as if following the demonstration of a calisthenics instructor. Church's course permanently improved the rigor of our reasoning.

The course began with the axioms for the propositional calculus (those of Russell and Whitehead's *Principia Mathematica*, I believe) that take material implication as the only primitive connective. The exercises at the end of the first chapter were mere translations of some identities of naive set theory in terms of material implication. It took me a tremendous effort to prove them, since I was unaware of the fact that one could start with an equivalent set of axioms using "and" and "or" (where the disjunctive normal form provides automatic proofs) and then translate each proof step by step in terms of implication. I went to see Church to discuss my difficulties, and far from giving away the easy solution, he spent hours with me devising direct proofs using implication only. Toward the end of the course I brought to him the sheaf of papers containing the solutions to the problems (all problems he assigned were optional, since they could not logically be made to fit into the formal text). He looked at them as if expecting them, and then pulled out of his drawer a note he had just published in *Portugaliae Mathematica*, where similar problems were posed for "conditional disjunction", a ternary connective he had introduced. Now that I was properly trained, he wanted me to repeat the work with conditional disjunction as the primitive connective. His graduate students had declined a similar request, no doubt because they considered it to be beneath them.

Mathematical logic has not been held in high regard at Princeton, then or now. Two minutes before the end of Church's lecture (the course met in the largest classroom in Fine Hall), Lefschetz would begin to peek through the door. He glared at me and at the spotless text on the blackboard; sometimes he shook his head to make it clear that he considered me a lost cause. The following class was taught by Kodaira, at that time a recent arrival from Japan, whose work in geometry was revered by everyone in the Princeton main line. The classroom was packed during Kodaira's lecture. Even though his English was atrocious, his lectures were crystal clear. (Among other things, he stuttered. Because of deep-seated prejudices of some of its members, the mathematics department refused to appoint him full-time to the Princeton faculty.)

I was too young and too shy to have an opinion of my own about Church and mathematical logic. I was in love with the subject, and his course was my first graduate course. I sensed disapproval all around me; only Roger Lyndon (the inventor of spectral sequences), who had been my freshman advisor, encouraged me. Shortly afterward he himself was encouraged to move to Michigan. Fortunately, I had met one of Church's most flamboyant former students, John Kemeny, who, having just finished his term as a mathematics instructor, was being eased — by Lefschetz's gentle hand — into the philosophy department. (The following year he left for Dartmouth, where he eventually became president.)

Kemeny's seminar in the philosophy of science (which that year attracted as many as six students, a record) was refreshing training in basic reasoning. Kemeny was not afraid to appear pedestrian, trivial, or stupid: what mattered was to respect the facts, to draw distinctions even when they clashed with our prejudices, and to avoid black and white oversimplifications. Mathematicians have always found Kemeny's common sense revolting.

"There is no reason why a great mathematician should not also be a great bigot," he once said on concluding a discussion whose beginning I have by now forgotten. "Look at your teachers in Fine Hall, at how they treat one of the greatest living mathematicians, Alonzo Church."

I left literally speechless. What? These demi-gods of Fine Hall were not perfect beings? I had learned from Kemeny a basic lesson: a good mathematician is not necessarily a "nice guy."

WILLIAM FELLER

His name was neither William nor Feller. He was named Willibold by his Catholic mother in Croatia, after his birthday saint; his original last name was a Slavic tongue twister, which he changed while still a student at Göttingen (probably on a suggestion of his teacher Courant). He did not like to be reminded of his Balkan origins, and I had the impression that in America he wanted to be taken for a German who had Anglicized his name. From the time he moved from Cornell to Princeton in 1950, his whole life revolved around a feeling of inferiority. He secretly considered himself to be one of the lowest ranking members of the Princeton mathematics department, probably the second lowest after the colleague who had brought him there, with whom he had promptly quarreled after arriving in Princeton.

In retrospect, nothing could be farther from the truth. Feller's treatise in probability is one of the great masterpieces of mathematics of all time. It has survived unscathed the onslaughts of successive waves of rewriting, and it is still secretly read by every probabilist, many of whom refuse to admit that they still constantly consult it, and refer to it as "trivial" (like high school students complaining that Shakespeare's dramas are full of platitudes). For a long time, Feller's treatise was the mathematics book most quoted by nonmathematicians.

But Feller would never have admitted to his success. He was one of the first generation who thought probabilistically (the others: Doob, Kac, Lévy, and Kolmogorov), but when it came to writing down any of his results for publication, he would chicken out and recast the mathematics in purely analytic terms. It took one more generation of mathematicians, the generation of Harris, McKean, Ray, Kesten, Spitzer, before probability came to be written the way it is practiced.

His lectures were loud and entertaining. He wrote very large on the blackboard, in a beautiful Italianate handwriting with lots of whirls. Sometimes only one huge formula appeared on the blackboard during the entire period; the rest was handwaving. His proofs — insofar as one can speak of proofs — were often deficient. Nonethless, they were convincing, and the results became unforgettably clear after he had explained them. The main idea was never wrong.

He took umbrage when someone interrupted his lecturing by pointing out some glaring mistake. He became red in the face and raised his voice, often to full shouting range. It was reported that on occasion he had asked the objector to leave the classroom. The expression "proof by intimidation" was coined after Feller's lectures (by Mark Kac). During a Feller lecture, the hearer was made to feel privy to some wondrous secret, one that often vanished by magic as he walked out of the classroom at the end of the period. Like many great teachers, Feller was a bit of a con man.

I learned more from his rambling lectures than from those of anyone else at Princeton. I remember the first lecture of his I ever attended. It was also the first mathematics course I took at Princeton (a course in sophomore differential equations). The first impression he gave was one of exuberance, of great zest for living, as he rapidly wrote one formula after another on the blackboard while his white mane floated in the air. After the first lecture, I had learned two words which I had not previously heard: "lousy" and "nasty." I was also terribly impressed by a trick he explained: the integral

$$\int_0^{2\pi} \cos^2 x \, dx$$

equals the integral

$$\int_0^{2\pi} \sin^2 x \, dx$$

and therefore, since the sum of the two integrals equals 2π, each of them is easily computed.

He often interrupted his lectures with a tirade from the repertoire he had accumulated over the years. He believed these side shows to be a necessary complement to the standard undergraduate curriculum. Typical titles: "Gandhi was a phoney," "Velikovsky is not as wrong as you think," "Statisticians do not know their business," "ESP is a sinister plot against civilization," "The smoking and health report is all wrong." Such tirades, it must be said to his credit, were never repeated to the same class, though they were embellished with each peformance. His theses, preposterous as they sounded, invariably carried more than an element of truth.

He was Velikovsky's next-door neighbor on Random Road. They first met one day when Feller was working in his garden pruning some bushes, and Velikovsky rushed out of his house screaming: "Stop! You are killing your father!" Soon afterward they were close friends.

He became a crusader for any cause which he thought to be right, no matter how orthogonal to the facts. Of his tirades against statistics, I remember one suggestion he made in 1952, which still appears to me to be quite sensible: in multiple-choice exams, students should be asked to mark one wrong answer, rather than to guess the right one. He inveighed against American actuaries, pointing to Swedish actuaries (who gave him his first job after he graduated from Göttingen) as the paradigm. He was so vehemently opposed to ESP that his overkill (based on his own faulty statistical analyses of accurate data) actually helped the other side.

He was, however, very sensitive to criticism, both of himself and of others. "You should always judge a mathematician by his best paper!," he once said, referring to Richard Bellman.

While he was writing the first volume of his book he would cross out entire chapters in response to the slightest critical remark. Later, while reading galleys, he would not hesitate to rewrite long passages several times, each time using different proofs; some students of his claim that the entire volume was rewritten in galleys, and that some beautiful chapters were left out for fear of criticism. The treatment of recurrent events was the one he rewrote most, and it is still, strictly speaking, wrong. Nevertheless, it is perhaps his greatest piece of work. We are by now so used to Feller's ideas that we tend to forget how much mathematics today goes back to his "recurrent events"; the theory of formal grammars is one outlandish example.

He had no firm judgment of his own, and his opinions of other mathematicians, even of his own students, oscillated wildly and frequently between extremes. You never knew how you stood with him. For example, his attitude toward me began very favorably when he realized I had already learned to differentiate and integrate before coming to Princeton. (In 1950, this was a rare occurrence.) He all but threw me out of his office when I failed to work on a problem on random walk he proposed to me as a sophomore; one year later, however, I did moderately well on the Putnam Exam, and he became friendly again, only to write me off completely when I went off to Yale to study functional analysis. The tables were turned again 1963 when he gave me a big hug at a meeting of the AMS in New York. (I learned shortly afterward that Doob had explained to him my 1963 limit theorem for positive operators. In fact, he liked the ideas of "strict sense spectral theory" so much that he invented the phrase "To get away with Hilbert space.") His benevolence, alas, proved to be short-lived: as soon as I started working in combinatorics, he stopped talking to me. But not, fortunately, for long: he listened to a lecture of mine on applicatons of exterior algebra to combinatorics and started again singing my praises to everyone. He had jumped to the conclusion that I was the inventor of exterior algebra. I never had the heart to tell him the truth. He died believing I was the latter-day Grassmann.

He never believed that what he was doing was going to last long, and he modestly enjoyed pointing out papers that made his own work obsolete. No doubt he was also secretly glad that his ideas were being kept alive. This happened with the Martin boundary ("It is so much better than *my* boundary!") and with the relationship between diffusion and semigroups of positive operators.

Like many of Courant's students, he had only the vaguest ideas of any mathematics that was not analysis, but he had a boundless admiration for Emil Artin and his algebra, for Otto Neugebauer and for German mathematics. Together with Emil Artin, he helped Neugebauer figure out the mathematics in cuneiform tablets. Their success give him a new harangue to add to his repertoire: "The Babylonians knew Fourier analysis." He was at first a strong Germanophile and Francophile. He would sing the praises of Göttingen and of the Collège de France in rapturous terms. (His fulsome encomia of Europe reminded me of the sickening old Göttingen custom of selling picture postcards of professors.) He would tell us bombastic stories of his days at Göttingen, of his having run away from home to study mathematics (I never believed that one), and of how, shortly after his arrival in Göttingen, Courant himself visited him in his quarters while the landlady watched in awe.

His views on European universities changed radically after he made a lecture tour in 1954; from that time on, he became a champion of American know-how.

He related well to his superiors and to those whom he considered to be his inferiors (such as John Riordan, whom he used to patronize), but his relations with his equals were uneasy at best. He was particularly harsh with Mark Kac. Kitty Kac once related to me an astonishing episode. One summer evening at Cornell Mark and Kitt were sitting on the Fellers' back porch in the evening. At some point in the conversation, Feller began a critique of Kac's work, paper by paper, of Kac's working habits, and of his research program. He painted a grim picture of Kac's future, unless Mark followed Willy's advice to master more measure theory and to use almost-everywhere convergence rather than the trite (to Willy) convergence in distribution. As Kitty spoke to me — a few years after Mark's death, with tears in her eyes — I could picture Feller carried away by the sadistic streak that emerges in our worst moments, when we tear someone to shreds with the intention of forgiving him the moment he begs for mercy.

I reassured Kitty that the Feynman–Kac formula (as Jack Schwartz named it in 1955) will be remembered in science long after Feller's book is obsolete. I could almost hear a sigh of relief, forty-five years after the event.

Emil Artin

Emil Artin came to Princeton from Indiana shortly after Wedderburn's death in 1946. Rumor had it (*se non è vero è ben trovato*) that Indiana University had decided not to match the Princeton offer, since during the ten years of his tenure he had published only one research paper, a short proof of the Krein–Milman theorem in "The Piccayune [sic] Sentinel," Max Zorn's *samizdat* magazine.

A few years later, Emil Artin had become the idol of Princeton mathematicians. His mannerisms did not discourage the cult of personality. His graduate students would imitate the way he spoke and walked, and they would even dress like him. They would wear the same kind of old black leather jacket he wore, like that of a Luftwaffe pilot in a war movie. As he walked, dressed in his too-long winter coat, with a belt tightened around his waist, with his light blue eyes and his gaunt fact, the image of a Wehrmacht officer came unmistakably to mind. (Such a military image is wrong, I learned years later from Jürgen Moser. Germans see the Emil Artin "type" as the epitome of a period of Viennese Kultur.)

He was also occasionally seen wearing sandals (like those worn by Franciscan friars), even in cold weather. His student Serge Lang tried to match eccentricities by never wearing a coat, although he would always wear heavy gloves every time he walked out of Fine Hall, to protect himself against the rigors of winter.

He would spend endless hours in conversation with his few protégés (at that time, Lang and Tate), in Fine Hall, at his home, during long walks, even via expensive long-distance telephone calls. He spared no effort to be a good tutor, and he succeeded beyond all expectations.

He was, on occasion, tough and rude to his students. There were embarrassing public scenes when he would all of a sudden, at the most unexpected times, lose his temper and burst into a loud and unseemly "I told you a hundred times that..." tirade directed at one of them. One of these outbursts occurred once when Lang loudly proclaimed that Pólya and Szegő's problems were bad for mathematical education. Emil Artin loved special functions and explicit computations, and he relished Pólya and Szegő's "*Aufgaben und Lehrsätze*," though his lectures were the negation of any anecdotal style.

He would also snap back at students in the honors freshman calculus class which he frequently taught. He might throw a piece of chalk or a coin at a student who had asked too silly a question ("What about the null set?"). A few weeks after the beginning of the fall term, only the bravest would dare ask any more questions, and the class listened in sepulchral silence to Emil Artin's spellbinding voice, like a congregation at a religious service.

He had definite (and definitive) views on the relative standing of most fields of mathematics. He correctly foresaw and encouraged the rebirth of interest in finite groups that was to begin a few years later with the work of Feit and Thompson, but he professed to dislike semigroups. Schützenberger's work, several years after Emil Artin's death, has proved him wrong: the free semigroup is a far more interesting object than the free group, for example. He inherited his mathematical ideals from the other great German number theorists since Gauss and Dirichlet. To all of them, a piece of mathematics was the more highly thought of, the closer it came to Germanic number theory.

This prejudice gave him a particularly slanted view of algebra. He intensely disliked Anglo–American algebra, the kind one associates with the names of

Boole, C. S. Peirce, Dickson, the late British invariant theorists (like D. E. Littlewood, whose proofs he would make fun of), and Garrett Birkhoff's universal algebra (the word "lattice" was expressly forbidden, as were several other words). He thought this kind of algebra was "no good" — rightly so, if your chief interests are confined to algebraic numbers and the Riemann hypothesis. He made an exception, however, for Wedderburn's theory of rings, to which he gave an exposition of as yet unparalleled beauty.

A great many mathematicians in Princeton, too awed or too weak to form opinions of their own, came to rely on Emil Artin's pronouncements like hermeneuts on the mutterings of the Sybil at Delphi. He would sit at teatime in one of the old leather chairs ("his" chair) in Fine Hall's common room, and deliver his opinions with the abrupt definitiveness of Wittgenstein's or Karl Kraus's aphorisms. A gaping crowd of admirers and worshippers, often literally sitting at his feet, would record them for posterity. Sample quips: "If we knew *what* to prove in non-Abelian class field theory, we could prove it"; "Witt was a Nazi, the one example of a clever Nazi" (one of many exaggerations). Even the teaching of undergraduate linear algebra carried the imprint of Emil Artin's very visible hand: we were to stay away from any mention of bases and determinants (a strange injunction, considering how much he liked to compute). The alliance of Emil Artin, Claude Chevalley, and André Weil was out to expunge all traces of determinants and resultants from algebra. Two of them are now probably turning in their graves.

His lectures are best described as polished diamonds. They were delivered with the virtuoso's spontaneity that comes only after lengthy and excruciating rehearsal, always without notes. Very rarely did he make a mistake or forget a step in a proof. When absolutely lost, he would pull out of his pocket a tiny sheet of paper, glance at it quickly, and then turn to the blackboard, like a child caught cheating.

He would give as few examples as he could get away with. In a course in point-set topology, the only examples he gave right after defining the notion of a topological space were a discrete space and an infinite set with the finite–cofinite topology. Not more than three or four more examples were given in the entire course.

His proofs were perfect but not enlightening. They were the end results of years of meditation, during which all previous proofs of his and of his predecessors were discarded one by one until he found the definitive proof. He did not want to admit (unlike a wine connoisseur, who teaches you to recognize *vin ordinaire* before allowing you the *bonheur* of a *premier grand cru*) that his proofs would best be appreciated if he gave the class some inkling of what they were intended to improve upon. He adamantly refused to give motivation of any kind in the classroom, and stuck to pure concepts, which he intended to communicate *directly*. Only the very best and the very worst responded to such shock treatment: the first because of their appreciation of superior exposition, and the second because of their infatuation with Emil Artin's style. Anyone who wanted to understand had to figure out later "what he had really meant."

His conversation was in stark contrast to the lectures: he would then give out plenty of relevant and enlightening examples, and freely reveal the hidden motivation of the material he had so stiffly presented in class.

It has been claimed that Emil Artin inherited his flair for public speaking from his mother, an opera singer. More likely, he was driven to perfection by a firm belief in axiomatic *Selbsständigkeit*. The axiomatic method was only two generations old in Emil Artin's time, and it still had the force of a magic ritual. In his day, the identification of mathematics with the axiomatic method for the presentation of mathematics was not yet thought to be a preposterous misunderstanding (only analytic philosophers pull such goofs today). To Emil Artin, axiomatics was a useful technique for disclosing hidden analogies (for example, the analogy between algebraic curves and algebraic number fields, and the analogy between the Riemannian hypothesis and the analogous hypothesis for infinite function fields, first explored in Emil Artin's thesis and later generalized into the "Weil conjectures"). To lesser minds, the axiomatic method was a way of grasping the "modern" algebra that Emmy Noether had promulgated, and that her student Emil Artin was the first to teach. The table of contents of every algebra textbook is still, with small variations, that which Emil Artin drafted and which van der Waerden was the first to develop. (How long will it take before the imbalance of such a table of contents — for example, the overemphasis on Galois theory at the expense of tensor algebra — will be recognized and corrected?)

At Princeton, Emil Artin and Alonzo Church inspired more loyalty in their students than Bochner or Lefschetz. It is easy to see why. Both of them were prophets of new faiths, of two conflicting philosophies of algebra that are still vying with each other for mastery.

Emil Artin's mannerisms have been carried far and wide by his students and his students' students, and are now an everyday occurrence (whose origin will soon be forgotten) whenever an algebra course is taught. Some of his quirks have been overcompensated: Serge Lang will make a *volte-face* on any subject, given adequate evidence to the contrary; Tate makes a point of being equally fair to all his doctoral students; and Arthur Mattuck's lectures are an exercise in high motivation. Even his famous tantrums still occasionally occur. A few older mathematicians still recognize in the outbursts of the students the gestures of the master.

Solomon Lefschetz

No one who talked to Lefschetz failed to be struck by his rudeness. I met him one afternoon at tea, in the fall term of my first year at Princeton, in the Fine Hall common room. He asked me if I was a graduate student: after I answered in the negative, he turned his back and behaved as if I did not exist. In the spring term, he suddenly began to notice my presence. He even remembered my name, to my astonishment. At first, I felt flattered until (perhaps a year later) I realized that what he remembered was not me, but the fact that that I had an Italian name. He had the highest regard for the great Italian algebraic geometers, for Castelnuovo, Enriques, and Severi, who were slightly older than he was, and who were his equals in depth of thought as well as in sloppiness of argument. "You should have gone to school in Rome in the twenties. That was the Princeton of its time!" he told me.

He was rude to everyone, even to the people who doled out funds in Washington and to mathematicians who were his equals. I recall Lefschetz meeting Zariski, probably in 1957 (while Hironaka was already working on the proof of the resolution of singularities for algebraic varieties). After exchanging with Zariski warm and loud Jewish greetings (in Russian), he proceeded to proclaim loudly (in English) his skepticism on the possibility of resolving singularities for all algebraic varieties. "Ninety percent proved is zero percent proved!," he retorted to Zariski's protestations, as a conversation stopper. He had reacted similarly to several other previous attempts that he had to shoot down. Two years later he was proved wrong. However, he had the satisfaction of having been wrong only once.

He rightly calculated that skepticism is always a more prudent policy when a major mathematical problem is at stake, though it did not occur to him that he might express his objections in less obnoxious language. When news first came to him from England of Hodge's work on harmonic integrals and their relation to homology, he dismissed it the work of a crackpot, in a sentence that has become a proverbial mathematical gaffe. After that débacle, he became slightly more cautious.

Solomon Lefschetz was an electrical engineer trained at the *École Centrale*, one of the lesser of the French *grandes écoles*. He came to America probably because, as a Russian-Jewish refugee, he had trouble finding work in France. A few years after arriving in America, an accident deprived him of the use of both hands. He went back to school and got a quick Ph.D. in mathematics at Clark University (which at that time had a livelier graduate school than it has now). He then accepted instructorships at the Universities of Nebraska and Kansas, the only means he had to survive. For a few harrowing years he worked night and day, publishing several substantial papers a year in topology and algebraic geometry. Most of the ideas of present-day algebraic topology were either invented or developed (following Poincaré's lead) by Lefschetz in these papers; his discovery that the work of the Italian algebraic geometers could be recast in topological terms is only slightly less dramatic.

To no one's surprise (except that of the anti-Semites who still ruled over some of the Ivy League universities) he received an offer to join the Princeton mathematics department from Luther Pfahler Eisenhart, the chairman, an astute mathematician whose contributions to the well-being of mathematics have never been properly appreciated (to his credit, his books, carefully and courteously written as few mathematics books are, are still in print today).

His colleagues must have been surprised when Lefschetz himself started to develop anti-Semitic feelings which were still lingering when I was there. One of the first questions he asked me after I met him was whether I was Jewish. In the late thirties and forties, he refused to admit any Jewish graduate students in mathematics. He claimed that, because of the Depression, it was too difficult to get them jobs after they earned their Ph.D.s. He liked and favored red-blooded American boyish Wasp types (like Ralph Gomory), especially those who came from the sticks, from the Midwest, or from the South.

He considered Princeton to be a just reward for his hard work in Kansas, as well as a comfortable, though only partial, retirement home. After his move he did little new work of his own in mathematics, though he did write several books, among them the first comprehensive treatise on topology. This book, whose influence on the further development of the subject was decisive, hardly contains one completely correct proof. It was rumored that it had been written during one of Lefschetz's sabbaticals away from Princeton, when his students did not have the opportunity to revise it and eliminate the numerous errors, as they did with all of their teacher's other writings.

He despised mathematicians who spent their time giving rigorous or elegant proofs for arguments which he considered obvious. Once, Spencer and Kodaira, still associate professors, proudly explained to him a clever new proof they had found of one of Lefschetz's deeper theorems. "Don't come to me with your pretty proofs. We don't bother with that baby stuff around here!" was his reaction. Nonetheless, from that moment on he held Spencer and Kodaira in high esteem. He liked to repeat, as an example of mathematical pedantry, the story of one of E. H. Moore's visits to Princeton, when Moore started a lecture by saying: "Let a be a point and let b be a point." "But why don't you just say 'Let a and b be points'!" asked Lefschetz. "Because a may equal b," answered Moore. Lefschetz got up and left the lecture room.

Lefschetz was a purely intuitive mathematician. It was said of him that he had never given a completely correct proof, but had never made a wrong guess either. The diplomatic expression "open reasoning" was invented to justify his always deficient proofs. His lectures came close to incoherence. In a course on Riemann surfaces, he started with a string of statements in rapid succession, without writing on the blackboard: "Well, a Riemann surface is a certain kind of Hausdorff space. You know what a Hausdorff space is, don't you? It is also compact, o.k.? I guess it is also a manifold. Surely you know what a manifold is. Now let me tell you one nontrivial theorem: the

Riemann–Roch Theorem." And so on until all but the most faithful students dropped out of the course.

I listened to a few of his lectures, curious to find out what he might be saying in a course on ordinary differential equations he had decided to teach on the spur of the moment. He would be holding a piece of chalk with his artificial hands, and write enormous letters on the blackboard, like a child learning how to write. I could not make out the sense of anything he was saying, nor whether what he was saying was gibberish to me alone or to everyone else as well. After one lecture, I asked a rather senior-looking mathematician who had been religiously attending Lefschetz's lectures whether he understood what the lecturer was talking about. I received a vague and evasive answer. After that moment, I knew.

When he was forced to relinquish the chairmanship of the Princeton mathematics department for reasons of age, he decided to promote Mexican mathematics. His love/hate of the Mexicans occasionally got him into trouble. Once, in a Mexican train station, he spotted a charro dressed in full regalia, complete with a pair of pistols and rows of cartridges across his chest. He started making fun of the charro's attire, adding some deliberate slurs in his excellent Spanish. His companions feared that the charro might react the way Mexicans traditionally react to insult. In fact, the charro eventually stood up and reached for his pistols. Lefschetz looked at him straight in the face and did not back off. There were a few seconds of tense silence. "Gringo loco!" said the charro finally, and walked away. When Lefschetz decided to leave Mexico and come back to the United States, the Mexicans awarded him the Order of the Aztec Eagle.

During Lefschetz's tenure as chairman of the mathematics department, Princeton became the world center of mathematics. He had an uncanny instinct for sizing up mathematicians' abilities, and he was invariably right when sizing up someone in a field where he knew next to nothing. In topology, however, his judgment would occasionally slip, probably because he became partial to work that he half understood.

His standards of accomplishment in mathematics were so high that they spread by contagion to his successors, who maintain them to this day. When addressing an entering class of twelve graduate students, he told them in no uncertain terms: "Since you have been carefully chosen among the most promising undergraduates in mathematics in the country, I expect that you will all receive your Ph.D.s rather sooner than later. Maybe one or two of you will go on to become mathematicians."

Halsey Royden has had a lifelong connection with Stanford University. His maternal grandmother enrolled with Stanford's first class in 1892, and two uncles graduated before 1920. After receiving a B.S. from Stanford in 1948 and an M.S. in 1949, Royden went to Harvard University to study under Lars Ahlfors' direction, obtaining a Ph.D. in 1951. At that point he joined the Stanford faculty. He has since served as dean of the school of humanities and science. He is known for his research in complex analysis and differential geometry, and is the author of a widely-used textbook, Real Analysis.

A History of Mathematics at Stanford

HALSEY ROYDEN

The older history of the Stanford Mathematics Department may be divided into a number of periods: Early Years, the Blichfeldt Era, the Szegő Period, and the later fifties and early sixties. I transferred to Stanford as an undergraduate in 1946, received my B.S. in 1948, my M.S. in 1949, and returned to Stanford as a faculty member in 1951, after a brief period of exile at a small college on the banks of the Charles. Memory in human institutions is transmitted not only by the written archive but also by direct transference of living memory from one generation of colleagues to the next. Those long associated with an institution come to have memories of the events before their time that are only slightly less vivid than those they have actually experienced. Although I arrived at Stanford in the middle of the Szegő era, memories were still current from the earlier time, when H. F. Blichfeldt dominated the Stanford department. My account of the course of Stanford's mathematics department utilizes material gleaned from the archives, together with memories of my own experiences, as well as those recounted by my colleagues, particularly Harold Bacon, Don Spencer, and Jack Herriot. I have taken some of the descriptions of former faculty members from Memorial resolutions of Stanford's Academic Council. I am especially indebted to Harold Bacon for his assistance in preparing this history.

My professional life for the last forty years has been involved with Stanford affairs, and I hope the reader will pardon me if this history sometimes assumes the character of a personal memoir. It is often difficult to view recent events with a proper historical perspective, and I have therefore not included events of the last twenty years, although this has meant neglecting the rise of differential geometry at Stanford with the appointments of Yau, Simon, Siu, and recently of Schoen and numerous younger people.

1. The Early Years

Instruction at Stanford began with the Academic Year 1891–1892. The senior faculty in mathematics, both in the early days and for a long time thereafter, consisted of Professors Robert Allardice and Rufus Green. They were both in their late twenties when they came to Stanford, and though neither had a Ph.D., they were established mathematicians.

Allardice was a professor in Stanford's original faculty of 1891–1892. He had received his A.M. from Edinburgh in 1882 and was a student of Chrystal. His assistance is noted in the preface to the second edition of Chrystal's *Algebra*. At Edinburgh, Allardice was Baxter Scholar in Mathematics '82–83, Drummond Scholar in mathematics for '83–84, and assistant professor of mathematics from 1884 until he came to Stanford.

Rufus Green received his B.S. from Indiana in 1885 and his M.S. in 1890. He was an instructor in mathematics at Indiana in '85–86, a graduate student at Johns Hopkins for '86–87, and a professor of pure mathematics at Indiana from 1887 to 1893, when he came to Stanford as an associate professor. Stanford's first president, David Starr Jordan, had been a professor at Cornell and president at Indiana before coming to Stanford. He recruited most of Stanford's first faculty from among his former colleagues at Cornell, and many of the rest from Indiana.

Hans Frederik Blichfeldt, who was to play such a large role in the history of mathematics at Stanford, began his undergraduate work at Stanford in 1894, and he received one of the three A.B.'s awarded by the Stanford mathematics department in 1896. He was appointed instructor in mathematics, and received his A.M. in 1897. He was reappointed instructor again in 1898, after receiving his Ph.D. from Leipzig. Blichfeldt was born in Denmark in 1873. His father emigrated to the United States in 1888, but before leaving, the young Blichfeldt had taken, and passed with high honors, the general preliminary examination conducted at the University of Copenhagen. In his early school career he had constantly excelled in mathematics, and by the time he took this examination at the age of fifteen, he had discovered by himself the solutions of the general polynomial equations of the third and fourth degrees — a remarkable performance for a schoolboy.

Coming to the United States, he found employment in Nebraska, Wyoming, Oregon, and Washington where he worked at various jobs in the lumber industry. During the four years from 1888 to 1892 he "worked with my hands doing everything, East and West the country across" (quoted by L. E. Dickson). Then for the two years 1892–1894 he worked as a draughtsman for the engineering department of the City and County of Whatcom, Washington, where his unusual mathematical talent began to come to the attention of his employers and fellow workers. Although he had not pursued a formal high school education in this country, he was persuaded in 1894 to apply for admission to Stanford. His application for admission was supported by a letter to Stanford's President Jordan from the County Superintendent of Schools of Whatcom County, who wrote, "He is about 21, of most exemplary physique and morals, evidently cultured in his native tongue and fairly proficient in English. He is a real genius in mathematics — working intuitively, to all appearance, abstruse integral calculus problems ... In short all who know him here look upon him as a genius — if he might have advantage of training ... " The Stanford Registrar was uncertain how to credit the examination Blichfeldt had taken at Copenhagen in 1888, but, nevertheless, admitted him as a special student in September of 1894. In January of 1895 he was granted "full entrance standing, except for English 1b, on the basis of work done before entering the University," and a month later he was granted an additional credit of sixty hours toward graduation.

At this time Stanford followed a "free elective" system somewhat similar to Harvard's, and this was particularly well adapted to a young man of great mathematical ability and originality. He received his A.B. degree in mathematics in 1896 followed by the A.M. in 1897. Apart from courses in English, German, and physics all of Blichfeldt's courses were in mathematics and included: calculus, quaternions, higher plane curves, differential equations, analysis, solid geometry, and invariant theory as an undergraduate, and projective geometry, curve tracing, vector analysis, theory of functions, and theory of substitutions as a graduate student.

At that time German universities were centers of great mathematical activity, and Blichfeldt was determined to go to Leipzig and study under Sophus Lie. His Stanford friends, and particularly Professor Rufus Green, urged and encouraged him to do this. Although his three years at Stanford (which at that time charged no tuition fees) had been financed by savings painstakingly accumulated during his earlier years of hard work, he held a teaching assistantship in 1896–1897, and Professor Green saw to it that he could borrow what was needed to make the European venture possible. He spent the year working with Lie, mastering the "Lie Theory" of continuous groups, and writing his dissertation "On a certain class of groups of transformations in

space of three dimensions," (*Amer. J. Math.* **22** (1900) pp. 113–120). He
was awarded the Ph.D., summa cum laude, in 1898.

He returned to Stanford as an instructor in mathematics in 1898 and remained a member of Stanford's faculty until his retirement in 1938. He was
assistant professor of mathematics, 1901–1906; associate professor, 1906–
1913; professor, 1913–1938; Professor Emeritus, 1938; until his death in
1945. He was executive head of the department from 1927 until 1938. In
1911 he was associate professor of mathematics at the University of Chicago,
summer quarter, and professor of mathematics at Columbia for the Summer
Sessions of 1924 and 1925.

Blichfeldt made contributions of lasting and fundamental importance to
the theory of groups and to the geometry of numbers. In the former he
solved the problem of finding all finite collineation groups in four variables, a
problem whose solution eluded Klein and Jordan. In the latter he determined
the precise limits for minima of definite quadratic forms in six, seven, and
eight variables. In addition to some two dozen research papers of importance,
he was the author of *Finite groups of linear homogeneous transformations*,
which forms part two of the book *Theory and Application of Finite Groups,*
by G. A. Miller, H. F. Blichfeldt, and L. E. Dickson, 1916. He was also the
author of the book, *Finite Collineation Groups,* 1917.

Blichfeldt was a member of the American Mathematical Society and served
as its vice-president in 1912. He was also active in the affairs of the Mathematical Association of America. He represented the United States officially
at two International Mathematical Congresses, was elected to the National
Academy of Sciences in 1920, and served as a member of the National Research Council in 1924–1927.

When Blichfeldt was promoted to an assistant professorship in 1901, he
was joined as assistant professor by George Abram Miller, who had been at
Leipzig while Blichfeldt was there. Miller later had a distinguished career at
the University of Illinois. Miller's curriculum vitae indicates what the career
of a mathematician was like in those days:

George Abram Miller:
> A.B., Muhlenberg College, 1887;
>
> Ph.D., Cumberland University, 1892;
>
> Professor of Mathematics, Eureka College, 1888–1893;
>
> Instructor in Mathematics, University of Michigan,
> 1893–1895;
>
> Student, Universities of Leipzig and Paris, 1895–1897;

Instructor, Cornell University, 1897–1901;

Instructor in Mathematics, University of Chicago, summer 1898;

Assistant Professor, Stanford, 1901–1902;

Associate Professor, Stanford, 1902–1905.

The department awarded its first Ph.D. to William Albert Manning in 1904. The department also awarded a B.S. that year to Eric Temple Bell, who, two years earlier, had been admitted and granted sixty hours toward graduation on the basis of previous work at the University of London. Except for several courses in education during his last year, all of his courses were in mathematics. The courses he took, some of which were taught by Blichfeldt and Miller, seemed to be somewhat more substantial than those offered to Blichfeldt eight years earlier. They included: determinants, advanced co-ordinate geometry, calculus, differential equations, solid geometry, theory of equations, theory of functions, theory of numbers, and history of mathematics in his junior year, and advanced calculus, theory of groups, continuous groups, and a second course in the theory of functions his senior year.

A department of applied mathematics had been established in the meantime with Professor Hoskins as its only member. In 1902 he was joined by Halcott Cadwalader Moreno and William Albert Manning as instructors. By 1907 they had become assistant professors and were joined by S. D. Townley, who had received his Sc.D. degree in astronomy from the University of Michigan in 1897. By 1925 all three had risen to the rank of professor. At this time Hoskins retired, and applied mathematics was merged with pure mathematics to form a single department.

The year 1921–1922 also saw the arrival of the first formal summer visiting professor with the appointment of G. D. Birkhoff. The faculty roster for the two departments was the following:

Mathematics:

Allardice, Green, Blichfeldt, *Professors*;

G. D. Birkhoff, *Visiting Professor (Summer)*;

H. W. Brinkman, *Instructor (Summer)*;

Dorothy Crever, W. W. Wallace, *Assistants in Instruction.*

Applied Mathematics:

Hoskins, Manning, Moreno, Townley, *Professors;*

L. G. Gianini, *Instructor;*

F. E. Terman, H. E. Wheeler, *Teaching Assistants.*

The 1921–1922 list of teaching assistants in applied mathematics contains the name of Frederick Emmons Terman. Terman was later to become chairman of electrical engineering, dean of engineering, and, ultimately, provost at Stanford. He was always interested in the affairs of the mathematics department, especially the teaching of mathematics to engineering students. When I was an associate dean of humanities and sciences in the early sixties and Fred Terman was provost, I sometimes found it difficult to get on his crowded calendar. If I mentioned anything about our engineering calculus course, however, Terman would hold forth for an hour or so with advice and comment, drawn partly from the time when he taught sections of it as a teaching assistant.

2. The Blichfeldt Era

With the retirement of Hoskins in 1925, the department of applied mathematics was discontinued, and Professors Manning and Townley became members of the department of mathematics. Professor Townley was an astronomer, who had been at Stanford since 1907. Professor Manning, who had been Stanford's first Ph.D. in mathematics, was, like many of his colleagues at Stanford, interested in group theory. Several of his children received Ph.D.s from Stanford: His son Lawrence was for many years a professor of electrical engineering at Stanford. His daughter Rhoda was a professor of mathematics at Oregon State, and I took a course from her on modern algebra during my junior year, when she was a visiting faculty member at Stanford. Another daughter, Dorothy, was a professor of mathematics at Wells College.

A long tradition at Stanford in mathematical statistics and mathematical economics began in 1925, when the roster of the department listed Harold Hotelling as "Junior associate, Food Research Institute." The Food Research Institute had been established at Stanford a few years earlier by Herbert Hoover, and its faculty and staff consisted mostly of economists and statisticians. Hotelling, one of the founders of mathematical statistics in America, received his A.B. from Washington in 1919; his M.S. in 1921; and his Ph.D. from Princeton in 1924. He was an instructor at Princeton, 1922–1924; a junior research associate at Stanford's Food Research Institute, 1924–1926; research associate, 1926–1927; and associate professor of mathematics at Stanford until 1931, when he left to become a faculty member at Columbia. He went to North Carolina in 1946.

In 1928 my long-time colleague Harold Bacon received his bachelor's degree from Stanford. Professor Bacon, who provided the principal direction for Stanford's undergraduate program in mathematics until his retirement in 1972, always seemed to his students and fellow faculty members as the embodiment of Stanford ways and history. Through his teaching of the calculus over many years, he probably taught more Stanford undergraduates than anyone else in the history of the University. Often in my travels I would meet a Stanford alumnus and mention that I taught mathematics at Stanford. Invariably, the response would be, "Oh, then you must know Professor Bacon."

The following year Bacon received a master's degree, writing his master's thesis under the direction of Harold Hotelling. He spent the year after working for an insurance company in the mistaken belief that he wanted to be an actuary. He returned to Stanford the next year and began work on his dissertation under the guidance of Harald Bohr, who was a visiting professor at Stanford that year. The dissertation was completed a few years later under the supervision of Professor Uspensky. Bacon was an acting instructor during this time, and in 1933 he became one of the three instructors in mathematics at Stanford. He was promoted to assistant professor in 1936.

The year 1929–1930 saw the appointment of James Victor Uspensky as an acting professor of mathematics. He was an acting professor again in 1930–1931 and a professor of mathematics from 1931 until his death in 1946. He graduated from the University of St. Petersburg in 1906, receiving his Ph.D. there in 1910. He was a Privat-Dozent in 1912–1915, a professor in 1915–1923, and a member of the Russian Academy of science from 1921 on. In Leningrad (then Petrograd) he was the teacher of the famous Russian number theorist I. M. Vinogradov. I am told that Uspensky visited America for a year in the early twenties. Upon his return to Russia he was interviewed by an agent from the NKVD (a predecessor of the KGB), who asked him how he liked America. Uspensky disarmed his interviewer by saying, "I loved it. It is a place of great opportunity, and if only I were a young man I would emigrate. But I am a member of the Academy of Science, and my career is established here. I am too old to start over again." The NKVD agent evidently reported that Uspensky was reliable and sound in his views. Thus, when Uspensky did decide to come to America a few years later, he came in style on a Soviet ship with his passage paid for by the government. This was in marked contrast with the case of Besicovitch and Tamarkin, who walked long distances through the woods to cross an uncontrolled stretch of border into Latvia.

Uspensky was conservative in his views. Once during a student's oral examination for the Ph.D. Uspensky asked the student to prove the existence of transcendental numbers. The student responded by showing that the algebraic numbers are countable, while the real numbers are not. Uspensky replied, "Yes, yes, these set-theoretic considerations are all very nice, but can you *prove* the existence of transcendental numbers." Fortunately, the student was familiar with Liouville numbers and could show their transcendence.

After Blichfeldt became chairman in 1927 the Stanford mathematics department had a steady stream of major mathematicians as visiting faculty, mostly for the summer quarter. A charming lecture given by Harold Bacon at a recent meeting of the Northern California Section of the MAA recaptures the flavor of the area's mathematical life during the thirties. I quote verbatim from Harold's remarks:

> When I was asked if I would make a little talk at this luncheon, it was suggested that, perhaps, I might tell you something about what mathematics was like at Stanford 'in the olden days', or that I might give you a brief account of the organizing of the Northern California Section. It occurred to me that I might do a little of both, so I shall take the liberty of 'setting the stage' of the early thirties as a background for the organization of the Section in 1939. As a start I checked the Monthly for 1933 and found in the membership list of the Association that there were 31 individual members in Northern California and Nevada distributed as follows: Atascadero 1, Berkeley 12, Chico 1, Davis 1, Fresno 2, Morgan Hill 1, Oakland 1, San Francisco 3, Stanford–Palo Alto 6, Stockton 1, and Reno 2. A similar check in the 1939 monthly shows a total of 30 individual members, similarly distributed. Clearly, this was not a period of spectacular growth! In fact our mathematical community was pretty much scattered with a concentration in the San Francisco Bay Area. There were occasional meetings of the American Mathematical Society in the neighborhood, but even these were not heavily attended. Our situation is illustrated by a remark made by Stanford's Professor Uspensky. He, Professor Blichfeldt, Max Heaslet, and I were riding in Blichfeldt's car to one of the mid-thirties meetings in Berkeley. We speculated on the number of colleges, universities, and other organizations that would have members at the meeting. One of us guessed ten, another a dozen. It was remarked that, if the meeting were in New York, there would be a hundred or so. Whereupon Uspensky said very solemnly, 'Yes, we must recognize that we live in a remote province.'

> But to look ahead for a moment, I can tell you that in 1949 there were 91 individual members of the Association in our area, and by 1960 there were over 500.

> Stanford in the 1930's was Stanford during the Great Depression, and funds were hard to get to finance such things

as visiting lecturers, faculty, and conferences. All of our educational institutions were on short rations. But some skillful and sympathetic administrators somehow managed to squeeze out some modest sums for such purposes. For instance, at Stanford we were fortunate to have Harald Bohr for the full year 1930–1931, and in the summer quarters we were able to have as visiting faculty Edmund Landau (1931), Gordon Whyburn (1932), Marshal Stone and George Pólya (1933), Dunham Jackson and Dick Lehmer (1934), Gabor Szegő (1935), Rudolph Langer (1936), Lawrence Graves (1937), Gordon Whyburn (1938), Emil Artin and Arnold Dresden (1939), and Artin again (1940). Our regular full-time faculty was 9 in 1930–1931 and 7 in 1938–1939; last year (1984–1985) it was 31. In those pre-war depression years times were tough, but we had good work and good fun just the same. I could tell you dozens of stories, but I shall spare you all but a few.

I mentioned our Professor Uspensky a moment ago. He was a kind and gentle, soft spoken man; quite formal in manner. But he liked to seem quite 'tough' — Once I was at his home at a small gathering of graduate students, and he was making a vigorous argument upon some political theme. Suddenly he drew himself up and announced, 'I have Tartar blood in my veins. That is why I am so fierce!' And once he gave me reprints of two of his papers that he had written in Spanish and published in an Argentine journal. I thanked him, but had to confess that I could not read Spanish. 'Well', said he sternly, 'learn it!'

Bohr was a very kind man. For instance, I remember my being in professor Blichfeldt's office shortly after I returned to Stanford in 1930 to continue my graduate work after my master's degree and a year's absence working for an insurance company under the mistaken impression that I wanted to become an actuary. Blichfeldt and I were discussing my getting started on work that might lead to a dissertation. Just then Bohr came into the office. Blichfeldt turned to him and, indicating me, said 'Here's a man who is looking for a thesis topic. How would you like to suggest one, then be his adviser?' Bohr bowed, smiled and very courteously replied, 'I should be honored.' He generously acted as my supervisor for the remainder of the year he was at Stanford. When he left, I was most fortunate to have Uspensky take over and see me through to the completion of my work on the dissertation. It was indeed a great privilege to have two such inspiring men as my friends and advisers at the beginning of my career. You may sense that in those days procedures and academic 'red tape' were minimal!

Landau was a man of commanding presence with a real sense of humor, an enthusiastic lecturer, meticulously dressed in a somewhat formal fashion. He was particularly annoyed by chalk dust. In those days the blackboards in our department classrooms were of black slate, and we had rather soft chalk — white, yellow, red, green and other colors. Landau would write in unusually large script, quickly filling the front blackboards. He would sometimes dart about the room and write on the sidewall blackboard — once he even climbed over a couple of chairs to get to the board on the back wall. But then, the boards must needs be erased so that the writing could go on. Landau abhorred the usual felt erasers — too much dust. So, on the first day of his 8:00 and 9:00 classes, his assistant brought in a granite-ware kettle in which were a sponge and some water. Since she adamantly refused to use the sponge on the blackboard, Landau himself (shades of Göttingen with assistants who did the erasing!) would grasp the sponge, wring the water out on the floor, make some passes at the board, and then call on one or two students or visitors to his lecture to come up and dry off the slate with paper towels. A very ineffective method of drying! The lecture would continue. But the slate was still slightly wet, so half the chalk marks didn't show. Eventually the board dried, however, and normal conditions returned. But the paper towels usually got on the floor where they mingled with water and various scraps of white, yellow and red chalk. After two hours of being walked on, these additions to the bare wood floors produced a, shall we say, cluttered appearance. As it happened, the 10:00 class that followed these first two classes in this classroom was a course in Education — something like 'The administration and care of the School Building and Classrooms.' I fear that those students had a rather spectacular illustration of the neat and orderly classroom! Incidentally, on the last day of classes, Landau made a graceful and humorous farewell speech in his heavily accented English. His last 'goodbye' ended with the request, 'Please preserve the sponge to remember me by!'

Landau's MWF 8:00 class was a graduate level course, 'Selected topics from the Theory of Functions', while the MTWThF 9:00 class was primarily intended for high school teachers and other interested students on 'Foundations of Arithmetic' — essentially Landau's *Grundlagen der Analysis.* Towards the end of the quarter Landau and his wife gave an evening party to all his students in the faculty home they had rented on the campus. It began at about 8:00 P.M., and there were excellent and bountiful refreshments, good conversation, amusing 'parlor games', good fun. As the time passed rapidly, and the clock neared midnight, and even after, people decided that they really ought to go home, so they gradually said their thanks and goodbyes. It turned out that Landau feared that people had not had a good time because they did not stay on until 4:00 A.M. or thereabouts. That seems to have been the time the Göttingen parties usually broke up.

Besides being a distinguished mathematician and teacher, Dunham Jackson (University of Minnesota) was an inspired composer of Limericks of the quite respectable sort! In fact, he and I had a fairly extensive correspondence in Limerick form, each Limerick written and mailed on what we used to call a 'penny postcard.' I had purchased Jackson's book, *The Theory of Approximations,* Vol. XI of the American Mathematical Society Colloquium Publications, and I took it into his office and asked him to autograph it. I suggested that, in doing so, he write a Limerick for me. He immediately picked up the book, and without lengthy cogitation wrote on the flyleaf:

> There was a young Fellow named Bacon
>
> Whose judgement of books was mistaken
>
> In a moment too rash
>
> He relinquished some cash
>
> And his faith in the Author was shaken
>
> (August 17, 1934)

I must add that my faith in the author was by no means shaken; it was greatly reinforced!

In 1936 Emil Artin was offered a professorship at Stanford. Artin wanted to accept but was refused permission to leave Germany. As he later told Dave Gilbarg, who was his student at Indiana, "My work was considered too valuable to the Fatherland for me to be allowed to leave. The next year they kicked me out of the country."

The development of mathematics in northern California was much strengthened in 1934 when G. C. Evans came from Rice to become chairman of the mathematics department at Berkeley. Even before his arrival at Berkeley he had arranged for the appointment of C. B. Morrey as an instructor beginning in 1933–1934, and Hans Lewy was appointed lecturer beginning in 1935. With the arrival of Gabor Szegő as head of the Stanford department, Stanford and Berkeley began their development of a major faculty group in analysis.

The roster of the department in 1938 when Blichfeldt retired was the following:

> *Professors:* H. F. Blichfeldt, W. A. Manning, J. V. Uspensky.
> *Assistant Professor:* Harold Maile Bacon.
> *Instructors:* Helen Glover Brown, William H. Myers.
> *Acting Instructors:* Charles R. Bubb, Carl Douglas Olds.

3. The Szegő Period

Gabor Szegő was appointed professor of mathematics and executive head of the department of mathematics at Stanford upon the retirement of Blichfeldt in 1938. Szegő was a distinguished Hungarian mathematician. He studied in Berlin, Göttingen, and Budapest, and received his Ph.D. from Vienna in 1918 while in military service. He was an assistant in Budapest in 1919 and 1920 before becoming a Privat-Dozent at Berlin in 1921. The dissertation for his habilitation was a fundamental paper on the development of an arbitrary function in orthogonal polynomials. He was an "extraordinary" professor at Berlin from 1924 until he succeeded Knopp as professor in Königsberg in 1926. In 1934 he left Germany for the United States as a result of the rise of Hitler. He was a professor of mathematics at Washington University in St. Louis from 1934 until he came to Stanford in 1938. He and George Pólya were the authors of the famous problem book: *Aufgaben und Lehrsätze aus der Analysis.* While at Washington University he wrote a volume, *Orthogonal Polynomials,* for the Colloquium series of the American Mathematical Society.

Before coming to the United States, Szegő established a number of fundamental theorems in complex analysis, potential theory, Toeplitz forms, and special functions. His thesis contained a limit theorem which has been fundamental for the study of Toeplitz forms. He showed that the 'transfinite diameter' of a compact subset of the plane is just its logarithmic capacity and that the Hausdorff and capacitary dimensions of a set are the same. Szegő was the first to introduce the study of orthogonal polynomials on the unit circle, and for twenty years he was the only person studying them. Later his work on recursion methods for them was fundamental for work in the electronic synthesis of speech (*cf.* the note by Kailath in **[Sz]**). Related to this work is his development and use of the Szegő kernel, which was later studied by Garabedian and which plays an important role in contemporary work in several complex variables. One of his theorems on best approximation by the boundary values of holomorphic functions on the unit circle became a cornerstone of Beurling's theory of invariant subspaces of H^2. Szegő had extreme technical facility: Hans Lewy once sought to prove the positivity of the coefficients of a certain multinomial series. He wrote to Szegő, who responded a week later with a beautiful three page proof involving triple integrals of Bessel functions. Carl Loewner, who was a contempory of Szegő's at Berlin, called Szegő a virtuoso and related that I. Schur referred to him as "der begabte Szegő."

During Blichfeldt's time much of the emphasis of the mathematics department had been on algebra, group theory, and number theory. It is reported that Szegő was met on his arrival by Fred Terman, head of the electrical engineering department, and Felix Bloch, a professor of physics who later received the Nobel Prize for his work on nuclear magnetic resonance. They wanted to make sure that Szegő would arrange for the type of mathematics courses that were important for students in engineering and physics. Szegő was just the man for their purposes, and when I arrived as an undergraduate transfer student in 1946, there were many beautiful courses in analysis, useful for engineering and physics. One was an undergraduate course with the nondescript title "Advanced Calculus." Although this course was then taught by other faculty, it was clearly designed by Szegő, and the contents reflected his elegant style. Besides a thorough treatment of ordinary differential equations, it provided a magnificent treatment of the classical partial differential equations of mathematical physics. Topics included Bessel and Hankel functions, Legendre and associated Legendre Polynomials, spherical harmonics, orthogonal expansions in eigen-functions, boundary and initial value problems.

I later took courses from Szegő on the calculus of variations, mathematical methods in physics, and transform theory. Gabor Szegő was the best classroom teacher I have ever had the pleasure of taking courses from. His lectures were elegant, and covered the important material in what looked easy. He always used a direct approach to his topics that seemed natural, however, rather than some clever shortcut which might be a quicker way to prove things. My course in functions of a complex variable was taught by George Pólya for the first term and by Gabor Szegő for the second. Virginia Voegeli (later Virginia Royden), Lincoln Moses (who was beginning his graduate work in statistics, and who was later professor of statistics and graduate dean at Stanford), and I used to sit in the back row of this class. Sitting next to us were Fred Terman, then dean of engineering, and Hugh Skilling, then head of electrical engineering.

Szegő believed that one built a strong department by first building strength in a specific area, and he concentrated the majority of his appointments in classical analysis, particularly in complex function theory. His first appointment made at Stanford was that of Albert Charles Schaeffer as instructor in mathematics for the year 1939–1940. Al Schaeffer had received his B.S. from Wisconsin in 1930, his Ph.D. from MIT in 1936, and had been an instructor at Purdue before coming to Stanford. The following year saw the appointment of T. C. Doyle as instructor. By the time I arrived at Stanford, he had left to begin his career at the Naval Research Laboratory, but "D-" Doyle was already a legend among students and faculty.

George Forsythe, who had just received his Ph.D. from Brown, was appointed instructor in 1941. He was a conscientious objector during the war and left the following year to perform his alternative service. He was replaced as instructor by John G. Herriot, who had been a fellow graduate student with Forsythe at Brown. Forsythe returned to Stanford as a professor of mathematics in 1958. When a formal division of computer science was formed as a subunit of the department of mathematics in 1962, he and Herriot constituted its original faculty, and Forsythe was the architect and first chairman of the department of computer science when it was established in 1964.

The Register for the year 1942–1943 lists the appointments of Donald Clayton Spencer and George Pólya as associate professors, and the promotion of A. C. Schaeffer to associate professor.

George Pólya holds a special place in twentieth century mathematics, not only for his original and lasting contributions to pure and applied mathematics, but also as a great teacher of mathematics and for his contributions to the teaching of mathematics through his seminal work in heuristics and the methods of problem solving. He studied at Budapest and Vienna, receiving his doctorate from the former in 1912. He was at Göttingen from 1912 to 1914. Then, at the invitation of Adolph Hurwitz, he took his first teaching position at the ETH in Zurich, where he was to stay until 1940 and to which he returned for frequent visits thereafter.

At the suggestion of G. H. Hardy, Pólya was awarded the first international Rockefeller Fellowship in 1924. This was used to spend the year at Oxford and Cambridge with Hardy and Littlewood. Thus began the long friendship and collaboration with these mathematicians, one outcome of which was the famous book *Inequalities* by Hardy, Littlewood, and Pólya. While Pólya was at Cambridge, Hardy was in the midst of his campaign to reform the mathematics Tripos and asked Pólya to take the exam unofficially. Hardy expected Pólya's poor showing would demonstrate that most of the questions on the Tripos were irrelevant to "modern continental mathematics." Unfortunately for Hardy's plan, Pólya's performance was the best on the examination, and he would have been named Senior Wrangler if he had been a student.

In 1940 Pólya left Switzerland, and, after two years at Brown and Smith, came to Stanford, where he remained for the rest of his academic career. Pólya was one of the most popular teachers at Stanford. When Pólya became emeritus in 1953, Terman, who became provost shortly thereafter, and who had attended some of Pólya's courses, used the excellence of Pólya's teaching as an argument to modify strict rule that emeritus faculty no longer taught. Thus Pólya became the first Professor Emeritus at Stanford recalled to active duty. He taught nearly full-time for a decade, and part-time for many years thereafter. The last course he taught was combinatorial analysis for the computer science department in the Autumn Quarter of 1977. He also celebrated his ninetieth birthday that quarter.

Pólya's doctoral dissertation was on probability. Since there was no one at Budapest in this subject, he wrote without an advisor. He continued his study of probability, and early papers explored aspects of geometrical probabilities. He may have been the first person to use in print the term "Central Limit Theorem" to describe the normal limit law in probability. Pólya also worked on characteristic functions in probability theory, for which there is a "Pólya criterion." One example of his work is the "Pólya urn scheme," which is often used as a model to describe the spread of contagious diseases. An offshoot of this model is the "Pólya distribution." He was also the first person to investigate "random walk," a phrase he originated. In 1921 he showed that a random walk in a plane almost surely returns to its starting point, but in three dimensions it almost never returns.

Pólya's most profound and difficult work is in the theory of functions of a complex variable. He was one of the pioneers, along with Picard, Hadamard, and Julia, of the modern theory of entire functions. It is an indication of the level of Pólya's contribution that the language of the subject contains such phrases as "Pólya peaks," "the Pólya representation", "the Pólya gap theorem", "the Pólya-Carlson theorem," "Pólya's 2^z theorem," etc. Some of Pólya's most interesting work in this area concerns the zeros of entire functions. One paper of Pólya's in 1926 came close to proving the Riemann hypothesis. Although it failed to do so, it led to further developments, including some in statistical mechanics.

Pólya was much interested in geometry and geometrical methods, especially those involving symmetry. In 1924 he described the 17 types of symmetry in the plane. The Dutch artist M. C. Escher studied this paper, and soon after, some of the additional symmetries found by Pólya began to appear in Escher's etchings and prints. Pólya and Escher corresponded with each other prior to the second world war. Pólya's interest in symmetry emerged again in

1935 in a series of papers on isomers in chemistry, culminating in his monumental paper in 1937 on groups, graphs, and molecular structures. One of the high points in the history of combinatorics, this paper showed how to count essentially different patterns, patterns that could not be changed into each other by geometrical transformation such as rotation in space. Pólya's work was accessible and comprehensive, and the principal theorem is now called the "Pólya Enumeration Theorem." Found in any combinatorics text, it provides a powerful and subtle technique for counting graphs, geometrical patterns, and, not surprisingly, chemical compounds.

In his later years Pólya became much concerned with problems of the teaching of mathematics. Even before coming to America he had started a manuscript for his book *How to Solve It,* which was published by the Princeton University Press in 1945. It proved to be very popular and has now sold more than a million copies. It has been translated into fifteen languages. After this came the two-volume set, *mathematics and Plausible Reasoning* (1954), again illustrating some of the heuristic principles set out earlier in *How to Solve It,* and in some of his articles. That was followed by a more elementary set of books, *mathematical Discovery,* in 1962 and 1965. These works established him as the foremost advocate of problem solving and heuristics in his generation. Though he had distinguished antecedents from Descartes to Hadamard, the latter having also written about heuristics and the psychology of problem solving, Pólya nevertheless is the father of the current trend toward the emphasis on problem solving in mathematics teaching.

Donald Spencer had been a premedical student as an undergraduate at Colorado, and began simultaneous studies in medicine at Harvard and in aeronautics at MIT. He soon found that he preferred engineering and mathematics to medicine. After receiving his master's degree in aeronautical engineering from MIT in 1936, he studied at Cambridge University, receiving his Ph.D. in 1939 under the direction of J. E. Littlewood. He was an instructor at MIT from 1939 until he came to Stanford in 1942. In 1944–1945 Spencer worked with Max Shiffman in the Applied Mathematics Group at NYU, a research group established by the applied mathematics Panel, a subsidiary of the Office of Scientific Research and Development. They were concerned with some problems on impact and splash, developing a mathematical description of the behavior of the shape of the surface of a liquid into which a solid sphere is dropped.

His dissertation with Littlewood had been on mean p-valent functions, and at Stanford he began his fruitful collaboration with A. C. Schaeffer on variational methods in conformal mapping. This led to the determination of a number of coefficient and other regions of variation for the class of schlicht functions. He and Schaeffer received an ONR contract at Stanford to support their work on conformal mapping. They explicitly calculated the region of variation of the second and third coefficients of a normalized schlicht function. They were able to have models of three dimensional cross-sections of this region cast and machined in aluminum by a local firm specializing in the precision casting and machining of blades for jet engines. This work gave a new proof of Loewner's result that $|a_3| \leq 3$. Their goal, of course, was to prove $|a_4| \leq 4$. They knew this was true for points in the coefficient body sufficiently close to those corresponding to the Koebe function, and they also had an estimate for the continuity of points on the boundary of the region. As Spencer told me, this would enable them to do the inequality for a_4 if "only we could integrate 10^6 differential equations numerically with sufficient accuracy." That was far beyond the possibility of computation at that time, but some years later Garabedian and Schiffer greatly improved the estimates Schaeffer and Spencer. They sufficiently reduced the number of equations that needed to be integrated so that Garabedian could integrate them by hand on a Marchant Calculator.

In 1943 Mary Virginia Sunseri was appointed to an instructorship in mathematics. She became an assistant professor in 1948 and later an associate and full professor, becoming Professor Emerita in 1986. For forty-three years she was to be one of the mainstays of our freshman-sophomore teaching. In her teaching career at Stanford she has taught more students than anyone else in mathematics except for Harold Bacon, and has been the much respected advisor of many generations of Stanford undergraduates. She has received many awards at Stanford for her teaching and university service.

Paul Rosenbloom earned a Ph.D. at Stanford in the early forties working under the supervision of Gabor Szegő. He was later to become a professor at the University of Minnesota, and sometime later at Teachers College, Columbia. He has written of his study at Stanford **[Ros]**, and I include a brief excerpt from his account:

> In September 1941, I started my graduate work at Stanford after graduating from Pennsylvania. Szegő met me once a week to discuss my progress. My assignment was to do problems in Pólya-Szegő and to read Titchmarsh's *Theory of Functions*. This weekly meeting gave me a feeling of responsibility to have something to report so as not to waste Szegő's time. ... Often Szegő would discuss the ramifications of the problems and related results in the recent literature. He would point out natural questions for further investigations. Szegő had a broad and deep knowledge of general theories, but he preferred to work on

concrete problems which test the power of these theories. He had an amazing technical facility.

We had a weekly seminar. Since I was the only serious graduate student in mathematics, the members were Szegő, Schaeffer, Hille (on sabbatical), Forsythe (then an instructor), Doyle (an assistant professor), and myself. Hille returned to Yale the following year, but Pólya and Spencer joined the department then. Hille lectured on the Nevanlinna theory and on the Gelfond–Schneider solution of Hilbert's problem (later written up for the Monthly). I was assigned to present Brun's twin prime theorem and Ahlfors's thesis. The second year Spencer lectured on multivalent functions, and Schaeffer on Schiffer's variational method in conformal mapping. That was when their collaboration on univalent functions began.

In January 1943, I received an offer of an instructorship at Brown, providing that I could start in February. Szegő arranged for me to take my final orals immediately, even though my thesis wasn't written yet. I was asked to outline my main results thus far, and then the committee probed me with general questions. Pólya began asking me to give the definition of Gaussian curvature in terms of the area of a spherical map by the normals. I protested that I had never studied it, but he insisted that I try to work it out at the blackboard. He said, 'It is not forbidden to learn something from an examination!'

A few years after Rosenbloom had left for Brown, I transferred to Stanford from a junior college. The mathematical student life in 1946 was much more active than in Rosenbloom's time. Veterans were returning en masse from the war, classes were well populated, and there were numerous graduate students in mathematics. Although Al Schaeffer had just left Stanford for Purdue and Uspensky died during my first quarter at Stanford, there were still many inspiring teachers of mathematics. I found the mathematical atmosphere quite stimulating.

The graduate students in mathematics included Albert B. J. Novikoff, Mike Aissen, Ken Cooke, Arthur Grad, Joseph and Betty Ullman, Burnett Meyer, and David Haley. Although I was an undergraduate, I took many graduate courses and came to know the graduate students even better than many of my fellow undergraduates. The distinction between undergraduate and graduate students was not so great then as usual, since most of the undergraduate men were returning veterans of the same age as the graduate students.

There was an active summer term in those years, and I attended the summer term of 1947. I took a course in non-Eulidean geometry from Harold Bacon which was particularly memorable for meeting at 7:00 AM. I also had a course of Farey Series and Continued Fractions from Hans Rademacher, who was a visiting faculty member that summer. I also got to know Peter Lax, who was often visiting at Stanford, and for the summer of 1949 he was accompanied by his new wife Anneli. I remember attending a seminar by Kurt Reidermeister together with Anneli and my wife Virginia, who was finishing her M.S. in physics.

It was my good fortune that Harold Davenport was a visiting professor throughout my senior year. I took undergraduate courses in group theory and number theory from him, much of which I have forgotten, but I remember vividly his graduate courses on continued fractions, geometry of numbers, and analytic number theory. Davenport was extremely friendly and encouraging to students, and he would lunch several days a week with me and Nesmith Ankeny, one of my undergraduate classmates, who later got his Ph.D. from Princeton and is now a professor at MIT. Our discussions ranged over mathematical topics, anecdotes about mathematicians, the differences between American and British ways of doing things, the philosophy of mathematics, politics, history, etc. I have often supposed that discussions at the High Table of an English college were like those we had then. Harold Davenport was the first established mathematician that I felt I knew on a personal basis, and I remember him warmly as a teacher and friend. The department wanted Davenport to stay as the successor to Uspensky. Davenport told me that it was an extremely attractive possibility, but he ultimately decided that he should remain department head at University College, London, because of his responsibility to the young mathematicians he had recruited there. He was a good friend and frequent visitor to Stanford thereafter.

As an undergraduate I had been interested in Hilbert's program for the foundations of mathematics, and remember Davenport telling me about a theorem (possibly Littlewood's theorem about the alternation of the number of primes of the form $4k + 1$ versus the number of those of the form $4k + 3$) that had been proved under the assumption of the Riemann Hypothesis and had later received a very different proof on the supposition that the Riemann Hypothesis was false. The question then arose whether the theorem had really been established. Fortunately for the number theorists' peace of mind, a constructive proof was soon found that was independent of the Riemann Hypothesis (using the Skewes number). Not long after I returned to Stanford as a faculty member, Davenport was again visiting for a quarter, and our discussions continued. He had looked into intuitionism as a result of a talk he had given for undergraduates at London. He found it appealing philosophically, but thought it unsuitable as a basis for mathematics because of the extreme limitations it put on the mathematics one could do.

Harold Bacon has given a vivid description of the mess in Landau's lectures caused by sponging off blackboards full of colored chalk. I was first exposed to this phenomenon during my senior year: It was announced that Marcel Riesz would give a series of four lectures on the wave equation, Clifford numbers, retarded potentials, and the Riemann–Liouville Integrals. The day of the first lecture was warm, and the good-sized lecture room was full of faculty and students. Gabor Szegő introduced Riesz, who promptly took off his jacket and proceeded to lecture in his shirtsleeves and suspenders. A bowl of water and sponge had been provided. After filling up the blackboard, Riesz motioned imperiously to Szegő, who jumped up and washed off the blackboard while Riesz stood by and watched! Now Szegő was very distinguished and autocratic, wore elegant tailor-made suits, and was always regarded with awe by the students and most of the faculty. To see him in the role of a young European assistant to Riesz was startling! After several repetitions of this performance, needless to say, blackboard and floor soon became quite a mess. Sitting directly behind me was George Pólya, who had brought Felix Bloch to hear a distinguished fellow Hungarian. Pólya was somewhat embarrassed by the performance and muttered apologies *sotto voce*. The lectures were brilliant, however, full of new insights into novel mathematics I had never seen before.

Marcel Riesz was a frequent visitor to Stanford in subsequent years, and I always found him friendly and helpful. One of my graduate students invited Marcel to his house for dinner. Besides Riesz there were only graduate students and their wives. The host was quite nervous about how the evening would go, but all went well. After dinner a full bottle of whiskey was brought out and put on the coffee table in front of Riesz. Riesz stayed and talked amiably with the students until the whiskey bottle was empty, whereupon he got up, graciously said good night, and walked home.

Richard Bellman became an associate professor at Stanford, beginning with my senior year in 1948. His degree was from Princeton a few years earlier. He had spent some time there as a junior faculty member and brought the aura of Princeton with him to Stanford. At that time he was interested in the qualitative theory of ordinary differential equations, *a la* Lefschetz, what would now be called dynamical systems. He gave a beautiful course on the subject from which I learned much. At the beginning this course was overflowing with students both from mathematics and from engineering and physics. After one look at the crowded classroom, Bellman talked about prerequisites and assigned a long, highly theoretical problem set involving existence theorems and the Arzela selection theorem. By the second meeting of the class all of the engineering and physics students had dropped the course, but there were still a good many students left. Bellman promptly assigned another lengthy problem set, this one involving very applied numerical problems requiring numerical calculations to a high degree of accuracy. By the third meeting most of the mathematics students were gone, and Bellman proceeded to lecture for the rest of the term to a few hardy souls, mostly auditors. No mention was ever again made of the assigned problem sets, for which no due dates had been specified, and no more problems were assigned. He did, however, require a paper analyzing a particular differential equation as a final exam. I confess that I have sometimes emulated Bellman's method when a course that ought to be a small informal advanced class starts out to be overcrowded.

Max Shiffman became an associate professor beginning in 1948. He had received his degree from NYU in 1938 as a student of Courant. He had been a faculty member there since then, and during the war he was with the Applied Mathematics Group at NYU. Shiffman was largely interested in the calculus of variations and hydrodynamics, but was well versed in a wide range of modern analysis. I learned some of the rudiments of differential geometry from him as well as the use of topological methods in analysis and variational theory.

Max Schiffer first came to Stanford for a short visit in 1947 as a guest of Spencer and Schaeffer. (Schaeffer was then at Purdue, but was frequently on the Stanford campus because of his ongoing research work with Spencer.) They had not previously met Schiffer, and there was speculation among the graduate students about the possibility of friction, since Spencer and Schaeffer considered Schiffer the "competition" in the development of variational methods in conformal mapping. Arthur Grad, who was writing his dissertation with Spencer, kept us apprised of the arrangements for the coming meeting with the great man. As soon as the historic meeting took place, we eagerly sought out Arthur for a blow-by-blow account. Arthur reported that Schiffer had turned out to have charmed everyone and that there were no fireworks. Arthur sounded disappointed!

Schiffer returned as visiting faculty member in 1948 and lectured on potential theory, a course I was fortunate enough to attend. Schiffer originally began his mathematical studies in Berlin with I. Schur and emigrated to Jerusalem in the thirties, where he received his M.A. and Ph.D. from the Hebrew University. In addition to his work on group theory with Schur, he was active in developing variational methods in mathematical physics and in conformal mapping. He was a senior assistant and lecturer there from 1938 to 1946. From 1946 until 1949 he was a research lecturer at Harvard, where he worked with Stefan Bergman along with Nehari, Garabedian, and Springer. He was a frequent visitor to Stanford, where he collaborated with

Szegő on problems in mathematical physics and hydrodynamics and with Spencer on variational problems in function theory. Following a year visiting at Princeton and a year as professor at the Hebrew University, he accepted an appointment as a professor of mathematics at Stanford beginning in 1952. He has been at Stanford ever since.

I spent the academic year 1948–1949 as a graduate student at Stanford writing a master's thesis under Spencer's supervision. The title of this thesis was "Loewner's Kappa Function when the Slit is Analytic." The kappa function appears in the parametrization of schlicht functions that Loewner obtained by growing a slit into the interior of the unit disk, and I was concerned with getting the first few terms of the expansion for it in terms of those for the slit being grown into the disk.

Spencer, as many later generations of students will attest, was an excellent and stimulating man to study with, always friendly, helpful, and quite interested in the work done by his students. His stories of mathematicians fascinated me, especially those of his student days at Cambridge with Hardy and Littlewood. I came to feel almost more like a very junior colleague than a student. This feeling was reinforced by the presence of Paul Garabedian, who had just gotten his degree with Ahlfors, and was spending the year 1948–1949 at Stanford working with Spencer as a National Research Council Postdoctoral Fellow. Paul was an assistant professor at Berkeley the following year, returning to Stanford in 1950 as an assistant professor, becoming an associate professor in 1952.

Shiffman and Garabedian conducted the departmental seminar for the year 1948–1949. Various topics in conformal mapping were treated, including variational theory and the conformal mapping of multiply-connected regions. There were no books on variational methods in conformal mapping at that time, and the only book on multiply-connected regions was that of Julia.

The departmental seminar was an institution of some standing at Stanford. In Paul Rosenbloom's time it was largely a faculty seminar, but by the time I arrived, it had become very much a student seminar. It met for two hours every Thursday afternoon with an intermission in between. All graduate students were obliged to attend and to present assigned talks developing aspects of the theme for the year. Most of the faculty were in regular attendance. For the previous year the seminar had been conducted by Davenport and Pólya on topics in irrational number theory and Diophantine approximation.

Fellow students told me that in earlier years it had sometimes been a harrowing experience for them. According to Albert Novikoff, one of our more flamboyant graduate students and now a professor at NYU, the first hour consisted of the student's attempt to present the assigned topic, while Pólya and Uspensky argued with each other about the adequacy of the student's statements. For the second hour Pólya or Uspensky would demonstrate how the lecture should have been given. The only mitigation for the unfortunate student was that, whichever of Pólya or Uspensky was critical, the other would be supportive.

During the intermission Pólya and Uspensky would bait Schaeffer and Spencer about the latter's ignorance of classical mathematics. Schaeffer was oblivious to this, but Spencer would sometimes rise to the bait and respond by asking Uspensky what the Betti numbers of the sphere were. Uspensky would indicate that this modern stuff was nonsense beneath his notice. Sometimes the argument descended to the personal level, with Uspensky maintaining that the younger generation (i.e., Schaeffer and Spencer) lacked the strength of character and fortitude exhibited by their elders. Spencer recalls only once that he or Schaeffer got the better of the exchange: Uspensky had been holding forth about the degeneration of the younger mathematicians and recounted the story of an ancient Roman who, becoming tired of the world, ordered his servants to construct a huge funeral pyre which he proceeded to walk into. Turning to Schaeffer, Uspensky said, "Would you do that, Schaeffer?" Schaeffer bowed and replied, "After you, Uspensky."

It had always been my intention to go East for doctoral work after finishing my master's degree at Stanford. Don Spencer and Paul Garabedian both told me to go to Harvard and work with Ahlfors. Although that may not have been the best advice I have ever been given, it was certainly the best advice I ever accepted. Paul also instructed me to introduce myself to Stefan Bergman and to ask him for a research assistantship. This I did upon my arrival at Harvard, and my acquaintance with Garabedian, Schiffer, and Spencer were sufficient to obtain an appointment as one of Stefan's assistants.

Stefan Bergman, who joined the Stanford faculty as a professor in 1951, studied engineering at Wroclaw and Vienna. Finding himself strongly attracted to the theoretical aspects of engineering and to problems in pure and applied mathematics, he entered (in 1921) the Institute for Applied Mathematics which had just been established by Richard von Mises at the University of Berlin. In 1930 Bergman was appointed Privat-Dozent at the University of Berlin. His scientific career at the university, however, was cut short in 1933 by the Nazi takeover of Germany. He left Germany and found refuge for some years in Russia. In 1934 he became professor at the University of Tomsk in Siberia. In 1936 he moved to Tbilisi, Georgia, where he stayed until 1937. The success of his stay in the Soviet Union is best appreciated if one observes that some of his students became leading mathematicians in their own right, such as Vekua, Fuks, Kufarev, etc.

In 1937 his position became precarious, because of the increase of Stalinism. An agent from the NKVD, with whom he was aquainted, told him that things would become extremely difficult for foreigners, and advised him to leave the country while he still could. This he did, escaping the fate of Fritz Noether (Emma's brother). He had been with Bergman at that time, remained in the USSR, and was never heard of again.

Bergman left for Paris, where he worked under most difficult conditions. He spent most of his time at the Institut Henri Poincaré, where he wrote a two volume monograph on the kernel function and its applications to complex analysis, which appeared in the series "Mémorial des sciences Mathématiques." Through the help of Hadamard he was able to immigrate to the United States, leaving France just before the outbreak of the second World War in 1939. He was at MIT and at Yeshiva College from 1939 to 1941, and in 1941 he went to Brown, which was at that time a real haven for refugee scientists from Europe. His assistants there included L. Bers and A. Gelbart, who worked out his lecture notes.

In 1945 he joined his old teacher and friend von Mises at the Harvard Graduate School of Engineering. There he directed various research projects until 1951. During his years at Harvard his research projects included an impressive array of associates and assistants. Before my time these included Max Schiffer, Zeev Nehari, Paul Garabedian, and George Springer, and while I was there, my fellow assistants included Philip Davis, Henry Pollak, and Bob Osserman.

My colleague Max Schiffer, who knew Stefan from the time when they were both in Berlin, Stefan as Dozent and Max as student, has written [Sch] of those days in Berlin and of Bergman and his work:

> Von Mises was one of the leading theoreticians in aerodynamics and probability theory, who believed that applied mathematics should be as precise as pure mathematics but that its methods should be feasible and practical. His ideas had an enormous impact on Bergman's scientific outlook. At the Institute for Applied Mathematics Bergman worked on such down to earth problems as the magnetic field in an electric transformer and the distribution of temperature in the stator of a generator. He studied boundary value problems of elasticity and various other problems of potential theory. To obtain a large number of harmonic functions he applied the Whittaker method for creating harmonic functions by means of integrals over analytic functions.

Another mathematician at Berlin who had great influence on Bergman's scientific development was Erhard Schmidt. Shortly after his arrival at Berlin, Bergman participated in Schmidt's seminar and was charged with giving a lecture on development of square-integrable functions in terms of an orthogonal set. As he told me, he misunderstood the task, and instead of dealing with real functions over a real interval, he attacked the problem for analytic functions over a complex domain. He found the task hard but carried it through. This was the genesis of his famous theory of orthogonal functions and the kernel function, which formed his doctoral thesis in 1922. Interestingly enough another student from Schmidt's seminar, Salomon Bochner, was also attracted to the problem of orthogonal systems but developed into a different direction of analysis.

Bergman applied his results on orthogonal analytic functions in fluid dynamics, conformal mapping and potential theory, but also developed the central concept of his theory which is now called the *Bergman kernel*. He soon realized that he could define his kernel function in this case of functions of several complex variables. The subject was still quite undeveloped in the Twenties, and he was one of the founders of this important branch of research. An impressive achievement in this field is his construction of a metric on domains in the space of two complex variables which is invariant under mappings by means of a pair of analytic functions.

Another fruitful idea was his discovery that for a large class of domains an analytic function of several complex variables is already completely determined by its value on a relatively small part of the boundary. He called this part the *distinguished boundary*, but it is now known as the *Bergman–Shilov boundary* of the domain. He then connected the theory of the boundaries of a domain with that of the kernel function by classifying boundary points in terms of the asymptotic behavior of the kernel under an approach to that boundary point.

In 1930 Bergman became a 'Privat-Dozent' at the University of Berlin. His thesis, which he had to submit for his official 'Habilitation,' dealt with the theory of boundary behavior of the kernel function. He was appointed simultaneously to the Institute of mathematics and the Institute for Applied Mathematics at the Berlin University. This was at that time a rare distinction. I was then a very young student present at his inaugural lecture and very impressed by his sponsors von Mises and Schmidt who presided in full academic dress on this occasion. The topic of the lecture was about the theory of a wing of an airplane.

I again became a colleague of Bergman's at Harvard in 1946. After I had shown the relation between the harmonic Green's function of a plane domain and its kernel function it was natural to extend the method of orthogonal solutions to problems of partial differential equations and obtain representations for their fundamental solutions. We worked in close cooperation at Harvard from 1946 to 1950, and I remember those days with nostalgia. In 1952 Bergman became again my colleague when we joined the mathematics department at Stanford University.

Gabor Szegő arranged for me to be a teaching fellow at Stanford for the summer term of 1950, and I earned my keep by teaching the three calculus classes that were offered that term. Visiting mathematicians for the summer included M. Fekete, W. Rogosinski, W. Fenchel, and Walter Hayman. Hayman and Fenchel were accompanied by their families. Szegő solved the housing problem for them by renting a fraternity house for the summer and putting them all there. One morning when Rogo met Fekete as they were both shaving, he asked Fekete if he had spent a good night. Fekete replied, "Not bad. I proved the following theorem ... "

Paul Garabedian had just returned to Stanford from Berkeley, and discussions about function theory abounded. Hayman was working on p-valent functions and successfully applied Pólya symmetrization to them. It was from Hayman that I learned about p-valent functions and the details of the work of Pólya and Szegő on symmetrization in function theory. Spencer was also at Stanford for most of the summer, and I remember a lunch at Rickey's Restaurant with Spencer, Hayman, and Rogosinski. We were talking about the beauties of Pólya Symmetrization, and during the conversation several of the people at lunch, principally Rogo and Walter, used it to give a two line proof of the Koebe "One-quarter Theorem." We all agreed that this was the simplest possible proof. I also learned from Walter about mean p-valent functions and the problem of establishing the "One-quarter Theorem" for them. I worked on this problem quite a bit that summer and kept finding "solutions," but when I would show them to Walter, they would turn out to be founded on a certain amount of wishful thinking. Nevertheless, some of these attempts indicated the proper direction, and a year or so later Paul Garabedian and I succeeded in finally proving the theorem.

In 1951 Carl Loewner, Max Schiffer, and Stefan Bergman came to Stanford as professors, and I arrived as an assistant professor.

Carl Loewner received his Ph.D. from the Charles University in Prague in 1917 and was an assistant there until 1922, when he became a Privat-Dozent in Berlin. In 1928 he left to be an assistant professor at Cologne for 1928–1930 and Prague for 1930–1933. He was professor at Prague from 1933 until the Germans came in 1939. Upon coming to America, he taught at the University of Louisville from 1939 to 1944 and was a research associate at Brown in 1944–1945. He was professor of mathematics at Syracuse until he came to Stanford in 1951. Loewner is probably best known for the work on univalent functions where he gave a method for generating a dense set of conformal maps of the unit disk by means of semi-groups. This gives a representation for the coefficients of a univalent function, enabling him to prove $|a_3| \leq 3$, and this method forms the basis of the recent proofs of the Bieberbach Conjecture. Carl was interested in and knowledgeable about a wide range of mathematical topics: differential geometry, lie groups and semi-groups, matrix theory, geometric topology, complex analysis, and differential and integral equations. He was a popular teacher of graduate students and an excellent dissertation supervisor. While at Stanford he probably directed more Ph.D. students than the rest of the department's faculty combined. Lipman Bers, who wrote his dissertation for Loewner at Prague, once remarked that any mathematics department containing Loewner was fully qualified to give the Ph.D. degree, even if he were the sole member! Carl treated his students as colleagues, inspiring the best to superior work, while exhibiting much patience and help for the slower student. Because of the generosity of his help when needed, it has been said of his students "the weaker the student, the stronger the dissertation."

During my early years on the Stanford faculty, I was closer mathematically to Loewner than to anyone else in the department. I learned a great deal of differential geometry in those days by working or trying to work out problems and theorems for myself. Whenever I got stuck in the process, I would ask Carl how to do it. As I remember, he always knew how. I also learned an enormous miscellany of mathematics by collaborating with Carl on a proseminar for graduate students. A decade later I suddenly realized that, although I was still close to Carl personally, I seemed to have less mathematical contact than previously. I wondered if his outlook was finally aging, but observed that he was still in active contact with students and our youngest faculty — it was I who had grown older in my ways, not Carl!

Shortly after I joined the faculty at Stanford, there was a movement to reform and modernize the mathematics curriculum led by the "Young Turks" consisting of me, Paul Garabedian, Max Shiffman, and Carl Loewner. Although Carl was the oldest in years, he was probably the youngest in viewpoint. Gabor Szegő demonstrated his skill and polish as an administrator by his handling of this reform effort: He promptly constituted a committee consisting of Paul Garabedian and me to rewrite the curriculum and degree requirements for the University Bulletin, although I am sure he thought our ideas were newfangled nonsense.

Szegő was extremely productive mathematically at Stanford and provided mathematical and intellectual as well as administrative leadership for the department. He worked on inequalities for geometrical and other quantities from mathematical physics and, together with Pólya, developed and applied the theory of symmetrization to them. This work appeared in a book written with Pólya, *Isoperimetric Inequalities in mathematical Physics*, Princeton University Press, Princeton, 1951. He collaborated with Max Schiffer on finding estimates and extreme values for the virtual mass of a body in hydrodynamics. In 1952 he published a fundamental paper "On Certain Hermitian Forms associated with the Fourier Series of a Positive Function," which has had major applications to mathematical physics. Not long afterwards he wrote a fundamental paper with Helson at Berkeley on prediction theory for weighted L^2 spaces, and a paper with Mark Kac and W. L. Murdock on "Eigenvalues of Hermitian Forms" which has been influential in problems of numerical analysis and partial differential equations. He also wrote a book with Ulf Grenander on *Toeplitz Forms and their Applications,* which appeared in 1958.

The evolution of the Stanford department of mathematics owes much to the skill and effort of Gabor Szegő. Upon his arrival at Stanford he found a department largely oriented towards number theory and group theory and whose strength, although typical of American departments of that time, was hardly of the caliber of the major European centers. In his tenure as executive head Szegő reoriented the department towards classical analysis and took advantage of the influx of distinguished European mathematicians to build a department of world stature. In the fifties Stanford, along with NYU, Berkeley, and MIT, had become one of the leading departments in classical analysis. With a faculty containing Pólya, Szegő, Loewner, Bergman, Schiffer, Garabedian, me, and later Osserman, and with such visiting faculty as Ahlfors, Bers, and Spencer, Stanford had become the leading center for complex function theory. Szegő presided over this development with old world courtesy and tact, but with an autocratic, almost aristocratic, firmness and certainty of purpose. His knowledge and evaluation of the mathematical activities of the department's faculty was unusual, and I am told that he would read all of the papers published by members of the mathematics department. He ran the department with a grace that seemed effortless and still found ample time to give beautiful, well-organized courses and to maintain an active high-level research program.

At the conclusion of Szegő's period as executive head of the department in 1953 the roster was as follows:

Professors: Gabor Szegő, Harold M. Bacon, Stefan Bergman, Charles Loewner, George Pólya, Menahem M. Schiffer, Max Shiffman.

Associate Professors: Paul Garabedian, John G. Herriot.

Assistant Professors: Gordon E. Latta, Halsey Royden, Mary Virginia Sunseri, Robert Weinstock.

Several of the Stanford mathematicians of this period, Loewner, Pólya, Spencer, and Szegő have had volumes of collected works published. These are noted in the bibliography. In addition, the Pólya Picture Album [P 5] contains a number of photographs of Stanford mathematicians of that period.

4. Formation of the department of statistics

Statistics and probability have a long history at Stanford. Harold Hotelling was a member of the mathematics department in the late twenties. Uspensky, one of whose fields was probability, was professor of mathematics from 1929 until his death in 1946, and the appointment of Pólya in 1942 added more strength to the field of probability. The statistical tradition had continued in economics and the Food Research Institute from the time of Hotelling, and its importance was becoming recognized in engineering. There was also a long history of research and teaching in statistics carried out in the department of psychology. In the middle forties there was a Committee on Probability and Statistics consisting of Allen Wallis from economics, Holbrook Working of the Food Research Institute, Eugene Grant of industrial engineering, and George Pólya from mathematics.

In 1946 Albert Hosmer Bowker accepted Szegő's offer to be an assistant professor of mathematical statistics in the mathematics department. Bowker had studied under Hotelling at Columbia and North Carolina. During the Second World War Allen Wallis had been the Scientific Director of the Statistical Research Group at Columbia, and Bowker had served as one of his assistant directors.

It was expected that Wallis would play a leading role in the establishment of statistics at Stanford. Before the department was established, however, he left Stanford to become chairman of the department of statistics at Chicago and later dean of the business school there. He suggested that Al Bowker would give leadership for the development of the new program in statistics. His suggestion was followed, and when the department of statistics was established in 1948, Al Bowker became acting head, and, subsequently, its first executive head.

M. A. Girshick, then at the Rand Corporation, was recruited to be professor of statistics. Abe Girshick had studied under Abraham Wald at Columbia, and had a distinguished career with the federal government as a statistician before coming to Stanford. Herman Rubin was appointed assistant professor in 1949. Herman Chernoff came from Illinois as an associate professor in 1951, and the following year Charles Stein came from Chicago.

Herbert Solomon, who had done his graduate course work at Columbia, received the first Ph.D. from the new department in April of 1950. Lincoln Moses and K. D. C. Haley had been doing graduate work in the mathematics department before the establishment of the statistics department, and they received Ph.D.s in statistics not long after, Moses in August of 1950 and Haley in August of 1952. Solomon, after receiving his degree, headed the statistics section of the Office of Naval Research and was later on the faculty at Teachers College, Columbia. He returned to Stanford in 1959 as professor and head of the department. Moses, after teaching for two years at Teachers College, returned to Stanford in 1952 to a joint appointment in statistics and the school of medicine. His long career at Stanford includes a term as dean of graduate studies. Haley has been a perennial visiting professor teaching in the department's summer program. The early group of students admitted for graduate work in the new department included Gerald J. Lieberman, who received his Ph.D in 1952 and was appointed to an assistant professorship jointly in statistics and industrial engineering. He originally taught quality control and sampling inspection, but soon began to establish a group in the newly developing field of operations research.

In its early years the department of statistics began the practice of making joint appointments with other departments that made major use of statistics: Quinn McNemar, a professor of psychology at Stanford with a distinguished career in psychological statistics, was made professor of psychology and statistics when the department was established. Kenneth J. Arrow, who had joined the economics department in 1949, was appointed associate professor of economics and statistics in 1950. In the fall of 1952 Lincoln Moses, who had received his Ph.D. from the new department of statistics in 1950 and had spent two years on the faculty of Teachers College, Columbia, was appointed to a joint assistant professorship in statistics and in the department of Community Medicine in the Stanford Medical School, where he began to build up a group in biostatistics.

Not long after this Sam Karlin was appointed professor in mathematics and statistics. Karlin received his Ph.D. at Princeton with Bochner, and spent his early years at Cal Tech. He was a frequent consultant for Rand, and became one of the leaders in the new field of operations research. By the time of his arrival at Stanford he had become interested in "birth and death" processes. This led him into his research on population genetics and to his collaboration with Stanford's department of genetics when it was established some years later.

The department's roster for the year 1968–1969 was the following:

Emeritus: Quinn McNemar

Professors: Theodore W. Anderson, Jr., Herman Chernoff, M. Vernon Johns, Gerald J. Lieberman, Rupert G. Miller, Lincoln E. Moses, Ingram Olkin, Emanuel Parzen, Herbert Solomon, Charles Stein, Patrick Suppes. *Consulting:* William G. Madow

Associate Professor: Bradley Efron

Assistant Professors: Richard Olshen, David O. Siegmund, Paul Switzer, George G. Woodworth

5. The Applied Mathematics and statistics Laboratory

On the recommendation of Al Bowker and with the support of Fred Terman, then dean of engineering, a laboratory for applied mathematics and statistics was established in 1950. The laboratory, under the directorship of Al Bowker, was a significant factor in the development of the mathematical sciences at Stanford in the next decade. Not only did mathematics and statistics prosper through involvement with the laboratory, but also computer science and operations research grew out of activities of the laboratory. In addition the laboratory's work influenced mathematical and statistical development in economics, psychology, and other social sciences. The Applied Mathematical and Statistics Laboratory provided direction and coordination for Stanford's activities in the mathematical sciences, was a channel for university support, and was instrumental in obtaining funding from the federal government. The availability of external funding in those days allowed the mathematics and statistics departments to grow more rapidly than they would have otherwise, since appointments could often be made well in advance of the time when university support was to be expected.

Much of the early government money came from the Office of Naval Research, where the mathematics program was directed by Mina Rees. She had been involved during the war in the administration of mathematical and statistical work with the applied mathematics Panel of the Office of Scientific Research and Development. Schaeffer and Spencer's work on the coefficient region for schlicht functions had received ONR support, and Pólya and Szegő had an ONR project on potential theory and capacity out of which came their beautiful work on symmetrization.

The Office of Naval Research supported basic research which could be useful to the Navy in the long run. This support was focused largely in hydrodynamics, partial differential equations, complex function theory, and other areas of classical analysis, although some support was distributed more generally over pure mathematics, since Rees and her deputy Joachim Weyl believed that the health of mathematics as a whole was important for the development of those parts with more direct application to engineering and physics. I was told by Mina Rees that in the early postwar years the major share of ONR's support of mathematics was concentrated in five centers: NYU, MIT, Stanford, and Berkeley in classical analysis, and Tulane in modern analysis and topology.

There was an amusing anecdote about ONR's support of mathematics in those days: The admiral in command of the Office of Naval Research was rehearsing the staff at ONR for an impending inspection by the Chief of Naval Operations. He asked Jo Weyl, who had succeeded Mina Rees as the director of the mathematics branch, "What do we tell the CNO when he asks why we spend all this money for research in mathematics?" Joachim responded with a typical Weyl metaphor: "The tree of science has many branches, but the trunk is mathematics," to which the Commandant said, "No, no, much too high flown! We must have practical examples of the usefulness of mathematics." Consequently, when the Commandant brought the CNO around to the mathematics branch, Weyl responded to the CNO's inquiries by talking about a number of research projects and pointing out their applications and usefulness to topics in physics and engineering. When Weyl had finished, the Commandant turned to the CNO and said, "Perhaps I can explain it this way, Admiral: The tree of science has many branches, but the trunk is mathematics!"

When Loewner, Bergman, Schiffer, and I arrived at Stanford in 1951, work was being completed on a remodelling of Sequoia Hall to house the Applied Mathematics and Statistics Lab. This provided offices the department of statistics and the new arrivals in mathematics. Paul Garabedian, who had returned to Stanford from Berkeley the year before had his office there, as did Patrick Suppes, then an instructor in philosophy with interests in logic and the philosophy of science. There was space for research associates, visitors, and graduate students. We were soon joined by new faculty, Gordon Latta, John McCarthy, Paul Berg, Bob Osserman, Harold Levine, and James McGregor in mathematics, and Herman Chernoff, Charles Stein, Lincoln Moses, and Gerald Lieberman in statistics, and Sam Karlin jointly in mathematics and statistics. Kenneth Arrow and a group of mathematical economists associated with him were affiliated with the laboratory and located in Serra House next door to Sequoia Hall.

Garabedian and Schiffer had a substantial contract from the ONR for work in hydrodynamics. Research associates on this project included Hans Lewy, who was in exile from Berkeley because of his refusal to sign the Loyalty Oath, and Julia Robinson, who was, in those days, precluded from an appointment at Berkeley by Berkeley's anti-nepotism rules.

One of the major projects from the earliest days of the Applied Mathematics and Statistics Laboratory was in quality control and sampling inspection. Originally headed by Al Bowker, it soon came under the joint directorship of Bowker and Gerald Lieberman, who received his degree from the Stanford statistics department in 1953, and became an assistant professor jointly in statistics and industrial engineering. I understand that procedures and tables devised by this project still form the basis for the Department of Defense handbook on procedures for testing and acceptance.

Those were radiant days, and life in the Applied Mathematics and Statistics Laboratory was permeated with excitement and expectation. Garabedian and Schiffer were doing $|a_4| \leq 4$. A number of people including Schiffer, Garabedian, and Szegő were engaged in hydrodynamics. Game theory and decision theory were in the air and applied to many fields. David Blackwell was a regular visitor in those days and was in the midst of his great collaboration with Abe Girshick in their development of statistical decision theory using game-theoretic ideas. Chernoff and Stein were each involved in basic work in mathematical statistics. Ken Arrow and Herb Scarf were applying the new ideas to mathematical economics. Logic and foundational investigations were vigorously pursued with Chen McKinsey, Pat Suppes, and Jean and Herman Rubin. Suppes also began his study of the methodology of the social sciences using some of the concepts from game theory. With Dick Atkinson, who was later director of the National Science Foundation, and others, Suppes succeeded in applying some of these concepts to psychology. Excitement from work in one field was infectious and affected those working in a very different area. In this atmosphere it was easy to believe in the "Unity of Science" and to expect mathematics to be the ideal tool for its understanding and unification.

6. THE MATHEMATICS DEPARTMENT IN THE FIFTIES AND SIXTIES

Gabor Szegő was succeeded as department head by Max Schiffer in 1954. Schiffer was assisted in that role by Al Bowker who served as co-head of the mathematics department, as well as head of statistics and director of the Applied Mathematics and Statistics Laboratory. It was at this time that I began my long involvement in Stanford administration by serving first as assistant head and later as associate head of the mathematics department. Work with Schiffer and Bowker in planning the department's programs and charting its future growth was an exciting educational opportunity for me.

During Szegő's time as head there was an emphasis on complex analysis, but the following years saw growth in other areas of analysis. The first new area in analysis to blossom was that of partial differential equations and its applications to problems in fluid dynamics and mathematical physics. I think Paul Garabedian was probably the strongest supporter for moving in this direction. His interests were shifting from pure function theory towards applied mathematics even then, and he had collaborated with Hans Lewy while Hans was at Stanford, and he began work with some of the aeronautical engineers coming to Stanford. Of course he was still involved in complex analysis: He collaborated with Schiffer to give the first proof of $|a_4| \leq 4$, with me to prove the 'One-quarter Theorem' for mean univalent functions, and with Spencer to discover an early form of the $\bar{\partial}$-boundary condition. Schiffer and Szegő were also interested in the partial differential equations of mathematical physics and had collaborated on some problems in fluid dynamics and given bounds for various quantities, including the virtual mass of a body in a fluid.

Thus it was natural to add appointments in this area, and senior appointments were soon made during the latter part of the fifties. These included David Gilbarg in partial differential equations and hydrodynamics, Harold Levine in differential equations and mathematical physics, Robert Finn in nonlinear partial differential equations, particularly fluid flow and the Navier–Stokes equation, and Erhard Heinz in theoretical aspects of differential equations. When Heinz left to return to Germany in 1962, he was succeeded by Lars Hörmander, who spent half the year at Stanford and half at the University of Stockholm.

The Stanford mathematics department also began to acquire strength in broader areas of analysis with the beginning of the sixties: The appointment of Ralph Phillips as a professor, and the appointments of Paul Cohen, Karel deLeeuw, and Don Ornstein as junior faculty members, gave representation in functional analysis, measure and ergodic theory, and harmonic analysis. Yitzhak Katznelson became a frequent faculty visitor, adding additional strength in these areas. Paul Cohen was also interested in partial differential equations and had given an example to show the failure of uniqueness for the Cauchy Problem. The appointment of Kai-Lai Chung, jointly with statistics, continued the tradition of strength in probability, complementing the activities of Karlin and McGregor in stochastic processes. The appointment of Hans Samelson in 1961 continued the tradition of work in Lie groups, as exemplified by Blichfeldt and Loewner. He also gave us representation in topology.

Complex analysis, although no longer in a preëminent position, continued to flourish. Robert Osserman and Newton Hawley were promoted to permanent positions, and there was a distinguished group of younger complex function theorists on short-term appointments, including Jim Hummel, Peter Duren, and Simon Hellerstein. Visitors and research associates in the earlier part of this period included Lars Ahlfors, Lipman Bers, Hans Bremmermann, Fred Gehring, and Albert Pfluger. In 1962 Don Spencer returned to Stanford from Princeton, and a few years later he was joined by his colleague Kunihiko Kodaira.

In addition to the activity in analysis and topology, this decade also saw the growth of activity in logic and foundations. This was an area that had its origins in the philosophy department with Suppes and McKinsey. Solomon Feferman began his career at Stanford with a junior faculty appointment in mathematics and philosophy in 1956, and the philosophy department appointed John Myhill and Georg Kreisel to professorships. Although these appointments were primarily in the department of philosophy, they also held rank in the mathematics department. Myhill was later replaced by Dana Scott, who, because of his background, had the larger share of his appointment in the mathematics department. At about this time Paul Cohen, as a result of a dinner conversation with Sol Feferman after a Joint Stanford–Berkeley Colloquium, became interested in consistency proofs for arithmetic and succeeded in giving one after an intense period of work. Of course it turned out to use a nonelementary argument at one point (induction up to ε_0) and had similarities with Gentzen's consistency proof. This led Paul to think about the axiom of choice and the continuum hypothesis and eventually led to his famous proof of their independence.

The Joint Stanford–Berkeley Colloquium was a regular institution in those years. It was started in the early fifties and met on alternate months at Stanford and Berkeley. The speaker was a faculty member (or occasionally a visiting faculty member) from the department at which the talk was *not* held, and a large percentage of the speaker's colleagues would make the fifty mile trip to the other institution. The dinners were eagerly anticipated as an opportunity to meet our Berkeley colleagues and to share mathematical news and gossip. I first met John Kelley at one of these dinners and discovered our common interest in function algebras. It was at one of these Joint Colloquia held at Stanford that Charles Morrey first announced and described his proof that every compact real analytic manifold can be real-analytically embedded in Euclidean space. My interest in the foundations of geometry first became serious as a result of a Joint Colloquium talk by Tarski.

The two departments were close in those days, with numerous collaborations between their members. Loewner and Pólya often gave courses at Berkeley, and for many years Kelley and I conducted a joint seminar on function algebras for many years, sometimes meeting at Stanford, but far more often at Berkeley. As the two departments grew in the later sixties, attendance dropped off. The newer faculty members never achieved the familiar contact of the older ones, and the Joint Colloquia were discontinued. I am pleased to observe that in very recent years there has been a tendency to have joint Stanford–Berkeley seminars in a number of disciplines, although not with the frequency and universality of the old days.

One term Pólya was giving a course at Berkeley and would go up to one morning each week, spend a night or two there, and return to Palo Alto for the rest of the week. Since I had one of my seminars with Kelley on the day he went up, we arranged to go together by taking a bus that left at 7:30 in the morning and arrived about 9:00. Sitting next to Pólya on a bus for an hour and a half each week was a priceless education. I learned much mathematics and even more about well-known mathematicians and their idiosyncrasies. Many of Pólya's mathematical anecdotes have later been published elsewhere. Our conversations were by no means limited to mathematical topics. Since Pólya had lived in Zürich for many years and I had recently spent a sabbatical there, I asked Pólya about the etiquette of using the familiar 'du' forms in German — a topic that native speakers of English find it difficult to comprehend. He told me, among other things, that Hungarian also has familiar forms and their usage is as subtle as the German. He said that since coming to America he and Szegő always spoke English because they could never decide about the use of the familiar when they spoke Hungarian or German. I particularly enjoyed Pólya's accounts of Switzerland and of his times at Cambridge with Hardy and Littlewood. Pólya told me that he felt that he never really knew Hardy personally and that most of his information about Hardy's personality came from Littlewood, with whom Pólya was quite close.

David Gilbarg succeeded Max Schiffer as executive head of the mathematics department in 1959. Although the leadership of the department had become more collective by this time with many of the senior faculty actively involved in departmental decision and planning, Gilbarg led the department with deftness during a period of major growth. He was in the forefront of our recruiting efforts and took the lead in setting high standards for appointment and promotion. The department prospered under his guidance and direction.

I conclude this section by giving the department's roster for the year 1967–1968:

> *Emeriti:* Stefan Bergman, Charles Loewner, William A. Manning, George Pólya, Gabor Szegő.
>
> *Professors:* Harold M. Bacon, Paul W. Berg, Kai-Lai Chung, Paul J. Cohen, Karel deLeeuw, Robert Finn, David Gilbarg, Samuel Karlin, Kunihiko Kodaira, Georg Kreisel, Harold Levine, James L. McGregor, Robert Osserman, Ralph Phillips, Halsey

Royden, Hans Samelson, Menahem M. Schiffer, Dana S.

Scott, Donald C. Spencer.

> *Associate Professors:* Solomon Feferman, Newton Hawley, Donald Ornstein.

Assistant Professors: Michael G. Crandall, Paul Rabinowitz, Mary V. Sunseri, Alan Howard, Amnon Pazy, Ngo Van Que.

Instructors: Robert O. Burdick, Mark A. Pinsky, John B. Walsh.

7. THE DEPARTMENT OF COMPUTER SCIENCE

During the Second World War the Statistical Research Group at Columbia had been calculating mathematical and statistical tables, and this activity was continued at Stanford in the Applied Mathematics and Statistics Laboratory. Gladys Rappaport (later Gladys Garabedian), who had worked for Bowker at the SRG, was in charge of this activity. In those days computers were young women with Marchant calculators, who calculated by following the steps of a program written by Gladys, who supervised and checked their work.

In 1952 Bowker and Fred Terman, then dean of engineering, decided that Stanford should have a computer jointly funded by the Applied Mathematics and Statistics Laboratory and the School of Engineering. Stanford's first machine was an IBM Card Programmed Calculator. The steps of a computation were controlled by a sequence of punched cards and a plug board. It had an electromechanical memory which would store all of 16 words. Although very primitive by of present day standards, it was a great improvement over manual computing. With the advent of this machine a Computation Center was formally established in 1953 with Jack Herriot from mathematics and Alan Peterson from electrical engineering as co-directors. The entire staff of the center consisted of the co-directors and a secretary. Stanford got its first real real computer in 1956 with the installation of an IBM 650. It had electronic operation and 2000 words of drum memory. In a few years this was replaced by a Burroughs 220.

The Computation Center had expanded considerably by 1963, and a new building was built to house it. At Bowker's instigation this building was called "Polya Hall" in honor of George Pólya. When Pólya was asked if the building could be for him, he replied that it was all right as long as everyone understood that he had not contributed any money for it. At that time the Computation Center acquired two new machines, an IBM 7094 and a Burroughs 5000. The program for the Burroughs (and I presume also for the IBM) was still entered on punched cards, but the program was written in the Burroughs' version of ALGOL, and then compiled by the machine. I don't know the memory capacity of this machine, but John McCarthy was considered a dreamer for talking about having a machine with a million words of memory one day.

With his long connection with computing projects Al Bowker believed it was desirable to have faculty members whose area of expertise was in computing. He was strongly supported in this view by Paul Garabedian, who had used a significant amount of hand calculating in showing $|a_4| \leq 4$ and already had some notion of the possibilities for computing in fluid dynamics. Accordingly, it was arranged in 1956 to create a new position in the mathematics department for someone in the field of computing.

The natural choice to fill this position was George Forsythe, and I believe he was the only one ever seriously considered for it. He had spent a year as an instructor at Stanford and was well known to everyone here and had been a classmate of Jack Herriot, who was then the director of the computation center. He worked in numerical analysis applied to partial differential equations and conformal mapping and had been involved in joint work with Schiffer on error estimates in conformal mapping. He came to Stanford from UCLA, where he was part of the group working on the Bureau of Standards machine SWAC (Standards West Automatic Computer).

Forsythe became the apostle at Stanford for the newly emerging discipline of "computer science." At that time most of us thought of computer science as dealing with mathematical or statistical computations and numerical analysis, but George was well aware of the new developments in such fields as programming languages, artificial intelligence, machine translation of languages, and he had a remarkably prescient vision of the future shape of the discipline.

A division of computer science was established in the mathematics department in 1962 with Forsythe as Director. The first year's faculty consisted of Forsythe and Herriot together with several visitors. The new division received strong support and encouragement from the provost, Fred Terman, and from Al Bowker, who was his assistant and later graduate dean.

The first appointment made specifically in computer science at Stanford was that of John McCarthy in 1963. He had been an assistant professor of mathematics at Stanford for three years after he got his Ph.D. (in differential equations) with Lefschetz at Princeton. Before his return to Stanford he was a professor at MIT, where he had developed LISP and was active in the original work on artificial intelligence. It was McCarthy, in fact, who coined the name "artificial intelligence."

This appointment was soon followed by those of Gene Golub and Niklaus Wirth as assistant professors. Golub is a numerical analyst specializing in numerical linear algebra. Niklaus Wirth was the originator of the programming language Pascal.

In 1964 the division became the department of computer science. Al Bowker, then graduate dean, and I, who had become an associate dean of the school of humanities and sciences, were involved at the administrative level with the formation of the new department, and Bowker and Terman had played a significant role earlier in establishing computing at Stanford, but it was George Forsythe who was the guiding spirit in the establishment of computer science at Stanford. He chaired the division and then the department of computer science and was also the director of the computation center from 1961 until he was succeeded by Ed Feigenbaum in 1965. Not only did he have a well conceived notion of what should be done, he also had the Quaker knack of building a consensus for getting it accomplished. Although he was a father figure to the members of his department, he considered himself a numerical analyst and sometimes remarked wryly that numerical analysts had gone from being those odd people in a mathematics department to become those odd people in a computer science department.

The following year William F. Miller was appointed jointly as a Professor in the new department and in the Stanford Linear Accelerator Center, and Edward Feigenbaum was appointed to an associate professorship. The year after saw the joint appointment of George Dantzig as professor of computer science and operations research.

I conclude this section by giving the department's roster for the year 1969–1970:

> *Professors:* George B. Dantzig, Edward A. Feigenbaum, George E. Forsythe, John G. Herriot, Donald Knuth, John McCarthy, Edward J. McCluskey, William F. Miller.
>
> *Visiting:* C. William Gear, James H. Wilkerson.
>
> *Associate Professors:* Jerome A. Feldman, Robert W. Floyd, Gene H. Golub.
>
> *Assistant Professors:* Zohar Manna, D. Rajagopal Reddy.

8. Operations Research

The discipline of operations research began during World War II with the application of mathematical techniques to solve various problems of optimization. Further research in this area was conducted at the Rand Corporation and other places, and Ken Arrow, who held a joint appointment in economics and statistics at Stanford, and Sam Karlin, who came to Stanford in 1956 as a professor of mathematics and statistics, were associated with some of the activity at Rand.

The first formal course in operations research at Stanford was taught by Gerald Lieberman in 1957. In 1960 the provost, Fred Terman, at the urging of Al Bowker, established a committee to consider the future of operations research as a discipline at Stanford. The committee was chaired by Lieberman, who held a joint appointment in statistics and industrial engineering, and included K. J. Arrow from economics and statistics, Sam Karlin from mathematics and statistics, James Howell from the Business School, Pete Veinott from industrial engineering, and Harvey Wagner from industrial engineering and statistics. A Ph.D. program in Operations Research was established as a result of this committee's study. This program had a committee in in charge with the authority to admit graduate students to the Ph.D. program and to control the curriculum for that program. The following year the program was authorized to make appointments, provided they were joint with an established department. The first such appointment was that of George Dantzig, one of the pioneers in the field and the inventor of the Simplex Method, one of the cornerstones of operations research. He was appointed jointly with computer science.

The program flourished under Lieberman's leadership, but the administrative arrangements, which involved reporting to three separate deans, were cumbersome. Hence the program welcomed the opportunity to become a department in the school of engineering in 1967. The new department's roster for the academic year '67–'68 was the following:

Professors: George B. Dantzig, Donald Iglehart, Rudolf E. Kalman, Gerald J. Lieberman, Alan S. Manne, Arthur F. Veinott.

Associate Professor: Frederick S. Hillier.

Assistant Professor: Richard W. Cottle.

REFERENCES

[L 1] Loewner, Charles. Collected Papers, Edited by Lipman Bers. Boston, Mass. Birkhäuser, 1988.

[P 1] Pólya, George. Collected Papers: v. 1: *Singularities of Analytic Functions.* Edited by R. P. Boas. Cambridge, Mass. MIT Press, 1974.

[P 2] Pólya, George. Collected Papers: v. 2: *Location of Zeros.* Edited by R. P. Boas. Cambridge, Mass. MIT Press, 1974.

[P 3] Pólya, George. Collected Papers: v. 3: *Analysis.* Edited by Joseph Hersch and Gian-Carlo Rota. Cambridge, Mass. MIT Press, 1984.

[P 4] Pólya, George. Collected Papers: v. 4: *Probability, Combinatorics, Teaching and Learning in Mathematics.* Edited by Gian-Carlo Rota. Cambridge, Mass. MIT Press, 1984.

[P 5] Pólya, George. *The Polya Picture Album.* Boston, Mass. Birkhäuser, 1987.

[Ros] Rosenbloom, P. C. Studying under Pólya and Szegő at Stanford. In Vol. 1 of [Sz]., pp. 12–13. Reprinted in *A Century of Mathematics in America*, Part II. Edited by Peter Duren, et al. American Mathematical Society, 1989.

[Sch] Schiffer, M. M., "Stefan Bergman (1895–1977) in Memoriam", *Annales Polonici Mathematici,* XXXIX (1981) 5–9.

[Sp] Spencer, Donald Clayton. *Selections.* Singapore. World Scientific, 1985.

[Sz] Szegő, Gabor. *Collected Papers.* v. 3: Edited by Richard Askey. Boston, Mass. Birkhauser, 1982.

Studying Under Pólya and Szegő at Stanford [1]

P. C. ROSENBLOOM

In September 1941, I started my graduate work at Stanford after graduating from Pennsylvania. Since there were few graduate students in mathematics then, and I had covered many of the elementary graduate courses by independent reading, I was permitted to register for a rather light course load. This consisted of a course in differential geometry with Uspensky, one in group theory with Manning, and advanced reading and research with Szegő. My teaching load was 6 hours of college algebra and trigonometry, under the supervision of Bacon, at a salary of $700 for 9 months. After Pearl Harbor that December, there was a tremendous increase in enrollment in mathematics. My load was increased to 9 hours, and Blichfeldt was pressed to come out of retirement to teach again.

It may be of some interest to note that I paid $25 a month for room and board which, because of inflation, was increased to $35 the following year. On Sunday evenings I would splurge with a full-course restaurant dinner for 45 cents. Still I saved enough money to pay my expenses to the American Mathematical Society meeting at Vassar in the summer of 1942.

Szegő met me once a week to discuss my progress. My assignment was to do problems in Pólya-Szegő and to read Titchmarsh's *Theory of Functions*. This weekly meeting gave me a feeling of responsibility to have something to report so as not to waste Szegő's time. I have found that such regular meetings have the same effect on my own students and so have followed this practice ever since. Often Szegő would discuss the ramifications of the problems and related results in the recent literature. He would point out natural questions for further investigations.

We had a weekly seminar. Since I was the only serious graduate student in mathematics, the members were Szegő, Schaeffer, Hille (on sabbatical), Forsythe (then an instructor), Doyle (an assistant professor), and myself.

[1] Reprinted with permission from *Gabor Szegő: Collected Papers* (Birkhauser; Boston–Basel–Stuttgart, 1982) Vol I, pp. 12–13.

Hille returned to Yale the following year, but Pólya and Spencer joined the department then. Hille lectured on the Nevanlinna theory and on the Gelfond–Schneider solution of Hilbert's problem (later written up for the Monthly). I was assigned to present Brun's twin prime theorem and Ahlfors's thesis. The second year Spencer lectured on multivalent functions, and Schaeffer on Schiffer's variational method in conformal mapping. That was when their collaboration on univalent functions began.

I was given a desk in Schaeffer's office and had many opportunities for informal discussions with him. I earned a little extra money by babysitting for him, and also for Spencer the next year.

Szegő gave an evening course for engineers, sponsored by the Army as part of the defense program, on functions of a complex variable. He gave me the job of writing up the notes. His lectures were models of clarity and well spiced with physical applications. It was instructive to observe his careful attention to motivation without any compromise on mathematical standards.

At that time at Stanford, the language examinations for doctoral candidates in the sciences were administered by the scientists. That year Ogg, of the chemistry department, was in charge. When I called him, he said I should just come over to his office with some mathematics books in French and German. At the sight of Kuratowski's book, he said, "I've never studied topology. Read me some of that." After hearing several pages of topology, he tested me in German with Bieberbach's function theory. I have never encountered such a rational procedure at any other university since then!

Every couple of weeks Szegő would invite me home for dinner. It was a charming family circle with his wife, his son Peter, and his daughter Veronica. Since Mrs. Szegő had a degree in chemistry and Peter, a high school student, was already interested in engineering, the conversation at the dinner table often concerned science and mathematics and frequently became quite technical. I first heard the proof of Picard's theorem, starting with the construction of the modular function by conformal mapping, over the Szegős' dinner table. Szegő had an orchard behind his house and would often invite me to pick a basketful of fruit for my landlady to preserve.

At my preliminary examination, which was then oral at Stanford, Szegő asked me to prove Picard's theorem. I had never expected such a question but couldn't deny acquaintance with it since I had heard it from his own lips. He pushed me, in Socratic style, to stumble through it.

My first paper, on Post algebras, written as an undergraduate, had been accepted by the *American Journal*. Szegő said I could submit it as my thesis, and he would arrange for an outside expert to judge it, since there was then no logician at Stanford. But I didn't want to miss the chance to learn analysis from him. I had started, in the winter of 1941, to try to do for the sections of the power series for the error function what Szegő had done for the exponential series in 1924. During my first week on this, I made some stupid mistakes which are still embarrassing to recall. Szegő was patient and kind, and encouraged me to persist.

When I broached the question of a thesis topic, he said I should wait until Pólya came. He visited in the spring of 1942 and, at Szegő's home, suggested that I try to prove the results on sections of power series for entire functions which Carlson had announced in 1924, but whose proofs had never been published.

Szegő continued our weekly meetings, but I met with Pólya more irregularly. At one point, when I was floundering, Pólya suggested that I try to apply problems 107–112 in the first volume of Pólya-Szegő, and these turned out to be crucial. I was soon able to handle entire functions of infinite order and the radial distribution of the zeros of sections of entire functions of finite order. But I was having trouble with the angular distribution. Then one day, Szegő had lunch with me and constructed a heuristic argument on a paper napkin. This hint enabled me to overcome my difficulties. When I told Pólya my results, he invited me to his home. The whole evening he pestered me with questions about why my proof worked, whether my solution was a special trick or an instance of some systematic method. I learned more from that one evening than from any other single experience in my career.

In January 1943, I received an offer of an instructorship at Brown, providing that I could start in February. Szegő arranged for me to take my final orals immediately, even though my thesis wasn't written yet. I was asked to outline my main results thus far, and then the committee probed me with general questions. Pólya began asking me to give the definition of Gaussian curvature in terms of the area of a spherical map by the normals. I protested that I had never studied it, but he insisted that I try to work it out at the blackboard. He said, "It is not forbidden to learn something from an examination!"

Szegő had a broad and deep knowledge of general theories, but he preferred to work on concrete problems which test the power of these theories. He had an amazing technical facility. The late C. Loewner used to call him a virtuoso.

It is perhaps impossible to impart this technique to others, but he did influence me by his taste, his ways of looking at and attacking problems, and his insights into the general significance of particular results. I hope he also influenced me by his broader concern for my progress as a teacher, my teaching load, my material welfare, and even such acts as giving me a ticket to Bartók's concert at Stanford. I don't know whether he was able to continue giving such personal attention to students when so many more began to come to Stanford in the late '50s, but I was lucky to have had this opportunity when it was possible.

Since 1981 Robin E. Rider has been head of the history of science and technology program in The Bancroft Library at the University of California, Berkeley. After a Stanford B.S. and a Berkeley M.A. in mathematics, she received her Ph.D. in history at Berkeley in 1980 with a dissertation on 18th-century algebra. As a Rockefeller Foundation Fellow in the humanities in 1989 she will spend several months at the University of Oklahoma working on a project entitled "Printing and the Book of Nature." Other research interests include the development of operations research in the United States.

An Opportune Time:
Griffith C. Evans and Mathematics at Berkeley

ROBIN E. RIDER

In May 1935, at the end of Griffith C. Evans' first year at the University of California, his colleague at Brown University, R. G. D. Richardson, commented on Evans' plans for the Berkeley department: "He hopes to build up [there] a great center in our subject comparable to Princeton, Harvard, and Chicago . . ." [1]. Evans' vision was one of a top-flight department strong in research as well as teaching, emphasizing both standard areas of mathematics and newer fields and applications, able to recruit talented students and postdoctoral fellows.

Evans' own accomplishments would have added luster to any department. His work ranged over a number of mathematical topics, from analysis to applied mathematics and mathematical economics. He earned his Ph.D. at Harvard in 1910 with a dissertation on integral equations; he later expanded on this work in a set of Colloquium Lectures for the American Mathematical Society, which appeared in 1918 as *Functionals and Their Applications: Selected Topics, Including Integral Equations.* In a series of papers beginning in 1920 he examined surfaces of minimum capacity; he published another AMS volume, *The Logarithmic Potential,* in 1927; and, in line with his abiding interest in the applications of mathematics, he helped to develop dynamic theories of economics. Shortly after the great crash of 1929 he published an influential textbook on mathematical economics.

View of the Berkeley campus in 1972. Evans Hall (dedicated in 1971)
appears immediately to the left of Sather Tower.
(Photograph, University Archives, The Bancroft Library.)

Griffith Conrad Evans (1887–1973), ca. 1939.
(Photograph, University Archives, The Bancroft Library.)

Evans enjoyed the respect of colleagues at home and abroad. The American Mathematical Society elected him vice president for 1924, and beginning in 1927 Evans served as an editor of the *American Journal of Mathematics*. In 1931 he was elected vice president (for mathematics) of the American Association for the Advancement of Science; five years later he would hold the same office for economics. His accomplishments were recognized in 1933 with election to the National Academy of Sciences [2] .

INSTITUTIONAL MANDATE

Forward-looking administrators, scientists, and engineers at Berkeley also envisioned a first-class mathematics department at the University of California. There was a clear institutional mandate for change. Shortly before World War I Berkeley began to invest heavily in both physical plant and personnel, with the intent of building a university with strengths in both research and teaching. The choice of Gilbert N. Lewis to head the chemistry department was an integral part of the program for change. In December 1911 Lewis made his initial visit to the Berkeley campus. A few days later he bombarded the university president with his ideas for change. "I believe that such a reorganization of the Department of Chemistry, as I have here suggested, would make this Department strong and efficient in instruction and research and that with the right men, whom I feel sure could be secured, this laboratory would assume high rank and reputation among the best Chemical Laboratories of the world." Lewis had in mind to build a department that rivaled the research institutes organized on the German model. His initial 2-year plan, for example, called for a new chemistry building plus a 75% increase in the department's annual budget in order to cover new appointments and equipment [3]. He got much of what he asked for, and before long the investment in salaries and research programs paid off, yielding a chemistry department of "national standing" [4].

In the 1920s the University of California applied the technique to physics. It instituted a physics department research budget, which grew to $13,000 per year by 1930, and constructed a new building to house the department's teaching facilities, research laboratories, and offices. With the new emphasis on research, less productive faculty members saw their job security evaporate; meanwhile, efforts at recruitment redoubled. Among the desirable prospects was a young experimentalist from Yale, Ernest O. Lawrence, who rejected Berkeley's initial offer in 1927. The department then turned to another candidate, who accepted. In 1928 Berkeley upgraded their offer to Lawrence, stressing the university's commitment to the support of research. This time the appeal succeeded. Within five years of his arrival at Berkeley, Lawrence obtained a patent for what he dubbed a cyclotron, established the Radiation

Laboratory to house larger versions of the machine, arranged for local industry to contribute obsolete equipment, attracted philanthropic support, and made use of G. N. Lewis' unique store of heavy water in significant experiments. The year 1929 saw two more appointments in physics, including a halftime post for the theorist J. Robert Oppenheimer, who would split his time between Berkeley and Pasadena [5].

As the mandate for change took effect in chemistry and physics, the campus administration and faculty members in science and engineering recognized the pressing need for a similar transformation in the mathematics department. One administrator commented that the department had "not been maintained at the proper standard"; it seemed to most observers "rather dead." The mathematics faculty was not known for its research productivity, and other science departments wanted their own programs matched and buttressed by a mathematics department at full strength [6].

The department needed new blood. There had been only two department heads in the preceding sixty years. The more recent of these, Mellen W. Haskell, had begun to function as department head in 1909 [7]. His retirement, projected for 1933, together with other unfilled professorial posts in the department, presented an appropriate moment for change. (Appendix A lists members of the mathematics faculty for the years 1930–1950.)

Though the time was ripe for significant institutional changes, the harsh economic climate interfered with the search for a new chairman. Since the income of American colleges and universities followed general business trends with about a two-year lag, university income fell to a low point in 1933–1934, and faculties and administrators grappled with harsh economic realities as best they could. Fear spread about academic unemployment, and junior faculty in many institutions waited for the axe to fall. The most common economic adjustments, however, were deep cuts in equipment and supplies, freezes in hiring and capital improvements, and salary reductions, rather than layoffs.

Salary reductions were especially common at colleges and universities in the western United States. In Sacramento, for example, the legislature panicked and ordered severe cuts in funding for the University of California: over the course of two years the budget was cut "from $17,000,000 down to $12,800,000." Raymond T. Birge, new chairman of the Berkeley physics department, predicted that the corresponding salary reduction would have to amount to 10%, while the expense and equipment funds for the department would have to be cut by a third. (He was only slightly too pessimistic: salary cuts at Berkeley were eventually held to 7%.) [8]

Although it faced economic constraints in the salary it could offer, the university committed itself in 1932 to recruiting a new chair for the mathematics department [9]. The projected departmental "reorganization," which would involve dismissals of several junior faculty, would free at least two instructorships as well. The plans for restructuring the department did not sit well with everyone. The retiring chairman not only viewed the emphasis on research as slighting the value of teaching; he also sympathized with the plight of junior faculty and regarded any dismissals "at the very depth of the depression [to be] most cruel" [10].

But University of California president Robert Gordon Sproul persisted, and appointed a blue-ribbon committee to assess the state of the mathematics department and make plans for the future. The committee he chose had an important stake in the promotion of scientific research at Berkeley. Chaired by G. N. Lewis, architect of the Berkeley chemistry program, the committee also included Raymond Birge of physics; Charles Derleth, Jr., dean of engineering; Joel Hildebrand of the chemistry department; Armin O. Leuschner, chairman of astronomy and of the university board of research; and B. M. Woods, chairman of mechanical engineering. The committee decided on a nationwide search. They assigned Hildebrand the task of seeking advice and sizing up the possibilities for new appointments; to fulfill his mission Hildebrand traveled around the country meeting with experts and prospects alike [11]. Acting on his advice the committee unanimously recommended in 1933 that the University of California offer the position of department chairman to Griffith C. Evans of Rice [12].

The salary offer was a generous one of $9000, less the yet unspecified reduction for all faculty salaries. The committee suggested an extra inducement — that Evans be allowed to fill "several additional positions" in the department [13]. The powerful budget committee of the University of California added its unanimous endorsement to the Lewis committee's recommendation to hire Evans, despite their preference for leaving vacancies unfilled as an economy measure: the need for "reorganization of the Mathematics Department, . . . long . . . projected on Professor Haskell's retirement, . . . was again approved even after the present financial prospects were realized." The budget committee underscored the need to invigorate a department "with intimate relations to the welfare of other important departments such as Physics, Chemistry, Engineering, Economics, and Astronomy" (a position in full agreement with that of Lewis' committee). Moreover, the budget committee insisted, reform had to come from outside. "The department is in vital need of leadership, which, in the judgment of the Committee, cannot be found in the present membership; and the availability of an exceptionally strong leader at this moment is an opportunity which . . . should not be allowed to pass, whatever the conditions may be." Sproul accepted these recommendations and sweetened the offer by making clear to Evans that the way was prepared for a complete reorganization of the department [14].

But Evans had strong ties to Rice and to Texas. He was also troubled by the vagueness of the financial arrangements, since he would likely suffer in extricating himself from the depressed housing market in Texas. He thus decided once again to decline a Berkeley offer. This turn of events occasioned anxious correspondence and consultation back in Berkeley. Provost Monroe Deutsch kept Sproul informed: "Hildebrand points out that after the three weeks [he spent traveling] throughout the country there was no doubt in his mind but that Evans was by far the best man for our situation." Hildebrand and Leuschner both expressed concern about settling on an appropriate second choice; certainly it was too late to secure someone for the coming academic year. "Moreover, if he is of the right timber we would probably have to offer him as much as we would offer Evans." And appointing even a temporary chairman from within the troubled department posed serious problems, especially with regard to repopulating the faculty. "[O]ne of the most desirable features in reference to calling a new chairman was that he should have the obligation of selecting the persons to replace those who were to be dropped." Leuschner's and Hildebrand's appeals echoed the budget committee's concern for "the key importance of the work in mathematics for the fields of Chemistry, Physics, Astronomy, etc." Campus scientists, who had "long looked forward to this opportunity to place the work in Mathematics on a high plane," feared that the chance might be lost [15].

Deutsch suggested to the budget committee that they firm up the salary offer to Evans. The committee followed his advice, and also proposed that Evans might delay his move to Berkeley until August 1934 should he consider that necessary. Sproul transmitted the enhanced offer; others chimed in. Professor Woods wrote Evans concerning his service on the special committee, and explained that some years ago he himself had been "for five years a member of the mathematics department." Woods assured Evans that other campus units, including physics, chemistry, and the several engineering departments, were eager to cooperate with him in developing the mathematics program. Woods also thought Evans should know "that conditions are especially favorable just now for the building of a great mathematics department at the University of California. I am confident that opportunities for additions to the staff will come fairly promptly. . . ." In fact, Evans had by May 1933 resolved to accept the revised offer. He took the university up on its suggestion that he might defer his appointment to 1 July 1934, and asked Sproul to keep the appointment confidential until the fall of 1933 [16].

Postponing Evans' move to Berkeley until the academic year 1934–1935 necessitated the appointment of a temporary chairman, Professor Charles Noble, from within the ranks of the department. Other posts also required filling. Again, Hildebrand's nationwide survey work guided the search. Hildebrand submitted to Sproul lists of "young mathematicians fully qualified, in my opinion, for Instructorships in the University of California." Among those Hildebrand had interviewed personally were Charles B. Morrey, with a specialty in real variables; Alfred Foster, interested in the foundations of mathematics and in mathematical physics; and Ralph D. James, a specialist in number theory [17]. Noble consulted with Hildebrand and recommended Morrey and Foster for instructorships. Evans agreed that both would be valuable additions to the department, and Morrey and Foster joined the faculty beginning in fall 1933. Noble also reminded Sproul that opportunities for more new appointments lay ahead: Noble and the elder Lehmer would reach retirement age in four years; Irwin, in five. And Florian Cajori's professorship was still vacant [18].

REPERCUSSIONS OF EUROPEAN EVENTS

Political events in Europe affected opportunities for building the Berkeley department. Interested parties in various departments, unaware of Evans' appointment, called attention to the availability of an unnaturally large pool of potential nominees for the chairmanship, as a consequence of a brutal policy just enunciated in Germany. On 7 April 1933, only three months after Hitler became chancellor, a key piece of the National Socialist program was enacted in Germany. In the name of the Reich, non-Aryans lost state positions, including faculty posts in universities and research institutes; the National Socialists also dismissed Aryans whose loyalty they questioned. Foreigners employed in German universities likewise faced possible dismissal. The *Manchester Guardian Weekly* listed in May 1933 the names of some 200 professors purged in the first month; the list included a dozen mathematicians and physicists. By mid-1935, 100 physicists and 60 mathematicians had been dismissed. In the course of just two years German science thus lost one-quarter of the nation's physicists and one-fifth of its mathematicians. Still others resigned their posts in protest and joined the ranks of the displaced. Where before the German academic system, anchored in the virtues of research, had inspired imitation, now it was left in shambles by racial and political purges.

If the National Socialists did not recognize the value of those whom they discarded, foreign colleagues did. As Monroe Deutsch put the point, here was the chance to "profit by the stupidity and brutality of the German government" [19]. Displaced scholars and scientists looked to Great Britain and the United States for assistance. The announcement of the founding of the Society for Protection of Science and Learning appeared in British newspapers on 24 May 1933; the first contributions arrived in the next day's mail. The SPSL sought to coordinate the placement efforts by individual British scholars and British institutions.

The Society's American counterpart was the Emergency Committee in Aid of Displaced Foreign Scholars, founded a few weeks later. Its membership included the presidents of seventeen American colleges and universities, among them Robert Gordon Sproul of the University of California; its program was the solicitation of philanthropic support to permit American institutions temporarily to add refugee scholars to their staffs.

As a member of the Emergency Committee, Sproul received copies of the lengthening lists of displaced scholars. The Emergency Committee list of June 1933, for example, included 25 mathematicians and 16 physicists, out of a total of 350 names. Two months later a list arrived with more than 500 names, including 59 mathematicians and 67 physicists. The Berkeley physics department annotated the second list for Sproul's office, identifying 6 young physicists of interest to the department [20].

The names of mathematicians also stirred considerable interest. Max Radin proposed appointing Professor Edmund Landau, at least temporarily. Others, including Noble, recommended importing Richard Courant, who had taught at Berkeley in the summer session several years earlier. Deutsch, though less than enthusiastic about the wisdom of appointing Landau, pursued the idea of securing Courant's services. Sproul, however, could not see how Berkeley could "permanently finance a permanent appointment" for Courant at the rank of full professor. He also explained to Noble the need to consult with Evans [21]. Noble appreciated these difficulties and proposed the temporary expedient of calling for one year or one semester "a man of the distinction and adaptability of Dr. Courant" — if not Courant, perhaps Konrad Knopp of Tübingen. Both Leuschner of the astronomy department and Woods of mechanical engineering concurred in the wisdom of offering Courant a temporary appointment for winter 1934 [22].

The invitation was issued in summer 1933 (shortly after Evans had accepted the Berkeley offer), but Courant had already taken up an offer to spend the year at Cambridge University. Berkeley renewed the invitation, this time for fall 1934; the invitation reached Courant just as he accepted a post, perhaps permanent, at New York University [23]. Word of the offer reached Evans, who remonstrated to Deutsch that it was not to the advantage of a university to appoint foreigners to major posts when no reciprocal appointments were possible abroad. "I fear that the result is to discourage the legitimate aspirations of our own young scholars, and that the process has already been carried too far" [24].

Evidently Evans had heard a misleading rumor concerning Courant. Deutsch hastened to inform Evans that Courant had been offered an appointment for one semester only; in fact, Deutsch added, it seemed likely that Courant would instead take a permanent position in New York. Deutsch, who was working hard to marshal support for displaced foreigners, gently chided Evans for his attitude toward their placement in American institutions: "You see, therefore, that there is no plan to fill major appointments with foreign professors. On the other hand, you will, I am sure, agree with me that we should recognize the solidarity of professors and do a little at least to assist scholars who were deprived of their posts for reasons which cannot

commend themselves in the slightest degree to their colleagues in other lands"
[25].

BUYER'S MARKET

By the end of his first year at Berkeley, Evans had a change of heart about
bringing in Europeans to replace retiring faculty and build the Berkeley mathematics department. In spring 1935 R. G. D. Richardson, leader of the mathematics program at Brown University, wrote to Evans about Brown's experience with displaced European mathematicians, among them Hans Lewy,
a specialist in the theory of differential equations, who had come to Brown
for 1933–1934 and 1934–1935. Changes in the curriculum requirements for
Brown University students meant that its mathematics department could not
keep Lewy on. Richardson assured Evans that, "With his fine sensitiveness,
[Lewy] wishes to avoid replacing some American mathematicians; but there
are sure to be places where in graduate work and upperclass work he can
occupy a niche that would otherwise be difficult to fill." Richardson put the
point more forcefully in a letter to Edward R. Murrow of the Emergency
Committee: "Lewy would fit excellently" into Evans' plans to build a great
center at Berkeley [26].

Encouraged by Richardson's optimism that funds might be found elsewhere to tide Lewy over while he sought a permanent position "one or two
years hence," Evans recommended a temporary appointment at Berkeley for
Lewy. "[His] work is already well and very favorably known to my colleagues
at Berkeley and to myself." Indeed, Courant had already talked up Lewy
during his visit to Berkeley in 1932. Lewy's name had also figured on the
Emergency Committee list received in Sproul's office in June 1933. There
Lewy was identified as "one of [the] two best mathematicians living of about
his age" (sic) [27]. Evans thought it "fair to let [Lewy, then 31 years old]
compete and cooperate here with other young men in our department, and I
recommend that we offer him the possibility of a regular appointment. . . on
the retirement of Professor Noble. In the meantime he might be designated
as 'lecturer.'"[28]

The university administration pondered this suggestion, and its implications for the growth of the mathematics department. They queried Evans:
"If only one appointment were to be made to replace these three [Noble,
then D. N. Lehmer in 1937, and Irwin in 1938], would Dr. Lewy be your
choice?" Might not young men now in the department and deserving of salary
increases have some "prior claims" on available funds? Sproul also doubted
the wisdom of making current financial commitments based on impending
retirements [29].

The queries spurred Evans to outline his plans for the department. Though he agreed that "retirement of senior members should in general be compensated by appointments at the rank of instructor," in accordance with Sproul's general policy, Evans also recalled the commitment to a strong mathematics program underlying his own appointment. Evans argued that "[o]ne major appointment of a mathematician with an already established reputation might be defended" on the grounds that it would help the mathematics department "to improve its reputation for original investigation with more than usual rapidity." He proposed trying Lewy out for a year without obligation or expense — a solution made possible by the fund-raising efforts of the Emergency Committee and the Rockefeller Foundation. After the grants ran out, "the stricture of the budget" could be invoked as "an excuse that will be in no way humiliating to him not to keep him on if he does not live up to our highest expectations." Out of fairness to young Americans, Evans envisioned a slower schedule of promotion for Lewy than for indigenous personnel [30].

Evans' plan to appoint Lewy met with administrative approval, and Berkeley then sought grants in partial support of his salary for 1935–1936. The university applied to the Emergency Committee, which had already helped fund temporary positions at Berkeley for other displaced scholars, and to the Rockefeller Foundation for matching funds. Both came through with grants for Lewy's salary for 1935–1936 (and in fact for 1936–1937). By spring of 1936 it was the unanimous recommendation of Evans and the other senior members of the department that Lewy be offered a position as assistant professor at the end of the two-year period. As Sproul later explained to the Rockefeller Foundation, Lewy's "extremely rapid rise in rank [at Berkeley] tells its own story as to how [he] has contributed to the department" as a teacher and as a creative scholar [31].

Economic improvement in the second half of the decade, combined with retirements of senior personnel, also assisted Evans in implementing his plans for change and growth. The department hired a local product, Raphael M. Robinson, in 1937 [32]. In the next three years Berkeley secured the services of Anthony P. Morse, a specialist in real variables, and the number theorist Derrick H. Lehmer, as well as several instructors for one-year stints in the department.

TAKING STATISTICS SERIOUSLY

In these same years Evans began to polish another facet of his plans for the mathematics department. He later recalled that his conversations with the noted statistician R. A. Fisher of University College, London, planted in his mind the idea of a major statistics program. By the time he had assessed the situation at Berkeley, Evans "envisaged California as the place for a really outstanding statistician" — indeed, the place for a program "superior to anything in the West." The administration required some convincing. In 1936 Evans wrote to E. B. Wilson of the Harvard School of Public Health, "I am doing my best to induce the University of California to take theoretical statistics seriously, and thus provide in the Department Mathematics a center for advice and research with respect to the practical applications which are being made in other departments of the University. The ultimate [objective] would presumably be some sort of Board of Statistics with suitable laboratory equipment" [33].

Initially Evans suggested to Sproul that they hire Fisher's successor at the statistical department of Rothamstead Experimental Station, England. In the succeeding months Evans continued to investigate the "possible use of a theoretical statistician in the Department of Mathematics." He sent inquiries to knowledgeable colleagues in various fields, including physicist R. H. Fowler of the Cavendish Laboratory at Cambridge, mathematical physicist E. T. Whittaker of Edinburgh, mathematician Oswald Veblen of the Institute for Advanced Study, statisticians H. L. Rietz of the State University of Iowa and Harold Hotelling of Columbia, as well as Wilson and Fisher.

In order to enlist the support of his allies elsewhere on the campus, Evans recommended to Sproul that a committee be convened "to consider the situation," to include representatives from genetics, economics or commerce, agricultural economics, physics, psychology, and perhaps anthropology [34]. Evans' plan for statistics at Berkeley coordinated well with his work on the mathematical theory of economics; the plan also spoke to the growing importance of statistics to other fields in the natural and social sciences.

Evans added more names to his list of possibilities. By the following year the list included a versatile Polish statistician, Jerzy Neyman, then working at University College, London. Word of Neyman's talents had reached Berkeley via W. E. Deming, a U.S. government statistician, who had worked with Birge. Complicated negotiations eventually brought Neyman to the Berkeley campus in 1938–1939. As in 1933, the wide-ranging search for an appropriate candidate had involved representatives of campus departments — especially Raymond Birge of physics, whose enthusiasm for statistical studies carried considerable weight at Berkeley. As in Lewy's appointment, political events in Europe — in particular, Poland's plight — made available a specialist who fit into Evans' plans [35].

The confidence in Neyman was not misplaced. In his first few months at Berkeley, Neyman established many contacts with "biologists, physicists, economists, etc." He found that statistical applications involved in their research projects "almost invariably throw a new light on various sections of the mathematical theory of probability and frequently suggest new mathematical problems." One such application was the statistical study of mutations in genes bombarded with products of the cyclotron's operation [36]. Although Evans and Neyman would come to differ about the proper place of a statistics program (within the mathematics department or independent of it), Evans added his enthusiastic endorsement to Neyman's requests for space, equipment, and administrative support for the growing statistical laboratory. Evans had high praise for Neyman, noting his great success in working with natural and social science departments and his unique command of both theory and applications [37].

CONFRONTED BY WAR

Neyman was fortunate to escape the fate awaiting many Poles. The Nazification of the European continent, in Hermann Weyl's phrase, added urgency to the task of rescuing displaced scholars. By late 1940 assistance agencies responded with new programs to secure American posts for deposed "scholars of eminence." Sproul, informed of these programs, thought it possible that the University of California could absorb a few of the scholars thus available using either its own resources or this new source of funds. Sproul asked Evans to chair a campus committee to identify four or five foreign scholars "whose presence would be advantageous to the University." The committee, which did not in fact confine its consideration to the list provided to Sproul in late 1940, settled upon five names, including that of the eminent Polish logician and mathematician Alfred Tarski. With assistance from the Rockefeller Foundation's program of Aid for Deposed Scholars, Tarski joined the Berkeley faculty in 1942 [38]. He proceeded, with Evans' encouragement and campus support, to weld an active and prestigious program in logic and the foundations of mathematics. The year 1942 also brought to the Berkeley department another refugee from Hitler's Europe, Frantisek Wolf, a specialist in Fourier analysis and complex variables.

Hitler's march across Europe meant more direct challenges to American mathematics. As president of the American Mathematical Society in 1938–1940 Evans pressed for the establishment of a joint AMS–MAA committee on war preparedness, to address questions of mathematical research and training for the military. Evans also worked with the National Research Council in their compilation of a national scientific roster in 1940. He was convinced that American mathematics could play a significant role in the nation's war effort.

A few months after Pearl Harbor the Berkeley department was already involved in special training programs, and the staff of the Statistical Laboratory found themselves "heavily engaged" in war-related research. Evans joined the wartime Applied Mathematics Panel, part of the National Defence Research Committee (NDRC), which contracted with eleven universities to furnish mathematical service and advice to the Army, Navy, and the NDRC. One such contract for statistical research went to Neyman at the University of California. The statistical research groups funded by the panel tackled probabilistic and statistical aspects of bombing problems, damage studies, and quality control. Neyman's services were in such constant demand that he was often dispatched on lengthy missions elsewhere. In 1944, for example, he participated in the planning of bombing operations against Japan. During his absence, George Pólya, by then at Stanford University, took over Neyman's graduate courses at Berkeley [39].

ASSESSING THE IMPACT

By most measures the first few years after World War II witnessed dramatic growth in the University of California. Nine mathematicians had joined the faculty between 1933 and U.S. entry into the war; seven more came between 1942 and 1948. (See Appendix A.) By 1949, when Evans turned over the chairmanship of the department, the number of faculty members in mathematics was thus double the number when he arrived at Berkeley fifteen years earlier.

Graduate enrollment also swelled. Between 1929–1930 and 1939–1940 the number of graduate students in all fields at Berkeley increased gradually from 2515 to 3539; by 1949–1950 the total had reached 6066. The number of Ph.D.s granted in mathematics grew accordingly. In the fifteen years before Evans' appointment the department granted a total of only thirty doctoral degrees, while the physics department granted more than one hundred. During Evans' fifteen years as chairman, production picked up to 55 Ph.D.s in mathematics and statistics (an increase of 83%), compared to 114 Ph.D.s in physics. In the last year Evans served as chairman (1949), Berkeley granted ten Ph.D.s in mathematics and statistics [40].

In assessing Evans' impact on the mathematics department, it is natural to draw comparisons to the record of achievement in physics at Berkeley. Like the Berkeley research and teaching program in physics, the mathematics department both suffered and benefited from the Depression. Both fields coped with slashed salaries, faint job prospects, and increased enrollments. At the same time, good help was easy to find. E. O. Lawrence took advantage of highly-trained volunteers in populating the Radiation Laboratory; Evans heard in 1934 that "the present is an excellent time to purchase most able young men at a reasonable price." In building the mathematics department Evans recruited some of these young Americans, and profited from the availability of displaced foreigners and from the philanthropy that facilitated their absorption in American institutions. Those years, Evans recalled, were "an opportune time, when there were more fine mathematicians available than there were places for them to go" [41].

Evans had arrived at Berkeley with a mandate to build a research program in mathematics. By comparison to the scaled-up expenses and expectations of Berkeley physicists, the material needs of the mathematics department remained comparatively modest throughout the 1930s. But Evans' insistence on productivity and the infusion of new blood brought results.

In the first year after Haskell's retirement, the interim chairman reported "substantial" research activity in the mathematics department. Evans added his encouragement. In 1936 he reported that the department was devoting "principal effort" to increasing "its contribution to scholarship"; he had also introduced an element of flexibility in the subject matter of graduate seminars in order to follow "modern developments" in mathematics. Two years later Evans praised the "noteworthy" research accomplishments of department members. By the end of the decade Evans urged the rehabilitation of a mathematics publications series at the University of California, and noted the increase in the number of pages published by department faculty in national journals [42]. Ambitious plans for research at the Statistical Laboratory prompted appeals to the campus administration and the Rockefeller Foundation for additional space, equipment, and staff [43].

As they applied their knowledge and skills to matters of national defense, members of both departments learned valuable wartime lessons. Evans, like his colleagues in physics, sat on councils of war and helped shape his profession's response to military needs and demands. Mathematics and statistics at Berkeley contributed to the war effort and partook of the bounty; and in the immediate postwar years the department accustomed itself to look to the government both for research support — what Evans called the "gravy train of government contracts" [44] — and for jobs for its students. The Berkeley mathematics department thus entered the second half of the 20th century with an eminent and productive faculty, the promise of governmental research support, and the prospect of positions for a new generation of students.

A quarter-century before Berkeley had first looked to Evans to provide energetic leadership for its mathematics program. His arrival at Berkeley was engineered by scientists and engineers eager for a revitalized mathematics department; economic considerations at first constrained his plans for building the faculty, but the availability of European scholars and philanthropic support shortly offered unprecedented opportunities; under his leadership the department stood ready to contribute to the war effort and to benefit from the changes it brought to American science. Thus was realized the ambition of a major center of mathematical research and teaching at Berkeley, which long bore the stamp of Griffith C. Evans' personality and vision.

NOTES

1. Part of the research for this article was funded by a grant from the Rockefeller Archive Center. I am grateful to Sheila O'Neill, William Roberts, and Bruce R. Wheaton for their assistance and advice, and to Marjorie W.

Evans, the Rockefeller Foundation Archives, New York Public Library, and The Bancroft Library for permission to quote from manuscript materials. The following abbreviations are used in the notes:

BL   The Bancroft Library, University of California, Berkeley
CU   Uiversity Archives, BL
EC   Emergency Committee in Aid of Displaced Foreign Scholars papers, New York Public Library, Rare Book and Manuscripts Division
GCE Griffith C. Evans papers, BL
JN   Jerzy Neyman papers, BL
RF   Rockefeller Foundation Archives, North Tarrytown, New York
RTB Raymond T. Birge papers, BL.

Quotation, Richardson to Murrow, 3 May 1935 (EC, 109/University of California 1935).

2. Charles B. Morrey, Jr., "Griffith Conrad Evans," U. S. National Academy of Sciences, *Biographical Memoirs, 54* (1983), 127–155; Morrey, H. Lewy, R. W. Shephard, and R. L. Vaught, "Griffith Conrad Evans," University of California, *In Memoriam*, 1977, 102–103. Cf. draft biographical notices in GCE.

3. Lewis to Benjamin Ide Wheeler, 9 Dec 1911 (CU-30, 4/26).

4. "Joel Henry Hildebrand: Physical Chemistry at the University of California," interview by A. L. Norberg (BL, 1980), pp. 7, 14.

5. Robert W. Seidel, *Physics Research in California: The Rise Of a Leading Sector in American Physics* (Ph.D. dissertation, University of California, Berkeley, 1978) pp. 94–104, tables 1–3; J. L. Heilbron, Robert W. Seidel, and Bruce R. Wheaton, *Lawrence and His Laboratory: Nuclear Science at Berkeley 1931–1961* (Berkeley, 1981), pp. 9–13.

6. Monroe Deutsch to the university president, 1 Sep 1932 (CU-5, 1932/100-Mathematics). Even the department chairman recognized the problem. Mellen W. Haskell to Evans, 25 Oct 1928: "We have an excellent staff of teachers, but they have never been strong in research. I think we should have the approval of the administration in building up in that direction . . . . ." (GCE, carton 6/"Leaving Rice"). In R. G. D. Richardson's report in 1933 on five American Mathematical Society publications, the University of California ranked only fifteenth in the number of pages published in the period 1926–1932; it ranked even lower when only pages in the AMS *Transactions* were counted (RF 1.1/200/125/1542). On the level of investment in senior faculty salaries in mathematics, see *Departmental Budget Recommendations, Biennium 1933–35* (CU). Comparison of recommendations for mathematics and physics affords indirect evidence of the need for younger blood in the mathematics department.

7. *The Centennial Records of the University of California*, ed. Verne A. Stadtman et al. (Berkeley, 1967), pp. 90–91.

8. Monroe Deutsch to Eva Deutsch, 1 Nov 1933 (CU-5, 1933/800); Robin E. Rider, "Alarm and Opportunity: Emigration of Mathematicians and Physicists to Britain and the United States, 1933–1945," *Historical Studies in the Physical Sciences*, *15* (1984), 107-186, on 126.

9. Budget Committee [to Robert Gordon Sproul], 17 Sep 1932 (CU-5, 1932/100-Mathematics).

10. Haskell's opinion reported by Deutsch to Sproul, 25 Feb 1933 (CU-5, 1933/1-Mathematics). On teaching versus research in American mathematics in the 1930s, see Rider, "Alarm and opportunity," 134–135.

11. Lewis to Sproul, 7 Dec 1932 (CU-5, 1932/100-Mathematics).

12. Evans' name had come up before. In 1927, for example, W. W. Campbell, president of the University of California, had offered Evans the maximum salary paid at Berkeley (and indeed, more than any other salary in the mathematics department) and had held out the promise that Evans should "assume leadership" in the department at Haskell's retirement. Evans resisted, and cabled back: "Very much appreciate generous offer and magnificent opportunity. Decline regretfully." Campbell to Evans, 14 Dec 1927; Evans to vice-president W. M. Hart, 17 Dec 1927 (GCE, 6/"Leaving Rice"). Cf. Evans to Leuschner, 1 Apr 1933 (CU-5, 1932/100-Mathematics). In fact, the negotiations dragged on until early 1929, but to no avail.

13. Lewis to Sproul, 7 Mar 1933 (CU-5, 1933/1-Mathematics). Cf. draft in RTB ("Lewis").

14. Budget Committee recommendation, 20 Mar 1933; Sproul to Evans, 31 Mar 1933 (CU-5, 1933/1-Mathematics).

15. Deutsch to Sproul, 17 Apr 1933 (ibid.).

16. B. M. Woods to Evans, 5 May 1933; Evans to Sproul, 6 May 1933; Evans to Sproul, 23 May 1933 (ibid.). Evans' official acceptance was in fact dated 23 Oct 1933.

17. Hildebrand to Sproul, 7 June 1933 (ibid.).

18. Noble to Sproul, 12 June 1933 (ibid.). James joined the faculty in 1934, then moved on at the end of the decade.

19. Deutsch to Sproul, 25 July 1933 (CU-5, 320).

20. Lists in CU-5, 1933/800.

21. Deutsch to Hugo D. Newhouse, 16 May 1933 (CU-5, 1933/800); Sproul to Noble, 21 June 1933 (CU-5, 1933/1-Mathematics).

22. Noble to Sproul, 30 June 1933. Noble renewed the appeal concerning Courant in a letter to Deutsch of 9 July 1933 (ibid.).

23. On the Berkeley invitations, see Constance Reid, *Courant in Göttingen and New York. The Story of an Improbable Mathematician* (Springer-Verlag, New York-Heidelberg, 1976), 155–158; cf. Deutsch to Jacob Bellikopf of

the Federation of Jewish Charities of Philadelphia, 21 Sep 1933 (CU-5, 1933/800).

24. Evans to Deutsch, 22 Jan 1934 (CU-5, 1934/1-Mathematics).

25. Deutsch to Evans, 25 Jan 1934 (ibid.).

26. Richardson to Evans, 13 Mar 1935 (CU-5, 1935/1-Mathematics); Richardson to Murrow, 3 May 1935 (EC, 109/University of California 1935).

27. Evans to Sproul, 26 Mar 1935 (CU-5, 1935/1-Mathematics); Reid, *Courant*, pp. 133-134; list received 26 June 33 (CU-5, 1933/800). Out of more than 350 names on the list, 25 were mathematicians'.

28. Evans to Sproul, 26 Mar and 15 Apr 1935 (CU-5, 1935/1-Mathematics). Cf. Deutsch to Duggan, 25 Apr 1935 (EC 109/University of California 1935).

29. Sproul to Evans, 4 Apr 1935; cf. Budget Committee memorandum (CU-5, 1935/1-Mathematics).

30. Evans to Sproul, 15 Apr 1935 (ibid.). Evans projected for 1935–1936 a mathematics department composed of 7 professors, 2 associate professors, 5 assistant professors, 3 instructors (down from 5), and 4 teaching assistants, in addition to Lewy as lecturer. Of these, Evans and the other senior members of the department were keen to recommend promotion of Bing C. Wong, a specialist in algebraic and enumerative geometry, from assistant to associate professor.

31. Evans to Sproul, 13 Apr 36 (CU-5, 1936/1-Mathematics); Sproul to R. Fosdick of the Foundation, 20 Nov 1939 (RF 1.1/205/11/167). Cf. Deutsch to Duggan, 24 Jul 35 (EC, 109/University of California 1936).

32. Cf. Constance Reid with Raphael M. Robinson, "Julia Bowman Robinson (1919–1985)," *Women of Mathematics. A Biobibliographic Sourcebook*, ed. Louise S. Grinstein and Paul J. Campbell (Greenwood Press, New York, 1987), pp. 182–189, on 183.

33. Evans to E. B. Wilson, 25 Nov 1936 (GCE); drafts of "Brief history of the Department of Mathematics" (GCE, box 1).

34. "Brief history" (ibid.); Evans to Sproul, 20 Oct 1936 and 16 Dec 1936 (CU-5, 1936/1-Mathematics). Cf. Constance Reid, *Neyman — From Life* (Springer-Verlag, New York, 1982); "Report on statistics courses" [1 Jun 37], in CU-9, Committee on Courses minutes, vol. 9 (1934–1936 to 1938–1939), appendix for 16 Sep 1937. This was not the first time the question of a statistics center had arisen at Berkeley. E. B. Wilson advised Evans against putting into the mathematics department a "center for advice and research in respect to practical applications of statistics being made in other departments of the university." Wilson recalled that he had made the same statement during his half-year stay in Berkeley in 1929, when faculty and administrators had approached him for "some proposition that would

be attractive to me to come out there and be responsible for the teaching
of theoretical statistics throughout the university." Wilson to Evans, 1 Dec
1936. On the earlier discussion with Wilson, see CU-5, 1928/Mathematics.

35. Evans' offer to Neyman, 10 Nov 1937 (RTB, "Evans"); cf. JN and CU-
5, 1933/1-Mathematics. See the lengthy account in Reid, *Neyman — From
Life*. Evans' advocacy of the importance of theoretical statistics, as well as
his long-standing interest in mathematical economics, was underscored by his
appointment in 1938 as chairman of the section on "Probability, Statistics,
Actuarial Science and Economics" for the International Congress of Mathematicians scheduled for Harvard University in 1940.

36. Neyman to Warren Weaver of the Rockefeller Foundation, 19 Apr
1939 (RF 1.1/205/11/168; copy in CU-5, 1939/1-Mathematics and in JN).

37. Evans to Marston Morse, 28 May 1941 (GCE, carton 14/AMS Preparedness Committee); Evans to Sproul, 14 Nov 1938 (CU-5, 1938/1-
Mathematics).

38. Sproul to Duggan, 29 Oct 1940, and Duggan to Sproul, 7 Nov 1940
(EC, 109/University of California 1940); draft biography, ca. 1968 (GCE,
carton 1); Evans to Sproul, 25 Feb 1941; Sproul to Evans, 11 Feb 1941
(GCE); Thomas B. Appleget, "The Foundation's experience with refugee
scholars," 5 Mar 1946 (RF 1.1/200/47/545a); and the Alfred Tarski papers
at BL.

39. Evans to Sproul, 23 Apr 42 (CU-5.1, Reports to the President, 4/75,
Mathematics 1940–1942). Cf. GCE, carton 14; JN; and Harriet Nathan,
[unpublished] "Report of the President of the University of California 1942–
44" (CU).

40. University of California *Courses of instruction, General catalogue*,
Commencement Programs, and *The Centennial Record*, ed. Stadtman et al.
Doctorate Production in United States Universities 1920–1962, comp. Lindsey
R. Harmon and Herbert Soldz (Washington, D.C., 1963), table 3, gives a
national total of 747 Ph.D.s in mathematics for 1920–1934 and 1219 for
1935–1949; comparable totals for physics are 1240 for 1920–1934 and 2230
for 1935–1949. The Berkeley mathematics department thus outpaced the
national expansion.

41. Heilbron, Seidel, and Wheaton, *Lawrence and His Laboratory*, p. 13;
G. A. Bliss to Evans, 11 Jan 1934 (GCE, carton 1); Evans to Henry Helson,
12 Feb 1966 (GCE, box 1). Cf. Rider, "Alarm and opportunity," 127–128.

42. Reports to the president for mathematics, 1933–1940 (CU-5.1).

43. RF 1.1/205/11/168.

44. "Introductory remarks" (GCE, carton 6/Applied Mathematics).

Appendix A

Mathematics Department Faculty
Appointments for Academic Years 1930–1931 through 1949–1950*

Lecturer, instructor, or professorial ranks based on departmental lists in
Courses of Instruction:

Alder, Henry L.	1947–1948
Andrews, Anne D. B.	1930–1932
Apostol, Tom M.	1948–1949
Barankin, Edward W.	1947–1950
Bernstein, Benjamin	1930–1950
Bernstein, Dorothy	1942–1943
Brenner, Joel L.	1939–1940
Buck, Thomas	1930–1950
Burdette, Albert C.	1944–1945
Cajori, Florian	1930–1930 emeritus
Diliberto, Stephen P.	1947–1950
Dresch, Francis W.	1938–1946
Duncan, Dewey C.	1930–1935
Edwards, George C.	1930–1931 emeritus
Eudey, Mark W.	1944–1947
Evans, Griffith C.	1935–1950
Fix, Evelyn	1944–1950
Foster, Alfred L.	1933–1950
Garabedian, Paul R.	1949–1950
Goldsworthy, Elmer C.	1930–1949
Haskell, Mellen W.	1930–1932; 1933–1948 emeritus
Hayes, Charles A.	1946–1947
Hodges, Joseph L., Jr.	1948–1950
Horn, Alfred	1946–1947
Hughes, Harry M.	1949–1950
Hurst, John W.	1937–1938
Irwin, Frank	1930–1937; 1938–1948 emeritus
James, Ralph D.	1934–1939
James, Robert C.	1947–1950
Kelley, John L.	1947–1950
Lehmann, Erich L.	1946–1950
Lakness, Ralph M.	1949–1950
Lehmer, Derrick H.	1940–1950
Lehmer, Derrick N.	1930–1936; 1937–1938 emeritus
Levy [McDonald], Sophia H.	1930–1950
Lewy, Hans	1935–1950
Loeve, Michel	1948–1950
Mann, Henry B.	1948–1950

McDonald, John H.	1930–1945; 1945–1950 emeritus
Morrey, Charles B., Jr.	1933–1950
Morse, Anthony P.	1939–1950
Nelson, M. Lewis	1947–1948
Neustadter, Siegfried	1948–1949
Neyman, Jerzy	1938–1950
Noble, Charles A.	1930–1938; 1938–1950 emeritus
Owens, Owen G.	1945–1946
Pan, Ting K.	1949–1950
Pinney, Edmund	1946–1950
Pólya, George	1945–1947 visiting professor
Putnam, Thomas M.	1930–[1942]**
Riberito, Hugo B.	1948–1950
Robinson, Raphael M.	1938–1950
Roessler, Edward B.	1930–1933
Schaaf, Samuel A.	1945–1946
Sciobereti, Raymond H.	1930–1950
Scott, Elizabeth L.	1948–1950
Seidenberg, Abraham	1946–1950
Shephard, Ronald W.	1941–1942
Smith, Marianne F.	1948–1950
Sperry, Pauline	1930–1950
Stein, Charles M.	1947–1950
Swinford, Lee H.	1930–1950
Tarski, Alfred	1942–1950
Wakerling, Virginia W.	1944–1950
Walton, Lewis F.	1944–1946
Williams, Arthur R.	1930–1950
Wolf, Frantisek	1942–1950
Wong, Bing C.	1930–1945; 1946–1947 emeritus

*In many cases, appointments through 1949–1950 extended well beyond 1950.

**Faculty list lacking in *Courses of Instruction* for 1943–1944.

Cathleen S. Morawetz was born in Canada and earned a bachelor's degree from the University of Toronto, a master's from MIT, and a Ph.D. in 1951 from New York University, where she was a student of K. O. Freidrichs. She joined the NYU faculty as a research associate in 1952, then rose through the ranks to become a professor. She served as director of its Courant Institute from 1984 to 1988. Her research has focused on the applications of partial differential equations, especially on problems of transonic flow and scattering theory. She is the daughter of the Irish mathematician John L. Synge. Her hobbies are sailing in Canada and entertaining her four grandsons.

The Courant Institute of Mathematical Sciences

CATHLEEN S. MORAWETZ

One of the measures of the blossoming and maturing of mathematics in America since the conclusion of the Second World War is the rise to eminence of a large number of departments of mathematics. Before the war, mathematics in America was largely shaped by what went on at Harvard, Chicago, Princeton, and the Institute for Advanced Study; since the war one has had to reckon with research carried on at Yale, Illinois, Michigan, Wisconsin, Berkeley, Stanford, Maryland, to mention just a few of a happily growing list. Among these newly minted centers stands the Courant Institute; its birth was a most improbable event, since at its inception its parent institute, New York University, lacked financial resources, and was without a tradition of excellence in the sciences; it was the most unusual among the new centers, for it took — and still takes — as its agenda the mathematics inspired by applications. It was the creation of Richard Courant, a practical visionary, who held a large view of mathematics, not just as an art for art's sake but also as a vital way of making sense of the world. His story and the influence his Institute had — and has — on American mathematics is worth telling in this Centennial Year.

The origins of the Courant Institute go back to 1933, when Richard Courant was forced under the Nazi laws, from his position as director of the renowned Institute of Mathematics at Göttingen. In 1934 he was offered a position at New York University which he accepted with some reluctance since NYU was by no means at that time an institution of quality. In 1935, after one year in New York, that position was made permanent and he became the head of a graduate department of mathematics. He had observed on a visit to America in 1932 the low esteem in which mathematics was generally held at the larger universities and he wanted to change that, at least at NYU. Initially the only colleague at NYU who sympathized with this desire was Donald Flanders. In fact it was Flanders who had initiated through Veblen the first approach to Courant by NYU.

Courant's acceptance of the offer from NYU was influenced by two factors: the difficulty he had in helping others to find mathematical jobs in America and the opportunity to develop, in Abraham Flexner's words to Courant, "the great reservoir of talent" that lay in New York.

Courant's initial struggles were for space and positions at NYU but by fall 1938 he had achieved only a little: positions for Kurt O. Friedrichs and James J. Stoker, some fellowships, and some offices carved out of the girls' dormitory at NYU. But despite the odds against him, Courant worked on. Seeing World War II looming ahead with a great demand for science, he put forward a plan for a national institute of science but it failed in the scientific community, and Courant concentrated on his own vision of an institute at New York University.

From 1940 to 1945 the American scientific community was first preparing for and then working for the war effort. As Courant himself said years later, the god Mars helped his fledgling endeavor. Funds from the Office of Scientific Research and Development made it possible for Courant to expand his little group and to undertake to solve mathematically challenging problems arising from war projects. Mathematicians, especially the emigrés, turned to the task of winning the war (an early NYU report was Hermann Weyl's study of shock waves). But Courant's group maintained close touch with fundamental research and the instruction of graduate students.

After the war, many wartime collaborations ended and there was a great surge toward not immediately applicable mathematics: Friedrichs taught topology, for example. Courant retained and strengthened his connections to the government. The Office of Naval Research encouraged further growth of the group and supported the open, unrestricted mathematical research that Courant wanted. In 1946 the group was formally named the Institute for Mathematics and Mechanics and at the same time incorporated a separate applied mathematics group at NYU mainly doing electromagnetic wave propagation under Morris Kline. It also acquired somewhat more reasonable space rented from the American Bible Society, so the chattering of the dormitory was replaced by the singing of hymns. About this time plans were started for a journal, *Communications on Pure and Applied Mathematics*, which first appeared in 1948. Friedrichs and Stoker threw their energies into making its early issues a success. It was to be, unlike other journals, long on exposition and applications. The first year Friedrichs contributed no less than five papers, including his well-known article on the formation and decay of shock waves and his fundamental work on the perturbation of continuous spectra.

In the years following the war, many of the present older faculty completed their studies in the new institute. Courant and his group of advisers, now including Fritz John, found their best candidates for continuing positions among them.

In 1952 the Atomic Energy Commission (today a part of D.O.E.) selected NYU as the site for the first university large, high speed computer (Univac). Courant always said that his ambivalence on accepting this offer was wiped out by the enthusiasm of Friedrichs and Stoker.

The new machine demanded more space, more staff, and more positions, and the institute grew accordingly. By now Lipman Bers and Wilhelm Magnus were playing important roles especially in the broadening of the graduate education. Stoker took up the computing enterprise enthusiastically with some novel approaches to the flow of flooding rivers. New blood from the world of scientific computation was brought in.

Inevitably the success of the group led to tempting offers from outside and Berkeley tried to entice all of them to the west coast. But Courant decided to stay where "the reservoir of talent" was.

The mission of the Institute expanded. Harold Grad initiated and led for twenty years a group dedicated to the study of fusion energy and the related basic research fields of plasma physics and kinetic theory. Computer science was begun as a program within mathematics and spun off as a department in the late sixties.

In 1958, at the age of 70, Courant retired from his directorship and was succeeded by Stoker. Soon the Institute was renamed the Courant Institute of Mathematical Sciences. Stoker's greatest battle during his tenure as director was to win for the institute a measure of autonomy. Since his day, the Institute has reported directly to the president of the university, truly a tribute to the special role it has played in NYU's development, and also, one should add, to its ability to grow and manage its own affairs. Among the many appointments made while Stoker was director were Paul Garabedian, Michel Kervaire, and Kurt Symanzik and during this era Jurgen Moser returned from MIT.

In 1963, with a generous helping hand from Courant's friend and advisor, Warren Weaver, the Institute's needs for development were recognized by A. P. Sloan and through a substantial grant from the Sloan Foundation a new program for postdoctoral visitors, both American and foreign, new scientific programs in probability and mathematical physics and, above all, a new building were made possible. Thus ended the era of remodeled factory lofts and other penurious conditions. Along with the new building came a new "biggest" computer, the CDC 6600, second of its kind to work. It was installed when the building was still unfinished but when it came time for it to be laid to rest in 1980 it was so big it had to be chopped in pieces with an ax.

The new building was known in many parts as the Courant Hilton, it was so luxuriously laid out. The fine tradition of dancing parties, dating to Jim Stoker's early days, was carried over and a special floor built for the purpose in the lounge where to this day the well-known gregariousness of the Courant Institute flourishes.

In 1970, the Institute survived the wave of student radicalism that swept the world, suffering a week's occupation and the planting of a plastic bomb with burning fuse in the computer room. It was extinguished, not by the police bomb squad, but by two enthusiastic young instructors, Chi and Greenleaf, cheered on by Lax and Donsker.

The rest of the seventies were hard years for mathematics and the funding of both the NSF and the defense department agencies declined. NYU was having its own financial troubles and sold its Heights campus in 1972. Joe Keller's satellite graduate program in applied mathematics was wiped out and the Institute suddenly acquired a lot of extra faculty in mathematics. Computer science on the other hand grew with dramatic increases both in student enrollment and government support. By 1980 the building so generous in 1965 was beginning to be inadequate.

New ventures in parallel computing and robotics have expanded the activities of the Institute in the eighties, and the experimental activities are now again in remodeled loft space on Broadway, not so cheap as in Courant's day. The scope of mathematical activities has also broadened to meet the challenges of new applications, especially those tied to computer science.

In 1988, as the American Mathematical Society celebrates its 100th anniversary, the Courant Institute is celebrating the 100th anniversary of Richard Courant's birth. His goal has been achieved and a new young generation is now in place to carry forward his vision. The underlying philosophy of enthusiasm for all parts of mathematics remains. A good bit of Courant's philosophy and a perception of relevance has penetrated the world, and alumni and former postdocs bear influence at all levels. The mathematical community no longer views "the Courant" as just the "impregnable bastion of partial differential equations." One can reflect endlessly on the history of the

Courant Institute and wonder why it succeeded. Why did Lax and Nirenberg choose when they were young to stay at New York University when exciting offers attracted them elsewhere? Why did Friedrichs turn down an extraordinarily generous offer from Rockefeller in the early sixties? What were the elements that stimulated such an active and productive intellectual life? Why, when there were battles, and there were, did it not fall apart? It was not just Courant's skill with mathematical people nor even his ability to transmit his philosophy to those around him. Some of it was a happy juxtaposition of circumstances. But most of it remains a mystery.

The Princeton University Bicentennial Conference on the Problems of Mathematics [1]

FOREWORD

The forward march of science has been marked by the repeated opening-up of new fields and by increasing specialization. This has been balanced by interludes of common activity among related fields and the development in common of broad general ideas. Just as for science as a whole, so in mathematics. As many historical instances show, the balanced development of mathematics requires both specialization and generalization, each in its proper measure. Some schools of mathematics have prided themselves on digging deep wells, others on excavation over a broad area. Progress comes most rapidly by doing both. The increasing tempo of modern research makes these interludes of common concern and assessment come more and more frequently, yet it has been nearly fifty years since much thought has been broadly given to a unified viewpoint in mathematics. It has seemed to us that our conference offered a unique opportunity to help mathematics to swing again for a time toward unification. For this, could a better way be found than the bringing together of a small and diversified group of capable and active workers and their description to one another of their current research and, above all, the problems facing them, whether close at hand or in the middle distance? Mainly owing to the unremitting efforts of all concerned, the response has been most encouraging, our basic purpose has been amply fulfilled, and decided success achieved. It is our hope that other groups will carry the trend forward, particularly in the fields which we could not cover.

Time was finite, and we were forced to some omissions. Applied mathematics, because of its wide ramifications into many sciences, could not, we thought, be treated as one field; at any event, we would be concerned with

[1] Reprinted from *Problems of Mathematics*, Princeton University Bicentennial Conferences, Series 2, Conference 2.

its unifying spirit, pure mathematics. Yet, as the summary shows, applied mathematics still makes its vitalizing contributions.

Owing to the spiritual and intellectual ravage caused by the war years, it seemed exceedingly desirable to have as many participants from abroad as possible. As the list of members shows, considerable success was attained in this. Our conference became, as it were, the first international gathering of mathematicians in a long and terrible decade. The manifold contacts and friendships renewed on this occasion will, we all hope, in the words of the Bicentennial announcement, "contribute to the advancement of the comity of all nations and to the building of a free and peaceful world."

Summary

The account of the sessions which follows is informal and subjective—it tries to reproduce the atmosphere of each session, the current of the discussion and the main points of interest. We feel that this sort of report will be most helpful to mathematicians at large, particularly since the details will appear in book form under the same title. In some cases the requirements of space have kept us from reporting active and exciting discussion, yet we hope that this brief report will go far in giving to all mathematicians much of the flavor and spirit of the conference.

The completeness and clarity of the reports owes much to the Reporters for the individual sessions, the responsibility for the errors and omissions is mine.

Algebra

The discussion in this, the first session, could not find time to cover all of algebra. Two main lines can be seen, however, one the generalization of known results with an eye toward increasing their scope and learning more of their inner meaning—this going on at widely different levels of abstractness—and the other the continuation along classical lines, represented by Brauer's imposing advance.

The contrast between the present conference emphasizing discussion and Harvard's more formal Tercentenary of ten years ago was noted by Birkhoff. Defining algebra as dealing only with operations involving a finite number of elements, he classified the results of the most general algebra into three classes. The first contains "trivial" results. The second those involving the Axiom of Choice ("where the infinite gets into algebra"), such as that every algebra is a subdirect union of subdirectly irreducible algebras. These Birkhoff felt were becoming trivial. The third was illustrated by the theorem that in any algebra containing a one-element subalgebra, whose congruence relations are permutable, we can prove all known variants of the Jordan-Holder Theorem, Remak's "six-way isomorphism theorem" and (assuming finite chains) the theorem that in any two representations of A as a direct union, the factors are pairwise isomorphic (this enormously generalizes the Wedderburn-Remak-Krull-Schmidt Theorem).

This definition of algebra was challenged by Artin—"What about limits? In a valuation theory we have limits, and while the analysts may consider it a breach of faith ..." Birkhoff replied that "I don't consider this algebra, but this doesn't mean that algebraists can't do it." The value of the additional conclusions which can be added to the subdirect representation theorem in special cases by using additional tools, often topological in character, was emphasized by several members. Mac Lane stressed how topology helped to make the representation theorem for Boolean algebras more useful. Dunford stressed the importance of being able to characterize a representing subalgebra as that of *all* elements with a certain property—in analytical applications this often allows one to prove the existence of square roots, which led Stone to say that "It has seemed to me that the words 'all' and 'compact' are vaguely equivalent here." The advantages of having known, simple structures for the subdirectly irreducible algebras were pointed out; thus for Boolean algebras the fact that they are just two-element algebras is of decisive significance.

Rado suggested that the definition of congruence relation might be generalized to include cyclic transitivity and the conjugacy relation in a Peano space. Albert felt that it would be better not to generalize so far, and pointed out the possibilities of shrinkable algebras, a generalization of Jordan algebras where the powers of any element still commute with one another.

The generalized Jordan-Holder theorem was discussed in the light of the hypotheses and proof of Birkhoff on the one hand and Tarski and Jonsson on the other. The need of unifying results was clear to all who took part.

Brauer reported his proof of Artin's conjecture about induced characters, which asserts that characters known to be rational combinations of certain special characters are in fact integral rational combinations. The idea of the proof, and much technique was supplied by the theory of modular characters, but results from this field were not used. From this result it follows, for example: (1) that the zeta-function of a field K normal over k is divisible by the zeta-function of k (in the sense that the quotient of the functions is an integral function); (2) that every representation of a group G can be written in the field of the q-th roots of unity, where q is the least common multiple of the orders of the elements of G; (3) a conjecture of Siegel (1934) concerning the asymptotic behaviour of the class number. Brauer's result represents a decisive step in the generalization of class-field theory to the non-Abelian case, which is commonly regarded as one of the most difficult and important problems in modern algebra.

Artin stated that "My own belief is that we know it already, though no one will believe me—that whatever can be said about non-Abelian class field theory follows from what we know now, since it depends on the behaviour of the broad field over the intermediate fields—and there are sufficiently many Abelian cases." The critical thing is learning how to pass from a prime in an intermediate field to a prime in the large field. "Our difficulty is not in the proofs, but in learning what to prove."

A general Galois theory was characterized by Jacobson as beginning with "some sort of system in the field" and then setting up "some sort of correspondence between the intermediate fields and the subsystems of the system." He presented two such theories, one based on the notion of self-representation, which was introduced by Kaloujnine, and a second, applying to the noncommutative case, which combined self-representations, the allied notion of a relation space and the methods of the structure theory of rings. A problem of Artin concerning the equality of the right and left dimensionality of a division ring is closely connected to these theories, but remains unsettled in the general case. Further open questions are: "Can Hilbert's theory for normal extensions of an algebraic number field (which links up the arithmetical structure with the Galois group) be generalized to the case of an arbitrary finite extension by means of the self-representations of the extension? Can this be extended to include infinite extensions?"

ALGEBRAIC GEOMETRY

Here the discussion followed two lines — new, deeper problems for the classical algebraic geometry over the field of complex numbers contrasted with new methods for developing algebraic geometry over abstract fields. Lefschetz stated four problems, three along the first line and the fourth along the other: (1) Study the dimension of linear systems on algebraic varieties; extend the Riemann-Roch theorem to higher dimensions. (2) Can algebraic surfaces be parametrized by automorphic functions (in a way analogous to that of Poincaré for algebraic curves)? The Betti numbers of the fundamental domains of the automorphic functions were suggested as a fundamental tool for disproving this. (3) Is the cubic variety in four-space rational? The rumor that Fano has settled this in the negative was discussed. (4) Examine the standard configurations (e.g. the 27 lines on a cubic surface) over fields of characteristic p. "This one is baby talk. But there are some very deep problems in the same direction. The ambitious young fellows with lots of shoulder muscles could do much worse than to study Max Noether's and Halphen's work on the classification of algebraic curves in a 3-space and try to carry it over to characteristic p."

The next topic was the rank problem for cycles on the orientable manifold defined by an (irreducible) algebraic variety over the complex numbers. The rank of a cycle is determined by the dimension of the smallest algebraic sub-variety which bears a homologous cycle. Necessary conditions were obtained several years ago by Hodge in terms of the vanishing of the integrals of certain closed differential forms over the given cycle. The problem is to show that these conditions are also sufficient.

Hodge pointed out three component parts from which a proof of sufficiency would follow. The central part was the following: If Γ_m is an m-cycle of V_m and every integral (of type zero) of the first kind has zero period on Γ_m, then Γ_m is homologous to a cycle on an algebraic sub-manifold D_{m-1} of V_m. This is unsettled, except in the case $m = 2$ (Lefschetz) and is a problem of classical algebraic geometry. Hodge feels that any solution of this part will, through its technique, provide solutions of the other parts. This initiated an active discussion of the possibilities of giving meaning in the abstract case (including characteristic p) of the topological ideas, such as integral and periods, which play such a central part in the present theory for the classical case. A. Weil's definitions of residues and sums of residues were agreed to represent the best progress to date but the field seemed comparatively untouched.

With the aid of the entire theory of numerical invariants, Castelnuovo and Enriques showed that every algebraic surface (except ruled surfaces) was birationally equivalent to a minimal model—one in which no birational transformation can further shrink a curve into a point. If this were provable without the aid of the theory of invariants, that theory could be greatly simplified. Zariski outlined such a proof, which also applies to the case of characteristic p. "By slipping in exceptional curves in all possible ways we obtain what is probably the worst offender in this respect—the Riemann surface of the field where there is a point for each of the places, each of the valuations of the field. Since I have used this model in my work, I felt that I should work toward a model with the opposite property." The revision of theory allowed by this result seemed to Zariski to offer a possibility of founding a theory for higher dimensional varieties.

The discussion here led Lefschetz to say that "To me algebraic geometry is algebra with a kick. All too often algebra seems to lack direction to specific problems." Birkhoff countered with "If the algebraic geometers are so ambitious, why don't they do something about the real field?" Lefschetz stressed the resemblance to number theory before the utilization of analytical methods. "There are only scattered results and no indication of general methods." An instance of our ignorance is Hilbert's sixteenth problem about the nested ovals, still unsolved.

DIFFERENTIAL GEOMETRY

Here the discussion was concerned with isolated topics from differential geometry in the small and a more integrated discussion of differential geometry in the large.

Hlavaty discussed some recent developments in the differential geometry of the lines of a flat projective three-space. These are expounded in full in his recent book ("Differentielle Liniengeometrie," Noordhoff, Groningen, 1945).

Thomas discussed the problem of determining all quadratic first integrals of the differential equations of a dynamic system. The purely geometric analog of this problem was solved by Thomas and Veblen in Thomas's doctoral thesis. They found a finite process to determine the exact number of independent quadratic first integrals. This can be applied to determining when two dynamical systems have the same trajectories (without regard to time parametrization). Thomas suggested that the simultaneous use of several quadratic forms would be useful in attacking other problems in differential geometry. Rado asked for a definition, saying "You spoke of trajectories independent of parameters. I have often tried to find out from physicists or differential geometers what notion of curve they use. Can you tell me here?" This was not followed up, perhaps because there are so many detailed definitions in the literature. Synge asked Thomas whether, given a concrete dynamical system, he could tell whether there is another first integral. "Yes, in theory. In practice I think it could be done without too much trouble—at least to count the solutions, if not to find them."

The discussion then passed to problems "in the large." Here the objective is to relate the local differential properties of a space to the topological properties of the space as a whole. Results in this theory are still fragmentary and few general methods have been developed. Bochner presented certain new relations between the Ricci curvature of a compact Riemann space and the characteristics of vector fields defined over the space. For example he showed that if the Ricci tensor is positive definite, there exists no vector field for which the divergence and curl both vanish. From this it follows that the first Betti number of the space is zero. Many of the results have more extensive form on a complex analytic manifold on which is defined a positive Hermitian metric. Whitehead commented that "This is the first time that anyone has got a topological result out of the Ricci tensor, without using the full curvature tensor." Veblen said that "This is the direction in which we want to see differential geometry go."

Hopf called attention to little-known but very suggestive results of Preissmann (1943) and Cohn-Vossen (1935–1936). From Preissmann's results we learn that a three-dimensional torus cannot have curvature everywhere < 0 (from Bochner's result it cannot be everywhere > 0). Also the product of two closed surfaces of genus at least two cannot have curvature everywhere < 0 although the curvature can be everywhere ≤ 0. Cohn-Vossen has made an essential extension of the Gauss-Bonnet equality for closed two dimensional surfaces to an inequality for open two-dimensional surfaces. These inequalities relate the total curvature of the surface (a local property) to the Euler-Poincaré characteristic (a topological property of the surface as a whole). The extension of the Gauss-Bonnet equality to closed n-dimensional Riemann surfaces has been accomplished by Allendoerfer and Weil; but similar extension of Cohn-Vossen's inequality remains to be carried out.

MATHEMATICAL LOGIC

Here the discussion revolved about a single broad topic—decision problems. The notion of a decision method is a formalization of the classical notion of an algorithm. The known equivalence of Turing's "computability," defined in terms of a very general computing machine. Herbrand and Gödel's "general recursiveness," and Church's "λ-definability" has led to general agreement on the natural formalization of the notion of an algorithm. More and more decision problems are being shown to be unsolvable, in the sense that there exists no algorithm with the required properties. While the general mathematician has to regard this as somewhat negative progress, it is real progress, not only for mathematical logic but for other fields of mathematics as well.

The nearest approach to a proof of unsolvability for a problem of importance in other fields is the recent theorem of Post that the word problem in semi-groups is insoluble. Such results have encouraged mathematical logicians to go further and try to show that problems of a more standard mathematical character are insoluble. In his summary, Church suggested the word problem for groups and the problems of giving a complete set of topological invariants for knots and for closed simplicial manifolds of dimension n as likely possibilities. Theorems are needed to characterize or provide criteria for the distinction between solvable and unsolvable decision problems (though to find a decision method is no doubt itself an insoluble decision problem).

Closely related to the decision problem for theories is the question of whether or not particular questions are decidable in a given theory. It was pointed out that in particular the Riemann Hypothesis might be undecidable in a particular theory, but that the flexible position of the general mathematician would prevent its ever becoming demonstrably undecidable for him.

The analogy between generally recursive sets of integers and Borel sets of real numbers was stressed by Tarski, who pointed out the possibilities of proving analogous theorems and of developing a single theory to include both as special cases. Tarski then surveyed the status of the decision problem in various logical fields: sentential (propositional) calculus, predicate (functional) calculus, many-valued systems of sentential calculus, number theory, analysis, general set theory, and various abstract algebraic systems. In all of these, even in two-valued sentential calculus, where we would like to be able to decide when a set of formulas is an adequate axiom system, he pointed out open, important problems.

No formal system, with the usual restrictions, which is strong enough to deal with the arithmetic of integers can be complete (Gödel 1931), it must contain undecidable propositions. This led Gödel to propose a particular tremendous enlargement of the notion of a formal system—which would allow uncountably many primitive notions and allow the notion of an axiom to depend on the notion of truth. "I do not feel sure that the set of all things of which we can think is denumerable."

This led to a spirited discussion led by Church and Gödel, which centered on the non-mathematical questions of what could reasonably be called a "proof" and when a listener could "reasonably" doubt a proof. From a psychological point of view the discussion resembled the classical debates on intuitionism to a remarkable degree; Church arguing for finiteness and security and Gödel arguing for the ability to obtain results.

Kleene discussed the limitations which general recursiveness may place on quantitative proof. McKinsey discussed the criticisms of general recursiveness as the formalization of the notion of an algorithm.

Quine proposed to evade the undecidability of arithmetic by studying a restricted arithmetic without quantifiers, which might have a decision method. It was pointed out that, in particular, Fermat's Last Theorem could be expressed in such a system. The corresponding problem for a partial system of real numbers without quantifiers was proposed as an open problem.

The subject of non-classical logic recurred throughout the session, with relatively favorable words for the quantum-mechanical system of Birkhoff and von Neumann. Tarski felt that "The system of von Neumann and Birkhoff seems to me to be the most interesting of these (non-classical logics), and the only one which has any chance to replace our customary two-valued logic, since it is the only one which has arisen from the needs of science." Rosser discussed the problems involved in applying a many-valued logic, for example Reichenbach's, to all the steps which lead up to quantum mechanics, including truth-function theory (propositional calculus), quantification theory (functional calculus), set theory, theory of the positive integers, theory of real numbers, theory of limits and functions, theory of Hilbert space, theory of quantum mechanics. The initial steps have shown a great tendency of the theory to ramify single notions in the classical theory corresponding to more and more distinct notions as one passes to more and more complex theories. Gödel proposed using the two kinds of logic, each in its place. Church would accept this proposal for consideration, only if a single logistic system were constructed providing syntactical criteria by which the place of each kind of logic is fixed. He felt that all non-classical logics faced great difficulties in such applications.

TOPOLOGY

The discussion centered around two main topics: (a) groups of transformations, (b) classification of homotopy classes of maps, fibre bundles and related questions. The detailed discussion was extremely active and so wide-ranging that it cannot be covered here in every detail.

The discussion of (a) was initiated by an account by Montgomery. "Let G be a topological group and M a topological space." "A function $f(g; x) = g(x)$ defined and continuous on the product of G and M with values in M gives rise to a transformation group of M (or a representation of G) if (a), for a fixed g, $g(x)$ is a homeomorphism of M onto itself and (b), $g_1[g_2(x)] = (g_1 g_2)(x)$ for all g_1, g_2, and x." "The transformation group is called effective if only the identity leaves all of M fixed." "The group G is usually taken to be compact or locally compact, and M is taken to be a manifold." There are problems on various levels of generality mainly singled out by whether continuity alone, or a certain amount of differentiability, or analyticity is assumed for $f(g; x)$. If only continuity is assumed, some of the main questions are closely related to the fifth problem of Hilbert, a central conjecture being that any compact group which acts effectively on a manifold must be a Lie group. This is true for a compact connected group acting on a three-dimensional manifold and is true for any dimension with the additional assumption that each individual transformation is of class C^1. The conjecture also holds for a locally compact group if each transformation is of class C^2. In the discussion several problems which lie on the path towards a solution of the main conjecture were stated. Can a p-adic group operate effectively on a manifold? Does every periodic transformation of a Euclidean space admit a fixed point?

The discussion of the second main topic was initiated by Steenrod. The general problem consists in finding algebraic methods for the enumeration of the homotopy classes of continuous maps $K^n \to L^m$ where K^n and L^m are polyhedra of dimension n and m respectively. If $L^m = S^n$, the n-sphere, the problem has been solved by Hopf. If $L^m = S^{n-1}$, the $(n - 1)$-sphere, the problem has been solved by Pontrjagin and Steenrod using various multiplications in the cohomology theory. The case where $K^n = S^n$ leads to the definition of the n-th homotopy group of L^m (Hurewicz). These two main trends overlap when both K^n and L^m are spheres. We then have the problem of computing the homotopy groups of spheres. Although many efforts were and are directed towards this problem, only very fragmentary information is available. "Here is a complete unknown morass of groups."

The classification of fibre bundles and finding their invariants is closely related in nature and methods. Recent progress and some problems for the future were discussed by Hopf, who also asked whether every abstract cohomology ring is realized by a complex.

On the methodological side two facts were stressed. First, the constant use of obstruction cochains and cocycles with coefficients in relevant homotopy groups. Second, the need of studying not only the groups involved but also the homomorphisms connecting them.

In a final summary J. H. C. Whitehead noted with satisfaction that one of the main trends in present day topology is towards the study of the deeper properties of relatively simple spaces, such as spheres, spaces of line elements on spheres, group spaces, etc. He classified the problems of topology under three broad headings. At one end there are the fundamental definitions and problems, particularly the Poincaré conjecture that the 3-sphere is the only closed 3-manifold whose fundamental group (first homotopy group) is the identity, the basic conjecture that topological equivalence of complexes implies identity of subdivisions and the "word problem." Our knowledge of these matters is practically nil. At the other end there is the theory of homology. Here some of the fundamental problems such as calculating the homology groups, intersection cycles, etc., are completely solved by finite algorithms. In between homology and the ultimate problems of topology comes the homotopy belt. Here there are in general no finite algorithms. However, it is often possible, as in the theory of knots, to find an algebraic expression of a general class of problems, which is so closely related to the geometry and is so amenable to manipulation that it gives a good chance of solving any concrete problem, not too complicated to be stated. Much of the work discussed at the session consists of pressing up from the homology to the homotopy region, partly by the application of the former to the latter. Whitehead also referred to combinatorial theory initiated by Newman, to Reidemeister's *Überlagerung* and to a combinatorial theory with its roots in these two which he is developing. He described this as approaching the problems of homotopy from the other direction.

NEW FIELDS

Although this session was entitled "New Fields" the discussion was principally of classical problems related to application, and of the need and feasibility of revitalizing work in these fields. An exception must be made for Wiener's discussion of communication problems. This is not to be taken as a sign that there are no new fields in mathematics, but rather as an indication that new fields grow as part of an old field, or of a particular application, until they are so large as to be no longer new.

Active discussion first took place about the importance of rigorous theorems in applicable potential theory. The fact that all these theorems held in their integral form satisfied Weyl and Courant, since one really needs only the force on a small body and not on a point. Birkhoff felt that "It is a worthwhile mathematical problem to show that what is obvious is mathematically deducible," while Weyl commented that "It depends on the kind of mathematical deduction."

Evans discussed the status of multiple valued harmonic functions in three-dimensional space, which are of distinct importance in the theory of diffraction. Murnaghan outlined a non-linear theory of elasticity with finite displacements. The relation of this theory to the theory of plasticity was discussed, and it was pointed out that it could not explain the observed phenomena of hysteresis. Synge pointed out how the consideration of vectors, whose components lie in a function space provides both geometric intuition and a workable method of computing the size of the error committed by certain approximate solutions of boundary value problems. The relation of this to the Rayleigh-Ritz procedure was discussed.

Wiener called attention to the problems of communication of information, and to their general nature, which has "nothing to do with electricity, passive networks, or electrical engineering." "The theories of control engineering and communication engineering are one and the same—(and) are indistinguishable from the theory of time series in statistics." He emphasized the problems arising in this field in modern computing machines and the human nervous system (mathematical neurophysiology), where many problems need solution. The relation between the notions of amount of information in this theory and in R. A. Fisher's theory of statistical inference was briefly discussed.

von Neumann pointed out that the success of mathematics with the linear differential equations of electrodynamics and quantum mechanics had concealed its failure with the nonlinear differential equations of hydrodynamics, elasticity, and general relativity. He emphasized the likelihood that we face singularities of new types, not foreshadowed by those of functions of a complex variable. In the general relativity theory of a mass particle, there is an apparent singularity near the particle whose position depends on the coordinate system. We do not know what the equivalent of the "objective" theory of function-element continuation is in this case. In non-viscous gas dynamics the "natural" boundary conditions generally do not allow solutions unless certain discontinuities are allowed, these on the other hand force one to violate certain integrals (entropy), and introduce difficulties with uniqueness.

Further, we have no adequate theory of the interactions of two shock waves! In turbulence, a symmetric problem is dominated by unsymmetric solutions!

MATHEMATICAL PROBABILITY

Cramér surveyed many of the problems related to sums of random variables. A first group of problems concerned finite sums of independent random variables and the pathological cases already known. He emphasized the lack of general knowledge of the situation. For example, which infinitely divisible distributions can be represented with an indecomposable factor?

The Central Limit Theorem for independent variables seems to be in satisfactory shape, but we lack much information about when it is valid for sums of dependent variables. This is the simplest case of the general problem of limiting distributions, about which we know so little.

Stochastic processes were discussed more actively than any other topic; Cramér went over the general state of the theory, pointing out many directions in which our knowledge is incomplete. Doob called attention to the advantages of actually considering the specific functions of the stochastic process, saying "In the good old days, a lot of people worked in probability, but they got out of this field and into differential equations, or integral equations as soon as possible. This is not a criticism, but only a description." Feller pointed out that the results already known for Markoff processes are valuable tools in studying the functional equations of many types, which are known to hold for such functions associated with the process as the probability of a given transition in a fixed time interval. These equations may be, in particular, (i) a system of infinitely many differential equations, (ii) a difference-differential equation, (iii) infinite systems of difference-differential equations. "It is theoretically possible to consider any stochastic process as a Markoff process, but in a perfectly crazy space. The question is, does it help us or hurt us?" The problems of multiple renewal theory seem to require much further work along the lines of classical integral equation theory.

Hotelling noted the absence of statistics, as contrasted with probability theory, from the discussion, and emphasized the need for a complete and careful study of the estimation problem. The fact that a finite number of samples of finite length must be used to estimate an infinite process is sure to lead to new difficulties and interesting results. Wiener's experience with actual estimation in stationary processes was closely parallel to the experience of statisticians in estimating distributions of a single variable—it was hard, but feasible, to estimate general parameters, but it was much more difficult to estimate detailed qualitative characters of processes, such as would be given by higher moments, for example.

Kac discussed some problems connected with the central limit theorem, and pointed out that, since for a stationary Gaussian process one can define the likelihood of a path, we could ask "Is there a path of maximum likeli-hood?" He also proposed the problem of the Brownian motion of a stretched string.

The difficulties of further advances in some fields of stochastic processes were sharply outlined by Kac's statement, partly in jest, that "It seems to me that Doob's discussion could be summarized by saying that people were using difference equations to calculate probabilities when they didn't know what probability was, and that now we know what probability is, but can't calculate it." The discussion repeatedly fringed on a discussion of the *ad hoc* nature of Doob's powerful methods, but this seemed to lie just outside the purely mathematical discussion.

Returning to more finite problems, Neyman raised a representation prob-lem for sequences, and Wald discussed the problem of the maximum abso-lute value of a partial sum of a random series. This problem, of considerable applicational interest, is just being effectively attacked. The use of higher moments and analytical techniques was proposed by Wiener.

ANALYSIS

The first group of problems discussed arose from the study of partial dif-ferential equations. Classical potential theory is concerned with solutions of Laplace's equation, a special equation of elliptic type. Riesz described the modern development of potential theory, first for Euclidean space, and then for the so-called Lorentz space in which the square of the "distance" between points $(x_1, \ldots, x_m)$ and $(y_1, \ldots, y_m)$ is defined as $(x_1 + y_1)^2 - (x_2 - y_2)^2 \cdots (x_m - y_m)^2$. In the first, or elliptic, case, the classical Newtonian potential depending on the inverse $(m - 2)$th power of the distance, where the dimension m is at least 3, is replaced by one depending on the inverse $(n - a)$th power, where a is between 0 and 2. Boundary value problems in this generalized potential theory introduce new problems which did not ap-pear in the classical theory; the methods developed for their solution throw new light on classical potential theory. Riesz stressed the idea, which is of particular importance in the new theory, that it is more satisfactory to de-fine the Green's function as a potential than as the solution of a boundary value problem. In Lorentz space, the potentials appear as generalizations of the Riemann-Liouville integrals of fractional order. Riesz showed how the potentials lead in a very direct way to the solution of Cauchy's boundary value problem for the m-dimensional wave equation. In discussion, Hille em-phasized similarities between the Riesz Lorentz potentials and the Riemann-Liouville integrals. He described some of their properties, regarding them as operators belonging to semigroups of operators over Banach spaces.

The second group of problems were inspired by the general problem of rep-resenting functions by series. Zygmund described the two major outstanding problems of the theory of trigonometric series. For Fourier series in one variable the problem of representing a given function may be considered solved in the sense of summability (the present popularity of summability problems seems unfounded); however, even this is far from true for multiple series. For convergence, on the other hand, even the problem of whether the Fourier series of a continuous function must have points of convergence is unsolved and almost no progress has been made since Riesz's report in 1913. A solution would give new insight into the structure of functions of a real variable. Several conjectures were mentioned, but "the theory of trigonometric series is one of the worst places to have conjectures. History teaches that almost all conjectures were proved false." The second problem is that of characterizing sets of uniqueness for Fourier series; sets such that two trigonometric series which converge to the same sum outside the set, they are necessarily identical. Rajchman's conjecture about sets $H\sigma$ is still unsettled. The properties of sets of uniqueness are known to depend in a sense on their arithmetic structure. In this connection the properties of a special class of algebraic integers are relevant. In discussion, Salem stated several further problems involving these algebraic integers. Walsh called attention to a problems analogous to Fourier series problems but related to a general closed curve as Fourier series are related to a circle. Wiener stressed the importance of these problems for open curves as in the Nyquist diagram. Birkhoff brought out the problem of sets of uniqueness for solutions of partial differential equations, in particular for axially symmetric harmonic functions.

The third group of problems centered about the distribution of the values of analytic functions of a complex variable. Ahlfors mentioned that since the advances resulting from the modern introduction of topological and differential geometric methods into the problems related to Picard's theorem, little progress has been made, either with results relating specifically to the problem of the distribution of values or with the so-called type problem, the problem of classifying Riemann surfaces with respect to their ability to be mapped onto either a circle or the finite plane. He stressed the importance for this field of the theory of the conformal mapping of multiply connected domains. Further progress in the theory of entire functions seems to depend on progress in such conformal mapping problems; the first step appears to involve the solution of some new extremal problems. Robinson discussed the fate of the maximum modulus theorem in a multiply connected domain. Boas gave an account of problems involving entire functions of exponential type. Here the slow rate of growth of the function allows one to make quite precise statements about the restrictions imposed on the function by its special behavior in various respects. Levinson called attention to a new technique introduced by A. Selberg in the study of the distribution of the zeros of the Riemann zeta function, which may have consequences in analytic number theory.

ANALYSIS IN THE LARGE

Here the discussion centered on the choice of method—many of the problems of analysis in the large can be attacked by at least two different techniques, and it is of central importance to choose the approach which will give the best results with the least difficulty. In opening the session, Morse emphasized that the division of the session into differential geometry in the large, applied mathematics in the large, and equilibrium analysis in the large was "all for convenience and has nothing to do with essential differences."

In outlining the problems facing differential geometry in the large, Hopf reemphasized the importance of the work of Cohn-Vossen (1935–1936) on open surfaces and the need of extending it to higher dimensions. "Can the result that on a complete surface of the type of the plane with $K > O$ every geodesic goes to infinity be generalized? Can the existence and uniqueness theorems for a complete surface with $K > O$ of the type of the sphere as subspaces of E^3 be generalized by omitting $K > O$?" "In which E^m can the hyperbolic plane be imbedded?" "Given a closed differentiable manifold of dimension n, for what k can we find k functions $x_1, x_2, \ldots, x_k$ whose gradients are of dimension n?" "On $x_1^2 + x_2^2 + \cdots + x_n^2 = 1$, there are given k continuous functions without common zero. Can they be extended to the interior of the sphere without common zero?" Recent progress has been made on this last topological problem. The necessary and sufficient condition can be given in terms of differential geometry (by the vanishing of an integral) when $n = k$, and, as J. H. C. Whitehead reported, when $n = 4, k = 3$.

Although he admitted that some may think it a dream, Allendoerfer said that a program of connecting differential invariants with topological invariants had been in the minds of many. For this program it seems that the invariants should be integers, although "we know that topologists like groups." Hedlund discussed the problems of geodesic flows on a manifold, which arose from the consideration of dynamical systems. "Is a metric on the torus either flat or such that each geodesic has two conjugate points?" "If there are no conjugate points, is there metric transitivity?" "If there is negative curvature (not necessarily constant) is there mixture?" Hodge considered a complex manifold of two complex dimensions with coordinates z_1, z_1 (all indices run through 1, 2) on which functions a_{ij} satisfy

$$\frac{\partial a_{ij}}{\partial z_k} = \frac{\partial a_{kj}}{\partial z_i}, \frac{\partial a_{ij}}{\partial z_k} = \frac{\partial z_{ik}}{\partial z_j},$$

which are the conditions of integrability of

$$\frac{\partial^2 \Phi}{\partial z_i \partial z_j} = a_{ij}.$$

"When does there exist a solution Φ, all of whose singularities lie on an algebraic surface?" He pointed out that few would realize the equivalence of this with the problem that he had stated in the algebraic geometry session.

The need of applied mathematics for existence theorems and uniqueness theorems as well as for computing procedures was stressed by Courant. "If you cannot prove the existence of the solution, then you are not certain that your mathematical model is at all a satisfactory representation of physical reality." "If you cannot prove uniqueness, but can prove the existence of solutions, then your results are unsatisfactory, since, at best, each individual solution may or may not be followed by physical reality." The history of gas dynamics and the phenomenon of shock waves gives much support to these results. The presence of shock waves indicates the utter failure of differential equations alone to describe an idealized gas without friction or heat conduction, for, contrary to Lord Rayleigh's belief, there are no shock-free solutions for even the elementary problems. "We are all used to problems where, when we let a parameter tend to zero, the solution will tend to the solution for the zero value of the parameter. But this is exactly what does not happen." Shiffman pointed out that the differential equations for imbedding the hyperbolic plane into three-space are nonlinear and the analogy is strengthened by Hilbert's proof of nonimbeddability, which shows that singularities must arise.

The calculus of variations and the Schauder-Leray method are the two broad tools in much of analysis in the large. There was much discussion of their properties and contrasts. Shiffman asked for a method of applying Schauder-Leray theory directly to a class of analytic functions. He also hoped that the two methods could be brought closer together "so that they may alter and improve each other, and also so that each may fill out the gaps in the scope of the other." Uniqueness is harder to establish than existence by Schauder-Leray methods—Weyl felt that this was distinctive. Morse felt that the distinction between fixed point and extremal problems was general, saying: "The Schauder-Leray theory treats more general problems, with less powerful tools, thus obtaining less complete results." Thus those problems which admit a variational form, e.g. those with symmetric kernels, are best treated by extremal methods.

Radó pointed out that the right restriction for all single integral problems was absolute continuity and that is exactly what was suggested by the length integral. This was as if "by looking at the first two terms of an infinite series, one determined its convergence." He suggested that essential absolute continuity might play the same role in two, and perhaps even in higher dimensions.

In opening the topic of equilibrium analysis in the large, Morse stressed that the study of the behavior of two or more functions can always be reduced to the study of one function with general boundary conditions. "I feel that this is a better way to do this." This way passes through seven problems, each broad enough to be a program, and all of these need solution for progress on a broad front. 1. Two-point connectivities and generalizations. So far found only for specific manifolds by analytic solution of analytic problems. 2. Incidence of homology classes. Only special cases known. 3. Lower-semicontinuity and upper reducibility. When you use compactness instead of upper reducibility "You are no longer abstracting, you are subtracting—you are throwing away the best part of the calculus of variations." 4. Finiteness of the number of minimal surfaces bounded by the given curve. "I conjecture that an analytic and regular curve in E^3 is spanned by only a finite number of minimal surfaces." (The higher dimensional analog is known to be false.) 5. The derived representation of a given boundary curve. 6. The minimal surface problem on a general R-manifold. 7. The topological foundation of planetary orbits. (This refers to the three body problem with one infinitesimal body.) "The cycles must be relative—they must be loops hanging over ridges and extending down to Dante's Inferno."

McShane discussed the application of L. C. Young's generalized curves, which are "as constructive as the Lebesgue integral, useful, but certainly not perspicuous." He pointed out that, at present, they do not solve the problems of the second variation, or of minimax curves of higher type.

Final Session

At the closing session, a dinner at Procter Hall, the dining hall of the Graduate College, the conference and its guests considered for the time the state of mathematics as a whole and its prospects in the century ahead. As the first speaker, Mac Lane discussed his impressions of the conference and of the state of mathematics in the following vein:

"In almost all the conferences we ran across the phenomenon of someone else moving in. I think this is a significant aspect of this conference. When you set out to solve problems in mathematics, even in nicely labelled fields, they may well lead you into some other field. To cite some instances: The logicians moved in on the algebraists, the topologists moved in on the differential geometers (and vice versa), and the analysts moved in on the statisticians.

"The problem is that of communication between different mathematicians in different fields. The day is no longer here when all mathematics can be done under the old ideal of the universal mathematician who knows all fields.

Problems must rather be done by mathematicians in different fields who understand what is being done in neighboring fields and who are able to communicate with each other in the presentation of their results, in the understanding of what that presentation means, and in the appreciation of the significance and purpose behind the problems which they are endeavoring to solve. It is my opinion that this conference at Princeton has contributed much to the solution of this problem of the communication of mathematics."

Weyl reviewed the development of mathematics in his lifetime, and its relation to his own work. He expressed his tendency toward intuitionism, saying:

"In his oration in honor of Dirichlet, Minkowski spoke of the true Dirichlet principle, to face problems with a minimum of blind calculation, a maximum of seeing thought. I find the present state of mathematics, that has arisen by going full steam ahead under this slogan, so alarming that I propose another principle: *Whenever you can settle a question by explicit construction, be not satisfied with purely existential arguments.*"

And again, more generally: "At a conference in Bern in 1931 I said: 'Before one can generalize, formalize, or axiomatize, there must be a mathematical substance. I am afraid that the mathematical substance in the formulization of which we have exercised our powers in the last two decades shows signs of exhaustion. Thus I foresee that the coming generation will have a hard lot in mathematics.' The challenge, I am afraid, has only partially been met in the intervening fifteen years. There were plenty of encouraging signs in this conference. But the deeper one drives the spade the harder the digging gets; maybe it has become too hard for us unless we are given some outside help, be it even by such devilish devices as high-speed computing machines."

In conclusion, Stone expressed his firm faith that mathematics would continue to grow as well and as fast as of old, now that the ten years under the shadow of war were past.

Judged from the interest of the conferees, these were the trends and results of most interest for the present and near future:

1. The discussion about general algebra—were limits and topological methods really a part of algebra? Many vital results seemed to need these other methods—if they were to be *defined* out of algebra, then algebra would lose much power.

2. Brauer's result on induced characters, which some feel to be the most important in algebra for the last ten years, and which may open the way to non-Abelian class-field theory.

3. The renaissance of interest in algebraic geometry over the field of complex numbers, side by side with a continuing interest in the abstract case.

4. The tendency toward a welding together of differential and algebraic geometric, topological, and Hermitian methods and problems—the trend toward a unified study of manifolds, vector fields, imbedding and critical points.

5. The liveliness of mathematical logic and its insistent pressing on toward the problems of the general mathematician.

6. The return of topology to the study of geometric and combinatorial problems of simple objects, in whose light the recent period of abstraction seems to have been a search for new tools and the natural, exuberant generalizations of a field engaged in finding itself.

7. The tendency for the applications to stimulate what may some day be new fields and to force the reawakening of old, incomplete fields.

8. The role of stochastic processes in the future of mathematical probability and statistics. This seems to be the main subject for advances in pure theory and a major subject for advances in the applications.

9. The absence in analysis of a central, guiding theme, or even of a few such themes. This field seems to be going through a period of wide ramification, perhaps in preparation for new syntheses.

10. The remarkable size and unity of analysis in the large, which yet may well absorb much of the activities and interests now thought of as belonging to topology, algebraic geometry and differential geometry, as well as those classically considered as belonging to analysis.

GUESTS OF THE UNIVERSITY
WHO PARTICIPATED

L. V. Ahlfors, Harvard University
A. A. Albert, University of Chicago
J. W. Alexander, Institute for Advanced Study
C. B. Allendoerfer, Haverford College
G. Ancochea, University of Salamanca, Spain
E. G. Begle, Yale University
G. Birkhoff, Harvard University
R. P. Boas, Mathematical Reviews, Brown University
H. F. Bohnenblust, California Institute of Technology
K. Borsuk, University of Warsaw, Poland
R. Brauer, University of Toronto, Canada
S. S. Cairns, Syracuse University
L. F. Chiang, Academia Sinica, Shanghai, China
I. S. Cohen, University of Pennsylvania
R. Courant, New York University
H. Cramér, University of Stockholm, Sweden

P. A. M. Dirac, University of Cambridge, England
J. L. Doob, University of Illinois
N. Dunford, Yale University
S. Eilenberg, Indiana University
A. Einstein, Institute for Advanced Study
L. P. Eisenhart, Princeton University
G. C. Evans, University of California
W. Feller, Cornell University
K. Gödel, Institute for Advanced Study
O. G. Harrold, Princeton University
G. A. Hedlund, University of Virginia
T. H. Hildebrandt, University of Michigan
E. Hille, Yale University
V. Hlavaty, Charles University of Prague, Czechoslovakia
G. P. Hochschild, Harvard University
W. V. D. Hodge, University of Cambridge, England
H. Hotelling, University of North Carolina
H. Hopf, Federal Institute for Technology, Zurich, Switzerland
L. K. Hua, National Tsing Hua University, Peiping, China
W. Hurewicz, Massachusetts Institute of Technology
N. Jacobson, Johns Hopkins University
M. Kac, Cornell University
S. C. Kleene, University of Wisconsin
J. R. Kline, University of Pennsylvania
N. Levinson, Massachusetts Institute of Technology
S. Mac Lane, Harvard University
W. Mayer, Institute for Advanced Study
J. C. C. McKinsey, Oklahoma Agricultural and Mechanical College
E. J. McShane, University of Virginia
D. Montgomery, Yale University
M. Morse, Institute for Advanced Study
F. D. Murnaghan, Johns Hopkins University
J. von Neumann, Institute for Advanced Study
M. H. A. Newman, Victoria University of Manchester, England
J. Neyman, University of California
W. V. Quine, Harvard University
H. Rademacher, University of Pennsylvania
T. Radó, Ohio State University
M. Riesz, University of Lund, Sweden
R. M. Robinson, University of California
J. B. Rosser, Cornell University
R. Salem, Massachusetts Institute of Technology
I. J. Schoenberg, University of Pennsylvania
M. Shiffman, New York University

P. A. Smith, Columbia University
D. C. Spencer, Stanford University
N. E. Steenrod, University of Michigan
M. H. Stone, University of Chicago
J. L. Synge, Carnegie Institute of Technology
A. Tarski, University of California
T. Y. Thomas, Indiana University
O. Veblen, Institute for Advanced Study
A. Wald, Columbia University
R. J. Walker, Cornell University
J. L. Walsh, Harvard University
J. H. M. Wedderburn, Princeton University
H. Weyl, Institute for Advanced Study
J. H. C. Whitehead, University of Oxford, England
H. Whitney, Harvard University
G. T. Whyburn, University of Virginia
D. V. Widder, Harvard Univesity
N. Wiener, Massachusetts Institute of Technology
R. L. Wilder, University of Michigan
J. W. T. Youngs, Indiana University
O. Zariski, University of Illinois
A. Zygmund, University of Pennsylvania

PROGRAM

PRINCETON UNIVERSITY BICENTENNIAL CONFERENCE

Problems of Mathematics

FIRST DAY – TUESDAY, DECEMBER 17, 1946

Opening of Conference
    Chairman: L. P. Eisenhart

Session 1: Algebra
    Chairman: E. Artin   Reporter: G. P. Hochschild
    Discussion leader(s): G. Birkhoff, R. Brauer, N. Jacobson

Session 2: Algebraic Geometry
    Chairman: S. Lefschetz   Reporter: I. S. Cohen
    Discussion leader(s): W. V. D. Hodge, O. Zariski

Session 3: Differential Geometry
    Chairman: O. Veblen   Reporter: C. B. Allendoerfer
    Discussion leader(s): V. Hlavatý, T. Y. Thomas

Session 4: Mathematical Logic
    Chairman: A. Church   Reporter: J. C. C. McKinsey
    Discussion leader(s): A. Tarski

SECOND DAY – WEDNESDAY, DECEMBER 18, 1946

Session 5: Topology
    Chairman: A. W. Tucker   Reporter: S. Eilenberg
    Discussion leader(s): H. Hopf, D. Montgomery, N. E. Steenrod,
                J. H. C. Whitehead

Session 6: New Fields
    Chairman: J. von Neumann   Reporter: V. Bargmann
    Discussion leader(s): G. C. Evans, F. D. Murnaghan, J. L. Synge,
                N. Wiener

Session 7: Mathematical Probability
    Chairman: S. S. Wilks   Reporter: J. W. Tukey
    Discussion leader(s): H. Cramér, J. L. Doob, W. Feller

THIRD DAY – THURSDAY, DECEMBER 19, 1946

Session 8: Analysis
        Chairman: S. Bochner   Reporter: R. P. Boas
        Discussion leader(s): L. V. Ahlfors, E. Hille, M. Riesz,
                        A. Zygmund

Session 9: Analysis in the Large
        Chairman: M. Morse   Reporter: M. Shiffman
        Discussion leader(s): R. Courant, H. Hopf

Final Dinner
        Chairman: H. P. Robertson
        Speakers: S. Mac Lane, M. H. Stone, H. Weyl

CONFERENCE COMMITTEE

E. Artin, V. Bargmann, S. Bochner, C. Chevalley,
A. Church, R. H. Fox, S. Lefschetz (Chairman),
H. P. Robertson, A. W. Tucker, J. W. Tukey, E. P. Wigner,
S. S. Wilks

Princeton Bicentennial Conference, 1946

The Problems of Mathematics

1. Morse, M., Institute for Advanced Study
2. Ancocbea, G., University of Salamanaca, Spain
3. Borsuk, K., University of Warsaw, Poland
4. Cramér, H., University of Stockholm, Sweden
5. Hlavaty, V., University of Prague, Czechoslovakia
6. Whitehead, J. H. C., University of Oxford, England
7. Garding, L. J., Princeton
8. Riesz, M., University of Lund, Sweden
9. Lefachetz, S., Princeton
10. Veblen, O., Institute for Advanced Study
11. Hopf, H., Federal Technical School, Switzerland
12. Newman, M. H. A., University of Manchester, England
13. Hodge, W. V. D., Cambridge, England
14. Dirac, P. A. M., Cambridge University, England
15. Hua, L. K., Tsing Hua University, China
16. Tukey, J. W., Princeton
17. Harrold, O. G., Princeton
18. Mayer, W., Institute for Advanced Study
19. Mautner, F. I., Institute for Advanced Study
20. Gödel, K., Institute for Advanced Study
21. Levinson, N., Massachusetts Institute of Technology
22. Cohen, I. S., University of Pennsylvania
23. Seidenberg, A., University of California
24. Kline, J. R., University of Pennsylvania
25. Ellenberg, S., Indiana University
26. Fox, R. H., Princeton
27. Wiener, N., Massachusetts Institute of Technology
28. Rademacher, H., University of Pennnsylvania
29. Salem, R., Massachusetts Institute of Technology
30. Tarski, A., University of California
31. Bargmann, V., Princeton
32. Jacobson, N., The Johns Hopkins University
33. Kac, M., Cornell University
34. Stone, M. H., University of Chicago
35. von Neumann, J., Institute for Advanced Study
36. Hedlund, G. A., University of Virginia
37. Zariski, O., University of Illinois
38. Whyburn, G. T., University of Virginia
39. McShane, E. J., University of Virginia
40. Quine, W. V., Harvard
41. Wilder, R. L., University of Michigan
42. Kaplansky, I., Institute for Advanced Study
43. Bochner, S., Princeton
44. Leibler, R. A., Institute for Advanced Study
45. Hildebrandt, T. H., University of Michigan
46. Evans, G. C., University of California
47. Widder, D. V., Harvard
48. Hotelling, H., University of North Carolina
49. Peck, L. G., Institute for Advanced Study
50. Synge, J. L., Carnegie Institute of Technology
51. Rosser, J. B., Cornell
52. Murnaghan, F. D., The Johns Hopkins University
53. Mac Lane, S., Harvard
54. Cairns, S. S., Syracuse University
55. Brauer, R., University of Toronto, Canada
56. Schoenberg, I. J., University of Pennsylvania
57. Shiffman, M., New York University
58. Milgram, A. N., Institute for Advanced Study
59. Walker, R. J., Cornell
60. Hurewicz, W., Massachusetts Institute of Technology
61. McKinsey, J. C. C., Oklahoma Agricultural and Mechanical
62. Church, A., Princeton
63. Robertson, H. D., Princeton
64. Bullitt, W. M., Bullitt and Middleton, Louisville, Ky.
65. Hille, E., Yale University
66. Albert, A. A., University of Chicago
67. Rado, T., The Ohio State University
68. Whitney, H., Harvard
69. Ahlfors, L. V., Harvard
70. Thomas, T. Y., Indiana University
71. Crosby, D. R., Princeton
72. Weyl, H., Institute for Advanced Study
73. Walsh, J. L., Harvard
74. Dunford, N., Yale
75. Spenser, D. C., Stanford University
76. Montgomery, D., Yale
77. Birkhoff, G., Harvard
78. Kleene, S. C., University of Wisconsin
79. Smith, P. A., Columbia University
80. Youngs, J. W. T., Indiana University
81. Steenrod, N. E., University of Michigan
82. Wilks, S. S., Princeton
83. Boas, R. P., Mathematical Reviews, Brown University
84. Doob, J. L., University of Illinois
85. Feller, W., Cornell University
86. Zygmund, A., University of Pennsylvania
87. Artin, E., Princeton
88. Bohnenblust, H. F., California Institute of Technology
89. Allendoerfer, C. B., Haverford College
90. Robinson, R. M., Princeton
91. Bellman, R., Princeton
92. Begle, E. G., Yale
93. Tucker, A. W., Princeton

Commentary on Algebra

B. GROSS AND J. TATE [1]

The problem of Artin concerning the equality of the right and left dimensionality of a division ring extension was settled negatively by P.M. Cohn and has recently been completely killed by A.H. Schofield, who gave examples in which the two dimensions are arbitrary integers > 1.

Brauer's results on modular characters, and blocks in the decomposition matrix, were useful in the classification of the finite simple groups — one of the great mathematical achievements of our time. Brauer's proof of the Artin conjecture on induced characters has remained a central result in finite group theory; as a technical tool it has also proved extremely useful in reducing various problems on Galois modules to the case where the Galois group is elementary or even cyclic. It was also an important step in the generalization of class-field theory to the non-Abelian case, though not perhaps a "decisive one." Artin's belief that "whatever can be said about non-Abelian class field theory follows from what we know now," and that "our difficulty is not in the proofs, but in learning what to prove," seems overly optimistic. In the late 1960s, Langlands recognized the connection between non-Abelian Galois representations and automorphic forms on reductive algebraic groups, and told us essentially what to prove. But now, twenty years later, the proofs have been carried out in only a few special cases. Using techniques of base change for the group GL(2), Langlands himself showed how to establish the holomorphy of Artin's L-series attached to 2-dimensional representations of solvable Galois groups, and thereby prove that such representations come from automorphic forms. A converse (over **Q**) was established by Deligne and Serre for holomorphic modular forms of weight one: every such form gives rise to a 2-dimensional Galois representation with odd determinant.

One important aspect of our present view is that the theory of ordinary complex representations of finite Galois groups, i.e., the theory of "Artin

[1] Benedict Gross and John Tate are both professors of mathematics at Harvard University.

motives", cannot be separated from the theory of general motives, i.e., from the theory of systems of ℓ-adic representations of Galois groups coming from Grothendieck's étale cohomology of algebraic varieties. This theory was not even dreamed of at the time of the Princeton Bicentennial Conference in 1946.

Commentary on Algebraic Geometry

HERBERT CLEMENS[1]

Of the four problems listed at the beginning of "Algebraic Geometry," only one which explicitly appears seems in retrospect to have been fundamental to the development of the discipline in the ensuing forty years, namely the first — the extension of the Riemann–Roch theorem to higher dimensional varieties. In fact, with the advent of index theory in the fifties and the relative formulation of the theorem by Grothendieck somewhat later, the Riemann–Roch theorem has been generalized and applied far beyond the wildest dreams of the participants at the 1946 conference! The profound nature of those subsequent successes make the other three problems, as stated, seem much more naive and superficial by comparison. Problem (2) is curious — there are indeed easy topological obstructions. More interestingly perhaps, there are "differential-geometric" obstructions to mapping complex lines into higher dimensional varieties; this is studied via a global geometric version of classical Nevanlinna theory [G]. One other problem hinted at in (3), that of determining when a variety is rational, is quite significant, and is beginning to be attacked successfully by Iskovskih and his school. A class of varieties which give many interesting and "controllable" examples is that of "conic bundles," that is, algebraic fiber spaces whose generic fiber is a rational curve (rational curves have a distinguished divisor class of degree two). This class has provided the way to an example of a "stably" rational variety which is not rational. An important related problem which remains outstanding is that of determining whether a smooth deformation of a smooth rational manifold need be rational. The prime candidate for a counterexample, the cubic fourfold, has resisted several assaults.

Further along in the problem list, the suggestion of looking at Halphen's work on classification of curves in P^3 by genus and degree was indeed taken up and straightened out in characteristic zero by Gruson–Peskine [GP]. (It turns out that even in characteristic zero Halphen's analysis was not complete.) Several techniques related to this problem, perhaps most notably the

---

[1] Herbert Clemens is a professor of mathematics at the University of Utah.

vast generalization of the notion of a Koszul resolution, have given algebraic geometers much deeper insight into the resolution of the ideal sheaves of curves in projective space and in more general settings.

With their discussion of Hodge's famous conjecture, the authors of the problem list were on target, though again in unexpected ways. This conjecture or problem (depending on whether one is a "believer" or not), remains a very important outstanding problem in complex geometry. Perhaps even more importantly, work in Hodge theory motivated by the conjecture has led to some of the deepest mathematics in complex geometry and related fields. For example, the theory of intersection cohomology of Goresky and MacPherson combines with Hodge theory to give the decomposition theorem of Beilinson–Bernstein–Deligne–Gabber for the Hodge theory of projective morphisms between complex varieties and Saito's theorem giving pure Hodge structures on intersection cohomology. See [S] for an overview of these results.

Finally, the citation of the problem of minimal models, central to birational classification theory, was again the correct insight. It is only in the last ten years, with the advent of Mori's theory [K] of classifying varieties by finding rational curves on them, that the centrality of the minimal model question has again been appreciated. The great technical stumbling block, that of handling exceptional sets of codimension greater than one, is finally being dealt with via the theory of "extremal rays" and "flips," thus allowing the construction of minimal models in dimension three.

REFERENCES

[G] Griffiths, P. "Holomorphic mappings into canonical algebraic varieties." Ann. Math. **93** (1971) 439–458.

[GP] Gruson, L. and Peskine, C. "Genre des courbes de l'espace projectif, II." Ann. Sci. École Norm. Sup. **15** (1982) 401–418.

[K] Kollár, J. "The structure of algebraic threefolds: An introduction to Mori's program." Bulletin of the AMS, **17** (1987) 211–273.

[S] Saito, M. "Introduction to mixed Hodge modules." RIMS-605, Kyoto Univ., Japan 1987.

Commentary on Differential Geometry

ROBERT OSSERMAN [1]

The discussion of differential geometry, at least as reported by Carl Allen-doerfer, is clearly more notable for what (and who) is left out than for what is included. To start with "who," the names Cartan, Chern, Weyl, and Myers are nowhere mentioned. Their absence is even more striking in the context of the discussion. J. H. C. Whitehead is reported to have said of a recent Bochner theorem, that "This is the first time that anyone has got a topological result out of the Ricci tensor, without using the full curvature tensor." However, in 1941 Sumner Myers had published his (now) well-known result that a complete Riemannian manifold with a positive lower bound on Ricci curvature must be compact and have a finite fundamental group.

The most glaring omission of all is Chern's intrinsic proof of the general Gauss–Bonnet theorem. The work was done in 1943 in Princeton's own backyard at the Institute for Advanced Study, where Chern was spending two years at the invitation of Veblen. It was published in two papers that appeared in 1944 and 1945 in Princeton's own house journal, the *Annals of Mathematics*. Its direct antecedents are the Gauss–Bonnet theorem of Heinz Hopf for hypersurfaces, the tube theorems of Herman Weyl which gave the form of the higher-dimensional Gauss–Bonnet integrand, and the theorems of Fenchel, Allendoerfer, and Weil. Yet with Veblen, Hopf, and Allendoerfer all present, and the subject of generalized Gauss–Bonnet theorems under discussion, there is no mention of Chern's work. (Again the *caveat*: there is no mention in the written summary, prepared by Allendoerfer.)

It would appear that the session opened with presentations by each of the discussion leaders, V. Hlavaty and T.Y. Thomas, of their own recent work. (In the case of Thomas, the subject was not differential geometry, but dynamical systems.) Bochner then described his own recent results, that have since acquired the name of "the Bochner technique." The responses indicate a clear appreciation of the value of Bochner's contribution, which subsequent

---

[1] Robert Osserman is a professor of mathematics at Stanford University.

developments have amply confirmed. In particular, the Bochner "vanishing theorem" to which Whitehead alludes in the comment cited above shows that certain geometric constraints imply the vanishing of harmonic forms which in turn leads to the vanishing of homology (or cohomology) classes via Hodge theory. This was the precursor of the renowned Kodaira vanishing theorems of the fifties, in addition to a steady stream of related results. As an indication of the breadth of that stream, two recent surveys — one by Pierre Bérard: "From vanishing theorems to estimating theorems: the Bochner technique revisited," in the *Bulletin of the American Mathematical Society*, Vol. 19, No. 2, October 1988, and the other a monograph by Hung-Hsi Wu: "The Bochner Technique in Differential Geometry," appearing in *Mathematical Reports*, Vol. 3, Part 2, January 1988 — have extensive bibliographies with almost no overlap.

The Bochner results lead naturally to the final circle of ideas discussed: the relations between curvature and topology. Specific references are to the theorems of Preissmann, Cohn–Vossen, and Gauss–Bonnet. All the results have in common with Bochner's that they relate curvature to topology, but whereas Bochner's conclusions regard homology, Preissmann looks at the fundamental group. Since the work of Gromoll–Wolf and Lawson–Yau in the early 70s, the subject of curvature and the fundamental group has been more and more intensively studied. Surveys may be found in recent articles by Eberlein, and the 1985 monograph by Ballmann, Gromov, and Schroeder.

As for the Cohn–Vossen theorem, subsequent results have been most interesting in the two-dimensional case. The original theorem is the bound

$$\int_M K\,dA \le 2\pi\chi$$

for the total curvature of a complete 2-dimensional Riemannian manifold M in terms of the Euler characteristic χ of M. Refinements by Robert Finn and Alfred Huber include the result of Huber that if $\chi = -\infty$, then also $\iint K\,dA = -\infty$. In particular, finite total curvature implies finite topology. A simpler proof of that result was recently obtained by Brian White. Another consequence of Huber's Theorem is that nonnegative curvature outside a compact set implies finite topology.

For complete minimal surfaces, a stronger version of the Cohn–Vossen inequality holds, and the strengthened version plays an important role in the recently elaborated theory (and examples) of complete *embedded* minimal surfaces.

In higher dimensions, only scattered results are known: for dimension 4, under certain curvature restrictions (Poor, Walter, Greene/Wu), and in arbitrary dimension, for a restricted class of manifolds (Harder). Recent work of Cheeger and Gromov has shed light on the possible values of Cohn–Vossen type integrals and other integrals of characteristic classes for complete

Riemannian manifolds of finite volume, with a bound on sectional curvature. Also, a number of recent papers have been devoted to studying the topology of manifolds whose sectional curvature is nonnegative (or even identically zero) outside a compact set.

The Gauss–Bonnet Theorem states that for a compact surface M,

$$\int_M K\, dA = 2\pi\chi.$$

Among its most important consequences is that if the curvature K is of one sign everywhere on M, then the Euler characteristic χ must have the same sign. In the higher-dimensional Gauss–Bonnet theorem, the integrand K is replaced by a fairly complicated expression in the components of the curvature tensor. An obvious question is the relation between the sign of the sectional curvature and the sign of χ. In four dimensions, it follows from Myers' theorem cited above that if the sectional curvature is positive, then the first Betti number $b_1(M) = 0$, and then by Poincaré duality $b_3(M) = 0$, so that $\chi(M) = 2 + b_2(M) > 0$. A different argument due to Milnor shows that in four dimensions, if the sectional curvature is either everywhere positive *or* everywhere negative, then $\chi(M) > 0$.

In the case of positive sectional curvature, there was the possibility that the Gauss–Bonnet integrand would be pointwise positive. However, Geroch showed in 1976 that for dimensions six and higher the integrand could be negative even if the sectional curvature is positive.

More important than these particular results were the broad developments leading from Chern's intrinsic treatment of the Gauss–Bonnet theorem to the general theory of vector bundles and characteristic classes, and to the Atiyah–Singer index theorem. Although the first steps in that direction had already been taken, it would have required enormous prescience to foresee where they might lead. On the other hand, the prediction that links between curvature and topology were the wave of the future has been fully borne out in the theorems of Rauch–Berger–Klingenberg, Gromoll–Meyer, Gromoll–Cheeger, Lawson–Gromov, Schoen–Yau, the recent results of Ballmann–Brin–Eberlein–Burns–Spatzier, and others far too numerous to mention.

Commentary on Mathematical Logic

YIANNIS N. MOSCHOVAKIS [1]

According to McKinsey's account of the session in mathematical logic, "the discussion revolved about a single broad topic — decision problems." A decision problem calls for an "effective method" (an "algorithm") which will decide whether an arbitrary object from some set D has or does not have a given property P. It can be solved "positively" by producing an algorithm which does the required job. To solve it "negatively" and prove that the given decision problem is *unsolvable*, you must establish that the function which needs to be computed *is not computable by any algorithm*, and for this you need a precise, mathematical notion of computability. The pioneering work of Church, Gödel, Kleene and Turing in the thirties had produced just such a notion — *recursiveness* — and by 1946 essentially everyone in the field had accepted *Church's Thesis*, the claim that (formal) recursiveness and (intuitive) computability are identical.

In the natural course of events, the first unsolvability results were inside logic, including Church's fundamental result that *the problem of validity in predicate logic is unsolvable*. Just before the Princeton meeting Post had proved that *the word problem for semigroups is unsolvable*, and Church now suggested three more candidates for unsolvability results "of a more standard mathematical character": "the word problem for groups and the problems of giving a complete set of topological invariants for knots and for closed simplicial manifolds of dimension n."

The first of these questions has been answered by Boone and Novikov (independently) who proved in the late fifties that *there exist finitely presented groups with unsolvable word problem*. This was a very active research area in the fifties and sixties, one of its best results (in the experts' opinion) being Higman's (1961) Embedding Theorem: *a finitely generated group can be recursively presented if and only if it is isomorphic to a subgroup of a finitely*

---

[1] Yiannis N. Moschovakis is a professor of mathematics at the University of California, Los Angeles.

presented group. Markov proved in 1958 that *the decision problem for homeomorphy of closed, simplicial manifolds in any dimension n ≥ 4 is unsolvable,* and this is generally assumed to rule out any acceptable solution to the classification problem for these dimensions. In dimension 2 manifolds have been classified by classical methods, and nothing much is known now about the most difficult dimension 3. The problem of classifying knots is apparently still open — no matter how you make it precise.

What is quite astonishing about this extensive discussion of decision problems is the absence of any reference to Hilbert's 10th, the only decision problem in Hilbert's famous list: *to determine whether an arbitrary polynomial with integer coefficients has any integer roots.* Hilbert's 10th was proved unsolvable in 1970 by Matijacevič, building on a great deal of groundwork laid by Davis, Putnam, and Julia Robinson. It is generally regarded as the most impressive among "applications" of recursive function theory to fields outside logic.

McKinsey next makes a passing reference to the problem "whether or not particular questions are decidable in a given theory," accompanied by the suggestion that the Riemann Hypothesis may be a good candidate for such an example. Actually this is a bad example, since the Riemann Hypothesis has the property that its consistency with any minimally strong axiomatic theory implies its truth. Much better candidates for such examples are simple statements of set theory, like the axiom of choice or the continuum hypothesis which had already been proved consistent with Zermelo–Fraenkel set theory by Gödel and were generally believed to be independent of ZF. (Their independence was proved in 1963 by Paul Cohen, who introduced in the proofs the seminal new method of *forcing* and earned with his results the only Fields Medal ever given for work in logic.) Why were these not mentioned — particularly with Gödel in the group? Even now, however, when the question of independence of "natural, mathematical" statements comes up, logicians look for "logically simple" examples, as close to number theory as possible. The first simple and elegant example of this kind was a Ramsey-type partition theorem which can be expressed in the language of number theory, and was proved true and independent of the classical Peano axioms by Paris and Harrington in 1977, building on earlier work of Paris. A lot has happened in this area since then, particularly by Friedman and his school.

McKinsey reports a brief reference by Quine to "restricted arithmetic" and a long passage on "many valued logics," not among the most studied areas of logic today. Other than that, McKinsey's account mentions just three other topics which were touched upon, and which are more relevant to current work in logic.

"The analogy between recursive sets of integers and Borel sets of real numbers was stressed by Tarski, who pointed out the possibilities of proving analogous theorems and of developing a single theory to include both as special cases." Most likely Tarski was referring to the work of Mostowski, who was pushing then the stronger analogy between *recursively enumerable* sets of integers and *analytic* sets of reals, and I wish the text recorded whether Kleene responded and how. Kleene certainly learned of Mostowski's proposal — if not in 1946, shortly afterwards — and he published in 1950 (at Mostowski's suggestion) his result that *there exist disjoint, recursively enumerable, recursively inseparable sets*, which disproved the proposed strong analogy and suggested that the weaker analogy suggested by Tarski was "defective." It took about eight more years before the *hyperarithmetic sets* of integers were introduced (independently) by Davis, Kleene, and Mostowski and the "correct" analogy between *them* and the Borel sets was proposed by Addison and firmly established by Kleene's proof of what we now call *the Suslin–Kleene Theorem*. *Effective descriptive set theory* was born of these results, it grew enormously in the seventies under the influence of determinacy and large cardinal hypotheses and has certainly provided some version of Tarski's "single theory" of definability in the integers and the reals. It is one of the main research areas of logic today.

There is a tantalizingly brief remark that "Kleene discussed the limitations which general recursiveness may place on quantitative proof." Although somewhat cryptic, the reference must be to *realizability theory* which Kleene was developing just then. The gist of the idea is that a *constructive assertion* of some statement A can be understood (at least partially) as a *classical assertion* that a recursive function exists which *realizes* some constructive (or computable) aspect of A. In a seminal paper published in 1945, Kleene gave the first precise definition of *number realizability* and announced Nelson's result, that *every theorem of intuitionistic number theory is realizable*. The theory took off after that, different realizability interpretations of many intuitionistic theories were introduced, and they have provided one of the main tools for the metamathematical study of constructive mathematics. More recently, notions from realizability theory have invaded theoretical computer science, primarily in the theory of programming languages.

Finally, McKinsey devotes two paragraphs to the "spirited discussion" on Gödel's proposal of "a particular tremendous enlargement of a formal system — which would allow uncountably many primitive notions and allow the notion of an axiom to depend on the notion of truth." This threw me for a while, until I looked up Gödel's *Remarks before the Princeton Bicentennial* in Davis' most useful source book, *The Undecidable*, and recalled what it was about: Gödel talked about set theory, as one might expect, even though the word "set" never occurs in McKinsey's account. He introduced the notion of *ordinal definability* and announced that *the class of hereditarily ordinal definable sets is a model of ZF with the axiom of choice*, and he also talked about *axioms of infinity*. Intuitively, an axiom of infinity is an assertion that

very large sets exist, in some specific way of measuring "size." Quoting now
from the account of his remarks in *The Undecidable*,

> There might exist, e.g. a characterization [of axioms of infinity]
> of the following sort: An axiom of infinity is a proposition which
> has a certain (decidable) formal structure and which in addition is
> true ... the following could be true: Any proof for a set theoretic
> theorem ... is replaceable by a proof from such an axiom of
> infinity.

In his account, McKinsey focused on the *logical* aspect of Gödel's proposal — about nonconstructive axiom systems — which has not yet led to
anything substantial. For set theory, however, the idea that we should look
to plausible axioms of infinity in order to settle the classical open problems
about sets has been one of the dominant themes of the field and has motivated
some of its most substantial successes in the last forty years.

It is quite a sobering experience to read this account of McKinsey's and
try to relate it to what is happening in mathematical logic today. Ours is
a very young field, indeed: there was no *model theory* in 1946; Gödel was
practically the only person in the west interested in *set theory*; *recursion theory*
was viewed almost exclusively as a tool for applications outside logic and to
proof theory; and even in the parts of the discussion on *proof theory*, there
is no mention of the work of Herbrand and Gentzen, which has influenced
the theory of proofs since then much more than, say, many-valued logics.
After the subsequent work of some who attended the Princeton meeting and
many others later, we can find today work in all these research areas which
is as deep and as mathematically sophisticated as any in pure mathematics.
In fact, this increasing "mathematization" of the field has prompted recent
accusations that "we have lost our roots" — but this is not our topic here.

The first AMS Summer Research in Symbolic Logic was held in Cornell
in 1957. Its list of participants is an incredible who-will-be-who of young
logicians and its proceedings are full of exciting new beginnings. A lot had
happened in those ten years. Ward Henson has commented that in contrast
to the 1946 meeting, "the Cornell meeting we can recognize as one of ours."

Commentary On Topology

WILLIAM BROWDER [1]

The year 1946 and the Princeton Bicentennial Conference discussion of topology could be said to celebrate the birth of one field (compact transformation groups) and the puberty of another (homotopy theory, and algebraic topology).

The Fifth Problem of Hilbert commanded considerable attention in 1946, but a "central question" at the time is still unsolved: if a compact group G acts effectively on a manifold is G necessarily a Lie group? Many other related problems were soon solved such as the theorem of Montgomery–Zippin and Gleason: a topological group which is a manifold is a Lie group.

On the other hand, the focus of attention in this area soon moved toward problems of classifying actions, a typical question being: how much does an arbitrary compact group action on a familiar space (e.g., disk, sphere, euclidean space) resemble (e.g., homeomorphic, diffeomorphic, homotopy equivalent, or have similar fixed set) a familiar action (e.g., linear action), the point of view pioneered by Smith in the 30s. The specific question asked of this type: "does a cyclic group acting on a euclidean space fix some point?" was soon answered in the negative (Conner–Floyd). But this class of questions became the central focus in this area, and compact transformation groups became a testing ground and application area for every new development in topology, with great success.

The field of algebraic topology, on the other hand, was already in a golden age, with the development of homology and cohomology theory, homotopy groups, obstruction theory, and the emerging perception of a general context for algebraic topology, the functorial approach, and homological algebra. Steenrod had just begun to discover cohomology operations in calculating $\pi_n(S^{n-1})$, but the "mass production" methods made possible by the spectral sequence of Leray were not yet visible. A decade or so later the spectral sequence would have made itself felt, most notably in topology with

---

[1] William Browder is a professor of mathematics at Princeton University.

the Leray–Serre and the Adams spectral sequences, but also in applications to other areas.

In fact, the introduction of topological and, more specifically, *homological* methods into so many diverse areas of mathematics is a striking development hinted at in this conference in many areas, but not then realized. The introduction of sheaf theory into complex analysis, homological algebra, and cohomology of groups into number theory and algebraic geometry, and later the rise of K-theory in its multifarious guises has changed the shape of numerous areas of mathematics.

As algebraic topology developed after 1946 with its flourishing development using spectral sequences, cohomology operations and fiber bundle theory, a new theme began to be sounded strongly in the middle 1950s and 1960s: differential and geometric topology. With Thom's work on cobordism theory, a new approach to studying manifolds using homotopy methods emerged. With Milnor's discovery of the exotic differentiable structures on S^7 a whole new field opened up, the classification of smooth structures on a given topological manifold. And with Smale's discovery of the h-cobordism theorem and the Poincaré conjecture in dimensions greater than 4, the possibility opened of attacking many of the old geometric problems using the well-developed and powerful machinery of algebraic topology. These methods of homotopy theory — algebraic topology, characteristic classes, fiber bundle theory — became the everyday tools of geometric topologists in "surgery theory," the systematic, *algebraically controlled* method of cutting and pasting manifolds (initially smooth manifolds).

Pioneered by Milnor and Kervaire in studying smooth structures on $S^n, n \geq 5$, surgery theory soon was applied to studying how to characterize the homotopy type of smooth manifolds and classify (up to diffeomorphism) the ones in a given homotopy type (Browder and Novikov) always with dimension ≥ 5. Initially developed in the context of the trivial fundamental group, the generalization of the theory to the nonsimply connected case by Wall opened up a whole new connection of topology with number theory, through K-theory of hermitian forms related to surgery obstruction groups. Among the outstanding accomplishments of the theory were Novikov's proof of topological invariance of rational Pontryagin classes and Sullivan's analysis of the *Hauptvermutung* for manifolds of dimension ≥ 5 (surgery theory having been adapted to "piecewise linear" manifolds).

There was a parallel development of piecewise linear (PL) manifold theory in the 1950s and 60s, which was hinted at in 1946 in Whitehead's thinking, as was the development of algebraic K-theory which had its birth in Whitehead's simple homotopy theory. But nobody in 1946 could have anticipated that all these diverse methods would come together 25 years later to give Kirby and Siebenmann's comprehensive theory of PL triangulation of manifolds of dimension ≥ 5, and another decade later, with the addition of the decomposition space methods pioneered by Bing, lead to Freedman's topological classification of 1-connected 4-dimensional manifolds.

The reference of the conference to "the need of studying not only the [homology] groups involved, but also the homomorphisms connecting them" leads quickly (in retrospect) to exact seqeunces, homological algebra, the Eilenberg–Steenrod axioms for homology, and the general abstract approach, which became so popular. The Eilenberg–Mac Lane spaces, $K(\pi, n)$'s, the universal models for cohomology led to the Postnikov tower methods, calculations of Serre and Cartan, and to more general cohomology operations (of higher order) and similar methods in homotopy theory (Toda brackets). The crowning achievement of this approach was the Adams spectral sequence, a powerful tool for calculating stable homotopy from cohomology, which led to Adams' solution of the Hopf invariant 1 problem (e.g., there are no nonsingular pairings $\mathbf{R}^n \times \mathbf{R}^n \to \mathbf{R}^n$ if $n \neq 1, 3,$ or 7).

Into this increasingly formalized algebraic study came another dramatic development of the 50s (totally unanticipated in 1946), the Periodicity Theorem of Bott, the development of topological K-theory and interesting *extraordinary* cohomology theories, whose existence was suggested by the Eilenberg–Steenrod axioms. Bott's bizarre result, proved initially with Morse Theory, that the homotopy groups of the unitary groups were stably periodic, made possible the K-theory of stable complex vector bundles. Combined with analytic ideas, the first results on indices of elliptic operators from the Russian school, and no doubt inspired by topological methods in algebraic geometry, as in Hirzebruch's book, Atiyah and Singer created their Index Theory for elliptic operators, and another chapter in the reunification of topology with other fields of mathematics opened.

At the same time K-theory quickly found its uses in topology both in the geometric side (e.g., nonparallelizability of $S^n, n \neq 1, 3, 7$ and in surgery theory), but also as a powerful new tool in homotopy theory, spearheaded by Adams' solution of the vector fields on spheres problem, and Atiyah's K-theory proof of Adams' Hopf invariant one theorem. Bordism theory, another new, geometrically defined, extraordinary homology theory soon was put to use by homotopy theorists and these two theories and derivative theories are now central tools in the subject.

All the new influence of differential geometry, smooth manifolds, and analysis in topology and homotopy theory was something not envisioned in the 1946 conference, but if Pontryagin had been there, perhaps some hint of his theory of framed manifolds would have emerged.

The three-dimensional Poincaré conjecture was, in 1946, and remains today, an outstanding problem. Even though enormous progress has been made in low dimensional topology since 1946, it is still difficult to foresee a solution in the near future.

On the other hand, beginning with the classical theorems of Papakyri-akopoulos in 1956, and most recently the work of Thurston and his school in the 1980s, many deep problems in 3-manifolds have been solved. Again, in Thurston's approach, we see new geometric input into topology, by show-ing many 3-manifolds have "geometric" structures (e.g., metrics of constant curvature) so that geometric methods can be employed. The positive so-lution of the Smith conjecture (any periodic transformation of S^3 fixing a point is homeomorphic to a linear one) is the product of this "geometriza-tion" method together with differential geometric input (Meeks–Yau) and algebraic input (Bass). When there is no fixed point this problem is still open for most groups.

But the biggest surprise since 1946, and probably of this century, is the enormous qualitative difference between topology in dimension 4 and all other dimensions, and the methods of physics on which these results are based. Donaldson's remarkable first result in 1982, proved using Gauge Field Theory Methods, that the only positive definite intersection forms on closed smooth 1-connected 4-manifolds are diagonal, caused universal aston-ishment. The corollary which followed from this and Freedman's topologi-cal classification theorem mentioned above (proved at almost the same time) showed that the smooth structure on $\mathbf{R}^4$ is not unique, a statement false in every other dimension. Since then, it has been shown $\mathbf{R}^4$ admits infinitely many different smooth structures, as do many closed 4-manifolds. The the-ory of smooth 4-manifolds is of an entirely different nature from higher or lower dimensions. In particular the kind of algebraic topological invariants which describe the whole story, by surgery theory, in higher dimensions, are irrelevant to the rich phenomenology of dimension 4.

This new influence of physics has also made itself felt strongly in algebraic topology with many new points of view from quantum mechanics contributed by Witten.

One might look at the history of topology since the 1946 conference in two complementary aspects:

(1) The enormous development of all the themes represented there, homol-ogy and homotopy theories, fiber bundles, group actions, and combinatorial topology, and the spreading of the central parts of these methods (in partic-ular homology and fiber bundle theory) into other areas, until there is hardly an area of mathematics today in which topology does not play a prominent role, as well as in parts of theoretical physics and molecular biology.

(2) An influx of new ideas into topology, from differential geometry and calculus of variations, differential equations, algebraic geometry and number

theory, to the most recent striking input of ideas from physics. Coupled with the internal development of the subject, it has led to enormous progress.

One might say that topology has developed from being a field of mathematics in 1946, somewhat isolated in its subject matter and approach, into a meeting ground for many different areas of mathematics, an area in which interaction and cooperation with other fields has led to enormous progress for all sides.

In this, topology mirrors, on a smaller scale, the change in position of mathematics with respect to other sciences.

Herman Weyl's worries "... maybe it [mathematics] has become too hard for us unless we are given some outside help..." were prophetic. In topology, with respect to other fields of mathematics and physics, we have received abundant outside help. But topology has richly repaid its debt with an enormous contribution to other areas, including physics.

Commentary on Probability

J. L. DOOB [1]

The basic difference between the roles of mathematical probability in 1946 and 1988 is that the subject is now accepted as mathematics whereas in 1946 to most mathematicians mathematical probability was to mathematics as black marketing to marketing; that is, probability was a source of interesting mathematics but examination of the background context was undesirable. And the fact that probability was intrinsically related to statistics did not improve either subject's standing in the eyes of pure mathematicians. In fact the relationship between the two subjects inspired heated fruitless discussions of "What is probability?," and thereby encouraged the confusion between probability and the phenomena to which it is applied. (This confusion still plagues the subject.) Furthermore in 1946 there was no advanced probability text leading from definitions to basic theorems, although at the research level there were specialized articles and books written under the assumption that the readers would not worry about the lack of a rigorous axiomatic basis underlying the work. Nevertheless Kolmogorov had published the basis for mathematical probability in his (German) 1933 monograph, including the identification of random variables as measurable functions and expectations as integrals. Furthermore he had there introduced conditional expectations as functions, and had defined measures on coordinate spaces of arbitrary dimensionality, thereby providing models (the coordinate functions) of families of random variables. Unfortunately the significance of his work was not appreciated for years, and some mathematicians sneered that probability should not bury its spice in the bland soup of measure theory, that perhaps probability needed rigor, but surely not *rigor mortis*.

The carrying out of the program inspired by Kolmogorov's additions was delayed by a factor usually ignored by historians of mathematics, the surprisingly long time it takes for new mathematical ideas to be absorbed, for new points of view to be taken seriously. Although Lebesgue measure dates back to 1902 it was only noted (by Fréchet) in 1915 that for measures on Borel

---

[1] J. L. Doob is a Professor Emeritus of the University of Illinois at Urbana–Champaign.

fields of subsets of abstract spaces the corresponding measurable functions and integrals can be defined and manipulated in precisely the same way as in the context of Lebesgue measure on a linear interval. And even thereafter, for a long period the elementary manipulations of measurable functions and their integrals, now an early part of the graduate education of every student of analysis, were avoided, or rejustified as needed when the measure was not Lebesgue measure. "Measure" meant "Lebesgue measure" as against the more general "positive additive set function"; this convention, together with the use of (Riemann) Stieltjes integrals provided a terminological delaying action against the acceptance of general measure theory. Probabilists were perhaps more inhibited in applying measure theory than other analysts because of their private measure terminology and because their probability measures were usually not Lebesgue measure. It is not much of an exaggeration to assert that probabilists are the only mathematicians who ever evaluate an integral by other than the antiderivative method of elementary calculus. The ironic fact is that the role of measure theory in probability has embarrassed many probabilists and still embarrasses some who like to think that mathematical probability is not a part of analysis. It is noteworthy that Kolmogorov himself, in his 1933 monograph that made mathematical probability rigorous by basing it on measure theory, hindered the acceptance of his own ideas by being chary in his use of measure theory terminology. For example, instead of defining a real random variable as a real measurable function he avoided the use of the adjective, going back to the defining property that the inverse image of an interval under the function should be a set in his specified Borel field. Although in an introductory paragraph he stated that his random variables were described elsewhere as measurable functions he did not refer to the customary measure theory terminology in discussing almost everywhere convergence or convergence in measure, and went to the extreme of proving that the limit of a convergent sequence of random variables is a random variable. He obviously felt, and he was right, that the jump from Lebesgue to general measures was psychologically difficult.

With this background it becomes understandable that in 1946 only a few mathematicians were taking mathematical probability seriously as mathematics, that probabilists who were carrying on deep research, for example Paul Lévy, wrote in a mysterious style that discouraged outsiders from entering the field. Yet with the infallible help of hindsight it has become obvious that by the 1940s the level of analysis, in particular that of mathematical probability, was such that there was bound to be a flowering of the latter and its application to all areas involving averaging, some quite unexpected. It would be pointless to list those areas here. Suffice it to write that probability theory has on the one hand penetrated many parts of analysis and on the other hand still has its own special character. Every analyst must now be familiar with some aspects of mathematical probability although only specialists can be expected to follow specific probability research.

Commentary on Fourier Analysis

E. M. STEIN [1]

When the progress of Fourier series is viewed from the perspective of 1946 to the present — a considerable time span in the history of any branch of mathematics — one can better appreciate the paradoxical fact that the rates at which various parts of the subject developed were not at all in proportion to the interest they initially held. This point is brought out when we consider the advances made in the last 40 years in three areas alluded to in Zygmund's remarks, namely (1) the uniqueness of trigonometric series, (2) the problem of convergence of Fourier series, and (3) extensions of the theory to n-dimensions. 1. In the theory of uniqueness of trigonometric series there has been one major result: the theorem of Salem and Zygmund (ca. 1950) that a Cantorlike set of constant ratio of dissection r is a set of uniqueness if and only if 1/r is a Pisot–Varadarajhan number, namely an algebraic integer whose conjugates all lie strictly in the unit disc. Except for this striking result the field of uniqueness of trigonometric series has lain mostly dormant.

2. As for Lusin's problem of the a.e. convergence of Fourier series, the crowning result is Carleson's positive solution in 1966. For higher dimensions (except for Fefferman's observation that the analogue for rectangular convergence does not hold), much still remains to be done. A major problem resisting solution is that of the spherical convergence of Fourier series in dimension greater than 2, as well as the corresponding problem for the eigenfunction expansions for the Laplacian on a compact manifold. Connected with this is the sobering fact that, as subtle and as ingenious as Carleson's theorem is, the ideas in it have as yet not been significantly exploited in other parts of harmonic analysis.

3. It is, however, in the direction of extending other parts of the theory to higher dimensions that there has been in the last 40 years the greatest activity and success. Here we have witnessed a bountiful harvest of results. I have in mind the theory of singular integrals, the related extensions of the Littlewood–Paley theory, and the n-dimensional theory of Hardy spaces; in

---

[1] E. M. Stein is a professor of mathematics at Princeton University.

all of these important new points of view were developed, and significant relations within other parts of mathematics have been achieved.

What might the next 40 years hold in store for us? It is impossible not to be fascinated by this question. However, experience teaches us that our present interests may not be much help in wrestling with this puzzle.

Commentary on
"Analysis in the Large"

KAREN UHLENBECK [1]

In the opening paragraph on the discussion of analysis in the large, it is commented that the discussion centered on choice of method. Morse is quoted as emphasizing that the categorization into subdisciplines is regarded as "all for convenience, and has nothing to do with essential differences." The emphasis seems to be on existence and uniqueness of solutions. Choice of methods is apparently between calculus of variations and fixed point theorems, although mention is made of the relationship of differential and topological invariants, perturbation in a parameter and generalized curves. Today the major change would be an emphasis on content. The slightly puzzling division into differential geometry, applied mathematics, and something called equilibrium analysis has exploded into dynamical systems, differential topology, minimal surface theory, global differential geometry, analysis on Riemannian manifolds, complex geometry and several complex variables, nonlinear elliptic and parabolic equations, nonlinear hyperbolic equations and a whole realm of various subjects in mathematical physics and applied mathematics. Most mathematicians would not work seriously in very many of these subjects, but there are, of course, more mathematicians. The "choice of method" is less central, although various techniques, such as implicit function theorems (strangely absent from the 1948 article), Fredholm index theory, fixed point theorems, variational methods, complex dynamics, geometric measure theory, analysis of singularities, topological techniques, computer graphics and numerical methods are exchanged and shared by these disciplines. It is convenient in discussing this field to point to the halfway mark (1968). In this year, there was a grand American Mathematical Society Summer Mathematics Institute held in Berkeley on the subject of what had come to be called global analysis. A perusal of the three-volume proceedings (Proceedings of Symposia in Pure Mathematics, Volumes 14–16), or even just their tables of contents, gives one a very good idea of how far this optimistic

---

[1] Karen Uhlenbeck is a professor of mathematics at the University of Texas at Austin.

belief that all of nonlinear mathematics could be put in a proper framework by a universal method could be carried. These volumes are divided informally and less mysteriously into what is in fact dynamical systems (vol. 14), differential geometry and infinite dimensional differential topology (vol. 15), and partial differential equations (vol. 16).

In dynamical systems, various current fundamental concepts of structural stability, stable manifolds, entropy, zeta functions and attractors are well-developed, but little contact with physical, geometric, or computer-simulated problems is present. The theory of infinite dimensional manifolds and the study of functions, flows, and topology for them is almost too well-developed, and the techniques for the study of partial differential equations are already very numerous. I noted monotone operators, fixed point theorems, critical point theory, kernels and pseudodifferential operators, basic ideas of characteristics, and the notion of genericity.

Superficially those first twenty years of spectacular abstract technical development bear out the concern of Hermann Weyl, quoted in the final session of the 1948 report. Especially note the quote dating even further back from 1931, "I am afraid the mathematical substance in the formulation of which we have exercised our powers in the last two decades shows signs of exhaustion." In hindsight, it appears to me that global analysis in 1968 was an abstract theoretical technique in search of applications. Much intuition of a more specific mathematical substance did lie behind most of the abstraction, but it was certainly not the fashion to make this transparent.

Analysis in the large became global analysis, went out of favor, and became in its present state something the National Science Foundation calls geometric analysis. The most striking feature of these last twenty years has been a revolution from the general to the particular of which I feel sure Hermann Weyl would approve. Ideas have poured into "analysis in the large" from other parts of mathematics and related sciences. Computers of all sizes from personal to super have revolutionized dynamical systems, not only justifying its theoretical and topological bias, but making contact with complex analysis as well as more applied subjects such as fluid mechanics. Strange attractors and universal bifurcation phenomena are of central interest in many sciences. A book on chaos has been on the bestseller list of the *New York Times*.

Global differential geometry has been enriched by the successful study of a list of specific problems which interact with other areas of mathematics and theoretical physics. I hope I offend no one by citing as an example the Calabi problem, which solves Einstein's equation in general relativity on a complex geometric manifold, has important applications in algebraic geometry, and now provides vacua for string theories in high energy physics. Yau's solution is technically that of the unsophisticated "continuity method," or estimates and the implicit function theorem. Similar interconnections are observed for the Yamabe problem, harmonic maps, minimal surfaces, and Yang–Mills theory.

The calculus of variations plays a central role in the 1948 discussion. By 1968 it had been elegantly abstracted to calculus on Banach manifolds by Palais and Smale. In 1988, every variational problem in geometry or applications is first measured in difficulty by judging how far it deviates from the Palais–Smale criteria for Morse theory, and then studied for specific details. Estimates, hard and soft implicit function theorems, and constructive fixed point theorems are the tools of the trade, while the purely topological fixed point theorems have grown out of favor as being "nonconstructive."

I really cannot be fair here to the new advances which have occurred in all of nonlinear partial differential equations. Elliptic and parabolic equations are in such good shape that often one can ask what solutions actually look like! The last decade has brought a revolution in the understanding of hyperbolic problems, especially that of many physically relevant systems. It might be emphasized that there is in these developments a lot of contact with geometric ideas, computational algorithms, and specific applications intertwined with the fundamental functional analysis.

Finally, in returning to the 1948 discussion, we should note that some of the problems are still only partly solved. Allendorfer's comments connecting differential invariants with topological invariants refer to the theory of characteristic classes then being developed, although I first thought of them as looking forward to the Atiyah–Singer index theorem. From one point of view it was the "Dark Ages." From another point of view, the insight of theoretical physicists such as Witten into global topological invariants via quantum field theory reminds us of the rich possibilities of an unknown mathematical future.

In responding to the 1948 discussion, I have not had much space to point toward the future and the central open questions. However, although geometric analysis does not appear to me unified, the rest of the final paragraph concerning its remarkable size and tendency to absorb the rest of mathematics still seems true.

THE Ph.D. DEGREE AND MATHEMATICAL RESEARCH *
By R. G. D. RICHARDSON, Brown University

Recommendations regarding the training and utilization of advanced students of mathematics must be based on specific information concerning the present situation and on a knowledge of how the past has contributed to its upbuilding. No group of persons can be entirely certain that sound deductions are possible from data as incomplete as those now available. However, queries often raised regarding quality of personnel and regarding supply and demand can be answered with considerable assurance. In so complicated a problem any light that can be shed is undoubtedly welcome, and it is proposed in this report to set forth a variety of facts and to venture partial and tentative answers to certain questions.

We propose questions such as the following: How many doctor's degrees have been conferred on American mathematicians here in the United States and Canada? How many in Europe? What proportion of the present teachers of mathematics in colleges and universities have such degrees? Is there a sufficient number of competent students of mathematics now being enlisted and subjected to proper training by our graduate schools? What percentage of the Ph.D.'s have published considerable research? Is the record of publication improving with the newer crop of Ph.D.'s? What universities have the distinction of the largest average amount of publication by those to whom they have granted degrees? How does the record of the National Research Fellows stand?

Beginning with 1900, the Bulletin of the American Mathematical Society has made a practice of publishing an annual list of the doctor's degrees conferred in mathematics. In addition, each university has, on request, been kind enough recently to furnish lists of all doctor's degrees in mathematics conferred since the beginning of its graduate work and including the year 1934. From these sources† accurate data have been assembled. The corresponding information regarding Americans and Canadians taking degrees abroad is not so complete.

Each year, beginning with 1890, the Bulletin of the American Mathematical Society has printed an *Annual List of Papers Read Before the Society and Subsequently Published*. These lists have been drawn on for the period 1890–1933, and other sources for the period before 1890. It has been possible to make a complete tabulation for the sixty years ending with 1933,‡ giving the number of such papers and the number of pages printed. It should be pointed out, however, that papers of American authorship not presented to the Society are

* A report to the Commission on the Training and Utilization of Advanced Students of Mathematics presented at its session on December 31, 1935.

† Beginning with 1934, a list of doctors and thesis titles in all fields of advanced study is being published annually by H. W. Wilson Company under the title *Doctoral Dissertations Accepted by American Universities* compiled for the National Research Council and the American Council of Learned Societies by the Association of Research Libraries.

‡ For the National Research Fellows the tabulation has been extended to include 1934.

Reprinted with permission from *American Mathematical Monthly*, Volume 43, 1936, pp. 199–215.

Note: Although subsequent research has led to changes in the statistics that Richardson displays, so that his numbers should no longer be used without verification, the paper remains a model of the type of study that periodically increases our awareness of the status of the field. (**The Editors**)

R. G. D. Richardson

GRAPH I
Distribution of Degrees Conferred in Mathematics 1862–1934

The upper portion of each block represents American mathematicians with European degrees. The lower portion represents American degrees.

Period	American	Foreign
1862–69	3	2
1870–74	3	0
1875–79	7	4
1880–84	9	1
1885–89	23	10
1890–94	28	12
1895–99	56	11
1900–04	75	21
1905–09	80	13
1910–14	126	18
1915–19	125	3
1920–24	129	9
1925–29	227	9
1930–34	394	7

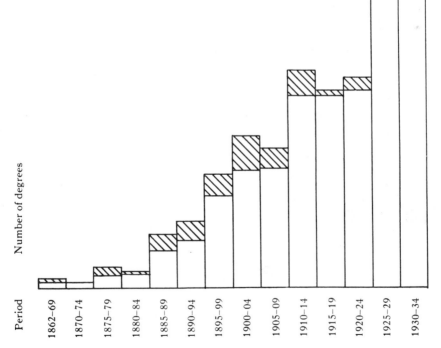

Number of degrees

Period   1862–69   1870–74   1875–79   1880–84   1885–89   1890–94   1895–99   1900–04   1905–09   1910–14   1915–19   1920–24   1925–29   1930–34

omitted from the tabulations; and this is true whether these papers appeared in American or foreign journals. Some mathematicians, especially among the more eminent, publish papers not read before the Society; but a study of the total publication of mathematical research for a typical year (1932) shows that only about 22% of such publication (1,158 pages out of a total of 5,290) is not included in our data. The data assembled seem, therefore, fairly adequate for the purpose in hand.

For each person it has been possible to compute an index of publication which is the annual average, measured in pages, of printed research during the period subsequent to receiving the degree, but in no case exceeding twenty-five years. The average of the indices for the group of doctors beginning with 1862 is 4.73, which then establishes a norm by which comparisons can be made. Otherwise stated, a doctor publishes yearly, on the average, 4.73 pages of research which has been presented to the Society. (If the year 1932 referred to above is a fair sample, this figure would be increased to 6 pages by including papers not presented to the Society. And it should be expressly understood that the list includes only papers published in research journals, and that separate monographs and books, which are so important in the development of research, are excluded from the count.) Reckoning eight dollars as the cost of printing a page, the average annual cost of publication per doctor is approximately fifty dollars, a relatively small item in the cost of production of research.

As a rough check on this figure of 4.73 pages, we note that the American Journal, Annals, Bulletin, and Transactions print about 3,500 pages of research annually and that other journals add perhaps 1,600 pages more. This makes a total of 5,100 pages as against approximately 1,200 doctors.

Number of Ph.D.'s Conferred in the United States and Canada

A careful investigation indicates that the number of Ph.D.'s in mathematics conferred by institutions in the United States and Canada during the period 1862–1934 is 1,286, 168 of which were conferred on women. Graph I exhibits the number of degrees conferred during five-year periods, this number following in general an exponential curve, though interrupted during the world war. The totals by years for the period 1930–34 are 84, 79, 69, 71, 92, respectively.

The figures naturally vary somewhat with the inclusion or exclusion of degrees taken in applied fields such as mathematical physics, mathematical astronomy, mechanics, theory of statistics, etc.; but they are accurate enough for our purposes.

Besides this number of American degrees, the information available indicates that there have been, during the period 1862–1930, 114 degrees conferred by foreign universities on mathematicians* who have been active in America.†

---

* The University of Göttingen accounts for 34 (Henry Blumberg, Maxime Bôcher, Oskar Bolza, Ann L. Bosworth, W. D. Cairns, A. R. Crathorne, H. B. Curry, Edgar Dehn, T. M. Focke, D. C. Gillespie, Charles Haseman, M. W. Haskell, E. R. Hedrick, W. A. Hurwitz, Dunham Jackson, O. D. Kellogg, A. J. Kempner, S. D. Killam, Luise Lange, Heinrich Maschke, Max Mason,

Included in this list are 40 mathematicians of European birth who subsequently took up residence in America and thoroughly established themselves by residence of several years.

Yale University, the first institution in America to confer the Ph.D. degree in course (1861), awarded the degree in mathematics in 1862 to J. H. Worrall. William Watson, later professor at Harvard University, received from Jena in 1862 the first foreign degree of which we have record. The earliest degrees conferred on women were granted to Winifred H. Edgerton by Columbia University in 1886 and to Charlotte A. Scott by the University of London in 1885.

As will be noted from Table I, more than one-sixth of the 1,286 degrees conferred in America have been awarded by the University of Chicago alone. Six institutions—Chicago, together with Cornell, Harvard,* Illinois, Johns Hopkins, and Yale—are responsible for more than half. Of the remaining 53 institutions, there have been 7 which have each conferred 25 or more degrees (California, Clark, Columbia, Michigan, Pennsylvania, Princeton, and Wisconsin) and 14 that have each conferred from 10 to 24 degrees (Brown, Bryn Mawr, Catholic, Cincinnati, Indiana, State University of Iowa, Massachusetts Institute, Minnesota, Missouri, Ohio State, Pittsburgh, Rice, Texas, and Virginia). The remaining 94 degrees were given by 32 institutions. During the past ten years only one-half the universities on this list (29 out of 59) have conferred five or more degrees and can thus be considered an important present factor. In the five years 1930–34, Chicago maintains its lead in the number of degrees conferred with 52, while Michigan is second with 35, followed by Cornell and Harvard each with 28.

C. A. Noble, L. W. Reid, W. B. Smith, Virgil Snyder, Elijah Swift, J. H. Tanner, E. J. Townsend, E. B. Van Vleck, Paul Wernicke, W. D. A. Westfall, H. S. White, Mary F. Winston, F. S. Woods); the others are distributed as follows: Berlin 4 (Harris Hancock, Ludwik Silberstein, A. G. Webster, E. J. Wilczynski), Bonn 2 (J. L. Coolidge, W. C. Graustein), Bordeaux 1 (P. L. Saurel), Cambridge 4 (E. W. Brown, R. C. Maclaurin, Frank Morley, H. M. Tory), Charlottenburg 1 (E. P. Wigner), Dublin 2 (James McMahon, J. L. Synge), Edinburgh 2 (Alexander Macfarlane, J. H. M. Wedderburn), Erlangen 4 (Henry Benner, W. F. Osgood, E. D. Roe, H. W. Tyler), Fribourg 1 (R. F. Schnepp), Ghent 1 (N. A. Court), Greifswald 1 (F. J. Dohmen), Heidelberg 2 (R. R. Fleet, T. E. Hart), Jena 1 (William Watson), Kazan 1 (G. Y. Rainich), Kiel 1 (Arthur Schultze), Königsberg 1 (J. B. Chittenden), Leiden 1 (D. J. Struik), Leipzig 9 (H. F. Blichfeldt, C. L. Bouton, H. B. Fine, A. G. Hall, J. M. Page, B. O. Peirce, D. A. Rothrock, W. E. Story, A. F. Wintner), London 1 (Charlotte A. Scott), Munich 9 (G. N. Armstrong, C. H. Ashton, P. S. Epstein, L. A. Howland, W. W. Küstermann, H. W. March, G. W. Myers, W. S. Seidel, Edwin R. Smith), Oslo 1 (Oystein Ore), Oxford 2 (W. R. Burwell, F. V. Morley), Paris 2 (Pierre Boutroux, Jacques Chapelon), Rome 1 (Oscar Zariski), St. Anthony 1 (J. A. Caparó y Peréz), St. Petersburg 4 (Alexander Chessin, J. A. Shohat, J. D. Tamarkin, J. V. Uspensky), Salzburg 1 (Eberhard Hopf), Stockholm 1 (Einar Hille), Strasbourg 5 (R. C. Archibald, J. W. Bradshaw, Myrtie Collier, E. V. Huntington, L. C. Karpinski), Szeged 1 (Tibor Radó), Tübingen 2 (S. C. Davisson, J. E. Manchester), Upsala 1 (T. H. Gronwall), Utrecht 1 (Hermance Mullemeister), Vienna 3 (F. W. Doermann, J. L. Gibson, James Pierpont), and Zurich 5 (W. H. Butts, S. R. Epsteen, Lulu Hofmann, William Marshall, A. B. Pierce).

† When persons have taken two doctors' degrees, the first only is listed.

* Throughout this paper, statistics for Harvard include Radcliffe.

TABLE I

NUMBER OF PH.D. DEGREES IN MATHEMATICS CONFERRED BY AMERICAN UNIVERSITIES

Institution	1862–69	70–79	80–89	90–94	95–99	00–04	05–09	10–14	15–19	20–24	25–29	30–34	Tot.
Boston U.								3					
Brown											1	13	1
Bryn Mawr			1	1	2			1	1	2	3	2	1
Calif. Inst. of Tech.											5	3	
California						2	1	5	7	8	11	11	4
Catholic									3		3	9	1
Chicago					4	15	20	25	31	41	49	52	23
Cincinnati								1		1	1	7	1
Clark				4	7	7	2	4	2				2
U. of Colorado							1					1	
Columbia U.			4	1	3	6	5	11	10	5	6	11	6
Cornell U.		1	3	1	1	7	8	8	8	8	16	28	8
Cumberland				1									
Dartmouth		1											
Duke												3	
Fordham											2		
Geo. Washington				1								2	
Harvard and Radcliffe		2	3		5	5	7	11	15	11	16	28	10
Haverford						1							
Illinois						1		6	8	10	25	23	7
Indiana								2	3		1	5	1
Iowa State												1	
State U. of Iowa										3	5	9	1
Johns Hopkins		2	12	7	9	8	10	14	7	8	15	11	10
Kansas					1			2			2	1	
Kentucky												1	
Lafayette			1	1									
Marquette												2	
Mass. Inst. of Tech.										1	4	12	
Michigan								4	5	3	8	35	
Minnesota										2	4	4	
Missouri								1	1	2	4	2	
Moravian									1				
Nebraska					2	1					1		
New York										1	2	2	
U. of N. Carolina						1						1	
Ohio State U.												16	
Ohio Wesleyan				1									
Otterbein				1									
U. of Pennsylvania			2	7		3	6	5	7	2	9	11	5
Pittsburgh											1	12	
Princeton			2	1		1	2	8	3	9	4	18	4
Purdue				1									
Rensselaer										1		2	
Rice									1	1	6	5	
St. Louis U.												2	
Stanford						1		1			2	3	
Syracuse U.			1	1		1		2	1	3		1	
U. of Texas									1	1	5	4	
U. of Toronto										1	1	6	
Tulane				1									
Vanderbilt			1				1						
U. of Virginia			1		3	1	3			1			
Washington U.			1									2	
U. of Washington											2	3	
West Virginia U.												2	
Wisconsin					2	1	2	1	3	5	5	20	
Wooster			1										
Yale	3	4	3	2	11	11	12	11	4	2	8	8	
Totals	3	10	32	28	56	75	80	126	125	129	227	395	128

In 1933 a committee of the American Council on Education made a study of institutions conferring Ph.D. degrees. As the result of a ballot in each field, a group of institutions was designated as approved, and from that list the most distinguished were singled out by stars. Of the 395 degrees in mathematics granted in 1930–34, 157 were conferred by institutions starred in this field, 187 by other institutions approved in the field, and 51 (about 12%) by institutions not on the approved list.

Of the 32 institutions now members of the Association of American Universities, 28 have conferred the degree in mathematics during the five years 1930–34, the total being 348, or 88% of all such degrees conferred.

Foreign Influence on American Mathematics

Grateful acknowledgment should be made of America's tremendous debt to foreign universities. The original inspiration for work in a large number of the fields now cultivated here had its source on the other side of the Atlantic, especially in Germany. As will be noted from Graph I, the number of foreign degrees during the period 1885–1914 was a considerable proportion of the whole; not only is this group important numerically but, as will be seen later in this discussion, it is significant scientifically. During the period in question, many of the ablest of our students went abroad to obtain degrees. Harvard in particular followed a very generous policy in awarding fellowships to persons wishing to take degrees in Europe. Up to and including 1913, Harvard sent abroad 26 mathematical fellows, of whom 16 took European degrees.*

Beginning with 1914, however, the number of Americans taking their degrees abroad declined rapidly. By that time the custom had been established of taking degrees in this country, even when the students wished to pursue further study abroad. The number of American mathematicians attending foreign universities in recent years is by no means so great as it was a quarter of a century ago, but scholars in this country are still indebted to universities in Europe for much stimulation. There seems to be no doubt that by our connection with foreign countries we have avoided and are avoiding the inbreeding which is apt to lead to sterility.

During the past decade the international fellowships, supported by the Rockefeller Foundation, have been of great importance in establishing contacts between American and European mathematicians. Other recent factors are the importation of some European scholars through the founding of the Institute for Advanced Study, and the recent infiltration of other displaced German scholars into our mathematical group. Besides the 40 foreign-born mathematicians referred to above as having entered the United States and Canada before 1930, there have been approximately 20 mathematicians who have more recently transferred to these countries from Europe, thus further raising the volume of our research activity.

---

* This includes B. O. Peirce and A. G. Webster whose fields bordered on physics.

While the tide of students venturing forth to Europe has abated since the world war, the number of foreign students arriving on our shores has rapidly increased, drawn partly by the fame of our leaders in scholarship and partly by the hope that a career might open for them here. Leadership in many fields is definitely passing to America, and the tides of students to and from Europe need to be regulated by an informed opinion with a view to best serving the cause of American scholarship.

Number of College Teachers of Mathematics

Data collected in the autumn of 1935, based on information furnished by the institutions themselves, indicate, as tabulated in Table II, that the number of persons teaching mathematics in colleges, universities, junior colleges, and degree-granting normal colleges in the United States (with its outlying possessions) and Canada is approximately 4,500. This includes some persons who are teaching descriptive geometry, mechanics, and methods in mathematics, as well as some who are teaching part time or who are largely in administrative work or who are emeriti; but on a conservative estimate, 4,000 persons are actually engaged full time in the teaching of mathematics of college freshman grade or higher. Similar figures were collected in 1932, and there seems to have been a considerable increase in the interim, due chiefly to the growth of junior colleges.

Table II is a statistical study by states of the number of teachers of mathematics in junior colleges, teachers colleges, and other colleges and universities, of the number of men and women teachers, of the number of teachers holding doctor's degrees, and of the number of those who are members of the American Mathematical Society or of the Mathematical Association of America or of both. The best information obtainable indicates that probably somewhat less than 1,300 of the present teachers of mathematics have the Ph.D. degree! Many of the 1,400[*] listed as having obtained degrees in America or abroad are deceased or have entered fields of work such as government service, banking, or industry. It should be remarked also that mathematics has furnished more than its share of administrative officers to colleges and universities. There have been some doctors who have drifted out of mathematics into other fields of science, and probably more who have correspondingly drifted in.

Of the 4,444 teachers of mathematics listed, 1,292, or 29%, hold the degree of Ph.D., while slightly less (1,263) are members of the American Mathematical Society and slightly more (1,333) are members of the Mathematical Association. In the Summary at the end of Table II the states have been grouped by sections of the country, and we note that 35% of the teachers in the northeast section from Illinois to Maine hold the doctor's degree. Only about 22% of those in the south central states from Kentucky to Texas hold the doctor's degree. In the remainder of the country about 26% hold that degree.

---

[*] This figure does not include the 1935 crop of Ph.D.'s, most of whom are doubtless included in Table II.

TABLE II

STATISTICS REGARDING TEACHERS OF MATHEMATICS IN AMERICAN AND CANADIAN COLLEGES AND UNIVERSITIES, 1935

	Number of Teachers						Holders of Ph.D.		Membership in				
	(a)	(b)	(c)	(d)	(e)	(f)	No.	%	A.M.S.	%	M.A.A.	%	Both %
Alabama	66	5	11	50	48	18	13	20	7	11	13	20	5	7
Alaska	3	3	—	—	3	0	0	0	0	0	0	0	0	0
Arizona	10	2	2	6	9	1	6	60	4	40	5	50	3	30
Arkansas	44	9	5	30	34	10	13	29	9	20	5	11	3	7
California	262	126	17	119	231	31	64	24	60	23	55	21	40	15
Canal Zone	2	2	—	—	2	0	1	50	0	0	0	0	0	0
Colorado	60	6	10	44	48	12	17	28	14	23	23	38	13	22
Connecticut	45	7	3	35	38	7	27	60	26	58	22	49	21	47
Delaware	7	—	—	7	6	1	2	28	5	71	4	57	3	43
Dist. of Columbia	48	13	8	27	30	18	18	37	16	33	11	23	8	17
Florida	38	5	—	33	37	1	12	31	10	26	13	34	9	24
Georgia	84	23	2	59	70	14	20	24	16	19	23	27	9	11
Hawaiian Islands	5	—	—	5	4	1	0	0	1	20	0	0	0	0
Idaho	19	9	—	10	19	0	3	16	3	16	1	5	1	5
Illinois	222	41	16	165	183	39	87	39	81	36	80	36	53	24
Indiana	96	4	8	84	87	9	33	34	25	26	42	44	15	16
Iowa	128	42	4	82	104	24	34	26	33	26	38	30	22	17
Kansas	92	24	10	58	71	21	22	24	25	27	41	44	21	23
Kentucky	77	22	19	36	56	21	19	24	14	18	28	36	12	15
Louisiana	49	7	4	38	37	12	11	22	9	18	21	43	8	16
Maine	25	2	1	22	23	2	6	24	6	24	10	40	5	20
Maryland	92	3	2	87	80	12	29	31	28	30	35	38	20	22
Massachusetts	170	14	16	140	139	31	53	31	66	39	43	25	38	22
Michigan	157	27	15	115	144	13	49	31	44	28	52	33	32	20
Minnesota	83	19	10	54	65	18	26	31	16	19	31	37	13	16
Mississippi	52	22	5	25	37	15	8	15	7	13	12	23	5	10
Missouri	122	31	21	70	97	25	30	24	27	22	29	24	12	10
Montana	19	2	1	16	18	1	4	21	7	37	5	26	4	21
Nebraska	55	5	10	40	45	10	14	25	13	24	19	34	12	22
Nevada	3	—	—	3	2	1	1	33	2	67	1	33	1	33
New Hampshire	28	1	3	24	26	2	13	46	11	39	11	39	9	32
New Jersey	107	17	7	83	102	5	35	33	44	41	29	27	24	22
New Mexico	20	3	4	13	19	1	6	30	8	40	7	35	5	25
New York	402	9	8	385	341	61	143	35	179	44	139	34	112	28
North Carolina	124	30	5	89	97	27	28	23	26	21	27	22	13	10
North Dakota	30	5	10	15	25	5	0	0	1	3	6	20	1	3
Ohio	195	4	5	186	167	28	74	38	67	34	90	46	52	27
Oklahoma	100	32	20	48	84	16	13	13	10	10	27	27	8	8
Oregon	31	4	—	27	26	5	6	19	7	22	9	29	5	16
Pennsylvania	283	12	28	243	239	44	106	37	110	39	104	37	67	24
Philippine Islands	38	—	—	38	31	7	4	10	0	0	0	0	0	0
Porto Rico	4	—	—	4	4	0	3	75	4	100	2	50	2	50
Rhode Island	32	1	2	29	29	3	13	41	19	59	14	44	13	41
South Carolina	50	3	3	44	40	10	7	14	5	10	6	12	2	4
South Dakota	27	5	2	20	24	3	6	22	7	26	6	22	3	11
Tennessee	82	10	12	60	69	13	21	26	9	11	14	17	4	5
Texas	202	62	18	122	169	33	47	23	42	21	55	27	27	13
Utah	28	13	—	15	26	2	6	21	6	21	4	14	3	11
Vermont	26	2	—	24	23	3	4	15	4	15	5	19	3	11
Virginia	98	20	6	72	74	24	23	23	26	26	29	29	21	21
Washington	47	10	5	32	44	3	19	40	18	38	13	28	11	23
West Virginia	40	6	9	25	34	6	17	42	9	22	7	17	6	15

TABLE II—*Continued*

Wisconsin	108	22	24	62	92	16	28	26	38	35	34	31	27	25
Wyoming	5	—	—	5	4	1	1	20	0	0	4	80	0	0
Canada	202	—	2	200	194	8	47	23	39	19	29	14	22	11
	4444	746	373	3325	3750	694	1292	29	1263	28	1333	30	828	19

SUMMARY

Geographic Division	Number of Teachers						Holders of Ph.D.		Membership in					
	(a)	(b)	(c)	(d)	(e)	(f)	No.	%	A.M.S.	%	M.A.A.	%	Both	%
New England	326	27	25	274	278	48	116	36	132	40	105	32	89	27
Middle Atlantic	792	38	43	711	682	110	284	36	333	42	272	34	203	26
East North Central	778	98	68	612	673	105	271	35	255	33	298	38	179	23
West North Central	537	131	67	339	431	106	132	25	122	23	170	32	84	16
South Atlantic	581	103	35	443	468	113	156	27	141	24	155	27	91	16
East South Central	277	59	47	171	210	67	61	22	37	13	67	24	26	9
West South Central	395	110	47	238	324	71	84	21	70	18	108	27	46	12
Mountain	164	35	17	112	145	19	44	27	44	27	50	30	30	18
Pacific	340	140	22	178	301	39	89	26	85	25	77	23	56	16
Territories	52	5	—	47	44	8	8	15	5	10	2	4	2	4
Canada	202	—	2	200	194	8	47	23	39	19	29	14	22	11

(a) Total; (b) Junior colleges; (c) Teachers colleges; (d) Other colleges and universities (e) Men; (f) Women.

New York State has the largest number of mathematics teachers of college grade, followed by Pennsylvania, California, Illinois, and Texas. Arizona and Connecticut are the states with the highest percentage of doctors on their faculties, with New Hampshire, West Virginia, Rhode Island, Washington, and Illinois following; at the other end of this scale is North Dakota with no doctors listed.

Whereas for all institutions above the rank of junior colleges (universities, colleges, and degree-granting teachers colleges) the proportion of persons with a doctor's degree is 33%, in the group of 285 colleges and universities on the approved list of the Association of American Universities it is 44%.

The decline of growth in the degree-granting institutions has been to a considerable extent offset by the establishment of emergency institutions of college freshman grade or of junior college grade. While the number of college teaching positions in mathematics probably doubled during the period 1918–1930, the unprecedented year-by-year increase in the college student population on which this increase was based was slackening rapidly even before 1929, and it seems improbable that, even if a period of economic stress had not ensued, there would have been many more academic students in the colleges of the country than there are at present. The decided drop in the number of additional appointments to the staffs of degree-granting institutions appears to have been inevitable. Without doubt the unemployment problem for college teachers is aggravated by the financial depression, but in main outlines it might have been foreseen by competent executives.

It is probably true that the number of persons attending institutions of grade beyond the high school will still increase greatly. Registration in junior colleges is increasing by leaps and bounds. The situation seems predictable, and it may be sufficiently accurate to estimate at 4,500–5,000 the number of professionally trained mathematicians who will be employed in college teaching during the next decade.

GRAPH II

COMPARISON OF PRODUCTIVITY FOR PERSONS TAKING AMERICAN AND FOREIGN DEGREES IN DIFFERENT PERIODS

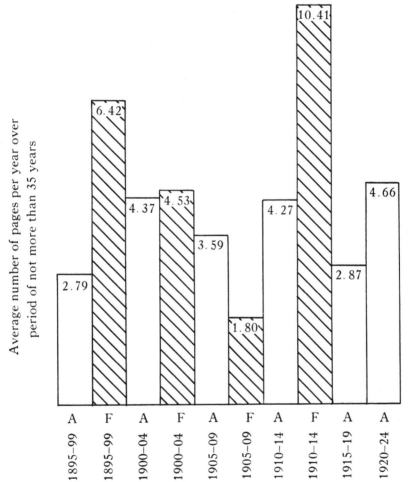

Of the 60,000 persons employed in teaching all subjects in the junior colleges, colleges, and universities (not including professional schools), about 7.5% are in the departments of mathematics. In this group of institutions there are approximately 800,000 students, or about one teacher of mathematics to 175 students. In the junior colleges alone, there are approximately 6,000 teachers, of whom 12.5% are in mathematics; the total enrolment is approximately 110,000, or one teacher of mathematics to 150 students.

If we consider twenty-five years as an average period of service, it would seem that, as soon as the situation becomes a little more normal, there will be need for at least 175 new persons to be added to the staffs each year. We have arrived at the point where the universities are granting about half that number of Ph.D. degrees annually. Less than one-third of the present staffs have such degrees, and, if one-half the new appointments are made from those holding the doctorate, the situation will probably be as satisfactory as we can hope for at present. The standards in this regard need elevation, and doubtless the next decade or two will see rapid advancement. There are many college teachers with pitifully meager preparation; the institutions must look forward to their gradual replacement by well-trained men and women.

Proportion of Ph.D.'s Publishing Research Papers

It goes without saying that the number of papers and the number of pages printed is not an adequate criterion for measuring the influence of a person on mathematical thought. Often the ideas of a scholar appear in papers published

TABLE III

ANALYSIS OF NUMBERS OF PAPERS PUBLISHED BY AMERICAN MATHEMATICIANS

	Persons taking degrees 1862–1933		Persons taking degrees 1895–1924	
	Number	%	Number	%
No papers	549	46	232	39
1 paper	227	19	109	18
2 papers	100	8	58	10
3–5 papers	131	11	66	11
6–10 papers	70	6	41	7
11–20 papers	69	6	50	9
21–30 papers	20	2	17	3
More than 30 papers	22	2	17	3
Total	1188	100	590	100

by his pupils or colleagues. But a study of the amount of publication is the easiest (perhaps the only) means that is available from a statistical standpoint. There is a great deal of information contained in Graph II and Tables III–V concerning this fundamental matter of the amount of publication, and, in spite of the reservations just made, the data have real significance.

Table III gives the numbers and percentages of persons who have published no research articles, one, two, three, etc. research articles; the first division of the table gives figures for all Ph.D.'s granted in 1862–1933 and the second (we note an improvement) for those during the central period 1895–1924, which was selected as being far enough in the past so that men have had an opportunity to get something into print, and not extending back far enough so that lack of publication facility and of stimulus enters into the calculation.

A perusal of the figures in the various tables indicates that not more than one-third of the persons taking doctor's degrees have made as substantial contributions to research as would be evidenced by the publication of three or more research articles; and that not more than one-fifth have really been consistently productive in their contributions. About 60 (or 5%) of the doctors are responsible for half of the published pages of research.

TABLE IV

STATISTICS REGARDING THE NUMBER OF PAPERS PUBLISHED BY AMERICANS WHO
TOOK DEGREES 1862–1933

Group	No papers	1 paper	2 papers	3–5 papers	6–10 papers	11–20 papers	21–30 papers	30+ papers
	%	%	%	%	%	%	%	%
American degree								
–1894	38	18	8	18	5	8	2	3
1895–1899	54	7	19	5	4	4	2	5
1900–1904	30	22	3	7	13	10	7	6
1905–1909	29	20	7	19	5	12	3	4
1910–1914	37	17	13	12	7	8	3	3
1915–1919	40	19	8	11	10	7	2	1
1920–1924	46	19	9	11	3	7	2	2
1925–1929	45	18	6	14	10	4	1	1
1930–1933	56	23	9	9	1	1	0	1
Foreign degree								
–1894	16	5	0	5	32	11	21	11
1895–1899	8	25	8	0	0	25	17	17
1900–1904	8	15	0	23	0	31	15	8
1905–1909	0	30	20	30	10	10	0	0
1910–1914	10	0	0	10	10	50	10	10

A more careful analysis of this situation in Table IV shows salient facts as follows: the percentage of persons doing no publishing has in recent years slightly increased; the percentage of persons publishing a large number of papers has remained fairly constant in the group of those taking American degrees as well as of those taking European; the proportion of prolific authors is greater among those taking degrees abroad. Contrary to the general opinion, America seems in recent years to be adding to the quantity of personnel, but not improving the average quality as judged by the number of papers published.

It would be exceedingly interesting to know what proportion of the men and women with mathematical ability of high order are now being drawn into the graduate schools; in other words, how efficiently the nation is using this human material. Are there more persons of mathematical talent in the nation than can well be utilized? But such a study, which would have to begin with high school students, is entirely beyond the power of any single organization such as the Mathematical Association of America or the American Mathematical Society.

TABLE V

AMOUNT OF PUBLICATION BY GROUPS OF FIVE YEARS AFTER RECEIVING DEGREE

Period	First 5 years	Second 5 years	Third 5 years	Fourth 5 years	Fifth 5 years	Sixth 5 years	Seventh 5 years	Total	
	pages	pages	pages	pages	pages	pages	pages	papers	pages
American degree									
−1894	14.91	8.92	7.38	8.75	4.51	7.03	6.25	6.32	57.75
1895–1899	19.43	16.24	17.50	9.72	15.74	7.39	11.70	7.02	97.72
1900–1904	33.40	26.67	30.80	17.58	9.31	13.33		7.15	131.07
1905–1909	34.07	30.30	12.78	13.80	8.94			5.76	89.89
1910–1914	30.18	19.70	18.31	17.29				6.63	85.47
1915–1919	18.61	10.75	13.66					3.57	43.01
1920–1924	27.40	19.18						3.65	46.58
1925–1929	34.02							2.97	34.02
Foreign degree									
−1894	13.95	44.32	43.74	23.63	30.68	18.26	25.37	14.45	199.95
1895–1899	50.00	47.17	60.00	25.25	19.00	15.00	8.37	16.75	224.79
1900–1904	45.54	23.15	21.85	15.77	15.92	13.69		11.70	135.92
1905–1909	24.10	4.50	4.70	8.50	3.90			3.80	45.00
1910–1914	39.20	33.40	45.10	90.60				17.20	208.30

In Table V the amount of publication is analyzed by periods of five years after receiving the degree, and it should be noted that figures are given for a five-year period and that no average per year is tabulated. It should further be observed that in this table the figures pertain to pages instead of to papers published. As an example of how they are to be interpreted, we note that for all the doctors receiving American degrees for the five-year period 1895–99, the average number of pages which they published during the second five years after receiving degrees was 16.24, which implies that the index, or average number of pages published annually, was 3.25.

For statistical purposes those papers which were printed before the degree was acquired are included in the first period. The average length of paper over the whole period is approximately 13.48 pages (4,916 papers and 66,268 pages*) and does not seem to vary significantly either with the age of the writer or with the date at which he took his degree. The first period of five years seems to be the most prolific one; doubtless this is due to the fact that men printing only one paper contribute during this period only. It is noteworthy that, even after more than thirty years subsequent to the conferring of the degree, there is on the part of many research workers not much slackening in the rate of publication.

The fifth edition of *American Men of Science* (1933) contains brief biographies of 1,242 living mathematicians who are listed because they "have contributed to the advancement of science." In the earlier editions there were listed an additional 168 names of people since deceased. Of this total of 1,410 mathematicians, 934 hold the doctor's degree; 104 of these degrees were awarded abroad; and of the recipients, 59 were born abroad Since *American Men of Science* contains primarily the names of persons now in residence in the United States and only incidentally a few of those working in Canada, the information has not so wide a basis as that in the study presented here, which includes Canada also.

On the basis of extensive ballots, Professor Cattell, in the various editions of *American Men of Science*, selected a total of 182 mathematicians (136 were listed as alive in the fifth edition) as representing the leaders in research, and stars were attached to their names. Of this group all but 19 have obtained the doctor's degree; 40 were born in Europe; and 56 received their doctor's degrees abroad. Our present study indicates that these 182 mathematicians have an average index of 13.03 as compared to 4.73 for the whole list of doctors, and that approximately 45,580 pages, or 69% of the whole mathematical publication of America, is due to them.

Comparison of Productivity as Regards Universities of Origin

Since it was the more enterprising and able of our students who went abroad during the twenty-year period beginning with 1895, it is to be expected that the productivity of the group taking their degrees in European universities would be greater than that of the corresponding group awarded American degrees. Graph II, which gives the average number of pages printed yearly during a period of not more than thirty-five years subsequent to attaining the degree, indicates that, for a period of twenty years before the world war (except for a lapse during 1905–09), the European universities attracted a considerable number of our ablest men as candidates for degrees. So far as the home product is concerned, we note that a peak was reached with the group of men taking degrees in 1920–24.

---

* This total is for Americans holding doctor's degrees and does not include articles presented to the Society by foreigners or non-doctors.

A department in a university waxes and wanes. At one period an outstanding professor attracts an able body of students and inspires them to continue their research; at another the quality of the staff and students is mediocre. If we consider a group of ten institutions, each conferring a large number of degrees, we note from Table VI that there is very considerable variation in the matter of productivity. This is true both as between the universities and in a given university as regards the four decades covered by the table. The figures in parentheses represent the number of degrees conferred in the period named in the heading of the column, and the other figures represent the average number of pages per person per year subsequently published. It is a matter of consider-

TABLE VI

COMPARATIVE STUDY OF AMOUNT OF PUBLICATION PER GRADUATE
IN TEN INSTITUTIONS

Institution	1890–99		1900–09		1910–19		1920–29		Whole period	
1	(4)	26.63	(35)	6.69	(56)	3.31	(90)	4.28	(185)	4.92
2	(16)	0.95	(18)	4.62	(21)	1.91	(23)	0.96	(78)	2.06
3	(5)	1.50	(12)	4.59	(26)	6.33	(27)	13.96	(70)	8.63
4	(13)	0.56	(23)	2.55	(15)	2.15	(10)	4.86	(61)	2.41
5	(2)	0.00	(15)	4.87	(16)	1.01	(24)	4.06	(57)	3.28
6			(1)	0.56	(14)	1.87	(35)	1.21	(50)	1.38
7	(4)	6.36	(11)	0.69	(21)	8.21	(11)	5.45	(47)	5.65
8	(9)	0.02	(9)	2.34	(12)	1.29	(11)	17.34	(41)	5.55
9	(1)	0.44	(3)	0.92	(11)	8.71	(13)	12.58	(28)	9.38
10					(9)	1.15	(11)	9.05	(20)	5.50
Average of the ten	3.01		4.22		3.77		5.83		4.62	
General average of all Ph.D.'s			4.34		3.60		5.68		4.73	

able significance that the persons taking degrees at these large institutions (and constituting more than three-quarters of the total during that period) do slightly less publication on the average than those taking their degrees at the remaining group of institutions. It is noteworthy also that the average of the indices for some of the institutions not included in this list of ten is higher than for any in the list.

Persons Without Degrees

Many persons who have not attained to doctor's degrees have published papers read before the Society. The number of such persons is about 125, and a few of them are ranked among our leading mathematicians.

National Research Fellows

It is to be expected that the publication rate for the National Research Fellows in mathematics will be greater than that for the whole group of persons taking degrees, not only because they have been given special opportunities, but also because they are a picked body of men. Let us select as a basis for discussion the 49 Fellows starting on their fellowships during the first five-year period, 1925–29, in which the fellowships were awarded. The average number of pages published annually over a five-year period subsequent to their entering on the fellowship is 17.82, while for all 227 taking Ph.D.'s during the years 1925–29, it is 6.80. For the period of tenure (generally two years) and one year thereafter (to allow for lag in publication), the average for the Fellows is 25.89; this indicates what can be expected from young men if a proper selection for quality is imposed and opportunity for full-time study is made available.

For accurate comparison purposes, however, it is desirable to have more equitable bases than those just used. Let us eliminate from the 49 Fellows starting on their incumbencies in 1925–29 the 4 persons who took their degrees earlier than 1923 and who had already thus had opportunity to establish themselves in research careers. Let us allow a lapse of a year and a half after giving up the fellowship; the doctor is then professionally employed and has probably already published the main results of the research done while an incumbent. For the other group let us select those Ph.D.'s of the same period who were not awarded fellowships and let us allow a lapse of a year and a half in their cases also, so that the doctoral thesis will probably not be included in the publications counted; let us compare these groups during the subsequent years and down to the time of the last information available (including 1934). The figure for the Fellows is 10.06 and for the remaining Ph.D.'s is 2.87, which establishes a significant differential. However, using the same groups and the same periods, let us compare the bottom one-third of the Fellows with the top one-third of the non-fellows. The results are 0.93 for the 15 persons of the one group as against 8.49 for the 60 persons of the other, indicating that the selection of the Fellows carried with it a considerable factor of error. It should, however, be pointed out that not all the more talented students apply for fellowships.

Let us next make a comparison of the 30 most productive of the 45 Fellows considered above and the 30 most productive Ph.D.'s among the 180 non-fellows of the period 1925–29 by subjecting their records to the conditions of the preceding paragraph. The indices are 14.63 and 14.77; but it must be pointed out that these selections are for the upper two-thirds and upper one-sixth respectively.

Another fundamental comparison that can be made is that between the Fellows of the 1925–29 period and the corresponding persons of the 1920–24 period before the fellowships were available. Selecting the 30 from each group who have published the most, and using comparable periods beginning one and one-half years after fellowship and degree respectively and terminating with 1934 and 1929 respectively, the figure for the Fellows is 14.63 and for the others is 10.16. Since in this comparison the case for the Fellows is discounted by the fact that not all the best men are elected to fellowships, the figures indicate a decided stimulus from the fellowships.

Conclusion

The rapidity with which a mathematical school of high distinction has been built up in America is one of the most striking phenomena in the history of science. It should be borne in mind that the American Mathematical Society, which has been an important factor in this development, was not founded until 1888, and that only in isolated cases was research carried on before that date. Building on the splendid foundations already laid, great forward movements are possible if the spirit of cooperation now animating mathematicians is fostered. The challenge of the future inspires the mathematical fraternity to high endeavor.

Many of the statistics exhibited in this paper will be of interest to those who follow the development of teaching and research in the field of mathematics. Those concerned with the strategy of promoting mathematical thought and achievement in America can find in the assembled material several sign-posts for their further guidance. Other studies could be made from the data which have been collected and which are available to anyone; the investigations appear to be worth pursuing further.

Judy Green is an associate professor of mathematics at Rutgers University, Camden. She received her Ph.D. in 1972 from the University of Maryland, writing a dissertation in mathematical logic under the guidance of Carol Karp. Having published previously in model theory, she has recently shifted her research interest in logic to its history. Jeanne LaDuke is an associate professor of mathematics at DePaul University. She did her graduate work at the University of Oregon, receiving a Ph.D. in 1969 as a student of Kenneth Ross in abstract harmonic analysis. She has published in harmonic analysis and, more recently in the history of American mathematics. The authors have been collaborating for several years on the history of women in American mathematics. Since 1979 both have been honorary research associates in mathematics at the National Museum of American History, Smithsonian Institution.

Women in American Mathematics:
A Century of Contributions*

JUDY GREEN AND JEANNE LADUKE

Women were a part of the young and vigorous American mathematical community at the time of the founding of the Society. Indeed, throughout the past one hundred years, women have made contributions to mathematics through their research, their teaching, their service to the professional associations, and their participation in the mathematical activities of the larger society. While women have consistently maintained a presence within American mathematics, their numbers and their influence have not increased steadily over the years. Before World War II the growth of the community of American women mathematicians roughly paralleled that of the larger mathematical community, and the number of women who contributed to mathematics was substantial. In the 1940s the war had a dramatic effect on the professional lives of mathematicians, both men and women. Even though women continued to earn Ph.D.'s, the explosive growth of graduate education for men that followed the war eclipsed the participation of women. Thus, the proportion and visibility of women declined significantly during the early

*This article expands upon remarks made by the authors at a panel sponsored by the Association for Women in Mathematics at the Centennial Celebration of the AMS, 9 August 1988. The texts of the talks given at the panel appear in LaDuke et al. 1988.

Portrait of Winifred Edgerton Merrill, oil on canvas by Helena E. Ogden Campbell, presented to Columbia University in 1933.
(Photograph courtesy of Columbia University in the City of New York, Office of Art Properties

postwar decades. Finally, after again sharing in the growth of mathematics during the 1960s, women attained a self-awareness and cohesiveness in the early 1970s. Since then, it has been difficult not to recognize their numbers and their steady, and sometimes striking, contributions to mathematics.

Throughout the last one hundred years there have been so many women involved in mathematics, and in such diverse ways, that there is no possibility of a comprehensive treatment here. In particular, the names mentioned in this article are nowhere near an exhaustive roll of the women who have made significant contributions.

The First Fifty Years

By 1888, the year in which the Society was founded, American women were already active in mathematics. At Wellesley College there was a strong, and entirely female, mathematics department. One of its graduates, Winifred Edgerton Merrill, had received her Ph.D. *cum laude* in mathematics from Columbia University in 1886, the first American woman to be granted a Ph.D. in mathematics. Bryn Mawr College had been in existence for three years and from its outset had offered fellowships for graduate study in mathematics. Charlotte Scott, who had received a doctorate in mathematics from the University of London, was head of the Bryn Mawr department of mathematics.

While Scott and Edgerton were, in 1888, the only two women in the United States with doctorates in mathematics, Christine Ladd-Franklin had been admitted to Johns Hopkins University ten years earlier on the recommendation of J. J. Sylvester. She had published extensively, including three articles in the *American Journal of Mathematics*, and had completed a dissertation in the algebra of logic under Charles Sanders Peirce in 1882.[1] Because the Johns Hopkins trustees were unwilling at that time to confer a degree on a woman, Ladd-Franklin's Ph.D. was not conferred until 1926.[2]

For the first two and one-half years the New York Mathematical Society had no women members, but the desire to publish a journal, the *Bulletin*, provided impetus for a major membership drive. Hence, in 1891, the first six women joined the NYMS. The first, admitted in May of that year, was Charlotte Scott of Bryn Mawr, a distinguished geometer who became one of the most active and recognized women in the early history of the Society, serving on the Council (1894–1897 and 1899–1901) and as vice-president

---

[1] It was published in 1883 in *Studies in Logic by Members of the Johns Hopkins University*. Boston: Little, Brown & Co.

[2] She was offered an honorary doctorate by Johns Hopkins in recognition of her work in color perception, but insisted on receiving instead the doctorate she had earned in 1882. See "Woman Ph.D. ... " 1926.

(1906). The other five were: Mary E. Byrd of Smith and Mary Watson Whitney of Vassar, both astronomers; Susan Jane Cunningham, a professor of mathematics and astronomy at Swarthmore; Ellen Hayes, an early graduate of Oberlin who was professor of mathematics at Wellesley; and Amy Rayson, who taught mathematics and physics at a private school in New York City. [3] Women not only joined the Society, but they participated actively in its meetings. The first paper presented by a woman, "An orthomorphic transformation of the ellipsoid," was read at the October 1892 meeting by Ella C. Williams, a new member and teacher in a private school in New York, who had studied at Bryn Mawr, Cornell, Michigan, Newnham College at Cambridge, and privately with H. A. Schwarz at Göttingen.

In connection with the World's Columbian Exposition, an International Mathematical Congress was held in Chicago in 1893,[4] the year after the University of Chicago had opened. Among the forty-five whose names appear on the official register of the Chicago congress are four women. The listing includes Achsah M. Ely, B.A., professor of mathematics, Vassar College, and three graduate students: Charlotte C. Barnum and Ida M. Schottenfels, who had spent the previous year at Yale; and Mary Frances Winston, who had just completed a year as honorary fellow at the University of Chicago. Barnum, a Vassar graduate, remained at Yale to become, in 1895, the first of three women to receive Ph.D.'s in mathematics from Yale before 1900.[5] Schottenfels studied at the University of Chicago after the Congress and received a master's degree in 1895. Although she never received a Ph.D., she was, throughout the first decade of the twentieth century, a regular contributor to sessions at Society meetings and published, mainly in the area of finite groups, in the *Bulletin*, the *Annals*, and the *Monthly*. She continued presenting papers to the Society into the early 1930s. Before her year at Chicago, Winston had studied at Wisconsin, where she earned a bachelor's degree, and at Bryn Mawr. In 1897 she received a Ph.D. *magna cum laude* as a student of Felix Klein at Göttingen, having been one of the first three women who were officially admitted as regular students in a university administered by the Prussian government.[6]

In the first decade of the new century there were eighteen new women Ph.D.'s in mathematics. Included among these is Mary E. Sinclair, who in

---

[3] Biographical details about the twenty-two women who joined the Society in the nineteenth century can be found in Whitman 1983.

[4] This Congress was a precursor of the International Congresses that have been held on a regular basis since the 1897 congress in Zürich.

[5] For more information on early American women who earned Ph.D.'s in mathematics see Green and LaDuke 1987.

[6] Although Sofia Kovalevskaia received a Ph.D. from Göttingen in 1874, she was never officially a student there.

Members of the Chicago Mathematics Department Faculty, mid-1920s: (l to r) E. H. Moore; G. A. Bliss; Mayme I. Logsdon, Ph.D. Chicago 1921; L. E. Dickson; W. D. MacMillan; F. R. Moulton; H. E. Slaught.
(Photograph courtesy of the University of Chicago Mathematics Department.)

1908 became the first of twenty-two women to receive Ph.D.'s in mathematics from the University of Chicago before 1930.[7] Sinclair worked in the calculus of variations and spent her teaching career at Oberlin, becoming professor and head of the mathematics department there. Also among the earliest women to receive Chicago Ph.D.'s were three other mathematicians of distinction: Anna J. Pell (1910), Mildred Leonora Sanderson (1913), and Olive Cleo Hazlett (1915). Although Sanderson, a student of L. E. Dickson, died in 1914 just a year after completing her graduate work, her thesis, which was published in the *Transactions*,[8] was a fundamental contribution to the theory of modular invariants and covariants. In 1915 Dickson described her as his most gifted pupil. Both Pell (later Pell Wheeler) and Hazlett had productive careers that included significant participation in and recognition by the Society.

During the early years of the twentieth century membership in the AMS increased, from 347 in 1900 to 630 in 1910 (Pitcher 1988, 2). In 1912 an item in the Notes and News section of the *Monthly* compared American and European mathematical society membership:

> According to the latest Annual Register of the American Mathematical Society about 50 of the 668 members are women. It is interesting to observe that the American Mathematical Society has a much larger per cent. of women members than the leading mathematical societies of Europe. According to the latest register of the German mathematical society ... only 5 of its 759 members are women; and only one of these 5 members is a German woman, while three of them are Americans and the remaining one is a Russian. The French mathematical society has also very few women members. The numbers of the women members of the Circolo Mathematico di Palermo and of the London Mathematical Society are considerably larger but they are much smaller than in our own society. (*Amer. Math. Monthly* 19 (1912): 84)

When the Mathematical Association of America was organized in 1915 in Columbus, Ohio, women were welcomed and involved. About twelve percent of the more than one thousand charter members were women, and they represented a variety of institutions. The two largest contingents were seven women from Wellesley and six from Iowa State College of Agriculture. Helen A. Merrill, who had received her Ph.D. from Yale in 1903 and was chairman of the department of mathematics at Wellesley from 1916 until 1932, was particularly active in the new association. She served as associate editor of the *Monthly* (1916–1919), was an early member of the Executive Council (1917–1920), and became vice-president in 1920.

---

[7]The Chicago department, under the leadership of E. H. Moore, was a leading center of graduate education in mathematics in this country until the early 1930s.

[8]Volume 14, 1913, pp. 489–500.

During the decade of the 1920s the two women who were most visible as scholars and as members of the Society were Anna Pell Wheeler and Olive C. Hazlett. Pell Wheeler, whose work and career have been detailed in Grinstein and Campbell (1982), spent from 1911 to 1918 at Mount Holyoke College and the remainder of her career at Bryn Mawr, where she succeeded Scott in the leadership of the department. She published in the area of functional analysis, directed seven Ph.D. dissertations at Bryn Mawr, served on the Council of the AMS (1924–1926), and was the first woman both to give an AMS invited address (1923) and to give the Colloquium lectures (1927). Hazlett, like Sanderson a student of Dickson, worked in the area of modular invariants and in linear associative algebras. After spending a year as a postdoctoral fellow at Harvard, she took a position at Bryn Mawr in 1916, and in 1918 moved to Mount Holyoke at the same time that Pell Wheeler made the opposite move. Although an associate professor at Mount Holyoke, she accepted a reduction in rank in 1925 in order to move to the more research oriented environment of the University of Illinois. She was a cooperating editor of the *Transactions* from 1923 to 1935 and served on the Council of the AMS from 1926 to 1928. Of the many papers she published, one was contributed to the 1924 Toronto International Mathematical Congress and another, at the beginning of a two-year Guggenheim fellowship period in Italy, Switzerland, and Germany, to the 1928 Bologna International Congress of Mathematicians.[9]

From 1924 to 1935 Caroline E. Seely served with Hazlett as a cooperating editor of the *Transactions*. She also was an assistant editor of the *Bulletin* from 1925 to 1934. Seely, a 1915 Columbia Ph.D. who studied with Edward Kasner, published in the area of integral equations throughout her career. She was employed by the AMS from 1913 to 1934. As clerk to the secretary of the AMS, she was responsible for much of the operation of the Society. Her contributions were such that, in response to her tendered resignation, J. R. Kline wrote to the Council in the fall of 1934: "The committee used all efforts within the dignity of the Society to secure the withdrawal of this resignation, but without success" (Hildebrandt Papers).

Besides Scott, Hazlett, and Pell Wheeler, only two women participated on the governing boards of the Society during its first half century: Florence P. Lewis served on the Council of the AMS (1921–1923) and Clara E. Smith served on the 1923 Board of Trustees. Lewis received her Ph.D. from Johns Hopkins in 1913 in algebraic geometry under Frank Morley and spent her entire career at Goucher College. She was on the faculty throughout the period 1912 to 1928, during which time nine women graduated from Goucher

---

[9]For more details about the work and careers of Hazlett, Pell Wheeler, and several of the other women mentioned in this article see Green and LaDuke n.d.

who later received Ph.D.'s in mathematics.[10]   Clara E. Smith received her
Ph.D. from Yale in 1904. She spent her career at Wellesley and coauthored
textbooks with her colleague Helen A. Merrill. Like Merrill, Smith was active
in the MAA. She served on its Board of Trustees (1923–1926) and as vice-
president (1927). Although the number of women who participated in the
Association on a national level before World War II was not large,[11] the group
of women who participated in the governance of the sections was both large
and diverse.

Even though a substantial number of women in the 1920s attained satisfaction and influence through their research, their teaching, or their roles
within the professional societies, certain restrictions and limitations were
understood. Examples of the perceptions of some women with degrees in
mathematics may be gleaned from responses to questions in a study conducted in the late 1920s of women with Ph.D.'s. An assistant professor of
mathematics in a college reported that

> Nothing but the most earnest conviction that she could never be satisfied
> without a Ph.D. in mathematics would justify a woman's setting herself
> that end. It is a long, hard road and when the degree is obtained,
> she finds that all the calls for mathematics teachers are for men, and
> that when a woman is employed in one of the large universities she
> is practically always given long hours and freshmen work for *years*,
> with less pay than a man would receive for the same service. If all
> the women could fare as well as I have fared, I'd say "Go ahead," but
> alas! such unexpected good luck does not come to many in a generation.
> (Hutchinson 1929, 185–186)

A woman who had received her Ph.D. before 1915 had the following observations:

> I am convinced that women have a special aptitude for mathematics as
> often as men but practically it is a very difficult line for women to pur-
> sue with financial success or even with intellectual satisfaction unless
> she [*sic*] has private means. ... A few women's colleges and an occa-
> sional co-educational college seem to give the best chance of teaching.
> Certain universities are particularly unfair to women mathematicians.
> ... A woman desiring to teach mathematics would do well to study sur-
> veying and engineering in order to break the force of the tradition that
> women do not understand such branches. (Hutchinson 1929, 206–207)

---

[10]Another student of Morley, Clara L. Bacon, Ph.D. 1911, was also on the Goucher faculty
during this period.

[11]Elizabeth Carlson (1927–1931) and Helen B. Owens (1936–1938) served as associate ed-
itors of the *Monthly*; Elizabeth B. Cowley served on the Executive Council (1918–1921), Mary
E. Sinclair served on the Board of Trustees (1936–1939), and Mayme I. Logsdon served on the
Board of Governors (1940–1942).

Some aspects of the early 1930s were of significance for mathematicians in general and for women in particular. The depression had devastating consequences for those seeking jobs. The beginning of the School of Mathematics at the newly founded Institute for Advanced Study created a new type of research institute. Mabel S. Barnes, a 1931 Ph.D. from Ohio State University whose first position was a temporary one at a teachers college in Nebraska, recently commented about that period, "In 1933 there were no jobs at all for men or women." She continued:

> Even in remote Nebraska I heard about a place called the Institute for Advanced Study opening in far away Princeton. I applied for admission and was accepted. For some years the School of Mathematics was the only School of the Institute and was housed with the mathematics department in Fine Hall at Princeton. Soon after I arrived the Director of the School of Mathematics took me aside and warned me that Princeton was not accustomed to women in its halls of learning and I should make myself as inconspicuous as possible. However, otherwise I found a very friendly atmosphere and spent a valuable and enjoyable year there. . . .
>
> At the end of that year there were, of course, still no jobs. Six men and I from the Institute and from Princeton University took a special qualifying exam to be taken on as substitutes for mathematics teachers on leave from New York City high schools. (LaDuke et al. 1988, 7)

There were several women among the European refugees whose arrival in the 1930s enriched mathematics in America while temporarily exacerbating the job shortage. One of these refugees was Emmy Noether. From the time of her arrival at Bryn Mawr College in 1933 until her death in the spring of 1935, Noether was the focus for intellectual activity for some of the most promising young women in mathematics. Fellowships from Bryn Mawr provided postdoctoral support for Grace Shover (Quinn), Olga Taussky (Todd), and Marie Weiss to work with Noether there in 1934–35. Quinn, who had been a student of C. C. MacDuffee at Ohio State, later spent the major portion of her career on the faculty at American University. Taussky-Todd, who came from Europe to work with Noether, returned there at the end of the year. She emigrated to the United States after World War II to work at the National Bureau of Standards and later became professor at California Institute of Technology.[12] Marie Weiss, who had been a National Research Council Fellow from 1928 to 1930, worked actively in algebra and was a professor at Newcomb College, Tulane, until her death in 1952. Ruth Stauffer (McKee), already a graduate student at Bryn Mawr at the time of Noether's arrival, worked with Noether until her death. McKee received her Ph.D. in

---

[12]See Taussky-Todd 1985 for more details about her career.

Graduate students and postdoctoral fellows at Bryn Mawr, June 1935: (l to r) Grace Shover, Ph.D. Ohio State 1931; Frances Rosenfeld; Marie Weiss, Ph.D. Stanford 1928; Ruth Stauffer, Ph.D. Bryn Mawr 1935; Olga Taussky, Ph.D. Vienna 1930.
(Photograph courtesy of Grace Shover Quinn.)

1935 and later was employed by a research agency of the Pennsylvania state legislature.

A less well known aspect of the 1930s was the influx of a large number of women religious into graduate schools. Even before 1930 Sister Mary Gervase, in 1917, and Sister Marie Cecilia Mangold, in 1929, had received Ph.D.'s from Catholic University. During the 1930s there was a move to upgrade the quality of Catholic women's colleges. In that decade seventeen women religious earned Ph.D.'s in mathematics, and most subsequently taught in Catholic women's colleges, often assuming major administrative positions. Twelve of these seventeen women received their degrees from Catholic, all as students of Aubrey Landry in algebraic geometry.

During the 1930s many mathematicians found nonacademic employment. This sometimes utilized their mathematical backgrounds in surprising ways. For example, in 1938 the Mathematical Tables Project was begun in New York under the administration of the WPA. Two major figures in this project were Gertrude Blanch, a Cornell Ph.D. in mathematics, and Ida Rhodes. The original project required the direction of about 130 people, mostly high school graduates, who produced volumes of mathematical tables without the use of machines more sophisticated than calculating machines. The mathematics involved in this project was early, pre-computer, systems analysis. After working at the Institute for Numerical Analysis at UCLA, Blanch spent the last 14 years of her career as senior mathematician at the Aerospace Research Laboratories at Wright-Patterson Air Force Base. Rhodes moved with the Mathematical Tables Project to the National Bureau of Standards, from which she retired after years as an expert machine language coder and influential trainer of many who became pioneers in computer software.

Evidence of the presence and visibility of women mathematicians in the prewar years can also be seen by their participation in major professional meetings. The 1936 International Congress at Oslo attracted a strong contingent of women; the list of attendees includes 17 women among the 87 AMS members from North America. The semicentennial celebration of the Society at Columbia in 1938 again inspired attendance and involvement of women; 47 of the 419 members who registered for the event were women. While the work of women scholars was noted in some of the addresses, none of the major addresses were given by women.

World War II and the Postwar Period

The advent of World War II introduced major changes in the life and activities of mathematics departments in this country. For example, the total number of Ph.D.'s granted in the United States and Canada dropped from an average of just under one hundred a year in the late 1930s and early 1940s to a low of twenty-eight in 1945. Meanwhile, teachers of college mathematics were desperately needed for training programs for the Army and Navy. In 1943, the government estimated that approximately 250,000 trainees would be sent to a selected group of about 300 colleges and universities; and it was estimated that at least 2500 teachers of mathematics would be required for the various programs ("The Problem ... " 1943). In some cases women who had not been permitted to take positions because of anti-nepotism rules were pressed into service. For example, Helen Owens, a 1910 Cornell Ph.D., whose husband was head of the mathematics department at Pennsylvania State College, finally obtained a position there in the early 1940s.

Even though many women found teaching opportunities at this time, such employment tended to terminate towards the end of the war. In 1944 a department head at a women's college who was seeking a new faculty member was advised, "I suggest you write Professor Coble; he has several girls teaching mathematics in the emergency and these are being let out. There must be a good many institutions such as Minnesota, Wisconsin and Kansas where similar things have happened" (Richardson Papers).

The war also produced major changes in the careers of some individuals. For example, Grace Murray Hopper, a Yale Ph.D. (1934), had been teaching at Vassar when she joined the US Naval Reserve and was assigned to the Bureau of Ordnance Computation Project at Harvard. While there she developed programs for the Mark I computer. Subsequently, through her contributions to the development of higher programming languages, she became known as a pioneer in that field. After years of service with UNIVAC and the US Navy, she retired from the latter with the rank of rear admiral.

With some exceptions, the 1940s and 1950s were not a hospitable period for women in American mathematics. While the total number of mathematics Ph.D.'s awarded in the 1950s was three times the number granted in the 1930s, women did not participate in this postwar growth. In fact, roughly the same number of women earned degrees in the 1950s as in the 1930s. The percentage of Ph.D.'s going to women fell from more than fourteen percent during the first four decades of this century to a low of five percent in the 1950s.

While the number of women annually obtaining advanced degrees did not increase in the 1940s and 1950s, the visibility of women in the mathematical community generally declined dramatically. In the 1940s, for example, about three percent of the abstracts in the *Bulletin* were by women. In the 1940s and 1950s no women were on the Council of the AMS, and, with the exception of one who was on the Board of Trustees in the 1950s,[13] none were on major

---

[13]Mina Rees served as a trustee from 1955 to 1959. She was the first woman on the Board of Trustees since 1923, when Anna J. Pell and Clara E. Smith served as two of the original thirty-one trustees.

Grace Murray Hopper, Rear Admiral Ret., while still captain in August 1981.
(Photograph courtesy of the Smithsonian Institution.)

committees. On the other hand, half of the members of the entertainment committee for the International Congress in Cambridge in 1950 were women.

An indication of the extent of the influence of one woman on the field of mathematics during and after the years of World War II can be seen in a resolution adopted by the Council of the AMS at its annual meeting in December 1953. It reads in part:

> The very striking and brilliant contributions made by pure (non-military, non-applied) science, not least of these by mathematics, to the winning of World War II is well known. It was clearly seen by the government and those responsible for the armed services that a large scale fostering by the U.S. government of fundamental research, the basis of all research, was unavoidable. . . . Needless to say as the purest of all sciences, mathematical research might well have lagged behind in such an undertaking. That nothing of the sort happened is beyond any doubt traceable to one person — Mina Rees. Under her guidance, basic research in general, and especially in mathematics, received the most intelligent and wholehearted support. No greater wisdom and foresight could have been displayed and the whole postwar development of mathematical research in the United States owes an immeasurable debt to the pioneer work of the Office of Naval Research and to the alert, vigorous and farsighted policy conducted by Miss Rees. (*Bull. AMS* 60 (1954): 134)

Mina Rees, who received her Ph.D. from the University of Chicago in 1931, was on the faculty at Hunter College until she became technical aide to Warren Weaver on the Applied Mathematics Panel of the National Defense Research Committee during World War II. After her service with the Office of Naval Research she returned to Hunter as dean of the faculty and later became president of the Graduate Center of City University of New York. In 1962 the MAA honored Rees with its first Award for Distinguished Service to Mathematics, and in 1983 the National Academy of Sciences bestowed upon her one of its most prestigious honors, its Public Welfare Medal. These are only two of the many ways she has been recognized for her work.

In 1949 Evelyn Boyd Granville became the first black woman to receive a Ph.D. in mathematics. She earned her degree from Yale University as a student of Einar Hille. Marjorie Lee Browne, a student of G. Y. Rainich, received her Ph.D. from the University of Michigan early in 1950. The third Ph.D. earned by a black woman was awarded posthumously by the University of Pittsburgh to Georgia Caldwell Smith in 1961 (Kenschaft 1981, 593). Six others followed during the 1960s.[14]

---

[14]See Kenschaft 1981 and Vivienne Malone-Mayes's contribution to LaDuke et al. 1988, 8–10, for fuller discussions of the participation of black women in mathematics.

Women remained active in the national and sectional leadership of the MAA in the postwar years. In the decade following World War II four different women served on the Board of Governors and four served as associate editors of the *Monthly*. Starting in 1957 there have been only two years in which no women were on the Board of Governors, and as early as the 1960s there were usually at least two women serving as vice-president or governor and often three.[15]

In contrast to the MAA, the leadership of the AMS conferred little responsibility or recognition upon women. Cathleen S. Morawetz, an applied mathematician at the Courant Institute, was, in 1969, only the second woman to be invited to address any AMS meeting since Emmy Noether in 1934.[16] Although in 1971 Mary Ellen Rudin, a set-theoretic topologist at the University of Wisconsin, began a term on the Council of the AMS, she was the first woman member in over forty years.

In the 1960s the American mathematical community experienced dramatic growth. The number of mathematics Ph.D.'s granted annually by schools in the United States was almost four times as large at the end of the decade (over 1100) as it had been at the beginning (under 300) (National Research Council 1978, 12). The increase for women was not quite as rapid, growing from 19 in 1960 to 63 in 1969. The total number of Ph.D.'s in mathematics granted to women by American universities during the 1960s (about 400 or six percent of the total) was roughly equal to the total number granted to women before 1960.[17]

THE 1970s AND AFTER

The growth of the mathematical community that characterized the 1950s and 1960s continued only a few more years, peaking in 1972 when over 1300 Ph.D.'s in mathematics were granted (National Research Council 1978, 12). After a period of decline the annual number of Ph.D.'s granted stabilized at nearly 800 in the middle 1980s, a level first attained in 1965. The pattern for women was somewhat different. The growth was slower during the 1960s, but there was no subsequent decline. Moreover, the proportion of women among new Ph.D.'s has been above sixteen percent since 1983, a level previously reached only in the 1930s.[18]

---

[15] Only one woman served as an associate editor of the *Monthly* during the 1960s.

[16] Olga Taussky-Todd delivered an invited address in 1959.

[17] These figures were obtained by collating the results of the authors' research with data from the Doctorate Records File, which is maintained by the Commission on Human Resources of the National Research Council.

[18] Tables summarizing the sex, race, and citizenship of recipients of new doctorates appear in the October issues of the *Notices of the AMS* each year from 1973 to 1979 and in the November issues in subsequent years.

At the same time that the number of mathematicians was increasing, the status of women was becoming an issue in the academic disciplines. Women in professional organizations were beginning to organize caucuses and task forces. In the mathematical community the Association for Women in Mathematics was founded at the 1971 winter meetings after Joanne Darken, then an instructor at Temple University, suggested the formation of a women's caucus.[19] The purpose of AWM was to encourage the participation of women in mathematics. Particular goals included promoting the concerns of women within the Society and the Association and increasing the number of women among the leadership of the AMS.

In April 1971 the AMS Committee on Opportunities in Mathematics for Disadvantaged Groups, which had been formed in 1969, recommended the establishment of a separate Committee on Women in Mathematics. The committee was formed with the following charge:

> to identify and to recommend to the Council those actions which in their opinion the Society should take to alleviate some of the disadvantages that women mathematicians now experience and to document their recommendations and actions by presenting data. (Pitcher 1988, 292)

In 1973 this committee, chaired by Cathleen Morawetz, recommended several measures designed to encourage the participation of women in mathematics. The Council adopted some of them. One recommendation resulted in the publication in August 1973 of a Directory of Women Mathematicians. The committee remained in existence until 1974 at which time a joint AMS-MAA committee was formed.

Mary W. Gray, soon to become the first chair of AWM, was particularly active in a campaign to pressure the Society to open Council meetings to all members of the Society. She also advocated the adoption of nomination by petition of candidates for some offices of the Society. Council meetings were opened in 1971, and nomination by petition was adopted in 1972. Several candidates, both male and female, who had been nominated by petition, were elected to the Council during the 1970s. At the same time, the nominating committee of the Society began to nominate women mathematicians to the Society's elective offices with some regularity and has continued to do so. Thus, the virtual exclusion of women from leadership positions in the Society gradually came to an end. Mary Gray, who served a term on the Council from 1973 to 1975, was nominated in 1975, by petition, and was elected as the second woman vice-president of the Society. Nomination by petition has fallen into relative disuse, but the representation of women within the leadership of the AMS has remained significant. Since 1972 the number of

---

[19]In addition to Joanne Darken, those present at the inception of AWM included Mary W. Gray, Judy Green, Diane Laison, and Gloria Olive.

women officers or members-at-large of the Council has been at least three and as high as eight. At all times since 1976 one of the five trustees of the Society has been a woman.

Although women served as assistant or cooperating editors of AMS research journals from the mid-1920s to the mid-1930s, the first woman to serve as editor of an AMS journal was Barbara L. Osofsky, an algebraist at Rutgers University, who became editor of the *Proceedings* in 1974. During the period 1976 to 1987 the annual membership of editorial boards of AMS journals rose from 58 to 133. The number of women among these members fluctuated between 3 and 8.[20] The 1985 change in the bylaws of the Society that restricted the membership of the Council to exclude all future editors except for chairs of editorial committees has had the effect of reducing the representation of women on the Council since no woman has been designated the chair of an editorial committee since the change went into effect.

In 1973 Barbara Osofsky became the first woman to address a national meeting of the AMS by invitation since Anna Pell Wheeler gave the Colloquium lectures in 1927. In 1980 Julia B. Robinson became the second woman to deliver the Colloquium lectures; in 1985 Karen Uhlenbeck became the third. Cathleen Morawetz delivered the fifty-fourth Gibbs lecture in 1981, the first by a woman. At the AMS centennial celebration in 1988, Karen Uhlenbeck, the only woman among eighteen invited speakers, delivered the address, "Instantons and their relatives," at the Symposium on Mathematics into the Twenty-First Century. Despite such precedents, the incidence of women as invited speakers at AMS meetings and AMS sponsored conferences remains extremely low. AMS statistics show that there were only between twenty-five and thirty women who were invited speakers at any AMS meeting, either national or regional, in the period 1978 to 1987. ("Statistics on Women Mathematicians ... " 1988, 1065). Programs for the semiannual joint meetings for the same period show that the MAA invited about the same number of women to address national meetings as were invited to address all meetings of the Society.

The MAA has continued to appoint woman to editorial positions. Since 1972 at least one associate editor of the *Monthly* has been a woman; often there have been several. However, no woman has yet served as editor since the *Monthly's* inception in 1894. Similarly, women have been involved on the editorial boards of the Association's two newer publications, *Mathematics Magazine* and *College Mathematics Journal*. Although the *College Mathematics Journal* has not yet had a woman as editor, Doris J. Schattschneider was editor of *Mathematics Magazine* from 1981 to 1985.

---

[20]These numbers include both editors and associate editors. See "Statistics on Women Mathematicians ... " 1986–1988 for more detail.

The MAA inaugurated its first woman president, Dorothy Bernstein, in 1978 and will inaugurate its second woman president when Lida Barrett takes office in 1989. Bernstein, a student of J. D. Tamarkin at Brown, worked in the area of partial differential equations. She taught for most of her career at the University of Rochester and at Goucher College. Barrett, a topologist in the tradition of R. L. Moore and a student of J. R. Kline at the University of Pennsylvania, is now Dean at Mississippi State University.

Julia Robinson served as the first woman president of the AMS from 1983 to 1984. In 1948 Robinson earned her Ph.D. at the University of California, Berkeley, as a student of Alfred Tarski. Her work in recursive function theory, carried out while she held no regular position, provided a major component of the solution to Hilbert's tenth problem. She was elected to the National Academy of Sciences in 1975, the first woman mathematician to be so honored. Subsequently, she was appointed professor at Berkeley. In 1983 Robinson became the first woman mathematician to receive a MacArthur Foundation Fellowship.[21] She died in 1985.[22]

While much smaller than the Society or the Association, AWM has also contributed to the increased visibility of women as active research mathematicians. In January 1974 Louise Hay presented the first mathematical talk, "Indices of Turing Machines," at a national AWM meeting. The annual Emmy Noether Lectures began in 1980.[23] In 1982 the AWM sponsored a symposium to honor Emmy Noether's one hundredth birthday. Its second symposium was held in 1985 in cooperation with the Mary Ingraham Bunting Institute of Radcliffe College and dealt with the legacy of Sonia Kovalevskaia.[24] The AWM newsletter has featured numerous historical articles on women mathematicians and their mathematics. Several of these have been precursors of more formal contributions to the growing literature on the history of women in mathematics.[25]

In 1984, Julia Robinson and Linda Rothschild, the presidents of the AMS and AWM, respectively, accepted a citation from the MAA which reads in part:

> The struggle is not over: the Women's Movement has not achieved all of its goals. But the achievements have been so great and the benefits

----

[21]Karen Uhlenbeck followed Robinson as the second woman mathematician to receive a MacArthur Foundation Fellowship and the second woman mathematician to be elected to the National Academy of Sciences.

[22]See Reid 1986 for further details about Robinson.

[23]Short biographies of the first nine Emmy Noether lecturers and titles of their lectures appear in Association for Women in Mathematics 1988.

[24]The proceedings of these symposia appear as Srinivasan and Sally 1983 and Keen 1987.

[25]Most of the literature deals with individual women mathematicians. For example, Grinstein and Campbell 1987 contains selections on forty-three women mathematicians, including some of those mentioned in this article.

to society so obvious that it is right to pause to acknowledge and honor the many women who have blazed the trail.

The Women's Movement in mathematics has been especially strong. Many women — and more than a few men — have worked hard and effectively to convince women that they have potential for excellence in mathematics and that they should receive recognition and rewards commensurate with their achievements ...

Women have achieved prominence in research, teaching, writing, and editorial responsibilities, and have risen to the highest levels of leadership in mathematical organizations. Public recognition for these achievements has inspired other women to make full use of their abilities, in mathematics as in all affairs, with pride and confidence. The Board of Governors of the Mathematical Association of America recognizes and honors their many contributions. ("MAA Gives Special Recognition ... " 1984)

Although the citation explicitly addresses only the recent achievements of the women's movement in mathematics, those achievements are part of a continuous tradition of women in American mathematics that goes back more than a century.

REFERENCES

Association for Women in Mathematics. 1988. *The Emmy Noether Lecturers.* Wellesley, Massachusetts. Pamphlet presented to the American Mathematical Society on the Occasion of Its Centennial Celebration.

Dickson, L. E. 1915. "A Tribute to Mildred Lenora [*sic*] Sanderson." *Amer. Math. Monthly* 22: 264.

Green, Judy, and Jeanne LaDuke. 1987. "Women in the American Mathematical Community: The Pre-1940 Ph.D.'s." *Mathematical Intelligencer* 9 no. 1: 11–23.

———. n.d. "Contributors to American Mathematics: An Overview and Selection." In *Women of Science: Righting the Record*, eds. P. Farnes and G. Kass-Simon. Bloomington: Indiana University Press. Forthcoming.

Grinstein, Louise S., and Paul J. Campbell. 1982. "Anna Johnson Pell Wheeler: Her Life and Work." *Historia Math.* 9: 37–53.

———, eds. 1987. *Women of Mathematics: A Biobibliographic Sourcebook.* Westport, Connecticut: Greenwood Press.

Hildebrandt, Theophil Henry. Papers. Bentley Historical Library. University of Michigan. Ann Arbor, Michigan.

Hutchinson, Emilie J. 1929. *Women and the Ph.D.* Greensboro: Institute of Women's Professional Relations. North Carolina College for Women.

Keen, Linda, ed. 1987. *The Legacy of Sonya Kovalevskaya.* Providence, Rhode Island: American Mathematical Society.

Kenschaft, Patricia C. 1981. "Black Women in Mathematics in the United States." *Amer. Math. Monthly* 88: 592–604.

LaDuke, Jeanne, et al. 1988. "Centennial Reflections on Women in American Mathematics." *Newsletter AWM* 18 (November–December): 4–12.

"MAA Gives Special Recognition to the Women's Movement in Mathematics." 1984. *Focus* 4 (March–April): 3.

National Research Council. Board on Human-Resource Data and Analyses. 1978. *A Century of Doctorates*. Washington: National Academy of Sciences.

Pitcher, Everett. 1988. *A History of the Second Fifty Years. American Mathematical Society. 1939–1988*. Providence, Rhode Island: American Mathematical Society.

"The Problem of Securing Teachers of Collegiate Mathematics for Wartime Needs." 1943. *Bull. AMS* 49: 175–176.

Reid, Constance. 1986. "The Autobiography of Julia Robinson." *College Math. Jour.* 17: 2–21.

Richardson, R. G. D. Papers. University Archives. The John Hay Library. Brown University Library. Providence, Rhode Island.

Srinivasan, Bhama, and Judith D. Sally, eds. 1983. *Emmy Noether in Bryn Mawr*. New York: Springer-Verlag.

"Statistics on Women Mathematicians Compiled by the AMS." 1986–1988. *Notices AMS* 33: 572; 34: 700–701; 35: 1065–1066.

Taussky-Todd, Olga. 1985. "An Autobiographical Essay." In *Mathematical People*, eds. Donald J. Albers and G. L. Alexanderson, 311–336. Boston: Birkhaüser.

Whitman, Betsey S. 1983. "Women in the American Mathematical Society before 1900." *Newsletter AWM* 13 (July–August): 10–14; (September–October): 7–9; (November–December): 9–12.

"Woman Ph.D. at 78 Tells Life Story." 1926. *New York World*. (28 February): 9.

William L. Duren, Jr. received his Ph.D. in 1930 from the University of Chicago, where he was a student of G. A. Bliss. He then returned to his alma mater, Tulane University, where he became chairman of the mathematics department in 1948 and established the Ph.D. program. In 1952–1953 he became the first program director for mathematics in the NSF. In 1954 he persuaded the MAA to establish the Committee on the Undergraduate Program (later CUPM), serving as its first chairman. In 1955 he moved to the University of Virginia as dean of the College of Arts and Sciences, serving concurrently as president of the MAA. Later he helped to form a new Department of Applied Mathematics and Computer Science in the Engineering School, where he taught until his retirement in 1975. The MAA honored him with a Distinguished Service Award in 1967.

Mathematics in American Society 1888–1988
A Historical Commentary

WILLIAM L. DUREN, JR.

1. The Place of Mathematics in American Society

Two histories. For the publication of new mathematics by American mathematicians, the century that began in 1888 has been one of promise and fulfillment. It is appropriate to celebrate that achievement on the occasion of the centennial of the American Mathematical Society.

Then there has been another history: the story of how mathematicians have practiced their profession and earned their living in American society, and of how that society has treated its mathematicians. It is a story of research, scholarship, and teaching, of the practice of mathematics in science, industry, and government, of industry's use of mathematics, of society's support through education and grants of money, of mathematicians as citizens and citizens as mathematicians. For this history 1888 was also a time when mathematics had high status and great promise. But thereafter both status and position have eroded, except for a brief heyday from about 1950 to 1970. So, as a contribution to a complete story of American mathematics, let us proceed to a brief review of some aspects of this second history, the social history of American mathematics.

Institutions that include mathematics. In American society higher education and secondary education are two of those institutions that are allowed to govern their territory so long as they obey the law and stay within the budget. Mathematics itself is not such an institution, but it belongs to both higher education and secondary education. Much trouble, particularly in mathematics, has resulted from the separate, and almost independent, governance of these two institutions. However, the histories of these two institutions have been written; and they provide the best available framework for the social history of American mathematics, [1–7]. To a remarkable extent they fill in the gaps in the social history of mathematics, which is only sketchily recorded, or the history of a single mathematics department [8].

Mathematics also belongs to less well-defined national institutions called science and technology whose official interface with the federal government is the National Academy of Sciences–National Research Council (NAS–NRC). In practice this is the main interface of mathematics with the government. In the government, we have the National Science Foundation (NSF) and the National Institute of Health (NIH), while in the research community we belong to the American Association for the Advancement of Science (AAAS).

In these scientific institutions we again have special difficulties. For mathematics is not entirely a science, having a humanistic culture as well which resides more comfortably in higher education. Moreover, among the sciences, our basic research tends to have a long delay before providing visible social utility. And our applied mathematics tends to be a tool subject, a problem-solving technology for problems that arise outside of mathematics. Either way we seem to fit in more as "servant of the sciences" than as a science in our own right. Our own associations, AMS, MAA, SIAM, ORSA, NCTM, and the statistical associations, have no direct interface with the government.

2. UNLIMITED PROSPECT

Mathematics in the general studies curriculum. In 1888 mathematics had a secure place in American society [9]. This was based on its long-established position, along with the Latin and Greek classics, as one of the required general studies that had made up the curriculum since the earliest medieval universities [11,13]. It was one of the essential humanities. In 1888 Daniel Gilman, in his thirteenth report as president of the new research university, Johns Hopkins, said with justifiable hubris: "How to begin a university... Enlist a great mathematician and a distinguished Grecian; your problem will be solved... Other teachers will follow them..." [14, p. 29]. Gilman's "great mathematician" was J. J. Sylvester and his "distinguished Grecian" was the renowned Gildersleeve.

Secondary school mathematics requirement. The position of mathematics was further bolstered by its requirement in the secondary school curriculum. There were some public "academies" preparing students for college but in general this was the function of independent private academies. Although there was no consistent standard, their curricula were designed to prepare for the general studies curriculum of the colleges by a required general studies curriculum of their own, dominated by the entrance requirements of the colleges. Mathematics was required for graduation from the academy even though the substance of this requirement was somewhat ragged.

The great boom. The period after the American Civil War up to 1893 saw the greatest boom in higher education in the nation's history. New state universities were established, including the new land grant colleges, such as MIT and Purdue, that included schools of engineering and applied science. There were new colleges for women and for blacks. During the war Yale had taken the lead to establish a Ph.D. program; and more such German-style graduate degree programs followed. Science had finally made its way into the university curriculum over the opposition of the entrenched humanities, although in a separate B.S. degree program segregated from the traditional liberal arts. Hence the name of the central university college was changed from "the academical college" to "the college of arts and sciences." And the national learned societies came into being, including the AMS in 1888.

The position of mathematics was further fortified by its requirement in the curricula of the new B.S. degree programs and those of the new schools of engineering and applied science.

European resources. American mathematics still drew strength from European and English mathematics. Young Americans with ambitions for mathematical research went to Germany, particularly Göttingen, for their Ph.D. degrees. A variation of this pattern was emerging in which one got the Ph.D. in one of the new American graduate schools, and then went to Germany for a year or two of further research training. English or French mathematics texts were still the mainstay of instruction in both secondary school and the university. (The Germans wrote treatises, not textbooks.) Distinguished European mathematicians were still being attracted to bolster the faculties of the new American graduate departments.

Applied mathematics. Applied mathematics appeared to be in good shape in 1888. True, the old applied mathematics consisting of the application of trigonometry to surveying, mapping, and military science had run its course and was being taken over by civil engineering. But mathematics then embraced rational mechanics and positional astronomy. This was the applied mathematics of 1888. Only a few American mathematicians were doing significant research in celestial mechanics, but they taught the classical work of Newton, Lagrange, Laplace, and Gauss. Of course there was the brilliant and original work of Willard Gibbs in phase equilibria, chemical thermody-namics, and statistical mechanics. He had, however, little in the way of a research group to continue and develop stochastic applications of mathemat-ics. In fact, the whole field of rational mechanics had become too difficult technically for all but a very few graduate students in the new doctoral pro-grams. They were not prepared to work in the statistical mechanics of Gibbs or the new topological dynamics of Poincaré.

3. THINGS GO WRONG FOR AMERICAN MATHEMATICS 1893–1940

Retrenchment. We may date the end of the great boom in higher education as 1893, the year of a severe national financial panic in the second adminis-tration of President Cleveland. As we shall see, there were also other reasons, more specific to mathematics, for regarding this as the end of an era and the beginning of another. The years 1893–1940 saw a slow but substantial growth in American mathematical research. In this respect it was a period of great achievement. But for the position of mathematics in American society the unlimited prospect of 1888 was not to be realized. Very shortly after 1888 a series of disasters fell on mathematical education, on applied mathematics, and on the service of mathematics in American society.

Mathematics loses its traditional place in the required general studies. The traditional curriculum of required general studies had been carried over into the expanded colleges from the early colleges that trained students for the ministry, or for teaching. That difficult course of studies included Greek and Latin for their cultural values as well as for reading the scriptures in the original. It included mathematics, partly by sheer weight of tradition, and partly for "training the mind." By 1888 it was apparent that there would never be enough jobs as minister of the gospel, or teacher, to employ the flood of B.A.s then graduating. Nor were there jobs enough in the technically primitive industry to employ the graduates of the new science and engineering programs. And students in the traditional degree program were clamoring for a German-style free elective system to replace the difficult and outmoded general studies curriculum.

Charles W. Eliot, President of Harvard (1869–1909), had started his ca-reer as a mathematics tutor (1854–1863), before changing to chemistry and then to the presidency of Harvard. He was undoubtedly the most influential educator of the century [15]. He took the lead to give the students what they wanted, a free elective system. Other universities quickly followed Harvard's lead. It was the ruination of classics, replaced by a newly synthesized subject called "English." Mathematics lost a large share of its enrollment and could have suffered the same fate as classics if it had not been for its requirement in the new physical sciences and engineering. Also there was no popular substitute for mathematics, no computer science! Thereafter mathematics was even more strongly regarded in the national government as a subsidiary of the physical sciences and engineering.

The wide adoption of the free elective system created so much chaos in the college curriculum that some constraints had to be reintroduced to restore order. One of these was the requirement of a major subject, introduced by Woodrow Wilson as president of Princeton. Both Harvard and Yale also introduced systems of group requirements and academic advising that, together with the major, make up the system we have today. But this did not restore the old position of mathematics.

Mathematics loses out in secondary school. A major trouble with the great expansion of higher education that had occurred was that it was built on an inadequate foundation. There was no national system of secondary schools to prepare students for the new public universities. A blue ribbon "Committee of Ten", headed by none other than Harvard's Eliot, was commissioned to decide what the new schools would be. In a report in 1893, it rejected the English, French, and German models. The committee recommended a new American high school that would be under the same local school boards that controlled the universal public elementary education. The additional four years would be offered, but not required. The colleges, which had dominated the old academies that prepared students for college, were to have no authority over the new high schools. It was believed that the majority of the new students would not go to college but would only add four years to their terminal general education. Many would profit by vocational, rather than academic, education [16].

While the new high schools were to offer college preparatory mathematics courses, these courses were not built into the graduation requirements as they had been in the academies. The estimates at the time called for only 10 or 15 percent of public high school students to take the college-bound curriculum.

These arrangements left the school-college interface in some confusion, so a new Commission on College Entrance Requirements in 1899 defined the boundary and set up the system of Carnegie units. These describe the content of a student's record. The minimal college entrance mathematics content was 2.5 units, including one unit of deductive Euclidean geometry. In addition, the report called for the establishment of the College Entrance Board (CEEB) which would prepare and administer standard examinations to validate the graduation credits. This strengthened the position of the private universities that required the CEEB scores. Originally the state universities, and many private ones, did not require the CEEB scores but the system of Carnegie units was universally used [17,18].

These arrangements gave the major private universities at least the same control of their admissions that they had enjoyed before 1893. In subsequent years these universities would exert the only real force for maintaining quality of performance in secondary schools. In the state and private universities that admitted on a basis of the high school diploma, there was little control of admission standards. Of course this affected mathematics most of all. The mathematicians hated it, but there was little they could do about it. Moreover, at first it was true that the new public high schools were supplying only a minor fraction of college admissions; so the issue was slow in coming to a head.

PEB. James B. Conant [19] has praised the American high school for its democracy and adaptability to local conditions. As parents and citizens we can all appreciate the undeniable virtues. But it has not worked well for mathematics in American society. We recognize that there are other important educational and social objectives besides the teaching of mathematics. But we, as custodians of the mathematical culture, have not only the right but the responsibility to look after the quality of mathematics that the secondary schools provide. We have already seen one way in which the system was flawed for providing good mathematics. It was seriously flawed in yet another way.

The local school boards had no authority, or capability, to educate or certify teachers, to establish a national curriculum, or to acquire and distribute state and federal funds for education. The federal government was deliberately denied such powers. And the universities did not have these powers or responsibilities. Somebody had to run the public school system; Eliot's plan did not provide for it. In this power vacuum there arose an unofficial Professional Education Bureaucracy (PEB). PEB came to consist of the superintentents who act for the local school boards, the state and federal officials that control and distribute the public funds and certify teachers, and the deans of the schools of education that train teachers. The U.S. Commissioner (now Secretary) of Education has never been a member of PEB. He is a temporary federal appointee of the president, who is permitted to sound off somewhat impotently. Unofficial PEB has never had an official organization. It speaks, when necessary, through some committee of the National Education Association (NEA) as Eliot's Committee of Ten did. The leaders of PEB occupy politically vulnerable positions, especially the superintendents, who are the key members. But the actions of PEB, in setting the national curriculum, in training and certifying teachers, and in distributing state and federal funds, are not directly subject to democratic approval. In particular, they are insulated from the criticism of the intellectual, university, and scientific communities.

In 1913 NEA appointed the Commission on the Reorganization of Secondary Education. It clearly reflected the thesis of John Dewey that the function of the school is to foster the growth of the pupils along the lines of their own interests rather than to impart subjects, and that the aim of education is social. This philosophy avoided the European objectives of liberal education and culture. But the Commission consisted of 14 committees representing subjects. Only mathematics was not represented. The general report, issued in 1918, set forth the agenda of social development and personal fulfillment as the aims of secondary education, and relegated the mastery of subjects to low priority. These objectives gained wide acceptance without any formal adoption into national policy [20].

Isolation from European Mathematics. Not only was the new American high school different from its European counterpart; the implication was that the American college program would be different too. European textbooks, course structures, and teachers were no longer transferrable to America as they had been in the past. The first result was that American mathematicians had to write new textbooks to fit the new courses. The last of the French textbooks to be translated and widely used in this country was that of Goursat and Hedrick (1904).

One of the first significant new American texts was Granville's *Calculus* (1904). Since college algebra was a new course, textbooks had to be written for it. The subject was a kind of hodgepodge. About half of it had to be a review of high school mathematics. This was in effect remedial mathematics from the start. The rest of it advanced into the study of polynomial algebra, including Horner's method for the numerical solution of equations of any degree. H. B. Fine of Princeton and Wylczinski and Slaught of Chicago wrote significant textbooks that were considerably higher in level than later ones. Fine also wrote a distinguished analytic geometry textbook. Since synthetic Euclidean geometry was assigned to the high school, this course was geometry on its own merits, continuing Euclidean geometry by Cartesian methods, not just precalculus coordinate techniques. L. E. Dickson of Chicago wrote a junior level polynomial algebra text, called *Theory of Equations*. Like college algebra, the analysis to follow calculus was something of a medley, called advanced calculus. E. B. Wilson of Yale wrote one of the significant texts.

The effect of all this writing of American texts was to divert energies from research to teaching, and to isolate American from European mathematics. Graduate work also turned inward. Not so many Americans went to Europe, even for postdoctoral study. The faculties of colleges and state universities were now American educated. Moreover, since so much low-level and remedial teaching was required, the Ph.D. degree was not considered necessary for all the faculty. A number of the universities that had previously announced Ph.D. programs quietly let them become inoperative and became large undergraduate colleges with only a Master's program attached. Research was concentrated in relatively few universities; and even they had less contact with Europe than they had earlier.

American mathematics loses relevance in science and technology. Not long after 1888 American mathematics lost mechanics, which had been its applied field and its connection with science and technology. In part this was due to the new American research interests in such subjects as algebra and topology, which could be undertaken without so much technical background as mechanics had come to require. In part it was due to the collapse of the Newtonian program to construct an axiomatic mechanics as an extension of Euclidean geometry. Maxwell's electromagnetic field physics had put this out of reach, and the emergence of Einstein's relativity and the new quantum theory of Planck clinched it. Mach's *Science of Mechanics* [21] was a historical farewell to the Newtonian program. The teaching of mechanics was gradually taken over by physicists and engineers, who trashed it!

Research in astronomy, too, moved toward physics, that is, into the analysis of the spectral quality of light gathered by large reflector telescopes. It was called astrophysics. Astronomy departments divorced themselves from mathematics and taught the celestial mechanics themselves as part of their continuing, but secondary, interest in positional astronomy. Gibbs's statistical mechanics continued to be studied mainly in physical chemistry and thermodynamics. That left the continuation of Poincaré's program in topological dynamics, which was taken up by G. D. Birkhoff and a few of his students, as about the only part of mechanics left to mathematics. And at the time that had little contact with the physical sciences. All told, that left American mathematical research and advanced teaching with virtually no viable applications in science and technology.

Nobody realized what was happening. E. H. Moore, in his 1902 retiring address as president of the AMS, entitled "*On the foundations of mathematics*" [22], gave a remarkable summary of the state of American mathematics and mathematical education at the time. He did not see what was happening. He saw only the continuation of the old relationship to European mathematics and the continued leadership of lower school mathematics by the university mathematicians. And he was optimistic about American applied mathematics. He did not realize the consequences of the loss of mathematics's traditional requirement in the general studies curriculum. In fact nobody seems to have realized what was happening.

One of the first to speak about the parallel between the demise of classics and the near demise of mathematics was E. B. Wilson in 1913 [23]. He wrote:

> The decline of the Graeco-Roman empire over our collegiate
> studies is apparently extending almost to extinction. That culture
> and valuable intellectual discipline are best obtained by application
> to subjects which are neither useful nor interesting to the student,

and over which he never obtains even a mediocre mastery, is an idea which is losing ground despite the entrenchment of vested interests. The fact is that Greek and Latin do not make good.

We mathematicians, however, are in no position to gaze upon the motes in the eyes of our classical bretheren, to whom we can hardly compare ourselves favorably. For there has been a great decline in the sway of mathematics over our collegiate studies. We suffer by the presence in college of great numbers of fellows neither primarily nor seriously there for the sake of intellectual advancement. But our chief difficulty is that we do not make good.

Wilson went on to propose that college mathematics should begin with "early calculus" for its intellectual value and real applications. Elsewhere he modified this by saying that first-year mathematics should consist of equal amounts of calculus and "choice and chance." By the latter he meant something like what we would call finite or discrete mathematics, with an introduction to probability and its applications. If we may look ahead in this history, we must report that Wilson's proposal did not take hold very widely, although Harvard took the lead somewhat later to establish calculus as its standard freshman course [24]. And in the years after 1953 the MAA Committee on the Undergraduate Program in Mathematics (CUPM) undertook to install Wilson's program of calculus and choice and chance as the standard first year mathematics in American colleges. After some early success, that initiative was overwhelmed by the tide of remedial, precalculus mathematics but left a substantially wider adoption of calculus and elementary combinatorics as the normal first-year mathematics. But, going back to 1913, it is doubtful that Wilson's proposal would have satisfied the engineers; for at that time they used finite difference and sum methods that they did not recognize as calculus.

By 1916 the mathematicians became alarmed at what was happening in the NEA Commission on the Reorganization of Secondary Education, in which mathematics was not represented. Belatedly E. R. Hedrick, as first president of MAA, appointed a national committee to prepare a statement for mathematics. It had representation from the various mathematical and educational organizations involved in high school mathematics. It was called "The National Committee on Mathematical Requirements." It was chaired by J. W. Young of Dartmouth, and it included E. H. Moore, Oswald Veblen, and D. E. Smith, as well as similarly prominent teachers and administrators from the secondary school system. But World War I was on, and the committee did not report until after the war ended. Meanwhile other events intervened to aggravate the problem.

Expansion strikes again. Before America's involvement in World War I, a national agreement was reached to extend required public education from

8 years to 12. Previously the high school component was provided but not required [25]. When the plan was implemented after the war, the sudden demand for teachers could not be met while maintaining quality standards; and if the students had been deficient before, these compulsory registrants were worse [26]. There was near chaos. Soon these students swarmed into the colleges and universities. Administrators were eager to have them for the growth they propelled. The mathematicians, with their outmoded standards, were blocking progress. The attack on mathematics requirements intensified. PEB was now much stronger politically. The education schools to train teachers were well established, if strained to meet the demand. PEB was joined by the college deans, many of whom now came from the popular social sciences where they had little use of, or appreciation for, mathematics. But, in spite of these difficulties, the expansion provided an opportunity to establish new graduate programs in mathematics. So the expansion produced a lot of new Ph.D.s in mathematics, too. They oversupplied the weak demand for college mathematics teachers; hence salaries were low. Then came the Great Depression in 1932. All academic professionals were hit, but mathematicians fared especially badly because of the oversupply. It was the worst of times for American mathematicians.

The Committee on Mathematical Requirements. We return to the J. W. Young committee, which Hedrick had appointed in 1916. In 1923 it finally issued its 652 page report that was probably the most thorough study of school–college mathematical education ever done. It titled its report, "The Reorganization of Mathematics in Secondary Education," presumably to identify it as the mathematics statement for the 1918 report of the NEA Commission on the Reorganization of Secondary Education [27]. The Young committee also made an extensive survey of secondary school mathematics curricula and training of teachers in foreign countries, which showed the United States to be deficient. It examined many available national tests of ability and achievement in mathematics. It presented sample units of exposition for topics of critical difficulty for the student. In particular it offered an outline of an intuitive approach to plane geometry for weaker students while retaining the traditional deductive course for college-bound students. It stressed the importance of secondary school mathematics for science and industry in this country.

It is fair to say that these authoritative and thoughtful recommendations came to nothing. By this time PEB had powerful political influence and the mathematicians had virtually none [28]. Moreover mathematics had few allies in the universities. And there seems to be no record of its seeking support from the National Research Council.

The tide begins to turn for American mathematicians. It was 1940 before the American public became aware again that mathematics is important, but years earlier some things happened in mathematics to reveal that recovery was under way. Recovery from irrelevance came first. Significant new fields of applications of serious mathematics opened up, though the public was not yet aware of this. (We will look at these new applications in the account of the postwar period as they unfolded and came into use in American society.)

Another great new start occurred when the Bamberger–Fuld families made a major gift to establish a new research institute above the level of the existing graduate schools. They entrusted the development of the idea to Abraham Flexner [29]. His idea was to choose, as the initial field, one in which there would be universal public recognition of the eminence of the chosen scholars and their work. Consultation with intellectual leaders in this country and in Europe convinced him that that field was old mathematics, apparently somewhat to his surprise. So the Institute for Advanced Study was established in Princeton in 1930 with mathematics as its first field of study and with Einstein as its first "mathematician" [30]. Veblen came over from Princeton as its first chairman [31]. Einstein suggested von Neumann. Herman Weyl came. Kurt Gödel came as Einstein's assistant. And the Institute for Advanced Study was under way.

The distinguished European mathematicians who joined the new Institute were only the first of a wave of fine European mathematicians who soon followed, fleeing Hitler and oppression. But they came at the very depth of the Great Depression, a time when many American mathematicians had no jobs, or poorly paid teaching jobs. There really was not room for the Europeans. But Veblen and AMS Secretary, R. G. D. Richardson, appealed to mathematics departments in the United States and Canada to find places for them. Veblen assigned to Leon Cohen the task of raising funds by contributions from American individuals and businesses to subsidize new jobs for a list of fine European mathematicians prepared by Veblen. Somehow positions were found for them in the United States and Canada. These immigrant mathematicians not only broke the isolation of American mathematics from that of Europe, they became the leaders in research of a great new era of American mathematics [32]. It was more than a triumph for mathematical research in America. It was a great day in the social history of American mathematics as well [94–97].

In 1938, the semicentennial year of the AMS, R. C. Archibald's history of the AMS revealed little about the status and service of mathematics in American society except that the Society was growing and gaining support after difficult times [84].

But G. D. Birkhoff's address, although primarily a survey of American achievements in research, also spoke with Birkhoffian bluntness about the status of American mathematicians. He quoted a prewar survey by J. C. Fields of Toronto which concluded that among mathematicians throughout the world "the American professor is the worst treated of all." Birkhoff gave reason to think that, since the war, things had improved. "There are now many chairs where the salary is good and the duties not excessive." On the other hand he worried that so many of those chairs went to distinguished Europeans, leaving capable Americans "as hewers of wood and drawers of water," adding, "it should be strongly emphasized that twelve hours of instruction each week is about all that can be required if the best standards of scholarship are expected." Then, expressing gratitude to the public for having supported mathematics so generously "if unwittingly," he added: "On our part there is an inescapable deep responsibility to the nation. It is our duty to take an active and thoughtful part in elementary mathematics instruction as well as to participate in the higher phases" [85]. He meant it. The Birkhoff–Beatley textbook for a simplified, but still rigorous, school geometry was one of his contributions.

4. RENAISSANCE 1940–1970

War makes mathematics important again. The approach of war in 1940 changed the climate for American universities. Now, for the first time in many years, a male student had to compete for admission on a basis of his performance in science and mathematics. A good performance was rewarded by a delay in military service and a college education at government expense. That would qualify him for a commission in the Army or Navy when his time came to go to war.

Then, in the war, it turned out that science, physics especially, made major contributions to the military victory through such innovations as radar, sonar, gun- and bomb-guiding computers, and finally the atom bomb. But mathematics was involved both in the original design and development of those weapons and in their most effective use [33–34]. To cite one outstanding example, the statistician Sam Wilks and the physicist Philip Morse made daily best estimates of the position of the German submarine fleet on the basis of the previous day's sightings. The superiority of sophisticated mathematical analysis over guessing in such battle situations resulted in the attachment of operations research, or operations analysis, sections to the military commands [35]. Moreover, the great importance of a large, high speed, electronic computing machine was recognized for such applications as logistics and secret intelligence, as well as the design and development of new weapons [36].

Basis of the return from irrelevance. While the war brought mathematics to the public consciousness again, the basis of the renewed usefulness of advanced mathematics in American science, technology, and organization had been established earlier, unknown to the public. It was not a product of the war, and it continued to develop with renewed impetus after the war. One significant start was the discovery that complex analysis was essential to the new electronic engineering that developed out of radio in the years between

1900 and 1925. In particular, by that time the Laplace or Fourier transform was recognized as the basic mathematical tool for the theory. Another, entirely different, kind of mathematics found new uses in electrical engineering. It was popularly presented in T. C. Fry's *Probability and its Engineering Uses* (1928).

The mathematical basis of the new applications of probability was set up by uniting the statistical methods of R. A. Fisher (1932) and von Mises (1939) with Kolmogorov's measure–theoretic formulation of mathematical probability (1933) [37]. Influential first expositions of the new mathematical statistics that resulted from this union were Cramér (Princeton, 1945) and Feller (1950) [37]. From these roots mathematical statistics grew into an enormously powerful and pervasive technology.

The completion in the years 1925–1930 of the linear operator formulation of quantum mechanics opened the door to a vast array of applications in physics, chemistry, solid state engineering, and, surprisingly, biology. The full implementation of this revolutionary development had to await the availability of the large-scale computer after 1945 [36]. It led Weyl to remark: "...this branch of mathematics [linear algebra and analysis] crops up everywhere in mathematics and physics, and knowledge of it should be as widely disseminated as the elements of differential calculus" [38].

The success of operations analysis in World War II gave a great impetus to the mathematics of optimization: operations research, linear programming, game theory, input–output analysis, optimal control, stochastic control, etc. These mathematical methods found more and more users in businesss management.

It was not only branches of mathematics, finding new applications, that reversed the inward turning of American mathematics that had isolated it from science and society for 50 years; it was also mathematicians and mathematical scientists, many of them the European emigrés. There was Einstein. General relativity at first had little impact on society. Among intellectuals it was either a humorous or outrageous paradox. Among professionals in mechanics it was a complicated theory that made small corrections in, still valid, Newtonian mechanics. But the atom bomb in 1945 and the subsequent nuclear energy technology changed all that. Even the common man now recognized relativity's mass–energy equivalence, $E = mc^2$. Moreover, in the modern science of the cosmos, popularized on TV, it has turned out that relativity is the only valid mechanics [86].

And there was von Neumann. His theory of games won almost instant recognition in economic theory and business technology. But his 1948 paper, "The General and Logical Theory of Automata", has had a slowly developing, deep impact on both computer architecture and biological science [87–88].

Other mathematical scientists of the years before and after World War II who brought serious mathematics into new uses in science and technology included Hermann Weyl, Vannevar Bush, Norbert Wiener, Claude Shannon, Philip Morse, Kenneth Arrow, Wassily Leontief, John Kemeny, and others. The list is uneven in terms of magnitude of mathematical content; we are talking about mathematical scientists whose work was influential outside the mathematical community.

And there appeared on the scene administrators and doers who understood the social value of advanced mathematics and were able to bring public resources to the recognition, support, and utilization of mathematics. They included Abraham Flexner, Oswald Veblen, T. C. Fry, Warren Weaver, Richard Courant, and Mina Rees. Finally there appeared popular interpreters of advanced mathematics such as Warren Weaver again and, later, the incomparable Martin Gardner. One would have to go back centuries to find a comparable list of prophets.

But perhaps the most important factor in the renaissance of mathematics in American scientific society was the idea of a mathematical model. It seems to have caught on with scientists and engineers in all fields. The public is now familiar with such models as the weather prediction model, as seen on TV. It is difficult to pinpoint just where the idea originated. One source seems to have been a seminal paper by G. A. Bliss in 1933 [39]. The computer was an essential tool to put the idea into practice, especially the many models that involved partial differential equations. The public could comprehend the idea of the mathematical model and understand the predictions it generated, with little or no technical understanding of the mathematics involved.

Postwar idealism and excellence, 1945–1954. When the war ended in 1945 the spirit of the country called for a better world to be achieved through competitive excellence. The government had plans ready to send the returning veterans to college, or graduate school, under subsidy of the GI Bill. The returning veterans, superimposed on the normal class of high school graduates, overflowed the universities. That created competitive conditions for admission and graduation. In this atmosphere of tough idealism the difficulty of mathematics was acceptable; mathematics was important. And there were not enough mathematicians to meet the demand. The public saw science and mathematics as keys to a better world. Only the people of the humanities felt neglected.

The idealism of the time was also expressed by an act of Congress which established the National Science Foundation (NSF) in 1950. Before it got into full operation, about 1952, the Office of Naval Research (ONR) conducted pilot operations in support of basic research in science and mathematics [40]. Other military research organizations joined in support of basic research.

Within the mathematical community a curriculum reform movement sprang up spontaneously, without any grants or commissions. Textbook writers came forth with new humanistic, or cultural, versions of freshman mathematics, and professors adopted them eagerly. There were books by F. L. Griffin, Morris Kline, Moses Richardson, Carl B. Allendoerfer and C. O. Oakley, and E. P. Northrop. The themes of these books included the cultural heritage of mathematics, the real numbers as a complete ordered field, logic à la Venn diagram and truth table, history, or unification of the technical components into a philosophical whole. They all made lower technical demands than the traditional texts; and this was their ultimate undoing. These were supposed to be terminal courses, but students in them kept showing up later in calculus, physics, or engineering where they were technically deficient. Yet in the postwar world mathematicians felt the need of their richer intellectual content.

Normally the universities have an excess of capacity in graduate studies and research because universities see this activity as conferring precious prestige. Normally they compete for graduate–research prestige with inadequate resources in qualified faculty and students, and always with inadequate funds. But the postwar era of excellence was a rare time when it was possible to establish new graduate–research departments and facilities without acting at the expense of one's rivals. However, it turned out that, if you were going to establish a new graduate department or institute, the time to do it was before 1954. After that it was too late; the old constraints began to return, though not all at once.

Around 1954 three little-noted events signaled the end of the era of excellence. First, the GI Bill veterans were phasing out about then. But now their numerous children, the war babies, could be seen and counted swarming through the lower grades. Clearly there would have to be a big expansion of college education to accommodate them, beginning in about 1963. That expansion would call for quantity rather than quality. Finally there was the 1954 Supreme Court decision (Brown *vs.* Topeka Board of Education) that outlawed segregation in the schools and guaranteed the right of public education without restriction by race.

This decision helped to justify a gradual shift from competitive excellence to concern for the underprivileged and underperformers in society. Mathematicians certainly shared this concern but it created special difficulties for mathematics. "Excellence" was still a good word and it took some years for somebody to invent the appropriate counter word, "elitism," to justify and mask the shift in national educational policy.

Policy and practice in the support of basic research in the early days of the NSF. The idea of a government foundation for the support of basic research was very new as the NSF got under way in 1952. There were no direct precedents to serve as guides for either policy or practice. The operations of ONR came closest. And there were the long-established private foundations, such as the Rockefeller Foundation. The physicists and engineers had some experience in writing contract proposals for war research. But core mathematicians had none. Mathematics proposals, and therefore mathematics grants, were lagging behind when the first program director in mathematics was belatedly appointed [41]. His first task was to get more proposals for mathematics research that could qualify for support under the still undetermined working criteria. Getting the right reviewers was easier because *Mathematical Reviews* told who was doing research in each field and who was reviewing that work.

Although the National Science Act had directed the NSF to support research in basic science and mathematics, a number of crucial policy issues surfaced almost immediately. Those issues are still with us in 1988. They have only grown more complex, more relative, and more unresolvable. A history of the early NSF must recreate the innocence and the enthusiasm of those days, as well as the extreme caution imposed by the superlative first director, Alan Waterman. We were not aware in 1954 that, at that very time, the era of excellence and idealism, in which the NSF was established, was yielding to a new one emphasizing expansion and more expansion.

From the start much of the difficulty in both policy and practice came from the word, "basic." For "basic" is a relative term, not permitting a workable general definition. Without such a definition the discriminations for day-to-day operations in support of basic research are difficult, especially in mathematics.

More trouble came from the mathematician's need to buy time from teaching duties for his research. No assistant, or laboratory, can do his research for him. Scientists and engineers did not understand this. Their custom is not to pay the principal investigator's salary. Support comes in the form of laboratories, instruments, materials, and assistants. When exceptions were made for the mathematicians, the grants escalated to levels that brought out the cry: "greedy mathematicians!" And the NSF terminated payments of any portion of academic year salaries for principal investigators.

But more fundamental difficulties involved the principles governing the support of basic research in this new government foundation. Fortunately for mathematics, Chester I. Barnard, first chairman of the National Science Board, brought with him from his former position as president of the Rockefeller Foundation a passive policy of supporting proposals submitted by scientists. They were chosen for support only on merit, determined by peer review in the field. It was intended to forestall any NSF control of research through its money-granting power. Barnard's policy, which was almost religiously followed in the early days of the NSF, also helped to avoid questions about what was "basic research." It was good for core mathematics.

This passive proposal-support policy had its critics, including Abraham Flexner, himself a former head of the Rockefeller Foundation [29]. Flexner strongly advocated a mission-oriented policy such as the long-standing Rockefeller program to eliminate tropical diseases. And his philosophy had also led to the establishment of the Institute for Advanced Study, the choice of mathematics as its first subject, and the selection of eminent mathematicians for its first faculty. But such successes depended on having intelligent, wise, objective and idealistic minds to direct the mission, Flexner's and Veblen's, in the case of IAS, and then letting them direct it. With improper direction, or with political direction, mission-oriented research can go very far wrong. It is a policy better suited to a private foundation.

The argument was made moot by a decision of Congress to make a massive appropriation for an upcoming International Geophysical Year. And the NSF, which would administer it, wanted it because it would mean a big increase in the NSF's small budget. It dwarfed the program for the support of individual projects, even though that item was increased. In subsequent years a succession of such huge projects, some in space, some in national defense, were adjoined to the NSF budget. In mathematics this implied that money available for specific applications increased out of proportion to core mathematics support.

Before we leave this subject it should be remembered that there are other noninvasive policies, besides Barnard's proposal-support policy, that have been used to support basic research in mathematics. These have included the Sloan Foundation's support of promising young mathematicians, the NSF fellowship programs, and the support of AMS programs such as its research conferences and its publications. But the support of individual research proposals retains its crucial importance.

Educational expansion, 1954–1962. There was no sudden shift in policy in 1954. Just when the humanities might have been able to regain their lost domination of undergraduate education, the Russian satellite, Sputnik, launched in 1957, shocked the nation into a new national effort to compete with the Russians in space. Moreover the new National Science Foundation was just coming into full operation in 1952. But the main thrust of national effort had to be directed to expand colleges for the war babies. New state colleges were established, new multiple branches of state universities, and many new two-year community colleges. To teach the war babies would require that the production of new college teachers be tripled over the next ten years. That meant that we would have to dig three times as deep as normally into the relatively thin age class born between about 1927 and 1937 for graduate students. They did not have GI Bill financing. Since the humanities were expanding at about the same rate, science and mathematics could not, and in fact did not, get the extra graduate student manpower by luring students from the humanities. They had to come from the bottom of the barrel.

The NSF now extended its departmental grants program to the major state university departments which had existing graduate programs. Mathematics was now even more the area of greatest shortage of both faculty and graduate students. The grants that subsidized the expansion relieved the faculty for research, essential in a graduate program in mathematics. The teaching loads of the graduate students, actually teaching assistants, increased. So did class sizes; mathematicians preferred large lecture sections to small classes as a way of handling the requisite student load. Faculty research was promoted by these conditions; but graduate student research was not.

Mathematicians get back into school and college education. An unanticipated aftermath of the war was that PEB lost control of secondary education, at least partially, for the first time since 1900. American women had found jobs better paid than teaching during the war and now had other interests. In any case there was no postwar rush of women back to teachers' colleges. PEB could not supply enough trained teachers for the upcoming school expansion.

PEB had been ready with postwar plans which reasserted the old estimates: 20 percent college bound, 20 percent vocational education, and 60 percent in a curriculum now called "Education for life adjustment" [42]. It was a gross miscalculation. Both PEB's previous educational policies and the postwar plan were ill fitted for the national drive toward excellence through science.

Leaders in NSF felt that something had to be done to help the mathematics and science teaching corps. In 1952 the NSF funded a new venture, a Summer Institute for Mathematics Teachers at Boulder. It was led by Burton W. Jones, and E. Artin was the distinguished visiting mathematician. The next year there were two math institutes. Then the program spread to the other sciences. It became one of the most popular and successful of NSF programs, much liked by Congress [43]. Then the NSF expanded the idea to academic year institutes. These were associated with colleges of arts and sciences, with only apprentice teaching conducted by the school of Education. State legislatures suspended the requirement of a degree in Education for certification of teachers. A new degree was established in the graduate schools of arts and sciences, the Master of Arts in Teaching, MAT. These moves put mathematics and science departments, teamed with the NSF, full scale into the training of teachers.

Mathematicians were also leaders in curriculum studies to make better use of students' time and taxpayers' money than the now-decadent old curriculum and discipline did. While the main problem was clearly in the lower schools, the college curriculum was more accessible, and the college teachers did not have to be taught mathematics. That seemed the place to start, and the objective from the beginning was to eliminate as much remedial mathematics as possible so that college mathematics could start with calculus and/or discrete mathematics and probability.

CUP–CUPM. So in 1953, the MAA Committee on the Undergraduate Program (CUP) was organized, funded by a modest Ford Foundation grant. Its major published work was two sample textbooks for the first-year calculus and discrete mathematics, entitled *Universal Mathematics*, Parts I and II [44–45]. A Dartmouth group produced for CUP a version which was the precursor of the textbook by Kemeny–Snell–Thompson, *Finite Mathematics*, a pioneer for modern textbooks in this field [46]. In 1958 CUP was terminated, reorganized into special panels as CUPM, and funded by a substantial NSF grant as one of the several curriculum commissions in science that it supported [47–48]. The CUPM panels explored mathematics courses in support of the physical sciences and engineering, biological sciences, teacher training, computing, business applications, and pregraduate education for professional mathematicians. There was also a general curriculum in mathematics to support these special objectives [49]. One product of CUPM was an undergraduate textbook of contemporary engineering applications of mathematics [50]. The gathering of the material for this was a joint enterprise of CUPM and the Commission on Engineering Education; it could hardly have been done by private efforts.

CUP–CUPM never intruded much into the American public consciousness, and it was not to be expected that college mathematicians would accept the authority of any curriculum recommendations. Nevertheless, its influence was considerable, mainly from the concerted national effort towards curriculum reform that it generated.

Perhaps the best test of the success of CUP–CUPM is in the progress of early calculus and elimination of remedial mathematics in colleges. A study by the old CUP around 1953 showed that capable European students first study (not necessarily rigorous) calculus at about age 16. In the United States at that time few, even capable, students encountered it before age 19. By the mid-sixties great progress was being made to establish calculus, and/or finite mathematics, as the first college course, that is, at age 18. Then, around 1967, a breakdown began in high school preparation which ultimately forced the reintroduction of much remedial, precalculus mathematics in college. But even after that the amount of freshman calculus taught remained far above the 1953 level. And high school, or prep school, calculus is still (in 1988) on the increase, that is, calculus at age 17. A reasonable guess is that Americans who are going to study calculus at all now do so on the average at least a year earlier than in 1953. And many study it two years earlier. On the other hand, remedial mathematics in college is far from being eliminated. A large part of it is unproductive, a waste of students' time, of mathematicians' time, and of public funds for higher education.

SMSG. The School Mathematics Study Group (SMSG) was the official school mathematics reform commission formed by the mathematics division of NRC, and Mina Rees' old Policy Committee, in 1958. E. G. Begle was asked to serve as its director. From the start it had a substantial grant as one of the curriculum study commissions supported by the NSF. Its board of directors represented the leadership of the Society and other mathematics organizations. For independence it was deliberately set up in a leading private university, first Yale and then Stanford.

There were several others of these curriculum groups in school mathematics supported by the NSF, including one headed by the somewhat flamboyant Max Beberman at the University of Illinois. Some of these proclaimed "the new math," but SMSG never claimed such a designation. In fact SMSG stayed within the confines of the traditional curriculum, grade by grade. Its operating membership included leading teachers from the public and private secondary schools as well as professors from the mathematics departments of teachers' colleges. With its resources SMSG was able to generate a great deal of national participation in the form of speaking, writing, conferences, and experimental teaching. Its membership included consultants from the other sciences and engineering. Moreover, the NSF conducted semiannual coordination meetings in which the commissions in all subjects met together and reported on their activities. No such massive and well-coordinated national effort has ever been mounted, before or since. The principal published output of SMSG consisted of sample text materials which it wrote for each of the grades. Its histories include a report in 1968 by Begle entitled "SMSG: The First Decade" [51–54]. It was the last decade; SMSG ended in about 1970, along with most of these educational reform commissions. We will return to that part of its history later.

Curricula for younger children. Two other commissions of the time were important. John Mayor, working at AAAS, conducted a study of mathematics in the primary and middle school grades, where many "math blocks" and deficiences are established. He addressed the important issue of the certification requirements in mathematics for *primary* school teachers [55].

Paul Rosenbloom at Columbia directed a commission studying the mathematical learning of quite young children, including preschoolers [56]. Many working on reform in mathematics education believed that those who later appeared as "disadvantaged" acquired their difficulties at preschool age. Begle, in particular, believed that remediation at age 18 was unproductive and wasteful of effort and money. He thought that this country had the resources and could not afford not to make a massive effort to educate disadvantaged mothers and their babies [57]. These judgments are reinforced by the widespread success of the Head Start programs for young disadvantaged children.

Research Potential and Training in the Mathematical Sciences. In 1957 that was the title of the joint AMS–MAA–NRC "Albert Survey" [58], which studied the conditions, favorable and unfavorable, for research and teaching in mathematics. Although NSF was making grants for research in 1957, the Albert Survey did not make more money for research its principal issue. In fact one of its surprising findings was that heavy teaching loads, up to 15 hours per week and more, do not inhibit research production as much as expected. To be sure, the performance of young mathematicians in universities with light teaching loads was better. But those universities were usually able to select the best mathematicians for initial appointment. Perhaps the heavily loaded ones were better motivated!

A subcommittee of the Albert Survey considered the "small college effect," which is the observation that small colleges produce a disporportionate number of future research mathematicians. It took off from a classical NRC report by M. H. Trytten, entitled "Baccalaureate origins of the Science Doctorates" [59]. G. A. Hedlund, himself a product of such a college, wrote in the report: "There is one noticeable pattern which runs through the small colleges which have been highly productive of mathematicians. It is the concern for students as individuals and the opportunities open [to them] for growth. These can take many forms, but it seems clear that their effect is great" [60].

A 1988 retrospective, now that the small college has virtually disappeared, might call for an examination of college economics. In the College Blue Book, giving academic data on these colleges, one would find that the student/faculty ratio was almost uniformly 10/1 before World War II. Typically a student took five courses; so a teacher had a total student load averaging 50. A teacher taught these students in four small classes. The student got more individual attention than is possible today; and the teacher still had time to keep up a modest research program, without a grant, if he or she was so disposed. Many did, as the Albert Survey found. Tenure was usually informal, by custom rather than by formal action, and these colleges did not practice up-or-out at the tenure threshold. Even in the financially poorer colleges, where the student loads were higher, the student got much more teaching service than is considered possible today. These conditions undoubtedly had something to do with the productivity of the small colleges. But this kind of small college was fast disappearing when the Trytten report came out.

In the leading universities, with substantial endowment and grants income, the student/faculty ratio would be below 5/1, approaching 1/1. Only these institutions practiced up-or-out at the tenure threshold to control costs and maintain quality. Those dropped found ready jobs in secondary universities.

In the state universities with little endowment or grants income, where the students paid little or no tuition, the student/faculty ratio ranged from 12/1 to 15/1 and teaching loads of 15 hours per week were not uncommon. The same conditions obtained in the less well-endowed, tuition-sensitive, private universities. The Albert Survey showed that a mathematician could do some research without a grant under these conditions. And the students got reasonably good teaching service. These conditions had not eroded much under postwar conditions, where teachers were available, up to about 1954, because the GI Bill paid liberal, real tuition. The trouble in mathematics was that the teachers were not available after the local supply of adjunct teachers was exhausted. Therefore the mathematics student/faculty ratio, and class size, tended to be higher than the university norm, and higher than those of English, which had a similar flood of freshmen. Unfortunately such ratios tend to become fixtures in the mathematics budget. Moreover things grew rapidly worse after 1954.

Community Colleges. A new type of college arose during the expansion. States usually set up the new two-year community colleges outside the system of state universities, with local boards of their own; yet they were teaching two years of college work that was guaranteed credit on transfer to the university. Initially they seemed to be more an extension of high school, and in PEB's domain, since many of their teachers were former high school teachers, often upgraded by the NSF academic year institutes. Later they got more Ph.D.s in their faculties and seemed to move towards college status. Mathematics was heavily involved in their identity crisis. They can teach remedial mathematics better than anyone else. But later many of the community colleges tried strenuously to escape the remedial teaching function. That left the problem of remedial mathematics unsolved. Also it would turn out that they often teach calculus better than the state universities. At first they rejected vocational education because of its low prestige but later found that they have a natural mission to train skilled technicians for electronics, health care, banking, accounting, computer programming, etc. [91].

CUPM tried to set up a program on the community college mathematics teacher, but was flatly rebuffed. In 1988 the proper relationship of the universities to the community colleges with respect to mathematical education is still unresolved. The community colleges, where one year of precalculus mathematics appears to be normal, not remedial, seem to have a clearer idea about it than have the state universities.

Manpower shortages 1962–1970. The Kennedy administration's ambitions in space brought to a head a conviction that had been growing among leaders of American business and industry. It was that our future prosperity in the new high-tech world depended on highly trained scientists and technologists, and that an acute shortage of such manpower was developing. Something had to be done about it, or we were headed for national decline [61]. This had been the feeling for some time in the mathematics departments of the universities. The Albert Survey had referred to "the critical shortage of mathematicians and teachers." The issue reached the government as a report of the President's Science Advisory Committee, the "Gilliland Report," 1962 [62].

The report called for a massive expansion of Ph.D. production in the fields of engineering, mathematics, and physical science. It was especially critical of mathematics, which was deemed to be underproducing both in undergraduate majors and in Ph.D.s. It set a hoped-for 1970 goal for mathematics Ph.D.s of seven times the 1960 output of 303 Ph.D.s, or, more realistically, at least four times. It called for large government appropriations and for financial help from the private sector to meet these goals. In a preface to the report, President Kennedy appealed to the patriotic spirit of young Americans to continue their education to an advanced stage. He said in part: "...It is the students themselves who hold the key to this nation's strength. It is my earnest hope that each college student will consider how valuable additional study will be in enhancing his abilities and potential contribution to the nation, and in bringing him greater satisfaction and rewards."

The immediate response of the mathematicians was expressed in a 1963 conference conducted by the Conference Board of Mathematical Sciences (CBMS) [63]. We felt important but skeptical of those big numbers. We wanted to do what our President asked, but nobody knew where we were to find these Ph.D.s. The amazing thing is that the *maximum* goal of the President's Commission, 2200 by 1970, was almost reached by 1999 mathematical science Ph.D.s in 1970–71 [64]. It had been 303 in 1960.

One of the responses of NSF was to establish a program of support to "Centers of Excellence," that is, to universities that had not previously granted significant numbers of Ph.D.s. This further aggravated the real shortage of mathematicians who were qualified for doctoral research and advanced teaching. Salaries of mathematicians went sky high by previous standards. The numbers of service-free NSF fellowships was not increased in proportion to the new goals. Those graduate students were needed as teaching assistants. Often the teaching loads of these teaching assistants were quite high. Also classes became large lecture sections, previously considered not good practice in freshman courses. Finally, even the youngest of these American graduate students in 1963 were coming from the low birth rate years before 1941. That meant that we were reaching well below levels of ability previously considered minimal for graduate study, we were overloading them with teaching, and we were teaching them with professors who had not previously directed doctoral research.

Actually they turned out better than we had any right to expect. But there had been no planning for the type of training they would receive. Mathematicians were mathematicians. Of necessity these expansion Ph.D.s were educated in core mathematics, and not much else. That limited their usefulness and employability to the academic sphere. Even there they were too narrowly educated to be the best teachers.

The President's Commission in 1962 set off a great deal of monitoring of scientific manpower production besides that of the U.S. Office of Educational Statistics [66]. In mathematics the 1967 Young survey (John Jewett, Executive) did the most thorough reporting of quantity and quality in mathematical education and manpower production [67].

The mathematicians of this time have a good record for objectivity when the national mood screamed "shortage." The famous 1971 Allan Cartter re-port, just before the crash, praised the Young–Jewett survey for its "sober and objective analysis" [68]. Somewhat earlier CUPM had considered a possible special project on an alternative to the Ph.D. as qualification for college teaching. But it rejected the idea on the advice of its special committee's hard-headed report by Lowell Paige. And in 1969 W. L. Duren, speaking to the conference of academic deans on their problem of shortage of mathematics faculty, urged them to wait and not to start any more Ph.D. programs in mathematics. Mathematicians would soon be available. They did not believe him.

COSRIMS. In 1968 the Committee on the Support of Research in the Mathematical Sciences (COSRIMS), with Lipman Bers as chairman, issued its report [65]. It had been set up as a ten-year followup to the Albert Survey [58]. But it was more specific in its objectives, as the name indicates. Its survey of education and manpower was delegated to the Young committee [67]. And its own subcommittee on undergraduate mathematics issued a separate report. The COSRIMS report was the subject of sharp debates within the mathematical community. It is difficult to gauge its influence in seeking more public support for research in mathematics since it came out just when the mathematics boom was about to collapse. And much of the NSF money had to go for an expansion of the postdoctoral research program for unemployed Ph.D.s. One of COSRIMS's small, but lasting, contributions was the phrase, "core mathematics."

Alternative graduate education in mathematics. As the shortage of mathematicians built up in the fifties, both for education and for industry, grave doubts arose about the existence of enough young people to satisfy the demand for Ph.D.s. For they had to have both the mathematical talent and the motivation to meet the research requirement of the Ph.D. This feeling was strong years before President Kennedy's 1963 call for a huge expansion of Ph.D. production. An obvious solution was a doctoral level graduate degree for college teachers with some substitute for the research dissertation. Such degrees had been given before, both in Germany and the United States, at least as early as 1888, often called "Doctor of Science." For the proposed doctoral level degree for college teachers Don Wallace proposed the name, "Doctor of Arts." There was much emotional feeling both for and against the idea. The issue came to a head in an official conference at Yale on October 21, 1962, for which Edwin Moise had prepared the presentation [69]. Richard Brauer spoke for the opposition, saying that "it would water down the Ph.D." The opposition won and the D.A. was dead, for a time at least.

The 1961 rejection of the teachers' D.A. degree did not settle the issue. The D.A. that was rejected was too narrowly a vocational degree for college teachers, described mainly negatively as a relaxation of the research require-ment. It did not spell out the educational qualifications for either a fully educated college teacher or a teaching assistant. (These questions were ad-dressed in CUPM reports of 1967 and 1969 [49, p. 102, p. 113].) And the D.A. proposal did not admit that the education needed by a college teacher is very similar to the advanced education needed by mathematical practi-tioners in science, industry, and government. Combining these objectives could produce a graduate education to equip the student for college teach-ing, or practice as an applied mathematician, or later specialization in some research field of the mathematical sciences.

In 1969, with the prospect of an oversupply of Ph.D.s looming, suddenly the question was: Are there too many Ph.D.s? R. D. Anderson, in the first of many counts he made of supply and demand, concluded: Not yet, but.... W. L. Duren said in effect: Not if we maintain high standards and broaden their education to equip them for a wide range of jobs and usefulness in society [71]. Lipman Bers, in the COSRIMS report, held the traditional research degree line. Others proposed many variations on these themes.

And, if we may look ahead in this history, even after the Ph.D. glut hit in 1971, the discussion of reforming the content of national graduate education has continued with a variety of motivations [98]. A panel of CUPM, chaired by Alan Tucker, developed the idea some years later as an undergraduate major program called: A General Mathematical Sciences Program [70]. But that idea never got very far because such curricula for generalists are too difficult at the undergraduate level. It might have had more chance as a professional Master's degree competing with the M.B.A. in business, or the Master's in computer science. The surplus of the Ph.D.s continues to block the acceptance of such reforms.

It is well to look at how the problem has been resolved in other academic fields. In physics, chemistry, and biology, all sciences with distinct theoreti-cal, experimental, and developmental emphases, one Ph.D. suffices, although it represents for individuals quite different kinds of study and research, or development. In medicine the practitioner's degree is a doctorate, the M.D., and the academic research degree, the Ph.D., is an extension of it. A similar situation holds in law. These make research education very long and costly to finance. But in all these fields, as well as in engineering, the number of graduate practitioner's degrees conferred is always several times as many as for academic research and graduate professors. In mathematics the insistence that the only nominally three-year graduate degree education shall be the one represented by the traditional, academic research Ph.D. severely restricts not only the number of mathematicians but also the usefulness of mathematics in American society.

Suddenly from shortage to glut. The end of the mathematics boom in this country came with great suddenness in 1971. It had come a little earlier in physics. The trigger was the Allan Cartter report [68], based on demographic data which showed that the war baby birth rate had already topped out, that the previous linear projections for the future demand for Ph.D.s as science and mathematics teachers were excessive, and that in 1981 college enrollments would begin to decline. The argument applied in the humanities just as well, since they had experienced the same growth boom that mathematics had, except that they had not been artificially inflated to three times their stable equilibrium share of college enrollments. Even so, why so sudden? It was still ten years to 1981.

But it was sudden. The 1999 Ph.D.s in 1971 found few jobs of any kind, certainly not the university positions they had anticipated. Many found no academic jobs at all. In desperation some took to driving taxicabs. Trying to help them, the AMS formed the Committee on Employment and Educational Policy. CEEP studied the evidence and could only say that it would not go away; it was not a short-term recession that would be followed by a renewal of growth. The NSF closed down the Center of Excellence grants and all the curriculum study commissions. Apparently it did not ever decrease its budget in support of basic research. But more of it had to go as postdoctorals for unemployed Ph.D.s, leaving assistant professors unsupported. Not only did the universities stop the employment of new Ph.D.s; they terminated assistant professors at the tenure threshold where they could legally do so. The terminated ones could not find other academic positions, as such rejects had done in the past. They were 35 years old, many had families, and now they had to start over in a new job where their mathematical education might be inappropriate. University administrations also tried to accelerate retirements of senior professors. If they could terminate one of these, it would save a larger salary. The new buzzword became "tenured in," something to be avoided at all costs.

But why was the collapse so sudden and so severe? The Cartter report had predicted that maximum enrollment would be reached in ten years. The main reason was the difference in the definition of "demand" between President Kennedy's commission and the Cartter report. In 1962 "demand" was conceived as a national need, not associated with visible salaried jobs. In 1971 Cartter defined it narrowly as predicted jobs in college teaching. And the prediction was based on an assumed student/faculty ratio of 25/1 instead of the prewar standards of from 10/1 up to 15/1. It also left out industry and government employment because that could not be quantified. The difference was enormous and the result cataclysmic.

Another factor was the vanishing of marginal income. The faculty expansion had been financed largely by the marginal tuition revenue from enrollments, increasing each year. When the Cartter report proved that enrollments would level off, this marginal income did not level off; it vanished suddenly.

There were other contributing reasons. When the war babies got to college they did not share the parental generation's ideals of excellence through competition, in which they were now the competitors. They turned activist against the Vietnam War, the draft, nuclear arms and energy, and against the science and technology that implemented war. That included mathematics. In the universities they found friends in the teachers of the humanities, who now perceived a chance to regain some of their lost importance. The informal coalition was joined by PEB, which wanted and got the curtailment of science education grants by the NSF in favor of educational grant programs out of the U.S. Office of Education. These events released the pent-up resentment of millions who could not compete, or did not choose to compete, in this postwar technological world.

5. Retrenchment, 1970–1988?

Retrenchment in the university administrations. Before we look at retrenchment in mathematics departments we should look at what university administrations did to counteract the effects of the sudden termination of money for expansion. For one almost immediate result was that mathematics departments were forced back into dependence on their university budgets more than when the department and members had grants from an outside agency. The research grants were still there but the educational grants for expansion of graduate production in mathematics were not.

As we have seen, university presidents' first concern was to cut back on salary commitments to tenured faculty. There was a lot of false reasoning about the "tenuring in" issue. Mathematicians were able to bring some rationality to the problem of how to get the unbalanced faculty flow back into a steady state. J. P. LaSalle [72] and John Kemeny [73] produced Markov chain models of faculty flow that showed what management strategies would get the faculty back into the desirable steady state and with the least hardship.

After that, in a rational world, one would expect the presidents to close down costly and unproductive programs of graduate work and research in science and mathematics, quietly, with no public announcement. Generally speaking that is not what happened. Instead, presidents of universities with new marginal graduate programs clung to them. Having lost their Center of Excellence grants, they had to finance their graduate programs out of general budgets for undergraduate teaching, plus research grants that their professors could bring in. They increased pressure to publish and get a grant. Publication was made a necessary condition for tenure in universities that had never required it before. The teachers to be cut, or have their salaries and rank frozen, were the older professionals who had taken positions as undergraduate teachers before the new graduate program was installed. Their teaching was replaced by "graduate teaching assistants" who were usually just low-paid, locally available adjunct teachers. Moreover, these conditions obtained even in colleges that still did not have graduate programs. Many of them are still pressing to establish one.

Many state universities now moved toward open admission policies in order to maximize enrollments and therefore the subsidy paid by the state in lieu of tuition. They could make a "profit" by admitting as many first-year students as possible and teaching them cheaply in large sections with low-paid adjunct teachers or graduate teaching assistants. They could do this at cost-per-student-semester-hour below the state subsidy. And the profit could be used to subsidize expensive graduate programs. This policy implied a great deal of remedial instruction as the quality of performance of high school graduates deteriorated. Much of the burden of this remedial teaching fell on the mathematics and English departments. The profit from this remedial instruction is, of course, a fictitious one created by the practice of unit cost accounting. In fact, such remedial instruction adds to the total cost which has to be paid, and diverts educational funds to unproductive efforts, actually leaving less for advanced teaching and research. The state councils of higher education are beginning to find this out. For example, a recent study in Virginia makes public the $30 million that remedial instruction is costing each year in the state, even with a lenient definition of "remedial." So mathematics departments should be warned that this way of earning money probably will not last [74].

Privately endowed universities and some well-financed state universities pursued the opposite strategy of maintaining admission standards. To them, remedial instruction is obviously pure cost. This proved to be a winning strategy in the competition between universities for good students.

Retrenchment in the mathematics departments. Math departments had more than their share of young people of professorial rank when the PhD. glut struck. Those who did not have tenure were cut mercilessly, even many promising and productive ones, because the department already had more than was deemed its share of tenured positions held over from the days of the math boom. This applied even in the top departments. In all but the top ranking departments graduate students in core mathematics became very scarce. To keep their research seminars active the faculty talked to each other about their research, or brought in visiting mathematicians. Academic year relief from teaching duties was no longer provided by NSF grants. But free time for research is important to a mathematician, so mathematicians generally chose to do their teaching in a fewer number of large lecture sections, often with quiz sections attached that teaching assistants handled. The system is legitimate, if well managed, but is subject to abuse. In some state university math departments the student/faculty ratio rose far above the 25/1 that was the basis of the Cartter report calculations. This is poor teaching service, especially if the professor is devoting much time to research.

Mathematics departments feel that this system was thrust upon them by the events following the 1971 crash. The relief they feel they need is more grant money to support core mathematics research. But even if the money were available, there is a question whether this solution would improve the overall service of mathematics in American society. It is one of the tough problems remaining in the wake of the 1971 crash.

PEB resumes control. The recognition of the cresting of the war baby population took the pressure off the demand for high school teachers. It did not produce a glut like the Ph.D. glut because the women had not returned to the teachers' colleges. In fact, enrollments in these colleges would continue to decline, forcing many of them to convert to liberal arts colleges with only a department of education. But PEB was able to supply the reduced demand. PEB got $887 million of National Defense Education Act funds, formerly assigned to the NSF, transferred to the U.S. Office of Education control [75]. The NSF curriculum study commissions, including SMSG and CUPM, were terminated. The activity of mathematics departments in curriculum and teacher training came to an abrupt halt. PEB's new curriculum plan to replace "Education for Life Adjustment" still called for 20 percent college bound, 20 percent vocational, and 60 percent in a terminal liberal curriculum that was to be more like a free elective system. A number of new subjects with contemporary flavor were to be added to the old ones. To restrain the chaos of a free elective system in high school, a new position, guidance counselor, was established in each high school. The guidance counselors received some additional training and higher pay than the teachers. That was where the NDEA funds went.

If anyone proposes more required courses, PEB's answer is to challenge the speaker for daring to dictate what free American students should study. Yet this is exactly what the guidance counselors often do. This happened again recently when Secretary of Education, William Bennett, proposed a curriculum for James Madison High School in which every student would take a course of study that included the old academic subjects covering the 15 Carnegie units for admission to college [76]. It included three years of mathematics. His argument was that this was the best education our civilization has been able to offer. The immediate challenge from the president of the American Federation of Teachers reasserted the old PEB estimate that such a curriculum is for "15 to 25 percent" [77]. The actual experience as of 1985 is that by ages 20–24 only 16 percent have less than 12 years of school, 46 percent finished high school but did not attend college, 41 percent had some college, and 23 percent had four or more years of college [78].

These figures imply that thousands of students are entering college without the standard college preparatory course, but with the terminal PEB education instead. Thousands more are entering college, nominally from the college preparatory program, but with little mathematics or science, if they have expressed a preference for the humanities to their guidance counselors. On the other hand, those students that choose careers in science or engineering may come to college with extra credits in mathematics, including high school calculus. There are many more of these students from private preparatory schools. The total is still relatively small but they make a large fraction of those entering the leading schools of science and engineering. And this system is sending mathematical illiterates to top liberal arts schools and even to the science and engineering programs of the many state universities. The situation is similar to the disorder which existed in college education after the introduction of Eliot's free elective system around 1900 and had to be corrected by stronger curricular direction. This is the basis for Secretary Bennett's call for a return to a secondary curriculum in which about two-thirds is a required common core of traditionally proven value, including three units of mathematics.

Did SMSG fail? A minor skirmish in PEB's campaign to recapture control of primary and secondary education involved the termination and discrediting of SMSG and the new math. The older mathematics teachers, trained in education schools, were unable to adjust to the new text materials and protested in mounting furor. Other critics, including Morris Kline among the mathematicians, furnished support to the claim that the SMSG curriculum had failed. Parents who could not help their children with their homework joined the clamor. This made PEB's takeover easier, but the main factors that ended SMSG were PEB's national political moves that terminated all of the NSF curriculum study and teachers' institute programs, together with the demographic changes that enabled it to resume supplying the teachers from its own schools of education, temporarily at least. But members of the mathematical community have apparently taken the criticisms of SMSG as the reason for its downfall, and still feel that its errors must be corrected to restore the good name of mathematics.

Actually SMSG did not fail. It did not have a life long enough for a definitive test of its first-edition text materials, whose obvious flaws included an excessive enthusiasm for logical language. With minor, necessary, exceptions, SMSG stayed in the traditional curriculum. Its advisers were leading university mathematicians who ensured that the mathematical content of its program was sound. Its great success lay in the massive national reform effort that it mounted, in the lively corps of teachers it and the institutes produced, and the enthusiasm for mathematics that they generated in their students. After the PEB takeover the old math teachers made life very uncomfortable for the SMSG–NSF trained teachers. Many of them left to get better-paying jobs in business where their computer training was an asset. Others moved up to the faculties of community colleges. But others stayed, and many of them are teaching high school calculus now.

If there are any lingering doubts about SMSG as a cause of the deterioration in arithmetic skills among high school students, one need only look at what happened after PEB restored the old math. Then things really got worse! To be fair, the real cause of the breakdown in school mathematics performance can hardly be attributed to the math curriculum, old or new. PEB has more serious problems with failure of discipline, drugs, crime, gang warfare, teenage pregnancy, and general chaos.

PEB loses control again. The schools of education are still dissolving, or converting to liberal arts colleges. PEB no longer has its monopoly on teacher training. The teachers are taking control of the National Education Association and turning it into a union [92].

The American public is aroused by evidence that American workers, educated in our schools, are inferior to their foreign competition, are causing the country to lose its competitive strength, and to lose jobs. Quality education will be an issue in the next presidential election. The only reason that this has not come to a head earlier is that the private schools educate the children of business, political, and industrial leaders. At the moment the focus is on the teachers. Businessmen think we can afford to pay good teachers more if we just identify the good teachers. Actually, the teachers are the main strength of the system. Something more fundamental is wrong.

The history of American education indicates that the decision of 1893 has become a mistake and needs to be changed. Mathematics is especially involved. Recall that in 1893 the decision was made, setting up the American high school under local school board control with no quality control either by the universities or the federal government. That worked well enough when only a small fraction of public high school graduates went to college; and most preparation for college was done by private academies, whose quality could be monitored by the College Entrance Examination system. But now that about half of public high school graduates go on to college, and state universities are expected to accept them, that arrangement is working badly. A large fraction of public funds for higher education is going into unproductive remedial teaching. And these remedial students are being poorly served by universities using them to support unproductive graduate programs with their fictitious tuition profits. It might be tolerable if the education of the noncollege-bound students was successful, but their dropout rate and the illiteracy rate among those students denies that.

The time has come when we must change the system to provide a much stronger direction of studies, and much stronger quality control of college preparation, than PEB and the politically vulnerable local school boards can manage by themselves. As recounted above, we mathematicians tried unsuccessfully to get that stronger direction of study, through graduation requirements, discussed in the report of the J. W. A. Young committee of 1923, and in several lesser efforts. We tried, with only temporary and limited success, to do it with the weak instrument of curriculum recommendations by CUP–CUPM and by SMSG. Now the time appears to have come for us to go back into the national political arena, urging the establishment of a general studies curriculum, such as the one that Secretary Bennett proposed for Madison High. That includes sufficient mathematics so that high school graduates can begin college mathematics with calculus, or discrete mathematics with probability. It will serve the nation and the students better, whether they go on to college or not. We ourselves should speak for mathematics, not leaving it to the engineers and physicists to speak for us.

American mathematicians and political issues. The course of events since World War II has forced American mathematicians to modify their long-standing, somewhat apolitical, attitudes. After all, we say, the business of the American Mathematical Society is mathematics: the promotion of research, publication, and teaching in mathematics. Our majority attitude had been: If mathematicians want to play politics, that is their right; but let them do it in the open public domain, not in the mathematical organizations. Still, as a group, we mathematicians have a characteristic political makeup that differs from that of the average educated American. On the whole mathematicians are more ignorant and unsophisticated about politics and its arts. The traditional issues of academic freedom have not touched mathematicians very much. Not since Galileo has mathematics been regarded as a subversive doctrine, although there were times when the new math appeared to reach that status.

We tend to be conservative in the sense that we treasure the right of the individual to do his thing. Many of us are military-oriented because of the long involvement of mathematics in military science. But we also tend to be internationalists since mathematics is an international culture, independent of language and politics. This characteristic has sometimes got us into trouble, or at least made us suspect as security risks.

The trouble started during the Eisenhower Administration (1951–1959), when some mathematicians were brought before the McCarthy or Velde (House Un-American Activities) committees, charged with being Communist "fellow travelers" at least. Others were fired from their tenured, civilian, positions in state universities because they refused to sign a loyalty oath as a condition for employment. Thus the issues of academic freedom came to mathematics.

AMS and MAA formed some joint committees to investigate and defend those fired. In at least some cases the real charge turned out to be homosexuality, an unmentionable in those days. The investigating committee was supposed to understand why, when the facts were revealed confidentially, the charge had to be stated as communist affiliation. But these ad hoc committees could do nothing effective to reverse any such dismissals. So AMS–MAA formed the Committee to Prevent the Loss to Mathematics of those Dismissed for Political Reasons. This committee heard the victims' stories and tried to find jobs for those dismissed. It did not attempt legal defenses. Very few learned societies attempted to give financial support to litigation in these cases.

During this same period evidence came out about the persecution of particular mathematicians or scientists in the Soviet Union and in other totalitarian countries. Resolutions in their defense were presented in AMS business meetings. Members hotly debated these resolutions, not on the merits, but on the propriety of AMS involvement. Several such resolutions passed [93].

Then came the 1971 crash, followed by sharp reductions in tenured faculty commitments. This could be accomplished in several ways: by denying tenure to assistant professors, by forced early retirement, by breaking tenure on grounds of hardship, or by freezing rank and salary for selected individuals. The affected mathematicians often naively believed that academic freedom gave them a constitutional right like the First Amendment freedom of the press. Actually the courts had never given academic freedom the status of constitutional law. But, in the era of litigation that followed, it turned out that the AAUP Statement of Principles of Academic Freedom and Tenure [79] did have a form of legal status. The enforcement of federal laws, such as those governing breach of contract, fair employment practices, or race or sex discrimination in the workplace, all depend on an accepted definition of fair employment practices in each industry. For the academic "industry" the only generally accepted such statement was the AAUP Statement of Principles. Both AMS and MAA, as well as most other learned societies, colleges, and universities had subscribed to it. Thus the AAUP rules of tenure became an essential instrument in the enforcement of federal law in these cases, even if they themselves did not have the status of law. This launched a second stage in the use of the rules of tenure: the protection of job security in the courts, independent of traditional issues of academic freedom.

Once the rash of litigation after 1970 had cleared, and college administrations had legally established the commitment to the Statement of Principles, a third stage in the use of the tenure principles evolved. Colleges, as well as graduate universities, denied tenure, or advancement, to those who did not publish research. Thus the rules of tenure became an instrument of job insecurity, as we have seen above. (We will return below to the special political issues involving women as mathematicians.)

The emergence of NSF support of research, with its issues of distribution of finite funds, also introduced hot political issues into mathematical society. The political issues may involve funds for mathematics as opposed to physics, funds for core mathematics research as opposed to applied research, basic research as opposed to mission-oriented research, "regional development" as opposed to highest quality, or education as long-range research support as opposed to a particular current research project. NSF itself steers assiduously clear of such issues, but they are unavoidable in the advisory panels. The American mathematician's old-time innocence of politics appears to be lost forever. And now we must overcome our timidity and get into the battle over the national high school curriculum and its mathematical content.

How to organize a national effort in mathematical education or support of research. The social history of mathematics in the years 1945–1970 records efforts of unprecedented scale to improve our conditions and our national service. There are some lessons to be learned from that history about the alternative ways to proceed in such ventures. A brief listing of them with their advantages and disadvantages may be useful.

In the dual, private–public system of American education, the private sector offers some advantages of prestige and freedom from leveling political forces. So when SMSG was set up, it was located in a leading private university, first Yale and then Stanford. The disadvantage was that no private university has the financial or manpower resources to conduct such a big project. All of that had to be brought in and financed by a foundation grant. When the grant was terminated the project was dead.

CUP–CUPM chose to locate in the association as a standing committee. This has the disadvantage that professional associations are not unbiased; they represent the interests of their membership. Consequently foundations do not like to make grants to projects so located. CUPM was able to overcome this, and get an NSF grant, only by assuming a dual status as an NSF commission. But it had the advantage of the mathematical manpower resources of MAA, the audience at meetings, and the publication facilities. While the NSF grant was terminated, CUPM has continued to live as a standing committee of MAA, though its activities are limited by financial constraints.

Another possible location is in the mathematics division of the National Research Council. This has some unique advantages since NAS–NRC is the official interface between the institution of science–technology and the goverment. It enjoys the prestige of NAS and unquestioned objectivity in matters of science. Since the NSF was established, NRC has traditionally conducted the screens of fellowship applicants. Moreover it has permanent government financing of its own, relatively small, operations. It is the ideal place to initiate a public project that seeks NSF financing. For example, SMSG was initiated in NRC. Its disadvantage as a place to operate is that it is small, its appointed membership visits only at brief meetings, and it is more oriented to research than to education. It loses some authority in the field of education since it represents only science education. At the present time the mathematics division of NRC houses both our research grant advisory function and the Mathematical Sciences Education Board.

By analogy, and by structure, the proper interface with the government for mathematical education should be the American Council on Education. But ACE has developed in a different way. So far it has not served as a base for efforts in education involving mathematics.

Both the physicists and the chemists have established Washington institutes that represent the combined membership of their various professional associations. Such an institute can perform the function of a lobby, or represent the profession other ways, especially in dealing with the government. This representation can be frankly in the professional interest of the scientists, without the tradition of objectivity that NRC has. We mathematicians tried to establish our own such institute called The Conference Board of Mathematical Sciences (CBMS). Baley Price was its founder. It served well during its existence, but not all of the mathematical science organizations were willing to participate. AMS and MAA, combined, find themselves too small to afford it. CBMS was able to exist for a time on grants, but when these gave out CBMS folded. We now have only Kenneth M. Hoffman as our part-time representative in government matters. He has proved to be the best communicator we have had in this status.

The NSF itself has mounted educational efforts in mathematics, as well as other sciences. These have included the summer and academic year institutes for teachers. And the fellowship programs have to be included. But, unlike the research grants, the NSF has not invited individual proposals in education; it has pursued its own missions in education, advised as always by NRC, or ordered by Congress. Also, when it had funds for such projects, it has supported large projects proposed by mathematical organizations. These included several major CBMS projects.

It may come as a surprise to mathematicians, but the NSF is not the only foundation making grants. In particular there are private foundations; and projects funded by them enjoy some of the prestige and freedom from political influences that private universities have. It is true that, when the NSF was formed to support basic science, there was an informal understanding that the private foundations would support the humanities. But mathematics is in large part a humanistic culture. In fact the first grant of $75,000 to support the old CUP in 1953 came from the humanities program of the Ford Foundation. The Foundation would have granted much more but, in the quaint innocence of 1953, the Executive Committee of MAA thought that such a large grant might unduly influence mathematics. More important is the Sloan Foundation fellowship program, selecting promising young mathematicians for support without specific project competition. And in the specific areas of applied mathematics, the military agencies and others too numerous to mention, have been, and continue to be, active in research support.

Private corporations have also helped. For example, around 1950, General Electric printed and distributed millions of copies of a pamphlet by their W. E. Boring called: "Why Study Math?" More recently American Telephone's Henry Pollak has spent a lot of company time as a sort of roving ambassador to mathematics. These voices offer a different, and often more influential message than that of academic mathematicians. IBM has also been a friend of mathematics, as have the other computer companies. One gets the impression that corporations would have done more to advance study and research in mathematics if we could have found the right ways for them to do it.

The College Entrance Examination Board [7] is a nonprofit association with many colleges and universities, both private and public, as members. Its data provide the best objective assessment of the quality of secondary school performance. This capability will be even more important in the future if we are to reform secondary education to eliminate wasteful, demoralizing, remedial mathematics teaching in college. Beyond its testing program, CEEB set up a commission on Mathematics (1955–1958) with A. W. Tucker as chairman, to determine what they should be testing for in the mathematics achievement test. Its report had lasting influence on the content of mathematics in good secondary schools.

We have excellent sources of statistical data on education in this country. The Division of Educational Statistics of the U.S. Department of Education has a fine tradition in this field. Its C. B. Lindquist was especially active in helping the mathematicians. The Research Division of the National Education Association (NEA) has provided the data on teacher supply and demand. And the Department of Labor regularly studies the supply of trained labor and the jobs for them.

The unofficial Professional Education Bureaucracy (PEB), the administrative divisions of the U.S. Department of Education, and the central power structure of NEA have been like enemy territory to the mathematicians. SMSG and the NSF institutes' programs were able to enjoy friendly cooperation with NEA's math subsidiary, the National Council of Teachers of Mathematics (NCTM). But this relationship deteriorated after PEB resumed full control of public school education and forced the termination of SMSG, as well as the NSF institutes, in 1970. In the large grants program for research in education that the U.S. Office of Education set up after that, there were projects funded in mathematical education, but apparently not many to members of a university mathematics department. The recent political activism of the teachers, which got cabinet status for Education and turned NEA into a labor union, is a good omen for the future. The teachers can be counted on to be stronger for educational quality than their bosses in PEB were. If the teachers really control NEA, we mathematicians can work with them, as we did with NEA's subsidiary, NCTM, during the days of SMSG.

Recent realignment of our relations to American society. Although CBMS continues to exist as a council of presidents of the mathematical sciences organizations, Kenneth M. Hoffman is now our designated spokesman to the federal government. Our public representation, both for research policy and education, is now concentrated in NRC [80]. These arrangements are designed to develop our part of the recommendations of the 1984 David Report viewing mathematics as a "Critical Resource for the Future" [83]. It is too soon to write any history, but to this historian it looks good.

6. Some Unfinished Stories

Women in graduate mathematics. It is obvious that mathematics does not come in male and female genders; but women as mathematicians have always been a special category in the American mathematical community. Their status has changed back and forth in response to conditions in American society. Their complete social history has not been recorded; and a part of it is still conjectural. The following might be a reasonable, if oversimplified, scenario.

In 1888 there were not enough graduate students for all the new American graduate programs in mathematics. So women were most welcome, then as they are today. On the other hand the demand for the graduates in paying and satisfying jobs was much less urgent. They could teach in the elementary and, soon to be developed, high schools or in the female colleges.

Then, after the 1920 women's suffrage amendment to the Constitution was passed, there was a great surge of professionalism among American women in which many became advanced mathematicians. But by the end of the decade the difficulty of getting university positions in competition with the men became discouraging. The Great Depression clinched it. Meanwhile PEB had established a virtually closed shop control of teaching jobs in the schools. These provided more security than university positions and were better suited to place-bound women. So most women mathematicians settled for education degrees and school teaching. That was the state of affairs until World War II when women found better paying and more satisfactory jobs in war work. And after the war they were busy raising families.

The production of the war-baby generation was about complete in 1963 when President Kennedy called for a massive, subsidized expansion in advanced engineering, mathematics, and physical sciences,. Mathematically talented women eagerly responded to his call. These women, returning to graduate mathematics, were a major reason for the unexpectedly high quality of these expansion Ph.D.s. But they had come back, only to be caught in the retrenchment after 1970. After 1970 many mathematical women switched to computers, to M.B.A. degrees, and to professional positions in business. Others found jobs in community colleges. They still did not return to schools of education. A surprising number, however, stayed in graduate mathematics. In university faculty positions the ratio of women to men is still increasing. University regulations now more freely permit husband and wife to be members of the same faculty.

All this did not happen without organized struggle. In this phase the name of Mary Gray has been prominent.

In this changing picture there is one invariant principle. It may be stated in terms of the, still valid, concepts of 1963, viewing mathematical brain power as a precious national resource. If half of the mathematically talented minds are female, then, for the national welfare, half of the fully educated and fully utilized mathematical minds should also be female. That has never become a reality in this country.

Foreign graduate students. Besides the women, the other unanticipated source of mathematical talent that made the crops of expansion Ph.D.s after 1963 better than we had any right to expect came from abroad. Their numbers have been increasing year by year, relative to native-born Americans, until in 1987 more than half of American Ph.D. degrees in mathematics were awarded to foreign students [81]. This happened in spite of the exclusion of foreigners from most fellowships, and in spite of the language difficulties of supporting them as teaching assistants. They do not come for the humanities. They have their own humanities. They want engineering, science, mathematics, and computer science. There are differing views as to whether this foreign invasion is good or bad.

Mathematics departments want them. Outside the very top departments, there are not enough good graduate students to keep the graduate programs functioning in a healthy manner. These students are not only selected for ability from a world pool of mathematical talent (excluding only the Soviet Bloc countries), they also tend to be better trained in certain areas such as hard analysis and mechanics. This may make them better than Americans in applied mathematics. There are many more in engineering than mathematics, and their registration in graduate courses helps to maintain a full program of advanced mathematics. Finally, we want these students because of our feeling that we and they belong to the world community of mathematics.

Managers of medical research laboratories, industrial, and government laboratories want these graduates and want them to stay in this country. We could hardly operate medical research or hospitals, or our research and development establishments, without them.

On the other hand a congressman, a taxpayer, or an industrialist may regard this as a shocking giveaway of costly American science and technology at taxpayer expense. For even if they pay full tuition, these students pay only a fraction of the high cost of their graduate education. Then they may go home to set up high-tech industry in competition with the American companies and their workers, whose know-how and financial resources created the technology they studied.

Still another view of this phenomenon has been expressed by J. J. Servan-Schreiber, former French cabinet member and writer on international economics [82]. His view is that America must remain the world's graduate university for the sake of both U.S. and world economic, technological, and intellectual development.

Clearly American mathematics has a stake in this socioeconomic issue, however it may ultimately be resolved.

Another way to find support for basic research in mathematics. There are now many able, well-trained, young mathematicians whose research proposals cannot be supported by the NSF for lack of funds. It is obvious to mathematicians that the relief needed is more money to support proposals in basic research. What could be wrong with an appeal for more grant money to help us complete our personal mathematical work, to relieve us of teaching, and to increase our annual income?

In fact there are cogent public policy arguments against providing more grant money to relieve mathematicians of teaching while the student/faculty ratio remains as high as it is now. Since a grant does not pay academic year salaries, it does not directly fund more research. Indirectly, its payment to the university can provide a teaching assistant to relieve the researcher of some academic year teaching. But the student/faculty ratio is already too high; and the students are getting only thinly diluted teaching service from professional mathematicians. This situation could only be made worse by using more teaching assistants. The grant cannot provide what is really needed—another tenured faculty position. In short, as long as the professional mathematics faculty remains too small, more research can only be bought at the cost of poorer teaching service.

History suggests another way to support basic research in mathematics as a corollary of a better educational policy. It is an accompaniment, not an alternative, to grants in support of research proposals. This is simply to reduce the student/faculty ratio back to the traditional prewar levels ranging from a standard 10/1 up to 15/1 in teaching colleges and universities.

As the Albert survey of 1957 showed, mathematicians could do creditable research while teaching 12 to 15 hours a week [58]. But the total number of students they taught was low, based on prewar student/faculty ratios. During the postwar period of explosive growth in enrollments, faculty shortages, and price inflation, academic economics could not sustain the old quality ratios. Postwar staffing called for student/faculty ratios such as Allan Cartter's norm of 25/1, and up from there. But now that enrollment is stabilized, and faculty manpower is in abundant supply, state universities should return to traditional quality standards in college education. Other universities and colleges will follow them.

Graduate professors cannot teach 12 hours. "Graduate professor" does not mean one who teaches a "graduate course", but one whose principal occupations are doing and leading research and writing, directing doctoral dissertations, editing and refereeing research institutes. For such professors a teaching load of three to six hours is appropriate, as it was before World War II when the basic college teacher's norm was 12. A graduate professor's position is commonly called a "chair," whether its reduced teaching is accounted for by endowment, grants, or other budgetary provision.

A first-class graduate department needs several such chairs; and its student/ faculty ratio is below 10/1. It is therefore very expensive. They cannot be proliferated as political patronage in every town, like community colleges. A large state university may be able to combine a good graduate and undergraduate department, but not without sacrifice of quality in one, or both, components when the overall student/faculty ratio is 25/1 or higher. Much harm has been done to U.S. higher education, both in quality and runaway cost escalation, by the continuing of too many pseudograduate departments after the onset of the 1971 Ph.D. glut and the subsequent proper discontinuation of the federal Center of Excellence grants. These pseudograduate departments are propelled by local pride and financed by abuses of low-level service and remedial teaching.

The main burden of undergraduate teaching should be carried by well-qualified professors, instructors, or legitimate graduate teaching assistants, whose primary duty is teaching. For them the full-time basis of 12 hours is appropriate. And, for fairness, their expected performance in research and publication should be less that for chaired professors, but still greater than zero. The rules for these discriminations have to be worked out locally, within departmental or university government.

The cost to states to lower student/faculty ratios will be high, and it will take time to implement. But there is no reason why postwar society cannot sustain prewar standards of education for its youth. Mathematicians would do well to join in public pressure for it. In doing so we would find that we have many friends, including English professors and their humanities colleagues, as well as many in the general public, especially women. By contrast, in an appeal for more research grant money for mathematics, we are isolated, and few in number.

Although research support would be only a secondary objective in going to a larger faculty, let us consider the ways in which it could help mathematical research. First, it would make more jobs for fully trained mathematicians. More mathematicians could do more mathematics. And we have many good mathematicians, women especially, who merit university positions. Admittedly the prewar faculty policy was less efficient as stimulator of research than NSF-style support of reviewed proposals. But it was a policy that left the mathematician free to work on what he chose. Moreover, contemporary pressure to publish as a condition for tenure would improve its efficiency. And that pressure would be more fair than it now is, as applied to an overloaded faculty, where grant-getters have all the advantage. These are reasons, based on history, why mathematicians should now give a higher priority to enlarging the faculty than to increasing research support for the existing overloaded faculty.

The appreciation and status of mathematics. We have worried for a hundred years that we and our subject are not understood and appreciated by the American public. Only recently Murray Gerstenhaber has expresssed our plight eloquently and philosophically [89]. He is only the latest to do so. Over the years the diagnosed causes have been manifold: not enough money for basic research, poor public relations in the media, not enough applied mathematics, and so on. But the lesson of history points to one basic cause. As E. B. Wilson put it in 1913, "We do not make good" [for the average American out there].

Consider the many millions of Americans whose collective viewpoints comprise our public image. Their view of mathematics is primarily determined by their experiences in their last, most advanced, course. For the vast majority of them this was a terminal course, some version of high school or college algebra, taught as a service course to prepare them for something else [90]. The most turned off of them must be the ones who got to state universities and were herded into large sections taught cheaply to earn budget credit for the support of the department's research program. Consider, even, the smaller, elite fraction whose terminal mathematics was a huge, monolithic, three-semester calculus course loaded down with too many good things. It was also primarily a service course. Will one of these former students understand or appreciate mathematics?

It is not that we have not tried. As we have seen in this history, we have tried, and are continuing to try, to solve this problem. Some departments, and some fine teachers, have succeeded. But the numbers of their fortunate students are small in comparison with the multitudes that were batch processed. We have not solved the problem in the aggregate; and that is our public image problem. We can blame PEB and John Dewey for a lot of it. We can blame our university administrations for some of it. But we ourselves are still to blame for much of it. We could have learned something from John Dewey about the motivation of young students and not depended as much on compulsion as we still do. But until we have educated a whole generation of Americans for whom mathematics makes good in their own experience, we will not have solved our "image problem."

The public understanding and appreciation of mathematics is related to our status in society, but it is not the same. Status is more subjective. Is it the same as prestige? The history has less to say about status. Surely it is not to be expected that improved teaching service will solve our status problem, though it might help. Certainly our status in American society took a great leap from limbo in the years after World War II. Indeed, as Abraham Flexner found out earlier, old mathematics had never lost it in high intellectual circles. Lately, after 1971, we perceive ourselves to have lost some of our postwar gains in status.

If we insist on regarding our status as a problem to be solved, then history tells us that we have an identity problem which should be solved first. We are hanging between the humanities on one hand, and science and technology on the other. If we hitch our wagon to the technology star, then we are forever a service subsidiary. But if we claim our rightful place in the ancient humanities, we are put down as materialists, and excluded from the club by such Johnny-come-latelies as the English and history professors and the snobbish literati outside of academia, bent on enhancing their own precarious status. History might sigh and say that there is not much we can do about it.

AMS and mathematical education throughout life. We come last to the newest, the most exciting, and explosive developments of the century in mathematical education, and the service of mathematics to American society. These are in the field of education continuing throughout life. They are beyond, or outside, the framework of formal education with its curricula, courses, lectures, credits, degrees, and its faculty.

The idea of education continuing throughout life is very old. But educators in the academic institutions have treated it more with lip service than with imagination. They have relegated it to an "extension division" of low prestige and often questionable quality. In mathematics its main product has often been evening courses in college algebra. An occasional new idea would show up, such as the sixties idea of canning lectures on film or videotape. But these electronic extensions of the devices and personalities of formal education never got very far.

Over the centuries the time-honored instrument of education throughout life has been the book, and then the library to make books accessible. During this century in higher mathematical education the most important institution of continuing education has been the American Mathematical Society, joined later by its sister societies in mathematics, applied mathematics, statistics, logic and computer science. The AMS programs that promote research, publish research, and disseminate mathematical knowledge through meetings, have enabled mathematicians to grow and develop after their formal education was complete. Under NSF support we have developed summer conferences, or institutes, in which mathematicians from around the world can cooperate in research in a chosen field. Finally, we must remember that in 1930 the Institute for Advanced Study became a great institution of continuing education in mathematics, outside the framework of formal education.

These developments have special significance in graduate education for applied mathematics. The nature of the subject is such that an applied mathematician's formal education could never be either extensive enough, or intensive enough, to meet his future needs. Now, as never before, his formal education need not be definitive. Later, in practice, he can continue his education, keep up with progress in his field, and grow as his problems demand. This was always the ideal but was never so possible to realize as it is now. Some of the new developments that make this possible include not only the research conferences, but short courses such as those that were offered by the Mathematics Research Center at Madison, Wisconsin, and now Cornell, and desktop publishing with the aid of Donald Knuth's T_EX software.

But the most explosive developments in continuing mathematical education are coming at a more popular level, in vocational mathematics. Mathematicians have been slow to get into this field. Medicine, with its changing technologies and strong financing, has long engaged in continuing education to keep its practitioners up to date. Agricultural extension services have made new technologies available to farmers. Now, belatedly, the uses of the computer have drawn mathematicians into it. Every meeting of AMS has associated short courses or demonstrations of mathematical applications of the computer. Also, with the computer itself as teacher, and with the aid of commercial software, many thousands of Americans are using mathematics as never before. Users of "spread sheets" are doing matrix algebra at high speed, even if they do not know what a matrix is, or that they are doing matrix algebra. Where this all leads, what its implications are for mathematics in American society, no one knows. In any case, from the highest mathematical level to the lowest, informal education, continuing throughout life, is where things are happening. And that is where AMS teaches.

References and Notes

[1] Flexner, Abraham, *Universities, American, English, German.* 1930.

[2] Hofstadter, Richard and C. DeWitt Hardy, *The Development and Scope of Higher Education in the United States,* 1952.

[3] Kandel, I. L. *History of Secondary Education,* 1930. Also his article "Secondary Education" in Encyclopedia Brittanica (1966).

[4] Kandel, I. L. *The New Era in Education,* 1955.

[5] Conant, James B. *Education and Liberty,* 1953.

[6] Conant, James B. *The American High School Today,* 1959.

[7] Valentine, James A. *The College Board and the School Curriculum,* College Entrance Examination Board, New York, 1987.

[8] For example, in the years 1945 to 1965 life in a mathematics department was dominated by the returning veterans with their G.I. Bill, then the phasing out of the veterans by the advent of the National Science Foundation, then Sputnik, the coming of computers, and the expansion of education to accommodate the war babies, all events belonging to the history of higher education.

[9] The written histories of American mathematics before 1888 tend to be social, rather than intellectual, histories, because the main activities of this period were teaching and scholarship. Research was just beginning. On the university level we have Cajori's 1890 survey [10]. And, mainly on the school level, there is D. E. Smith and J. Ginsburg [11]. There is also an excellent recent article by Judith Grabiner [13] which has the flavor of social history.

[10] Cajori, Florian, *History and Teaching of Mathematics in the United States.* U.S. Government Printing Office, 1890.

[11] Smith, David Eugene and J. Ginsburg, *A History of Mathematics in America before 1900*, Chicago, 1934.

[12] Tarwater, J. Dalton (Editor), *Bicentennial Tribute to American Mathematics, 1776-1976*, Math. Assoc. of America, 1976.

[13] Grabiner, Judith V. "Mathematics in America, The First Hundred Years," in Tarwater [12].

[14] Flexner, Abraham, *Daniel Coit Gilman*, New York, 1946, 29.

[15] James, Henry, *Charles W. Eliot (1834-1926)*, 2 vols. 1930.

[16] Kandel [3] 472 ff.

[17] Kandel [3] 468 ff.

[18] Valentine [7] 3 ff.

[19] Conant [6]

[20] Kandel [3] 486-491.

[21] Mach, Ernst, *Science of Mechanics*, 9th edition, 1933, trans. by T. J. McCormack, Open Court Pub. Co. 1942. In the preface Ludwig Mach quotes E. Mach as saying: "I do not consider the Newtonian principles as a completed and perfect thing, yet in my old age I can accept the theory of relativity just as little as I can accept the existence of atoms and other such dogma."

[22] Moore, E. H. "On the Foundations of Mathematics," *Bull. Amer. Math. Soc.* **9** (1902-3), 402. Also *Science*, **17**, March 1903, 401–416.

[23] Wilson, E. B. "Let us have calculus early," *Bull. Amer. Math. Soc.* **20** (1913), 30–36.

[24] Birkhoff, Garrett, "Some Leaders in American Mathematics." In Tarwater [12].

[25] Kandel [3], 449. Public high school enrollments grew from 0.9 million in 1910, to 1.9 million in 1920, to 4.2 million in 1928. And that was before the World War I babies arrived.

[26] I can report first hand on the bad effects of that postwar expansion. In 1918 I entered a new expansion high school, where a rapid succession of inept college students "taught" me algebra. In 1920 my parents moved to New Orleans where I transferred to an old, but also expanding, high school. The teacher in advanced algebra promptly sent me back to first-year algebra where Mr. Theriot, an old professional with a Master's degree from Tulane, rehabilitated me, at no extra compensation I am sure.

Without his generous help I could not have become a professional mathematician. Mr. Theriot was working with me after class one day when a group of the new teachers passed by. He muttered his disdain for them. To him they had no class and did not belong. Then he explained what had happened and why these unqualified teachers were there. The place was going to the dogs!

Later, in college, I became a tutor and teaching assistant, teaching remedial mathematics to graduates of similar expansion schools. My students were not fortunate enough to have been rescued in time by Mr. Theriot. And I did not succeed in saving many, if any, of them by remedial teaching in college.

[27] Young, J. W. (editor), *The Reorganization of Mathematics in Secondary Education.* A report of the National Committee on Mathematical Requirements, under the auspices of the Mathematical Association of America, MAA 1923.

[28] In my first encounter with PEB, I went with my Chairman to Baton Rouge in 1926 to meet the Louisiana state officers of education. Professor Buchanan urged that Louisiana adopt the Young Committee recommendations for a mathematics requirement for high school graduation. The officials treated us with bemused tolerance but deaf ears. They let us know that we had no influence whatever, while they had the political power. So why had we come all the way from New Orleans?

[29] Flexner, Abraham (with Esther S. Bailey) *Funds and Foundations,* 1952. Also recounted in *I Remember: An Autobiography,* rev. ed., 1960.

[30] The Institute for Advanced Study, *A Community of Scholars, Faculty and Members, 1930–1960.* Princeton, 1980.

[31] Montgomery, Deane, "Oswald Veblen," *Bull. Amer. Math. Soc.* **69** (1963), 26–36. Reprinted in [94], 118.

[32] Lax, Peter, "The Bomb, Sputnik, Computers, and European Mathematicians." In [12].

[33] Rees, Mina, "The Mathematical Sciences and World War II," *Amer. Math. Monthly* (1980), 607–621. Reprinted in [94], 275.

[34] Rosser, J. Barkley, "Mathematics and Mathematicians in World War II," *Notices Amer. Math. Soc.* (1982), 507. Reprinted in [94], 303.

[35] Brothers, Leroy A. *Operations Analysis in World War II.* Philadelphia, 1946. Also Charles W. McArthur, *Operations Analysis in the U.S. Army Air Forces in World War II,* to be published.

[36] Goldstine, H. H. *The Computer from Pascal to von Neumann,* Princeton, 1972.

[37] Feller, William, *Introduction to Probability and Statistics,* vol. 1, 2nd ed.,. New York 1957, 6, 31, 138. Feller gives an account of these ideas in the introduction.

[38] Weyl, Hermann, *Group Theory and Quantum Mechanics,* 1928, xxii.

[39] Bliss, G. A. "Mathematical Interpretations of Geometrical and Physical Phenomena," *Amer. Math. Monthly* (1933), p. 472.

[40] Rees, Mina, "Mathematics and the Government, the Postwar Years." In Tarwater [12].

[41] This account is based on recollections of my service as first program director for mathematics in the NSF. Bob Stoll had represented mathematics before me when it was included in physics. But in late 1952, at the urging of Marston Morse, then on the National Science Board, mathematics was given its own program. My secretary and I made up the whole section. What little I knew about the job I had learned from Mina Rees, but now she was unavailable for counsel because she was program

director in that rival agency, ONR! We were all extremely cautious not to overstep our congressional authority. I used to read the National Science Act daily. Director Alan Waterman's staff checked everything I did, or wrote. But the program directors in physics, chemistry, and engineering could not have been more friendly. And the mathematics community was terrific! In spite of the constraints, I think that Leon Cohen, who succeeded me, and I had more freedom to launch new initiatives than the program directors who came after. It was a great time to be in the NSF.

[42] National Education Association, *Education for all American Youth*, Report of the NEW educational policy committee, 1944. Also *Vitalizing Secondary Education.* U.S. Office of Education, Bull. No. 3, 1957.

[43] Krieghbaum, Hiller and Hugh Rawson, *An Investment in Knowledge.* NYU Press, 1969. (A history of the NSF institutes for teachers.)

[44] *Universal Mathematics, Part I, Functions and Limits*, CUP–MAA Lawrence, KS, 1954; *Part II. Elementary Mathematics of Sets*, Robert L. Davis, Ed., CUP–MAA, Charlottesville, VA, 1958.

[45] Duren, W. L. Jr. "CUPM, The History of an Idea," *Amer. Math. Monthly* **74** Part II, 1967, 23–37.

[46] Dartmouth College Writing Group, *Modern Mathematical Methods and Models*, 2 vols. MAA, 1958.

[47] Gehman, H. M. "The Washington Conference," *Amer. Math. Monthly* **65** (1958), 475.

[48] Davis, R. L. ed., *CUP Collected Reports and Transition Report,* MAA, 1957.

[49] Mastrocola, William E., ed., *A Compendium of CUPM Recommendations*, 2 vols. MAA, 1975.

[50] Noble, Ben, *Applications of Undergraduate Mathematics in Engineering.* MAA and Macmillan, New York, 1967.

[51] Begle, E. G. "SMSG, The First Decade," *The Mathematics Teacher*, March, 1968.

[52] Wooton, William, *The Making of a Curriculum*, Yale Univ., 1965. (A history of SMSG.)

[53] Jones, Phillip S., Ed., *A History of Mathematical Education in the United States and Canada*, 32nd Yearbook of NCTM, 1970, specifically the article by Alan R. Osborne and F. Joe Crosswhite, "Reform, Revolution and Reaction," 268–281.

[54] Jones, Phillip S. and Arthur F. Coxford, Jr., "Mathematics in the Evolving Schools." In [53] 11–92, especially Reaction, 81–83.

[55] Mayor, John. See the reports of the AAAS Committee on the Teaching of High School Science and Mathematics. Also see the reports of Mayor's Maryland Project.

[56] I cannot find the record of Paul Rosenbloom's work on mathematics for very small children, in an NSF commission at Columbia. It was based on the work of Jean Piaget, particularly his *The Child's Concept of Number*, New York, 1952. Other follow-up studies to Piaget's work are reported in [53] and in a later NCTM yearbook.

[57] Based on recollections of several conversations with Begle on what SMSG should do about remedial education, particularly in underprivileged sectors of society. Also see E. G. Begle, "The Role of Research in Improvement of Mathematical Education," Proceedings of the First International Congress on Mathematical Education, Dordrecht, Holland, 1969, 106.

[58] Albert, A. A. *A Survey of Research Potential and Teaching in the Mathematical Sciences*, 2 vols. Univ. of Chicago, 1957.

[59] Trytten, M H. *The Baccalaureate Origins of the Science Doctorates Awarded in the U.S. 1936–1950*, NAS–NRC, Pub. 382, Washington, D.C. 1955.

[60] Hedlund, G. A. *Report of a Conference on Undergraduate Mathematics* (Held at Hunter College), New Haven, 1957.

[61] Proceedings of the Sixth Thomas A. Edison Foundation Institute, *The Growing Shortage of Scientists and Engineers*, N.Y.U. 1955.

[62] Gilliland, Edwin R., Ed. "Meeting Manpower Needs in Science and Technology," Report Number One of the President's Science Advisory Committee: *Graduate Training in Engineering, Mathematics, and Physical Sciences*, The White House, Washington, D.C. 1962.

[63] Cohen, Leon W., ed., *Manpower Problems in the Training of Mathematicians*. Report of a CBMS Conference, Washington, D.C. 1963.

[64] Annual Publication, 1974 edition, National Center for Educational Statistics, *Projections of Educational Statistics to 1983–1984*. Table 25. U.S. Department of Health, Education, and Welfare.

[65] COSRIMS, *The Mathematical Sciences, A Report*, Committee on the Support of Research in the Mathematical Sciences, Lipman Bers, Chairman. Also *The Mathematical Sciences, Undergraduate Education*, National Academy of Science, Washington, D.C. 1968.

[66] Annual Publication, U.S. Office of Scientific Resources, *Doctoral Recipients in U.S. Universities*, Washington. Also Annual AMS Survey, *Notices Amer Math. Soc.*, Providence, and various NSF reports.

[67] Jewett John, ed. CBMS Survey of the Mathematical Sciences, Gail Young, Chairman.

    I. John Jewett and C. B. Lindquist, *Aspects of Undergraduate Training in the Mathematical Sciences*, Washington, D.C. 1967.

    II. John Jewett, L. J. Paige, H. O. Pollak, and G. S. Young, *Aspects of Graduate Training in the Mathematical Sciences*, 1969.

    III. J. P. Lasalle, C. R. Phelps, and D. E. Richmond, *Aspects of Professional Work in the Mathematical Sciences*, 1970.

    IV. John Jewett and C. R. Phelps, *Undergraduate Education in the Mathematical Sciences*, 1970–71. Washington, D.C. 1972.

[68] Cartter, Allan M. "Scientific Manpower for 1971–1985," *Science*, **172** (1971), 132–140.

[69] Moise, E. E. "The Proposed Doctor of Arts Degree," *Notices Amer. Math. Soc.* **8** (1961), 112. Conference at Yale on the proposed Doctor of Arts degree, Oct. 21–22, 1961. *Resolutions.* Apparently not published.

[70] Tucker, Alan, Chairman *Recommendations for a General Mathematical Sciences Program*, CUPM Report, MAA, Washington, 1981.

[71] Issues concerning the Ph.D. *Amer. Math. Monthly*, **77** (1970). R. P. Boas, "The COSRIMS reports," p. 623. R. D. Anderson, "Are there too many Ph.D.s?," p. 636. W. L. Duren, Jr. "Are there too many Ph.D.s?" p. 641. Alex Rosenberg, "Some further thoughts on the Ph.D. in mathematics," p. 646. Rosenberg refers to an earlier article, I. N. Herstein, "On the Ph.D. in mathematics," *Amer. Math. Monthly* **76** (1969).

[72] LaSalle, J. P. "Appointments, promotions, and tenure under steady state staffing," *Notices Amer. Math. Soc.*, **19** (1972), 69.

[73] Kemeny, John, "What every college president should know about mathematics," *Amer. Math. Monthly*, **80** (1973).

[74] McCartin, Anne Marie, and Margeret Miller, Report: Costs of remedial education in Virginia, VA. State Council on Higher Education, March, 1988.

[75] Johanningsmeier, E. V. *Secondary Education*, Academic American Encyclopedia, Grolier, 1985, describes how funds from President Eisenhower's 1957 National Defense Education Act were used "to train guidance counselors and administer tests that would identify those students who had the appropriate aptitudes for such studies" [i.e., for the science, mathematics and languages specified by NSDA]. I cannot find the original sources documenting the new slogan of "relevance" and the guidance counselor program, and once more rely on memory.

When the USOE grants program was set up with NSDA funds, largely replacing NSF education programs, I felt that mathematicians should cooperate. I served on various USOE committees and as a "field reader", i.e., referee, on many proposals. I often worked with Kenneth Brown of USOE, whom I liked and trusted. The end came for me in a USOE conference set up ostensibly to advise USOE how the NSDA funds should be allocated. I found that it was a sham; and that the decision had already been made to use the funds on the new guidance counselor program. It was the first I had heard of it. I found the program director in charge arrogant in his power, as no NSF program director would have dared to be. I stormed out of the meeting in protest. Later I got a soothing letter from the Commissioner of Education himself, but he did not ask what my grievance was; and UDOE never again asked me to serve.

[76] Bennett, William, "The Ideal Curriculum for James Madison High School," *New York Times*, Dec. 30, 1987.

[77] Shanker, Albert, Pres. Am. Federation of Teachers, "Bennett's James Madison High School," *New York Times*, Jan. 3, 1988, p. E9.

[78] Digest of Educational Statistics, 1987, USDE, Washington. Table 9, Years of Education Completed by Age, U.S. Bureau of Census, Sept. 1986. The figures quoted are based on the age class, 20–24, including delayed admissions to college. We can get some idea of the fraction of 18-year-olds who entered college directly by comparing Table 69, for the numbers of 17-year-olds graduating from high school in 1984-85, with the numbers of entering freshmen that fall. Assuming that 25 percent of community college freshmen are in degree-credit curricula, the numbers are: 27 percent dropout or delayed, 35 percent temporarily terminated high school graduates, and 38 percent entered degree credit programs in college. These figures will converge towards the ones given for the age class 20–24 as these students grow older. Actually freshmen admissions in some kind of college in the fall of 1985 were 85 percent of high school graduates the preceding spring. These figures do not support Shanker's and PEB's claim that the standard college preparatory curriculum is appropriate for only 15 to 25 percent of students.

[79] "1940 Statement of Principles of Academic Freedom and Tenure," published annually in the AAUP Bulletin, Washington, D.C.

[80] Hoffman, Kenneth M. "Washington Outlook," Periodic columns published in *Notices Amer. Math. Soc.* and MAA *Focus*, especially *Notices* 35 (1987), 898.

[81] Connors, E. A. "Annual AMS Survey," *Notices Amer. Math. Soc.* 35 (1988), 525–534.

[82] Servan-Schreiber, J. J. and Herbert Simon, "America Must Remain the World's University," *Washington Post*, Outlook, Nov. 15, 1987.

[83] David, Edward E. Jr. *Renewing U.S. Mathematics: Critical Resource for the Future*, NAS–NRC, Jan. 1984, "The David Report."

[84] Archibald, R. C. *A Semicentennial History of the American Mathematical Society, 1888–1938*, AMS Semicentennial Publications, vol. 1, 1938.

[85] Birkhoff, George D. "Fifty Years of American Mathematics," in AMS Semicentennial Publications, vol. 2, 1938.

[86] Chandrasekhar, S. "Einstein and General Relativity; Historical Perspectives," *Amer. J. Phys.* **47** (1979), 212.

[87] Dyson, Freeman, *Infinite in All Directions*, New York, 1988, p. 62.

[88] See Goldstine [**36**], 279–286.

[89] Gerstenhaber, Murray, "On the Status of the Mathematical Profession," *Notices Amer. Math. Soc.* **31** (1984), 469.

[90] Digest of Educational Statistics, USDE, Washington, (Annual) Also [**67**] vols. I and IV.

[91] Conference Report, "New Directions in Two-Year College Mathematics," *Notices Amer. Math. Soc.* **32** (1985), 176.

[92] "Virginia Praised for Revamping Teacher Degree," *Charlottesville Daily Progress*, August 19, 1988. As I write this the state board of education has tentatively approved a proposal to replace the education bachelor's degree by the liberal arts degree for the certification of all Virginia teachers.

[93] Pitcher, Everett, *A History of the Second Fifty Years of the American Mathematical Society, 1939–1988*, AMS, Providence, pp. 297–306.

[94] Duren, Peter, ed. with Richard A. Askey and Uta C. Merzbach, *A Century of Mathematics in America, Part I*, AMS, Providence, 1988.

[95] Reingold, Nathan, "Refugee Mathematicians in America, 1933–1941; Reception and Reaction," In [**94**], 175.

[96] Lefschetz, Solomon, "Reminiscences of a Mathematical Immigrant in the United States," In [**94**], 20.

[97] Bers, Lipman, "The European Mathematicians' Migration to America" In [**94**], 231.

[98] Nohel, John A. "Some Thoughts on the Role of Applications in the Development of Ph.D.s in Mathematics," *Notices Amer. Math. Soc.* **22** (1975), 380.

Leon Cohen received a Ph.D. from the University of Michigan in 1928, as the first doctoral student of R. L. Wilder. After two years at Princeton, he joined the faculty of the University of Kentucky and later moved to Queens College. In 1953–1958, he served as the second program director for mathematics in the NSF. He then became chairman of the mathematics department at the University of Maryland, while continuing to serve in organizations such as CUPM, CBMS, and NRC. In 1976 the MAA honored him with an Award for Distinguished Service.

Recollections of a Mathematical Innocent in Washington

LEON W. COHEN

Thirty-five years ago, in 1953, I went to Washington as Program Director for Mathematical Sciences in the National Science Foundation, expecting to stay one year. I stayed five. It is an indication of how remote the mathematical community was from the national scene at that time that while I had had no administrative experience, I ended by becoming Executive Director of the Conference Board for the Mathematical Sciences and also Executive Secretary of the Division of Mathematics in the National Research Council, the operative arm of the National Academy of Sciences. During the five-year period it was on-the-scene training. The rapidly rising level of military technology during the two great wars in the first half of the century thrust the mathematicians with the scientists into close governmental contact. In 1950, Congress initiated federal fiscal responsibility for the advancement of science by establishing the National Science Foundation as an independent agency with a government budget. It was not yet clear how important the applications of science were to become in the competitive world of the industrialized nations.

My first year in NSF was marked by an ongoing debate over the manner in which mathematical research would be supported. The received wisdom was to provide support requested by a proposal from a single investigator with the possible addition of a graduate student as a research assistant for a period of one or possibly two years. Alternative requests appeared in proposals for the support of research seminars, consisting of several investigators with related interests, one or two postdoctoral associates as visitors, and several graduate students as assistants. I urged that NSF policy permit grants on such proposals. After the year of discussion it was decided that only one grant be made in the Program. It was for $30,000 to support a small seminar. It was also decided that no future seminar proposal would be accepted for a grant.

That there was no budget for the Program was characteristic of NSF procedure. Each year proposals for grants came to the Program. They were refereed and evaluated at the annual meeting of the Program Advisory Committee. Based on that advice, I submitted requests for grants. Near the end of the year the National Science Board authorized grants to the extent allowed by the NSF budget.

It should be noted that in 1952 during the term of my predecessor, Professor W. L. Duren, Jr., the Division of Education in the Sciences successfully recommended a grant of $20,000 to the American Mathematical Society for the support of 29 mathematicians at a Summer Institute on Lie Theory.

The restrictions on research grants noted above were subsequently removed. Now NSF supports at least two mathematical institutes — much more elaborate than seminars — attached to universities.

Another problem arose from the reluctance of the mathematical societies to deal with nonmathematical institutions, e.g., government agencies. Professor G. Baley Price, anticipating the future interaction between the mathematical community and the institutions of government and industry, founded the Conference Board of the Mathematical Sciences (CBMS), as an institution whose members would be the several mathematical societies, among them the American Mathematical Society (AMS), the National Council of Teachers of Mathematics (NCTM), the Mathematical Association of America (MAA), and the Society for Industrial and Applied Mathematics (SIAM). CBMS was to have an office in Washington and was to provide liaison between the mathematical community and agencies of government and industry. CBMS originally administered a grant to Professor J. Sutherland Frame to study the needs of mathematics for buildings and facilities which at that time were insufficient for the rate at which mathematics had been developing during the wars. The unwillingness of the mathematicians to come to grips with nonmathematical problems involving the mathematical community is represented by the fact that the meetings of CBMS were paralyzed; nothing was done and I was told, perhaps humorously, by one of the representatives of the member organizations that his instructions were to vote positively on only one motion, namely, approval of the minutes of the last meeting.

These conditions have changed considerably, of course. Now the *Notices of the American Mathematical Society* prints a regular column edited by a distinguished colleague reporting on political activities in the national Capital; also, a Fellow is supported as liaison with the staff and members of the

Congress, representing the needs and purposes of the mathematical community. The staff of NSF has of course grown appreciably, as has its budget for mathematical research. One of the difficulties which occurred during my tenure in NSF was the organizational separation of research and education — or as it was called in NSF — Education in the Sciences. There was a steady conflict over what is called "turf," in the language of the gangs, between the two offices and, as a result, the necessary relationship between research and education in mathematics was not advanced as efficiently as it should have been.

In addition to the institutional difficulties which I've outlined, the negative attitude of individual mathematicians toward participation in any form of activity of a not strictly mathematical nature was evident. Two somewhat amusing stories illustrate the general tenor, although both stories represent extreme opinions. The Mathematical Association of America had developed a program to film hour lectures by distinguished mathematicians in the hope of using such films to stimulate interest among high school students in the discipline. At the suggestion of the MAA committee, I put in a phone call to one of our most colorful colleagues in the northeast of the country and when he came to the phone, he scolded me rather vigorously for having the temerity to encroach upon his privacy with an uninvited phone call. I apologized, of course, even though he and I had had pretty good personal relations previously, and got to his secretary, who promptly arranged the matter. The other instance occurred when I visited a southwestern university in the interest of the NSF program and after discussing matters with several of the colleagues, I met one of their prominent members who, when he learned that I represented a federal agency, quite unhumorously, offered to put on the gloves in a boxing match with me. Of course, I wouldn't have accepted that invitation even if I had not known that in his youth this mathematician had some reputation as an amateur boxer.

In the fall of 1957, the Russians surprised us rather shockingly by putting up "Sputnik" into space. There was immediate stirring in the scientific community in the United States and the Science Advisor to the President urged the prompt increase in the output of scientists and engineers at the Ph.D. level. A conference was called with support from NSF, to consider a program in the mathematical sciences. During the meeting of this conference, a sharp difference developed over the effect on the future of mathematics in the United States of the program for the rapid increase in the number of Ph.D. mathematicians and engineers. It was asserted by some of the most important members of the conference that this would actually produce a decrease in the mathematical effectiveness of the country, because it would encourage mediocrity, and the standards of excellence that had been encouraged by the AMS would be neglected. That this was not the conclusion of the conference was due in large part to the skills of S. S. Wilks (a statistician from Princeton), who was able to convince the members of the conference that a program to double the output of Ph.D. mathematicians in the next five years would not harm the excellent character of mathematical research. The conference finally adopted this view. As a matter of fact, at the end of the five-year period the number of Ph.D.s in mathematics had been doubled, and as we know from the the Fields Medals and other evidences of the excellence of U. S. mathematics, no damage resulted to the level of mathematics in the country.

One of the problems which was faced at that time was the lack of interaction between departments of mathematics and industrial research laboratories in the United States. A retired engineer who had been attached to the automobile industry was in the District and I made his acquaintance. After discussing the problem, he suggested that we go to Detroit where he would introduce me to the directors of the research laboratories of the automobile industries and I could lay my proposal before them. My proposal follows: A member of the research staff of an automobile company should be given a year's leave, be attached to a university department of mathematics to work out a graduate course in problems arising in his research, then he would return to his laboratory, taking with him a graduate student interested in one of the problems for development as a Ph.D. thesis. Thus a connection would be established between the mathematical needs of industrial research and the resources available at the academic center with mutual benefit. The net result was zero. Not a single director of research in any of the three companies showed the least interest in this proposal. And so, perhaps we see one of the reasons why the Japanese were so successful in invading the U.S. automobile market some years later.

When I went to Washington, there was a small number of mathematicians who were of influence in our national affairs. Notably, there were people like Oswald Veblen, Marston Morse, Marshall Stone, Mina Rees, and one or two others. However, there was no infrastructure — there were no second-level people of considerable mathematical tolerance and accomplishment who could provide connections between the academic community and the nonacademic institutions which depended on scientific work. This situation, of course, has been remedied in the meantime and it is hopeful for the future.

I should mention two incidents connected with the National Research Council. The National Research Council, Division of Mathematics, had certain responsibilities in advising the relevant authorities on postdoctoral Fellowships, on the Fulbright program, and other relatively small operations as compared to the needs of the Federal Government. It also had a big role in international affairs, since it was the agency through which American mathematics was connected with the quadrennial International Congress of

Mathematicians. The incidents happened at the end of my stay in Washington. One was a proposal that certain statistical research come under the direction of the National Research Council. This was presented by Professor Kruskal of the University of Chicago. I suggested to Professor Saunders Mac Lane, who was present as a member of the NRC Governing Board, that the project be assigned to the Division of Mathematics. He assented and I made the request. The Division of Biological Sciences also made a strong pitch for this, and it was my one small political achievement in Washington that I managed to beat out the Executive Secretary of the Biological Division and have this statistical program located in the Division of Mathematics. The other and more serious matter was the interest of the National Academy in science policy. The Academy authorized the development of committees on the national level to produce reports outlining the basic philosophy of the discipline, an estimate of its current status and a projection as to its possible future development. At this time, Professor Mark Kac was chairman of the Division, and the committee to handle this assignment in the Division was in the hands of the very vigorous Professor Lipman Bers. Such a report was developed, was approved by the Academy, and now forms the groundwork for the future development of mathematics in the National Research Council.

The National Science Foundation has matured. The mathematical community has learned to work more effectively with the nonmathematical world. There are now a substantial number of mathematicians capable and willing to assume relevant nonmathematical responsibilities. It is gratifying to have had a part at the start of this process.

Born in Budapest, Peter Lax came to America at age 15. During his entire mathematical career he has been connected with New York University. He received his B.A. there in 1947 and his Ph.D. in 1949, under the guidance of Kurt Friedrichs. He then joined the faculty and rose to serve for 8 years as director of the Courant Institute. Among his many honors are a Chauvenet Prize from the MAA and a term as president of the AMS. His research in applied mathematics has focused on hyperbolic systems of differential equations and computational methods. Jointly with Ralph Phillips he is the author of two books on scattering theory. He is a member of the National Academy of Sciences and a recipient of the National Medal of Science. He served on the National Science Board from 1980–1986 and was a recipient of the Wolf Prize in 1987. He also received the Norbert Wiener Prize of the American Mathematical Society and the Society of Industrial and Applied Mathematics.

The Flowering of Applied Mathematics in America

PETER D. LAX

Mathematicians are notoriously bad historians; they describe the development of an idea as it should logically have unfolded rather than as it actually did, by fits and starts, often false starts, and buffeted by forces outside of mathematics. In this sketchy account of applied mathematics in America, I shall describe the twists and turns as well as the thrusts.

Applied mathematics is alive and well in America today; just look at the 18 lectures chosen to describe the frontiers of research: one is on physiological modeling, another on fluid flow and combustion, yet another on computer science; a fourth is devoted to the formation of atoms within the framework of statistical mechanics. The subject of one lecture and the starting point of several others are physical theories; the conclusions reached are of interest to physicists and mathematicians alike.

It was not always so; for a few decades, in the late thirties, forties and early fifties, the predominant view in American mathematical circles was the same as Bourbaki's: mathematics is an autonomous abstract subject, with no need of any input from the real world, with its own criteria of depth and beauty, and with an internal compass for guiding further growth. Applications come later by accident; mathematical ideas filter down to the sciences and engineering.

Most of the creators of modern mathematics — certainly Gauss, Riemann, Poincaré, Hilbert, Hadamard, Birkhoff, Weyl, Wiener, v. Neumann, — would have regarded this view as utterly wrongheaded. Today we can safely say that the tide of purity has turned; most mathematicians are keenly aware that mathematics does not trickle down to the applications, but that mathematics and the sciences, mainly but by no means only physics, are equal partners, feeding ideas, concepts, problems and solutions to each other. Whereas in the not so distant past a mathematician asserting "applied mathematics is bad mathematics" or "the best applied mathematics is pure mathematics" could count on a measure of assent and applause, today a person making such statements would be regarded as ignorant.

How did this change come about? Several plausible reasons can be discerned. But first a bit of selective history.

The second world war, a watershed for our social institutions, concepts and thinking, has permanently changed the status of applied mathematics in America. That is not to say that there was no worthwhile applied mathematics in America before 1945; after all, already in the 19th century, Gibbs' contributions to statistical mechanics as well as to vector analysis and Fourier series, and Hill's studies of Hill's equation, had put America on the applied mathematical map. The leading American analysts in the twenties and thirties were G. D. Birkhoff, renowed worldwide for his work in dynamics, and Wiener, a pioneer in the study of physical processes driven by chance influences, such as Brownian motion and homogeneous chaos. The elusive goal of the ergodic theorem was assiduously pursued in the thirties. The early forties saw the birth of Shannon's theory of information, and Pitts and McCollough's theory of neural networks. Nevertheless, it is fair to say that applied mathematics before 1945 did not fare well in departments of mathematics; it was a marginal activity.

A shift from the margin to the center began after the war; the trickle of applied mathematics swelled to a river. A recent survey of the substance and outlook of applied mathematics has been rendered by Garrett Birkhoff. In the brief span of this talk it is possible only to indicate the broad areas of advance, and to select, somewhat arbitrarily, a number of highlights. If I fail to mention your favorite result in applied mathematics, that only underlines the *embarras de richesse* in this domain.

Partly because of the influential book by Courant and Friedrichs, *Supersonic Flow and Shock Waves*, fluid dynamics was one of the first fields to undergo a renaissance. The basic existence theorems of steady subsonic flow in two dimensions around fixed bodies were established by Bers and Shiffman in the early fifties; much excellent work has appeared since about flows with free boundaries. The problem of steady supersonic flow, and of one-dimensional time-dependent flow, turned out to be more difficult, because of the possible formation of shock waves; the only definitive existence theorem is Glimm's in 1966. Morawetz has studied smooth transonic flows and even those with shocks; there are some results, and many tantalizing open problems.

Perhaps the most exciting new development is computational fluid dynamics, the construction by elaborate numerical calculations of approximations to flow fields. The purpose is two-fold: first to provide engineers with accurate performance characteristics of devices that are in contact with moving fluids, such as pipe systems, aerodynamic shapes, turbines, etc., for purposes of design or control. This approach has been used in more and more complicated situations: combustive flows, magneto-hydrodynamics, etc. The second purpose for doing fluid dynamical calculations is to give theoreticians clues about the possible behavior of fluids, to jog their imagination, in short: to experiment. Such clues have been used to study the complete or partial breakdown of solutions of the Navier–Stokes and Euler equations, and for many other investigations.

The recent spectacular advances in computational fluid dynamics were made possible by increased machine speed, larger memories, and better software, but even more by the invention of clever new numerical methods and algorithms, such as Chorin's use of discretized vorticity, and the fast Fourier Transform of Cooley and Tukey.

Of course, pure mathematicians, too, perform numerical experimentations; that is how Gauss was led to surmise the prime number theorem. He would have loved the computing facilities available today to number theorists, students of dynamical systems, etc. Reliance on fancy computing today creates a strong bond between the pure and the applied.

Equally great advances have been made in other branches of mathematical physics. In the early fifties Kato succeeded in proving that the Schroedinger operator for the helium atom (and other heavier atoms) is selfadjoint. In the midfifties, he and Rosenblum proved the existence of the scattering operator for a pair of operators that differ by an operator of trace class.

Keller's work on diffraction of waves was also begun in the fifties. Using geometrical optics Keller and his coworkers were able to derive mathematically a large number of diffraction patterns; some of these were proved rigorously only much later by Melrose and Taylor, by means of specially designed microlocal operators.

The classical field of dynamics received a jolt in the early sixties, when Moser showed the existence of infinitely many closed curves invariant under area preserving maps of annuli; this shows that such mappings — which include many of physical significance — are not ergodic.

Starting in the fifties there were impressive advances in solving some of the basic problems of statistical mechanics — existence of thermodynamical limits, phase transition, stability of matter. Much of this work was done by physicists, many of whom deserve the title of honorary mathematician. It was an honorary mathematician, Mitchell Feigenbaum, who discovered the doubling of stable periods of selfmappings of intervals as the mapping is deformed, and the universal character of the transfer of stability, a highly unexpected result.

Even more unexpected was Kruskal's discovery of solitons, their curious interaction with each other, and their relation to the existence of infinitely many conserved quantities, and the complete integrability of systems with soliton-like structures. It is astonishing that there are so many completely integrable systems — KdV, sine-Gordon, nonlinear Schroedinger, Toda, etc. — unrecognized as such in the classical days of Hamiltonian mechanics. It is doubly astonishing that they all have a measure of physical significance. That one of them, the Kadomtsev–Petviashvili equation, arising in the study of water waves, has led Dubrovin, Arbarello, DeConcini, and Shiota to a solution of Shottky's classical problem of characterizing Riemann matrices in the theory of Riemann surfaces is truly mindboggling.

Another example of mathematical physics lending a hand to pure mathematics is Faddeev and Pavlov's use of the notions of scattering theory to study automorphic functions.

A great achievement of the last fifteen years is computerized tomography, a lovely combination of inversion of an integral transform, harmonic analysis, and construction of fast and effective algorithms.

The postwar period saw the rise to prominence of the theories of probability and of partial differential equations. Each field stands on two legs, one firmly planted on applications, the other in purely mathematical considerations. Before the war, they were regarded as specialties; today they play a central role in large parts of mathematics.

There also arose entirely new fields of applications, such as the theory of games, control theory, operations research, linear programming, dynamic programming, integer programming, etc. The general aim of these disciplines is to optimize; therefore they have much in common with the calculus of variations. But there are substantial differences as well: these modern theories of optimization often deal with discrete rather than continuous models, and their targeted applications are novel, usually some aspect of economics, business or finance. Equally novel are the algorithms used to achieve the desired optimum in the shortest possible time. A strikingly effective algorithm — simulated annealing — has been borrowed by Kirkpatrick from metallurgy and statistical physics. Annealing is a process applied to amorphous material, glasses of various kind, where the energy of configurations has many minima. The absolute minimum occurs in a highly ordered state, called crystalline; if an amorphous material is cooled very rapidly, it solidifies into a highly disordered state corresponding to a local minimum far from absolute minimum. If the material is cooled slowly, it settles into the crystalline state.

There is a large class of combinatorial optimization problems — of which the traveling salesman problem is typical — which resemble amorphous materials in the sense that the objective function has a superabundance of minima. In such cases any descent method is likely to steer the configuration to a local minimum that is far from the absolute minimum. Simulated annealing operates with a sequence of temperatures T getting smaller and smaller; for each temperature there is a corresponding Gibbs distribution, where the probability of the jth state is

$$\frac{e^{-E_j/T}}{Z},$$

where E_j is the energy of the jth state and Z defined by

$$Z = \sum e^{-E_j/T}.$$

The Metropolis algorithm is used to construct a sequence of states in equilibrium with the Gibbs distribution, as follows: the configuration is changed, according to a chosen recipe. If the new configuration has lower energy, accept the change; if the change in energy ΔE is positive, accept the change with probability $e^{-\Delta E/T}$; this part of the algorithm is implemented by a Monte Carlo method, employing a random sequence. After this algorithm has run for a certain time, the temperature is lowered and the algorithm sequence repeated. This procedure has been remarkably effective for finding excellent approximations to minima in a number of combinatorial optimization problems.

Computer science has been the source of much novel mathematics. It has focused attention on algorithms and has come up with many astonishingly efficient ones, such as the fast Fourier transform, fast matrix multiplication, the simplex method, and many more, described in Knuth's magnum opus. More recent are Karmarkar's algorithm, and Greengard and Rokhlin's method for the fast evaluation of potentials. It is often difficult to estimate the efficiency of an algorithm, especially if it works better in the typical case than in the worst case; see Smale's penetrating study of the simplex method.

An important problem is to design networks that perform efficiently sorting, parallel processing, and other such tasks. Such graphs, called expanders and concentrators, have the same number of edges issuing from each vertex and have good connectivity properties; the task is to construct concentrator graphs with as small a number of edges as possible. Sarnak and his coworkers Lubotzky and Phillips have explicitly constructed a family of expanders with nearly minimal number of edges, which they call Ramanujan graphs, since the proof that they have the desired property depends crucially on a conjecture of Ramanujan concerning the representation of numbers as linear combinations of four squares, as well as on delicate harmonic analysis on groups.

Computational complexity deals with the limits of cleverness, i.e., what are the fastest possible algorithms for evaluating a class of functions? There are few answers as yet to such deep questions; see, for example, Winograd's study of multiplication.

Perhaps the most fascinating area of computer science is artificial intelligence, with its implied threat to put out of business both pure and applied mathematicians of the human kind. But I am bothered by the widespread habit of some parts of the AI community to set their goals preposterously high, and to exaggerate past achievements. For instance, in his Gibbs lecture delivered in 1984, Herbert Simon described a computer program named BACON, designed to extract scientific laws from experimental data, without the benefit of theory, by a kind of curve fitting. He claimed three successes for BACON, the first the derivation of Kepler's third law of planetary motion, which Simon states as:

$$P = KD^{3/2}$$

where P is the period of revolution of the planet, D its distance from the sun, and K a constant that has the same value for all planets. This formulation is meaningful only for planets whose distance from the sun is constant, i.e., whose orbit is circular. Kepler's law on the other hand concerns *elliptic* orbits; he sets D equal to the arithmetic average of the closest and farthest distance of the planet from the sun:

$$P = K\left(\frac{D_1 + D_2}{2}\right)^{3/2}$$

That is, Kepler has found an expression for the period of any planet as function of the *two* parameters characterizing the planet's elliptic orbit. This is worlds away from finding the period of planets with circular orbits. If a mathematician proved a theorem in the spherically symmetric case, he wouldn't dream of claiming the general case; computer scientists must hold themselves to the same standard of precision. There are, to be sure, more profound objections to Simon's paradigm for AI — for example, those voiced by Grabiner and by Edelman.

The most curious — and controversial — of the new applied branches is catastrophe theory, the brainchild of the great mathematician René Thom. A sympathetic presentation of the epistemology of this theory is given in Ekeland's charming new popular book, *Mathematics and the Unexpected*, and in the treatise of Poston and Stewart. More jaundiced views are expressed, in deepening shades of yellow, by Arnold, Guckenheimer, and Sussmann.

Catastrophe theory already has some solid achievements to its credit, and I believe that more are to come. The hostility to the subject is a reaction to attempts to oversell it, like snake-oil medicine, good for whatever ails you. The applications touted by the popularizers were often flaky, and their novelty exaggerated. For instance, Zeeman, in his *Scientific American* article in 1976 wrote:

> For 300 years the preeminent method in building such models has been the differential calculus invented by Newton and Leibnitz. Nevertheless, as a descriptive language, differential equations have an inherent limitation, they can describe only those phenomena where change is smooth and continuous. In mathematical terms, the solutions to a differential equation must be functions that are differentiable. A mathematical method for dealing with discontinuous and divergent phenomena has only recently been developed.

This is strange talk in the age of the theory of distributions! Besides, the theory of discontinuous solutions is much older than the American Mathematical Society. The basic laws of shock waves, which are discontinuous solutions of nonlinear partial differential equations, were set down by Riemann 130 years ago.

Having described some of the achievements of applied mathematics, I would like to discuss briefly its methods. Some of them are organic parts of pure mathematics: rigorous proofs of precisely stated theorems. But for the greatest part the applied mathematician must rely on other weapons: special solutions, asymptotic description, simplified equations, experimentation both in the laboratory and on the computer. Out of these emerges a physical intuition which serves as a guide to research. Since different people have different intuitions, there is a great deal of controversy among applied mathematicians; it is a pity that these debates so often become acrimonious, shedding more heat than light.

We come back now to the question: What were the causes of the flowering of applied mathematics in America after World War II? Perhaps the most important factor was the war itself, which demonstrated for all the crucial importance of science and technology for such projects as radar, the proximity fuse, code breaking, submarine hunting, and the atomic bomb. Mathematicians, working along with physicists, chemists and engineers, made substantial — in some cases decisive contributions; without these developments, the United States might have lost the war. Those responsible for science policy after the war remembered this lesson well and applied it farsightedly and with imagination. They realized that applied science is a basic ingredient of technology, that applied mathematics is an essential component of applied science, and that all parts of mathematics, the pure and the applied, form an organic whole. Consequently, the U. S. Government started a vigorous program to support mathematics, for project-oriented work at government laboratories, for research at universities. A wide variety of subjects were encouraged, for a wide variety of reasons. Some, like numerical linear algebra, or the propagation of electromagnetic waves, were supported for use in immediate applications; others, like the theory of partial differential equations and statistics, because they were underdeveloped compared to their importance; others simply because they were part of the fabric of mathematics. The first agency to systematically support science and mathematics was the Office of Naval Research; it was followed somewhat later by similar offices of the Air Force and the Army, and the Department of Energy in its previous incarnation as Atomic Energy Commission. In addition to supporting a string of mathematicians whose names read like a Who's Who, these agencies were instrumental in the establishment of the School of Probability at Cornell, the School of Applied Analysis and Statistics at Stanford, and the Courant Institute at New York University. The National Science Foundation, coming somewhat later, took over many of the outlooks of its predecessors, as well as forming its own philosophy and point of view. Support of mathematics by the DOD continues to this day, supplementing and complementing support by the NSF and other agencies.

In view of the distinguished past and present success of this research program, it came as an utter surprise that a group within the AMS proposed to reduce support for mathematics by the DOD. Many who were supported by the DOD were deeply offended by the suggestion that they were accepting money from a tainted source, and that the support should have been given to worthier recipients.

The program to build up mathematics in general, and applied mathematics in particular could not have succeeded as well as it did without leadership in the mathematical community. Leadership in applied mathematics was largely provided by a remarkable group of immigrants, mostly refugees from Europe, such as Courant, Feller, Friedrichs, John, Kac, Kato, v. Karman, v. Mises, v. Neumann, Neyman, Prager, Schiffer, Synge, Ulam, Wald, Weyl and others. This group brought to these shores outlooks and styles that were radically different from the purity then prevailing, in particular a greater affinity for applications of mathematics to physics and engineering. Many of the newcomers were in their prime, and were able to put forward their ideas with vigor and confidence.

v. Neumann was a key figure among this illustrious coterie. There is hardly an area of applications that doesn't bear his stamp. In a prophetic speech in Montreal in 1945, when electronic computers were merely figments of his imagination, he declared that "many branches of both pure and applied mathematics are in great need of computing instruments to break the present stalemate created by the failure of the purely analytical approach to nonlinear problems." v. Neumann was a key figure in the American Mathematical

Society; his tragic, premature death has deprived applied mathematics and computer science of a natural leader, a spokesman, and a bridge to other sciences.

It is impossible to exaggerate the extent to which modern applied mathematics has been shaped and fueled by the general availability of fast computers with large memories. Their impact on mathematics, both applied and pure, is comparable to the role of telescopes in astronomy and microscopes in biology; it is a subject fit for another lecture; here I have time only for a few observations:

In the bad old days, when numerical work was limited to a few hundred, or few thousand, arithmetic operations, the task of the applied mathematicians called for drastic simplification, even mutilation, of their problems, to fit the available arithmetic capabilities; they had to cut every corner, exploit every accidental symmetry. Such expediency did not appeal to the mathematical mind, and probably had a great deal to do with the unpopularity of applied mathematics in the days before computers. Today we can render unto the computer what is the computer's, and unto analysis what is analysis', we can think in terms of general principles, and appraise methods in terms of how they work asymptotically for large n, rather than for $n = 8, 9, 10$!

There are many kinds of calculations carried out today, for many different purposes; the confidence we can place in them varies from case to case. Some of the most striking model truly chaotic phenomena such as multiphase flow, turbulent combustion, instability of interfaces, etc. Such calculations use discrete analogues of physical processes, and are very often fine tuned to resemble experimental results. For me, there is something unsatisfactory when a computational scheme usurps the place of a theory that ought to be independent of the parameters entering the discretization.

The applied point of view is essential for the much needed reform of the undergraduate curriculum, especially its sorest spot, calculus. The teaching of calculus has been in the doldrums ever since research mathematicians have given up responsibility for undergraduate courses. There were some notable exceptions, such as Birkhoff and Mac Lane's "Modern Algebra", and Kemeny, Snell, and Thompson's "Finite Mathematics", but calculus, in spite of some good efforts that did not catch on, has remained a wasteland. Consequently the standard calculus course today bears no resemblance to the way mathematicians use and think about calculus. Happily, dissatisfaction with the traditional calculus is nearly universal today; there are very few doubting Thomases. This welcome crisis was brought on by the widespread availability of powerful pocket calculators that can integrate functions, find their maxima, minima, and zeros, and solve differential equations with the greatest of ease, exposing the foolishness of devoting the bulk of the calculus course to antiquated techniques that perform these tasks much more poorly or not at all. We now have the opportunity to sweep clean all the cobwebs and dead material that clutters up calculus. We have to think carefully what we put in its place; I strongly believe that calculus is the natural vehicle for introducing applications, and that it is applications that give proper shape to calculus, showing how and to what end calculus is used. UMAP is an excellent source of such applications.

No doubt computing will play a large role in undergraduate education; just what will take a great deal of experimentation to decide. The brightest promise of computing is that it enables students to take a more active part in their education than ever before.

I would like to direct a comment at enthusiasts for discrete mathematics, a subject of great beauty and depth, which has gained enormous importance for applications because of the availability of computers. But it is mistaken to think that discrete mathematics should compete with or even replace calculus-based applied mathematics in the elementary undergraduate curriculum; this would disregard the explosive growth, thanks to computing, in our ability to bring calculus-based mathematics to bear on applications.

How can one resist the temptation to make guesses about the directions of future research? I am on the safest ground in surmising that computing will play an even bigger role in the next century than today. Mathematical modelers will explore their subjects in the manner of experimentalists. We shall enjoy routinely graphic display capabilities that would dazzle us today. We shall learn to use computations as an ingredient of a rigorous proof, a road already taken by Fefferman and Lanford. I am confident that fluid dynamics will be regarded as a central discipline, and that the elusive goal of understanding turbulence will be vigorously pursued. We will no doubt try to digest the large amount of chaos generated lately, by trying to extract information concerning average behavior. This can sometimes be done in completely integrable cases; the KAM theory gives reason to believe that such results have relevance for systems not too far from integrable ones.

I heartily recommend to all young mathematicians to try their skill in some branch of applied mathematics. It is a gold mine of deep problems, whose solutions await conceptual as well as technical breakthroughs. It displays an enormous variety, to suit every style; it gives mathematicians a chance to be part of the larger scientific and technological enterprise. Good hunting!

BIBLIOGRAPHY

[1.] Arbarello, E. and De Concini, C., "One a set of equations characterizing Riemann matrices", Ann. of Math., 120 (1984), 119-140.

[2.] Arnold, V. I., *Catastrophy theory*, Springer-Verlag, (1986), Berlin, Heidelberg, New York, Tokyo.

[3.] Birkhoff, G., *Applied Mathematics and its Future*, Science and Technology in America, R. W. Thomson, ed., NBS Publ. #465, 1977.

[4.] Chorin, A., "Numerical study of slightly viscous flows", J. of Fluid. Mech., vol. 57 (1973), 785-796.

[5.] Cooley, J. W. and Tukey, J., "An algorithm for the machine calculation of complex Fourier series", Math. of Comp., vol. 19 (1965), 297-301.

[6.] Courant, R., Friedrichs, K. O., *Supersonic Flow and Shock Waves*, Wiley-Interscience, Pure and Applied Mathematics (1948), New York.

[7.] Dubrovin, B. A., "The Kadomcev–Petviashvili equation and the relations between the periods of holomorphic differentials on Riemann surfaces", Math. USSR Izvestija 19 (1982), 285-296.

[8.] Eckmann, J. P., "The mechanism of Feigenbaum universality", Proc. Int. Congress of Math., Berkeley, 1986, vol. 2, 1263-1267.

[9.] Edelman, G. and Reeke, G. N., "Real Brains and Artificial Intelligence", Daedalus, 1988, 143-174.

[10] Ekeland, Ivan, *Mathematics and the unexpected*, The U. of Chicago Press, Chicago and London, 1988.

[11] Faddeev, L. D., Pavlov, B. S., "Scattering theory and automorphic functions", Seminar of Steklov Math. Inst. Leningrad, vol. 27 (1972), 161-193.

[12] Fefferman, C., "The N-body problem in quantum mechanics", CPAM 30 (5) Suppl. (1986).

[13] Glimm, J., "Solutions in the large for nonlinear hyperbolic systems of equations", CPAM 18 (1965), 95-105.

[14] Grabiner, J. V., "Computers and the nature of man, a historian's perspective on controversies about artificial intelligence, Bull. Amer. Math. Soc. 15(5) (1986), 113-264.

[15] Greengard, L. and Rokhlin, V., "A fast algorithm for particle simulations", J. Comp. Phys. 73 (1987), 325-348.

[16] Guckenheimer, J., "The catastrophy controversy", The Mathematical Intelligencer 1 (1978), 15-21.

[17] Keller, J. B., "One hundred years of diffraction theory," IEEE Trans. Antennas and Prop., Vol. AP-33, No. 2, 1985, 123-126.

[18] Kirkpatrick, S., Gelatt, C. D., Vecchi, M. P., "Optimization by Simulated Annealing", Science 220 (1983), 671-680.

[19] Lanford, O. E., "Computer assisted proofs in analysis", Proc. Int. Congress of Math., 1986, vol. 2, 1385-1394, Berkeley.

[20] Lax, P. D., Phillips, R. S., *Scattering theory for automorphic functions*, Annals of Math. Studies 87, Princeton Univ. Press, 1976.

[21] Lubotzky, A., Phillips, R. S., Sarnak, P., "Ramanujan graphs", Combinatoria 8, Issue 3 (1988), 267.

[22] Morawetz, C. S., "On the nonexistence of flows past profiles", CPAM 17 (1964).

[23] Melrose, R. and Taylor, M. E., "The radiation pattern near the shadow boundary", Comm. in P.D.E., 11 (1986), 599-672.

[24] Novikov, S., Manakov, S. V., Pitaevskii, L. P., Zakharov, V. E., *Theory of Solitons: The Inverse Scattering Method*, (1984) Consultants Bureau, Plenum Publ., New York.

[25] Poston, T. and Stewart, I., *Catastrophe Theory and its Applications*, Pitman (1978).

[26] Ruelle, D. "Is our mathematics natural? The case of equilibrium statistical mechanics", Bull. Amer. Math. Soc. (N.S.) 19 (1988), 259-268.

[27] Simon, H., "Computer modeling of scientific and mathematical discovery processes", Bull. Amer. Math. Soc. (N.S.) 11 (1984), 247-262.

[28] Smale, S., "Algorithms for solving equations", Proc. Int. Congress of Math., Berkeley, 1986, Vol. 1, 172-195.

[29] Sussmann, H. J., "Catastrophy Theory Mathematical Methods of the Social Sciences." Synthese, 31 (1975), 229-276.

[30] Thom, René, *Structural stability and morphogenesis*, Benjamin, 1974.

[31] v. Neumann, J. and Goldstine, H. H., "On the principles of large scale computing machines", J. v. Neumann, Collected Works, Vol. 5, 1-33, Pergamon Press, MacMillan, New York.

[32] Winograd, S., "Arithmetic complexity of computations", CBMS-NSF Regional Conf. Series in Applied Mathematics, SIAM.

[33] Zeeman, E. C., "Catastrophe Theory", Scientific American, April 1976.

Theodore von Kármán and Applied Mathematics in America[1]

JOHN L. GREENBERG AND JUDITH R. GOODSTEIN [2]

Applied mathematics is generally regarded as having become a distinct discipline in the United States during World War II. Brown University, under Roland G. D. Richardson, formally instituted a program in applied mathematics, the nation's first, in 1941. New York University, under Richard Courant, later established its own program (1). By that time, Theodore von Kármán (1881–1963), Hungarian-born engineer and applied scientist and the first director of the Daniel Guggenheim Graduate School of Aeronautics at the California Institute of Technology, had already spent more than 10 years struggling to make applied mathematics respectable in his adopted country. To him, the measures taken during the war represented the first concerted, nationwide effort to resolve a long-standing scientific gap in the United States.

Von Kármán figured prominently in the rise of Caltech's school of aeronautics in the 1930s and his experience in America in the 1930s helped define the issues that would lead to the organized development of applied mathematics in the next decade. Frequently pressed for his opinions on how to mobilize mathematicians for the war, von Kármán contributed the lead article "Tooling up mathematics for engineering," to the first issue of the *Quarterly of Applied Mathematics*, published in 1943 (2) under the auspices of Brown's program. Using the form of a dialogue, he eloquently stated the case for the applied mathematician in the service of science. He did not, however, wholeheartedly approve of the proposals for new applied mathematics institutes drafted just before Pearl Harbor, especially the "exaggerated" appeal to an "emergency" created by the war. In his review of one such proposal, he noted that the problem of applied mathematics could not be solved "through the ordinary process of supply and demand" (3,4). Indeed, an entirely different set of imperatives guided von Kármán in the 1930s.

[1] Reprinted with permission from *Science*, Volume 222, pp. 1300–1304, 23 December 1983, "Theodore von Kármán and Applied Mathematics in America." Copyright 1983 by the AAAS.

[2] John L. Greenberg is a research fellow and Judith R. Goodstein is Institute archivist and faculty associate. Division of Humanities and Social Sciences. California Institute of Technology, Pasadena 91125.

MATHEMATICIANS AND ENGINEERS

Shortly after he had completed his first tour of the United States in 1926, which included a visit to Caltech, von Kármán wrote to Courant, then the head of Göttingen's mathematics institute, that "what strikes me most in regard to mathematics...[in America] is the complete lack of 'applied' mathematicians . . ." (5,6). In one sense, von Kármán certainly erred; since the late 19th century, electrical and radio engineering had evolved into highly advanced branches of applied science in the United States, involving the use of a great deal of sophisticated mathematics, thanks to the efforts of several dozen applied scientists, including Charles P. Steinmetz, Michael I. Pupin, and Frank Jewett (7).

At the same time, there was some truth in von Kármán's perception: pure mathematics had developed by leaps and bounds in the United States during the early part of the 20th century. The contact that aspiring American mathematicians had with certain European mathematical schools, especially the German abstract school, which underwent rapid development during the latter half of the 19th century, provided the initial stimulus. As a result, American mathematicians were able, within a short time, to build and sustain research groups in American academic settings in the areas of analysis, number theory, and especially a new branch of mathematics, topology (8).

Contemporary historians of mathematics also emphasized the origins of pure mathematics. R. C. Archibald, for example, reported in 1925 that pure mathematical research in American universities began with Benjamin Peirce (1809–1880) (9). In fact, less than half his output is considered pure mathematics today; the applied mathematics Peirce did all but escaped Archibald's attention.

Except for electrical engineering, von Kármán's reading of the state of American applied science was by and large accurate. The science of the strength of materials, for example, von Kármán's first field of study, remained almost exclusively experimental in the United States. American engineers, generally pragmatic, distrusted the increasingly sophisticated theoretical and mathematical formulations overseas. Mathematically unsolvable problems had no place in turn-of-the-century American engineering practice (10). This may help explain why Russian-born Stephen Timoshenko, sometimes called the "father of engineering mechanics" in the United States, was virtually unknown to the Americn scientific community when he arrived in 1922 at the age of 44. In time, his reputation in the Old World reached the new one (11). According to Timoshenko, who went to work for Westinghouse in 1923, "all the jobs requiring any theoretical knowledge whatsoever were filled mainly by engineers educated in Europe" (11, p. 248). While designing machinery at Westinghouse, he also taught elasticity theory to the other engineers, probably the first such course in the country (11, p. 252; 12). After moving to the University of Michigan in 1927, Timoshenko initiated a program in engineering mechanics similar to one von Kármán would bring to Caltech. Von Kármán, in fact, insisted that Timoshenko had made "the first attempt to gather the applied mathematicians and to institute some activity in applied mathematics" in the United States (5). Both men complained about the attitude of American engineering students. In Timoshenko's opinion, they only wanted "the final result — a formula which . . . [they] can apply mechanically, without thought, to solve practical problems." He traced this attitude back to inadequate mathematics instruction in American high schools (11, p. 26). Indeed, during the 1930s, mathematics came under constant attack, especially at the high school level (13). Engineering educators well into the 1920s had seriously debated whether engineering students even ought to study calculus. Some thought that such courses were mere "cultural embellishments to the curriculum" (14).

Von Kármán did not find mathematicians at Caltech particularly helpful either. Number theorist Eric Temple Bell, for instance, was not interested in training engineers, and the mathematics learned in Bell's hands, according to von Kármán, was simply too abstract (15, p. 149). Von Kármán felt strongly that applied mathematics should be taught in graduate engineering schools, but this seldom happened. The mathematicians told the engineers to teach the course, but the engineers concentrated only on practical subjects (16).

Von Kármán had his feet planted in both worlds. He had done work on the buckling of columns, on the stability of vortex patterns that form behind stationary bodies in flowing fluids, and, with Max Born, on the lattice dynamics and vibrational frequencies of crystals, advancing work done earlier by Albert Einstein and Peter Debye on the heat capacity of solids, among other things. He brought a mathematically sophisticated point of view to all of these problems. Yet when he arrived in the United States, he found American engineers largely untutored in certain branches of mathematics, and quite unprepared for his unorthodox approach to the engineering sciences.

AERONAUTICAL TRADITIONS AND INNOVATIONS

Although aeronautics at Caltech is often said to have begun when von Kármán arrived in 1930, its roots go back to the formation of a committee on aeronautics at Throop College (later renamed Caltech) in 1917, as the United States prepared to join the war against Germany. Throop's science-minded trustee George Ellery Hale promoted aeronautics research as a way for the school to gain national stature. The college hired Harry Bateman, an English mathematical physicist, and Albert A. Merrill, an American inventor. Bateman was the theoretician, Merrill the tinkerer. After designing a small wind tunnel for testing models, Merrill began work on a plane design featuring a movable wing.

By the mid-1920s, Bateman had acquired several graduate students, including Clark Millikan, the son of Caltech physicist Robert A. Millikan. Following von Kármán's first visit to the campus in 1926, Clark Millikan kept him informed by mail of what was going on in aeronautics until von Kármán returned to Pasadena permanently in 1930. The letters told of the construction of the aeronautics laboratory, breaking in and experimenting with the 10-foot Göttingen-style tunnel von Kármán had urged the school to construct, Merrill's new airplane, and Bateman's recent work on airfoil theory (17).

Merrill, a self-taught inventor well versed in the practical side of aeronautics, had the field to himself at Caltech until Robert Millikan, the school's head, engaged Arthur E. Raymond, a member of the technical staff of Douglas Aircraft Company and an expert in designing planes, to teach a class in aircraft design. Merrill left Caltech before von Kármán became director of the Guggenheim Aeronautical Laboratory. The 1928 crash of Merrill's biplane, "the dill pickle" as his students called it, may have hastened his departure.

Bateman specialized in finding particular solutions to complicated equations used by physicists and applied mathematicians. In contrast to Merrill, Bateman was shy and unassuming. Indeed his Caltech colleague E. T. Bell, fearful that Bateman might shortchange his chances of election to the National Academy of Sciences by listing too little on his curriculum vitae, counseled, "Spread yourself; it pays, in our glorious country, to kick over the bushel and let your light to shine before men that they may see your good works . . ." (18).

In Merrill and Bateman's time, the aviation field still belonged to amateurs, and Merrill was high in their ranks. In contrast, von Kármán's students and co-workers attacked a host of theoretical problems related to airplane design and flying that industry used to good advantage (19). The presence of Raymond on the campus indicates that the southern California-based aircraft companies and Caltech had discovered each other before von Kármán took up permanent residence in the United States. There is little doubt that the companies profited even more from the creation of a first-class school of aeronautics in Pasadena. Indeed, records from the Guggenheim Aeronautical Laboratory reveal that of the 30 most prominent graduates in the 1930s nearly half — those who were theoretically oriented – joined universities and the others worked in industry, especially the local aircraft companies. In general, students who did their work in aerodynamics went into the aircraft industry, and those who specialized in fluid mechanics (studying problems such as turbulence and the boundary layer) became academics (20). The 10-foot wind tunnel at the Guggenheim Laboratory, designed to von Kármán's specifications, was used to test practically all the aircraft built by the companies on the West Coast during the 1930s, including the Douglas Company's

DC-3 series, the most successful commercial aircraft of the time. The aircraft companies also recruited Caltech's outstanding students. W. Bailey Oswald, who received his doctorate in 1932, was hired by Raymond as Douglas' chief aerodynamicist when the company began working on the DC-1. The relation between the Guggenheim Laboratory and local aircraft companies in the 1930s foreshadowed the rapid development during World War II of an academic-industrial complex.

Von Kármán's aeronautics school of the 1930s directly benefited from the European applied mathematics and mechanics movement of the 1920s and von Kármán's participation in it. That movement found expression in new organizations, journals, and academic departments. As head of Aachen's Aerodynamics Institute in Germany as well as professor of aerodynamics and mechanics, von Kármán took the initiative in organizing the 1922 conference on hydro- and aerodynamics in Innsbruck, Austria. The 4-day meeting, boycotted by French and British scientists, attracted 33 applied mathematicians and physicists from seven European countries. This informal post-World War I conference meeting, for which von Kármán personally divided the organizing costs with Italian mathematician Tullio Levi-Civita, succeeded in bringing together a number of people with similar scientific interests for the first time. Von Kármán saw that aerodynamicists like himself did not get the attention they deserved because there were not enough of them to stand out at ordinary scientific meetings. "And even among the group they are very split," he pointed out, "because the mathematicians attend mathematics meetings, the physicists attend physics meetings, and the technical people go only to technical meetings" (21). Von Kármán belonged to a group of scientists who decided to do something about the problem. Innsbruck was his solution.

A contemporary, Richard von Mises, founded in 1921 and edited a new journal for applied mathematics and mechanics (*Zeitschrift für angewandte Mathematik und Mechanik*). The head of Berlin's Institute for Applied Mathematics, itself a post-World War I development, von Mises had a flair for organizing like-minded scientists. According to von Kármán, it was von Mises who first mobilized physicists, mathematicians, and scientifically minded engineers working on applied problems to publish their results in the same place (22).

Von Kármán also disregarded the traditionally defined boundaries for aeronautics and aerodynamics in the United States. The range of problems he tackled encompassed more than either science usually did. Robert Millikan's criticism of the state of American engineering paved von Kármán's way at Caltech. Millikan had singled out "the 'ad hoc approach' to the practical problems to be solved" as the weak link in the nation's engineering schools (23). "If a man does not learn his physics, chemistry, and mathematics in college, he never learns it," he told a Caltech audience in 1920, adding, "the attempt to learn the details of an industry in college is futile. The industry itself not only can, but it must, teach these" (24). Although von Kármán did not succeed in converting all Caltech's engineers to his point of view, the institute's philosophy nevertheless provided him with the necessary freedom to pursue his own course.

In 1932, the Metropolitan Water District of Southern California asked von Kármán for help in designing pumps for its Colorado River aqueduct project. In petitioning the school to establish a hydraulics laboratory, von Kármán likened the state of hydraulics to that of aeronautics when engineers first began to turn away from purely empirical computations and started to embrace the methods of the applied mathematicians (25). In the "pump lab," as it came to be called, Caltech's engineers, von Kármán among them, studied a variety of water flow problems. Among other things, they designed and built a water "wind tunnel" to test the efficiency of various pumps. The work done in the hydraulics laboratory, von Kármán once said, "showed a generation of engineers how pure scientific ideas in hydrodynamics, aerodynamics, and fluid mechanics can be used to solve problems of practical design in related fields that at first seem remote" (15, pp. 205–206).

Von Kármán also had a hand in the Grand Coulee Dam project. When cracks appeared after the dam opened, von Kármán realized almost immediately that the forces on the dam exceeded the buckling limits for which it was designed. What the civil engineers had done, in effect, was to use standard design factors obtained from a handbook and then extrapolate to get the figures for building a dam the size of the Grand Coulee. Although they had taken into account static forces due to water pressure acting on the dam, they had failed to consider the special buckling conditions that would arise in such a large dam (15, pp. 207–208). Von Kármán advised the dam engineers to put in stiffeners, drawing on his experience with stiffeners in making sheet metal usable in aircraft design to solve a civil engineering problem.

In another nonaeronautical assignment, von Kármán and his co-workers solved the mystery of "Galloping Gertie," the collapse of the Tacoma Narrows Bridge in 1940. In characteristic fashion, von Kármán transformed a statics problem in civil engineering into a dynamic instability problem. The solution rested on an appreciation of a complex hydrodynamic phenomenon known as vortex shedding first explained by von Kármán in 1911 (26). In recalling the episode many years later, von Kármán noted that "the bridge engineers couldn't see how a science applied to a small unstable thing like an airplane wing could also be applied to a huge, solid, nonflying structure like a bridge" (15, p. 214). In all these instances, it was von Kármán, the applied mathematician, who was able to see the solution by cutting across the boundaries of the traditional engineering fields.

APPLIED MATHEMATICS VERSUS MATHEMATICAL PHYSICS

Often, neither the mathematicians nor the engineers grasped the role of mathematics in applied science. Von Kármán continually pointed out the difficulties mathematicians and physicists had in dealing with nonlinear problems, where intuition alone would not suffice (27). In some instances, the mathematics to deal with such problems had not yet been invented, as was the case with the solitary wave problem (28), the forerunner of solitons, the mathematics of which physicists struggle with today. Mathematicians preferred to deal in generalities, seldom taking, in von Kármán's words, "the pains to find and discuss the actual solutions," except in the simplest cases (29).

Von Kármán continually stressed the difference between mathematical physics and applied mathematics. Once, to make his point, he compared working in applied mathematics to shopping in "a warehouse of mathematical knowledge." The scientist could live in the warehouse and find uses for the equations on the shelf, or he could visit the place from time to time with a shopping list. Von Kármán saw himself as a shopper, not as the caretaker of the mathematics building (30). To a rigorous mathematical physicist like John L. Synge, however, von Kármán's style left something to be desired. Writing to H. P. Robertson, a colleague at Princeton, Synge said, "Kármán has a wonderful intuition, but to a mathematician his exposition is appalling; I think you know that already" (31).

Von Kármán and Courant, who emigrated to the United States in 1934, did not see eye to eye on the development of applied mathematics. Courant was fundamentally interested in mathematical physics (32, p. 226). He used mathematics to make the underpinnings of physics more rigorous. In lectures on this subject, he discussed mathematical problems which had their roots in classical physics. Unlike von Kármán, who used mathematics to solve physics problems, Courant stressed general theories. The algebra and analysis in the 1924 textbook on mathematical physics (33), coauthored by Courant, later provided physicists with tools for further developing quantum mechanics, despite the book's classical physics origins (32, p. 98 and pp. 113–114). Because physicists incorporated some of this mathematics into their own work. Courant saw his role in this work as that of an applied mathematician. But von Kármán would not have defined Courant in that way; Courant, in his view, was really preoccupied with the kinds of questions mathematicians ask, not those of applied scientists. More often than not, von Kármán and his colleagues had to devise their own mathematics of approximate solutions in working out specific technical problems (29).

Caltech's Bateman illustrates another aspect of mathematical physics. Bateman used his mastery of partial differential equations to push Maxwell's equations of electrodynamics to their limits. During the early 1920s, he applied his considerable mathematical skills to devise ingenious theories of radiation to account for the Compton effect, in an effort to save classical physics (34). Paul Ehrenfest, a visitor at Caltech in 1923, marveled at Bateman's uncanny ability but was not persuaded that the mathematician grasped the physics that underlay his calculations. In describing how they wrote a paper together, Ehrenfest remarked: "By my completely desperate questioning, I chased him around for so long in the primeval forest of his calculations that the thing grew clearer and clearer. The connections among his curious isolated results stood out ever more sharply (for him, too!!!)" (35). Bateman, to use von Kármán's metaphor, lived in the mathematical warehouse. He had little in common with applied mathematicians, whom he described on one occasion as mathematicians "without mathematical conscience" (36).

Von Kármán's approach to applied mathematics reflected those of people like von Mises and Hugh Dryden. He especially admired Dryden, who like Raymond, belonged to that small band of early American aviation enthusiasts with solid backgrounds in physics and mathematics. At the age of 20, Dryden earned his Ph.D. from Johns Hopkins in 1919, with an experimental thesis on airflow. As chief of the Aerodynamics Section in the National Bureau of Standards, Dryden continued to work on airflow problems, including turbulence and the boundary layer. In 1941 he succeeded J. C. Hunsaker as the editor of the *Journal of the Aeronautical Sciences* (37).

PUBLISHING

There was no American journal in the 1930s comparable to Mises' *Zeitschrift* to publish applied mathematics papers. Moreover, the banding together of engineers by specialty hindered the founding of interdisciplinary journals. Von Kármán described the problem in a letter to Harvard professor Den Hartog: "American engineers are organized in separate societies. Mechanical, civil, electrical, aeronautical, and automotive engineers have their own organizations, and very little contact exists between them" (38). Each engineering society had a separate journal. Applied mathematics issues, essentially interdisciplinary papers, had to find space in existing journals. A key problem was finding sympathetic editors to deal with manuscripts that straddled more than one discipline. In von Kármán's opinion, only the *Journal of the Aeronautical Sciences* had "the proper attitude for theory" and not a "panicky fear of mathematics" (39). This engineering journal only came into existence in 1933 as the publication arm of the newly founded Institute of Aeronautical Sciences.

Throughout the 1930s few publication outlets were available for topics in applied mathematics. "Many papers are undoubtedly misplaced," von Kármán wrote Brown University Dean Richardson in 1942, because of the way the societies and their journals were organized. Here was the proof, if any was needed, he told Richardson, of a niche "for such a [new] journal" (40).

During World War II applied mathematics turned into a subject for national debate over what the scope, objectives, and theory-to-practice ratio of the program at Brown University should be, as well as what to call the new journal to be published through the program. Some thought the journal should be the sequel to the von Mises *Zeitschrift*. Others thought the word "mechanics" ought to be dropped from the title. The final choice was the *Quarterly of Applied Mathematics*. Still one critic felt that the words "applied mathematics" had "no generally accepted meaning" (41) even in 1942. In some sense the difficulties that von Kármán faced in the 1930s had come to a head.

CONCLUSION

Von Kármán encountered many obstacles during the 1930s beyond a lack of appropriate journals. Some of his novel solutions to structural and civil engineering problems, for instance, were looked at askance by civil engineers, despite Timoshenko's pioneering work. Many older engineers were initially skeptical of von Kármán's proposal to build a water wind tunnel in connection with the Colorado River aqueduct project. When a number of experts, including several Caltech civil engineers, could not solve the mystery of the cracks in the Grand Coulee Dam, and von Kármán was called in as a last resort, some protested, "but he has no civil engineering experience" (15, p. 207). When von Kármán recommended testing a model of the new Tacoma Narrows Bridge in a wind tunnel, even the eminent civil engineer O. H. Ammann said: "You don't mean to say that we shall build a bridge and put it in a wind tunnel?" Von Kármán later noted that the builder of New York's George Washington Bridge "knew better, but long tradition was dictating his remarks" (15, p. 214). The structural engineers assigned to investigate the collapse of the bridge simply found it hard to get beyond their deeply held beliefs in static forces.

The obstacles placed in the path of applied mathematics were probably no greater than those placed in the path of other interdisciplinary endeavors. Subjects ranging from physical chemistry and astrophysics, at the turn of the century, to biophysics and bioengineering in more recent times, have successfully bridged several disciplines and become independent enterprises. In all cases, skepticism from co-workers in the traditional fields appears to be part of the natural-selection process. Under what circumstances some become independent disciplines and others do not, is poorly understood.

REFERENCES AND NOTES

1. World War II as a stimulus for applied mathematics is discussed by M. Rees *Am. Math. Mon.* **87**, 607 (1980)]. N. Reingold [*Ann. Sci.* **38**, 313 (1981)], and J. B. Rosser [*Am. Math. Soc. Not.* **29**, 509 (1982)].

2. T. von Kármán, *Q. Appl. Math.* **1**, 2 (1943).

3. Theodore von Kármán Papers, Milikan Library, California Institute of Technology, Pasadena.

4. T. von Kármán, letter of 13 September 1941 to M. Moore (3, box 70, 11).

5. T. von Kármán, letter of 14 February 1927 to R. Courant (3, box 6, 14).

6. P. A. Hanle [*Bringing Aerodynamics to America* (MIT Press, Cambridge, Mass., 1982), chapters 3–6] describes von Kármán's experiences in Germany.

7. See C. Susskind [*Twenty-five Engineers and Inventors* (San Francisco Press, San Francisco, 1976)] for biographical sketches of these and other applied scientists.

8. G. Birkhoff, in *The Bicentennial Tribute to American Mathematics*, 1776–1976. D. Tarwater, Ed. (Mathematical Association of America, Washington, D.C., 1977), pp. 25–78.

9. See P. S. Jones, in *The Mathematical Association of America: Its First Fifty Years*, K. O. May, Ed. (Mathematical Association of America, Washington, D. C., 1972), p. 5.

10. E. Layton, *Technol. Culture* **12**, 573 (1971).

11. S.P. Timoshenko, *As I Remember*, R. Addis. Transl. (Van Nostrand, Princeton, 1968), p. v.

12. C. R. Soderberg, *Biog. Mem. Natl. Acad. Sci.* **53**, 335 (1982).

13. C. B. Boyer, in *The Mathematical Association of America: Its First Fifty Years*, K. O. May, Ed. (Mathematical Association of America, Washington, D. C., 1972), p. 32.

14. E. T. Layton, *The Revolt of the Engineers: Social Responsibility and the American Engineering Profession* (Case Western Reserve Univ. Press, Cleveland, 1971), p. 4.

15. T. von Kármán with L. Edson, *The Wind and Beyond* (Little, Brown, Boston, 1967).

16. T. von Kármán, letter of 16 March 1937 to R. Courant (3, box 6.15).

17. The Clark B. Millikan Papers (Institute Archives, California Institute of Technology, Pasadena) are a rich source of information about the period before von Kármán's arrival.

18. E. T. Bell, letter of 4 July 1927 to H. Bateman, (Harry Bateman Papers, Institute Archives, California Institute of Technology, Pasadena), box 1.1.

19. W. R. Sears and M. R. Sears, *Annu. Rev. Fluid Mech.* **11**, 7 (1979).

20. See the *Bull. Calif. Inst. Technol.* **49** (No. 2) (1940) and Caltech aeronautics department statistics on employment of students after graduation.

21. T. von Kármán, letter of 12 April 1922 to T. Levi-Civita (3, box 18.8).

22. ____, letter of 8 September 1942 to R. G. D. Richardson (3, box 80.38); letter of 16 February 1945 to H. M. Westegaard (3, box 20.37).

23. ____, and F. L. Wattendorf, in *Miszellanneen der angewandten Mechanik, Festschrift Walter Tollmien* (Akademie-Verlag, Berlin, 1962), p. 103.

24. R. A. Millikan assembly address of 6 January 1920 (Robert A. Millikan Papers, Institute Archives, California Institute of Technology, Pasadena), box 20.9, pp. 4–5.

25. T. von Kármán, a proposal to establish a hydraulics laboratory at the California Institute of Technology, 11 January 1932 (3, box 72.10).

26. A. F. Gunns, *Pac. Northwest Q.* **72**, 168 (1981).

27. T. von Kármán, *Bull. Am. Math. Soc.* **46**, 617 (1940).

28. ____, *ibid.*, p. 60.

29. ____, *Q. Appl. Math.* **1**, 2 (1943).

30. ____, letter of 8 September 1942 to R. G. D. Richardson (3, box 80.38).

31. J. L. Synge, letter of 7 April 1941 to H. P. Robertson (Howard P. Robertson Papers, Institute Archives, California Institute of Technology, Pasadena), box 4.47.

32. C. Reid, *Courant in Göttingen and New York: The Story of an Improbable Mathematician* (Springer-Verlag, New York, 1976).

33. D. Hilbert and R. Courant, *Methoden der Mathematischen Physik* (Springer, Berlin, 1924).

34. R. H. Stuewer, *Proc. Int. Congr. Hist. Sci.* **2** 320 (1975); *The Compton Effect: Turning Point in Physics* (Science History Publications, New York, 1975), pp. 312–314.

35. P. Ehrenfest, letter of 16 February 1924 to T. Ehrenfest, quoted in M. J. Klein, "America observed: Paul Ehrenfest's visit in 1923–24," paper presented at the Conference on the Recasting of Science between the Two World Wars, Florence and Rome, 1980.

36. M. A. Biot, *J. Aeronaut. Sci.* **23**, 406 (1956).

37. T. von Kármán, letter of 24 February 1940 to J. P. Den Hartog (3, box 7.2); letter of 8 September 1942 to R. G. D. Richardson (3, box 80.38).

38. ____, letter of 24 February 1940 to J. P. Den Hartog (3, box 7.2).

39. ____, letter of 14 February 1940 to J. P. Den Hartog (3, box 7.2).

40. ____, letter of 8 September 1942 to R. G. D. Richardson (3, box 80.38).

41. W. Prager, letter of 9 December 1942 to H. Dryden, in (3, box 80.38).

42. Supported in part by the Haynes Foundation.

Brockway McMillan studied electrical engineering at Armour (now Illinois) Institute of Technology, then studied mathematics and physics at MIT, receiving a Ph.D. under Norbert Wiener in 1939. After postdoctoral positions at Princeton University, he served in the Navy at the Naval Proving Ground and at Los Alamos. In 1946 he went to Bell Telephone Laboratories as a research mathematician. In 1961–1966 he was assistant secretary of the Air Force for Research and Development, then under secretary of the Air Force. Returning to Bell Laboratories in 1966, he retired in 1979 as vice president, military systems. He was president of SIAM and chairman of CBMS. His technical interests include electrical network theory, random processes, and the theory of information.

Norbert Wiener and Chaos

BROCKWAY McMILLAN

The roots of Wiener's "Homogeneous Chaos" [W3] can be found in a series of papers from the period 1919–1922, of which "Differential Space" [W1] was the penultimate, and in the later paper "Generalized Harmonic Analysis" [W2] of 1930. The final paper of the 1919–1922 series recasts some of the arguments of [W1] and of the earlier papers. It is [W3] that introduced the term "chaos," a term that did not long retain Wiener's original intended meaning, and is used in a different technical sense today.

The contributions of [W3] over its predecessors were of course more than verbal. They included

(1) the definition of the *pure homogeneous chaos*, an extension to a many dimensional space of the random process on the time axis described in [W1], accomplished by a method different from that of [W1]

(2) a multidimensional ergodic theorem, later sharpened in [W4]

(3) extensions to multidimensional processes of some of the results of [W2] on time series

(4) the definition of the *discrete homogeneous chaos*, a multidimensional random point process, and

(5) a theorem to the effect that a certain class of functionals of the pure chaos is weakly dense within a much larger, but not clearly specified, class of random processes.

The term "chaos," as used by Wiener in [W3] and later, is a noun that, with qualifiers, indicates a kind of random process. A general definition is given in ¶2 of [W3], almost *en passant*: a chaos is a real- or vector-valued function $F(S; \alpha)$. S is drawn from a sufficiently rich class of subsets of n-dimensional Euclidean space E_n, a class later chosen to be a countable ring Ξ that generates the σ-ring of Borel sets. The function $F(S; \alpha)$, as a function of S, is finitely additive over Ξ and for each S is a Lebesgue measurable function of the real variable α on $0 \le \alpha \le 1$.

[W3] always makes the "random" label α explicit, and represents the complete probability space, of which α is a representative point, as the unit interval $0 \le \alpha \le 1$ with ordinary lebesgue measure.

[W3] does not adhere strictly to the initial definition of "chaos," using the term also to designate a random point function defined on E_n, now often called a random field. By specialization, then, the term also designates any function measurable on $0 \le \alpha \le 1$.

Throughout [W1] and its predecessors, there is emphasis on models of natural phenomena, particularly on models of Brownian motion. [W2] focuses on the analysis of observational data in the form of time series, but the element of randomness is not explicitly present. [W3] opens with a clear statement of intent to provide a mathematical basis for the modeling and study of a wide class of random phenomena in nature, mentioning specifically physics and statistical mechanics. Its opening paragraphs display Wiener's deep concern about ergodic theory — the problem of identifying the average of some local physical quantity over an *ensemble*, or random universe of states of a system, with the average of that same quantity over all localities in a particular sample state drawn from that universe. (In Wiener's terms, the average over an ensemble of states is called the "phase average.") This problem of identification is fundamental to the Gibbs approach to statistical mechanics.

Because of the ergodic issue, the paper lays great store by the uniform structure of the underlying Euclidean space E_n and by the invariance, not only of E_n itself but also of the mathematical structures to be erected thereon, under an n-generator group of translations. Hence the "homogeneous" in its title, and the definition, already in ¶2, of homogeneity: $F(\cdot; \cdot)$ is defined to be *homogeneous* if the distributions of the random variables $F(S; \cdot)$, $F(T; \cdot)$ are the same whenever T is a translation of S.

Paragraphs three, four, and five of [W3] deal with ergodic theory, proving an ergodic theorem based on iterates of translations drawn from the symmetry group of E_n, a theorem later sharpened in [W4]. This theorem establishes the existence of the spatial average, over a sample state (i.e., for a chosen α), of a fairly general kind of functional of a chaos. To identify this average with the ensemble average of that functional requires that the chaos be *metrically transitive*: a chaos is metrically transitive if, for each set S, given that S_v is the translation of S by the vector v, the joint distribution of $F(S; \cdot)$ and $F(S_v; \cdot)$ tends to independence as $|v| \to \infty$.

Paragraphs six and seven introduce one of the two random processes of interest in the present essay. In ¶6 we find the one-dimensional *pure homogeneous chaos*, known today as the Wiener process (an interesting synonymy in itself). This is exactly the process considered in [W1], now defined by explicitly introducing a measure into the space of additive set functions on the ring Ξ of subsets of E_1. In the companion papers to [W1], references to an underlying probability measure, in measure-theoretic terms, were absent or at best muted. These papers dealt with function spaces and the averages (expectations) of functionals defined on them. These averages were defined directly as Daniell integrals and the underlying probability measure per se given little explicit attention. [W3] made the probability measure a basic element by mapping elementary events—subsets of the function space called by Wiener *contingencies* and in [Ko] *cylinder sets* — into subsets of the interval $0 \leq \alpha \leq 1$ in such a way that the Lebesgue measure of the image equalled the desired probability of the elementary event.

Between the date, 1921, of [W1] and the publication of [W3], much had happened in the theory of random processes. In particular, Kolmogoroff's fundamental paper [Ko] had appeared. I have no evidence that Wiener ever read [Ko]. Certainly its basic result would have greatly simplified the proofs, in ¶¶6, 7, and 11 of [W3], as well as, later, in [WW], that a stochastic process with specified properties exists. Bochner [B] shows an elegant adaptation of the method of [Ko] specifically to spaces of additive set functions. Bochner told me that this latter paper was inspired by his impatience with Wiener's approach.

[W1] used Brownian motion as its motivation. It dealt with that motion in terms of the paths, or time-histories of displacement, of a particle subjected to random impulses. Accordingly, [W1] introduced a measure into the space of *paths*: functions on $0 \leq t \leq 1$ that vanish at $t = 0$. [W1] showed that this measure assigns outer measure unity to the class of continuous functions. This is now known as the Wiener measure.

A sample path can be thought of as the definite integral, from time zero, of the sample of "white noise" that constitutes the history of buffeting to which the Brownian particle was subject. This is indeed the basis on which [W1] defined the Wiener measure. An important contribution of [W3] then appears in ¶6, for here the point of view is shifted back from *paths* to their *increments*: the Wiener process of ¶6, an additive set function on the one-dimensional time axis, is the *indefinite* integral of a sample of white noise.

This change in point of view is fundamental, for it allows the extension of the theory, in ¶7, to several dimensions.

In ¶7 lies the heart of the paper. The argument of ¶6 is extended from the one-dimensional time axis to E_n and defines the multidimensional Wiener process or *pure homogeneous chaos*. This is a chaos $P(\cdot;\cdot)$ such that for every collection $S_1, S_2, \ldots, S_k$ of pairwise disjoint sets drawn from Ξ the random variables $P(S_1;\cdot), \ldots, P(S_k;\cdot)$ are independent and Gaussian, with means zero and variances respectively equal to $\mu(S_1), \ldots, \mu(S_k)$, where $\mu(\cdot)$ is Lebesgue measure in E_n. That a random process with these properties exists is in fact a theorem, proved from primitive hypotheses. For the proofs in both ¶6 and ¶7 Wiener uses the technique, mentioned above, of mapping measurable events into measurable subsets of the interval $0 \le \alpha \le 1$.

The pure chaos $P(\cdot;\alpha)$, as a set function, is neither countably additive nor of bounded variation, but it admits the definition of (stochastic) integrals. Examples are

(I1) $$\int f(x)P(dx;\alpha),$$

(I2) $$\iint g(x,y)P(dx;\alpha)P(dy;\alpha),$$

integration being over $x \in E_n$, $y \in E_n$. The objects (I1) and (I2) are of course random variables having values, and properties, that justify the suggestive notation. If the integrands here are *simple* functions, step functions measurable on Ξ and taking only finitely many values, the definitions of the corresponding integrals are obvious from the notation. (I1) is then extended from simple functions to integrands $f \in L^2(\mu)$ by continuity in $L^2(d\alpha)$. Multiple integrals such as (I2) are extended from the integrals of simple functions by a convergence in probability $(d\alpha)$.

From the definitions, Wiener shows that the phase average, that is, the expectation, $\int d\alpha$, of (I1) is zero, as is the expectation of any (Im) of odd order m. That of (I2) is given by

$$\int d\alpha \iint g(x,y)P(dx;\alpha)P(dy;\alpha) = \int g(x,x)\mu(dx).$$

More generally, the expectation of (Im) for even values of m is a sum of integrals each of order $m/2$ in μ, in which the integrand appears with its variables identified in pairs in all possible ways. The principle is simple enough but the combinatorics get complicated. One does well so to define his integrands that they vanish when any two variables coincide.

¶9 defines random fields—random functions of a point $z \in E_n$—from integrals such as (I1), (I2) or a general (Im), by replacing the integrands therein with $f(z-x)$, $g(z-x, z-y)$, etc. The discussion then returns in ¶10 to the generalized harmonic analysis of [W2]. By a careful extension of

[W2], it is shown that, except for a null set of labels α, to such a random field can be assigned a (random) multidimensional power spectrum, a nonnegative chaos on frequency space, frequency space being a copy E_n again. For such a random field $G(z; \alpha)$, the power spectrum is the generalized Fourier transform of the sample autocorrelation

$$H(x; \alpha) = \lim V(r)^{-1} \int_r G(x + y; \alpha) G(y; \alpha) \mu(dy),$$

in which "$\int_r$" denotes integration over a sphere about the origin, of radius r and volume $V(r)$, the limit is as $r \to \infty$, and "+" is vector addition in n-space.

The final issue, in regard to the pure chaos, appears in ¶12, identified as the weak approximation theorem. It asserts that a general chaos, as defined in ¶2, can be approximated weakly in distribution by a suitably chosen polynomial functional of the pure chaos. The statement of this theorem is somewhat vague and is inconsistent with its citations of earlier formulas. The subsequent argument is so obscure that I cannot deduce from it exactly what class of chaoses has been proved to admit this approximation in distribution. In fact, a much stronger result now stands. Taking off from the isometry between L^2 on E_n and $L^2(d\alpha)$, defined by the mapping of $f(\cdot)$ on E_n into the function (I1) (on $0 \leq \alpha \leq 1$), Kakutani [Ka], Ito [It1], and Segal [S] have developed alternative structures within which the results of ¶¶6–10 of [W3] are extended and, in particular, polynomials in the pure chaos appear as a dense set. A large class of random processes on the line are described by measures in function space that are absolutely continuous with respect to the Wiener measure.

¶11 introduces the other chaos of interest here, the *discrete chaos*, or multidimensional Poisson process. The argument starts from primitive assumptions and shows that there exists a random additive set function $D(S; \alpha)$ defined and Lebesgue measurable on $0 \leq \alpha \leq 1$ for each $S \in \Xi$, such that (i) if S_1 and S_2 are disjoint then the random variables $D(S_1; \cdot)$ and $D(S_2; \cdot)$ are independent and

(ii) $\text{Prob}\{D(S; \cdot) = k\} = \dfrac{\mu(S)^k}{k!} \exp\{-\mu(S)\}, \qquad k = 0, 1, 2, \ldots .$

Indeed, it is shown that the assumption (i), and the equality (ii) for $k = 0$, suffice to characterize the chaos $D(\cdot; \cdot)$. The discussion continues, defining the first order stochastic integral $\int f(x) D(dx; \alpha)$ by extension from simple functions $f(\cdot)$ to functions $f \in L^2(\mu)$ in analogy with the case of the pure chaos, and showing that

(A) $\displaystyle \int d\alpha \int f(x) D(dx; \alpha) = \int f(x) \mu(dx).$

These results, with ¶10, suffice for calculating the power spectrum of the reponse of a resonator (any linear time-invariant dynamical system) to a Poisson time series of unit impulses. The calculation results in a conclusion that has been known, at least in engineering terms, since 1909 [Ca]. It completes the discussion in [W3] of the Poisson process. Our discussion here of that process will resume later.

My first encounter with Norbert Wiener was in an undergraduate class. I had entered MIT, after two years at another school, without full third-year status, and was caught up in a quite irregular program. This put me, in the spring term of 1935, along with a few other misfits, into a hurriedly scheduled class in differential equations with Wiener as our teacher. It seems likely that this was the first undergraduate teaching he had done in some years. The text was that by H. B. Phillips, written for engineers and full of practical-looking problems, problems that could give one a feel for what the terms of a differential equation really meant. It was clear that Wiener enjoyed teaching this material and he taught it in the spirit of the book. He did not try to improve our minds with excessive rigor. With gusto he showed us all the tricks. He tackled the problems with enthusiasm, and brought in new ones of his own — for example, given the tensile strength of steel as then available, how long can you make a suspension bridge? (Answer: not long enough to bridge the Atlantic.) Wiener was always beautifully articulate but, before even a small group, he tended to adopt a somewhat oratorical style. Beyond this, however, his teaching manner was informal, chatty, even avuncular. He enjoyed a question that required a thoughtful answer. I liked him at once.

Differential equations met right after lunch. Wiener usually entered class with a cigarillo in his mouth. He would sneak a few puffs and then put the butt on the chalk rail. Later he would surreptitiously drop the butt into the side pocket of his jacket. We waited all term for him to drop a smoldering butt into that pocket but he never did.

Three years later, in the spring of 1938, it was arranged that I would do a thesis under Wiener, in the field of random processes. The obvious necessary reading of [W1] and [W2] was already under way or accomplished when a hint of something more specific turned up, in the suggestion by Wiener that I look into [Sch] on the shot-effect. Summer passed and the fall term began with two events, equally unforseen and having, to me, comparable immediate impact: New England's first great hurricane and, a few days later, the arrival of the galley proofs of [W3], handed to me by Wiener with the request that I proofread them. I decline to admit responsibility for the many typographical errors and inconsistencies that remain in the published version of [W3], for, by the time I had catalogued the ones I understood, Wiener had already returned the galleys!

Actually, the specific subject matter of [W3] scarcely figured in any discourse between Wiener and me during the roughly fifteen months that I worked under his tutelage. Throughout this period, his consuming interest was to apply the results of [W3] to the modeling of natural phenomena. He saw in the Wiener process (known to him of course as the pure homogeneous chaos) and in the related weak representation theorem a way to represent the distribution-over-states of fluids and fields. Similarly, though the working tools were not as fully developed, he saw in the Poisson process a way to represent the random states of a particulate system, such as a classical molecular gas or fluid, and as a means to model the shot-effect in electronic devices.

Wiener was anxious to get at both of these fields of application. He suggested that for a thesis I develop the calculus of stochastic integrals with respect to the discrete chaos, and of their ensemble averages, telling me to hurry — since "we" had more important problems to work on. As he had known, it was easy to do. I did hurry, and he accepted the result, [Mc1]. The task was easy for two reasons: the discrete chaos $D(\cdot\,;\alpha)$ is nonnegative, and is of bounded variation on Ξ with probability one, facts exploited in [Mc1]. Indeed, more strongly, the process can so be defined on a Euclidean space that with probability one $D(\cdot\,;\alpha)$ is a σ-finite measure on the σ-ring generated by the class of all bounded sets, so that stochastic integrals are unnecessary. This fact was unknown to me then. Whether it was then known to Wiener I am not sure. He never raised the question with me — to answer it would have made a *good* thesis. In hindsight, the fact explains why the Poisson chaos is easy to work with.

During that academic year, and into the summer of 1939, I saw Wiener's work with the Poisson process from the inside, so to speak, as an amanuensis and quasicollaborator. Along with several others, I also had the opportunity to observe from the outside his work with the Wiener process.

Wiener had scheduled for 1938–1939 a repeat of lectures given some years before on Fourier series and integrals. A good set of mimeographed notes was available from the original lectures, prepared by W. T. Martin and others. Exploiting the existence of these notes, Wiener sped through the entire subject matter of the original lectures in but a few weeks. That in itself was someting of an experience for his listeners, but more was to come. He immediately launched into a quick introduction to the Wiener process and then treated us for the remainder of the academic year to a research seminar addressing the problem of modeling or understanding fluid-mechanical turbulence.

None of his listeners were equipped in any way to participate actively in this research effort. It was a one-man show. Twice weekly he would lay before us his latest ideas for an attack on the problem, sometimes covering the blackboard with prodigious and clearly extemporaneous calculations, sometimes simply speculating, sometimes discussing why what he had just been attempting didn't work. The subject matter was interesting, the spectacle was entertaining, but almost invariably he would break off his line of attack, anticipating some mathematical difficulty, long before any difficulty became evident to his mortal listeners. Quite literally, he could see a shock wave ("Verschiebungstoss" was his word for it) coming long before we could. He regularly deplored the lack of a good existence theory for the Navier-Stokes equations.

This line of application did not lead far during Wiener's lifetime. Later developments, as of 1976, are summarized in [DMc]. At a more basic level, the impact of [W3] and of its predecessors on mathematics has been significant. See [It2] for a brief appreciation. [W3] has also, through the works of Kakutant, Ito, and Segal, had some influence on quantum field theory.

Directly upon the appearance of the galley proofs of [W3], Wiener also plunged into an attack on the statistical mechanics of fluids, work in which I was expected to participate. The working tool was of course the Poisson process. It was clear that Wiener found it stimulating to argue with, or perform before, some kind of audience, even a not very responsive one. I was the chosen audience for statistical mechanics, much as the class in Fourier analysis was that for fluid mechanics.

The work on statistical mechanics turned out to be more substantial and more concrete than that on turbulence. It is described in some detail in [DMc]. I was something more than straight man to the Wiener act, serving as amanuensis and working with the intricate and combinatorial calculations that soon dominated the enterprise. I had brief moments of glory before seminars; Wiener regularly left the exposition to me. He even had me write the abstract [WMc] and let me present the nonresults to the AMS meeting in February 1939. Occasionally I had the exquisite pleasure of contributing a computational trick.

The attack on statistical mechanics began with a natural idea: imagine an infinite cloud of points (molecules), a sample from the Poisson process randomly populating the phase space E_6 (three space coordinates, three velocity coordinates). At time $t = 0$ turn on the intermolecular forces and let this cloud evolve according to Newton's laws. In other words, consider the infinite system of differential equations (of first order, since we are in the phase space of one molecule) that governs the state of this cloud. Given a functional of the cloud such as

$$(F2) \qquad\qquad \Psi(f; \alpha) = \sum_x \sum_y f(x, y),$$

these equations will imply a differential equation that describes its growth. Here $f(\cdot, \cdot)$ is, say, a smooth function with bounded carrier such that $f(x, x) \equiv 0$ for all x, the sums are over all x and y in the cloud, and α, as you have guessed, marks the particular cloud at issue. Wiener would call (F2) a polynomial homogeneously of degree two in the chaos from which this particular cloud is a sample. It is in fact a double integral, (a stochastic integral, as of 1939):

$$\Psi(f;\alpha) = \iint f(x,y)D(dx;\alpha)D(dy;\alpha).$$

The differential equation for Ψ allows one to compute its time-derivatives at $t = 0$. The mth such derivative is a polynomial (in $D(\cdot\,;\cdot)$) such as (F2) is, of degree $m + 2$ in $D(\cdot\,;\cdot)$. It is linear in $f(\cdot,\cdot)$ and its derivatives, and is of degree m in the interparticle potential.

Since the cloud at time $t = 0$ is a sample from the Poisson process, one can calculate the averages ($\int\cdot\,d\alpha$) of the derivatives of Ψ at $t = 0$. These average derivatives describe a formal Maclaurin's series in the time for the average of Ψ at a later time. The key to the whole application of chaos theory is that integrals over α are reducible to integrals over phase space as in the display (A). By integrating these latter by parts to eliminate the derivatives of $f(\cdot,\cdot)$, one can recast the terms as functionals of $f(\cdot,\cdot)$ so that, formally,

(B) $$\int \Psi(f;\alpha)d\alpha = \iint f(x,y)\rho_2(x,y;t)\mu(dx)\mu(dy).$$

Here $\rho_2(\cdot,\cdot\,;t)$ is the second-order density of a point process that describes the distribution-over-states of the cloud at time t, given that it started as a Poisson cloud at time zero. Similar calculations lead to the other densities ρ_m, $m = 2, 3, \ldots$, (ρ_1 is trivial).

Wiener's hope was by this means to derive expansions for the ρ_m as series in powers of such physical parameters as density and temperature. Terms in such series would be sums of multiple integrals involving the interparticle potential. The calculations and results are in fact no less intricate than those found in other approaches to the problem, approaches that were generating an extensive literature at that very time: [MM], [BG] and others. After July 1939, little seems to have been done by Wiener along this line. Later, by quite another method, he did develop formulas for the multiplet-densities ρ_m. The work was submitted for publication, I believe, in the *Journal of Chemical Physics*, but the method and results had been anticipated by a paper already in press [MMo]. Wiener's manuscript has apparently been lost.

Though nothing of significance to statistical mechanics resulted from this work with the Poisson process, two mathematical problems emerged. For a convenient term, define a *true point process*, on Euclidean space E_n, as a complete probability measure on the class of all subsets of E_n that assigns unit probability to a certain subclass Γ. Γ consists of exactly those subsets $\gamma \subset E_n$ such that card$(\gamma \cap S)$ is finite for every bounded set $S \subset E_n$. Γ can be called the class of *locally finite* sets.

In defining the integral (I1) as a stochastic integral, [W3] explicitly avoids claiming that the Poisson chaos is a true point process. Whether or not it was such a process made no difference to the formal calculations of the work on statistical mechanics but the mathematical question remained open, albeit tacit at the time. Actually, during those calculations it became evident — it is already evident in the display (B) — that the Poisson process itself was irrelevant. What was relevant was some other point process that modeled the distribution-over-states of the physical gas, a distribution that depends of course upon the interparticle potential and depends upon it in a distressingly complicated way.

Display (B) suggests the existence of a distribution-over-states in which pairs near the point $(x, y) \in E_6 \times E_6$ occur with density $\rho_2(x, y; t)$. The analog of (B) for functionals Ψ of degree m similarly defines $\rho_m(\cdot, \cdot, \ldots, \cdot; t)$ as the density of m-tuplets in E_{6m}. A critical mathematical question then is: does there exist a true point process that exhibits this sequence $\{\rho_m\}$ of densities? This question subsumes that of the nature of the Poisson chaos, because the latter is characterized by a density sequence $\{\lambda^m\}$ in which the constant λ is the intensity of the process (λ is simply a scaling parameter, assumed $= 1$ in (A)).

Wiener and I subsequently and separately worked on these two problems. In the spring of 1940 I communicated to Wiener the fact that a chaos (not claimed to be a true point process) with well-behaved moments exists if and only if a certain set function $E\{S\}$, definable in terms of the given sequence $\{\rho_m\}$ of densities, is sufficiently regular and has the property of complete monotonicity. (See [Mc3] for complete monotonicity.) The chaos is then characterized by $E\{\cdot\}$ as a function on Ξ, in that $E\{S\} = \mathrm{Prob}\{\gamma \cap S = \varnothing\}$. (It is not then an accident that this latter probability alone, in the form of the function $\exp(-\mu(S))$, the archetypical "sufficiently regular" completely monotone function of S (see [Mc2], ¶6.32) sufficed to define the Poisson chaos in ¶11 of [W3].)

Wiener replied to this news by inviting me to join him and Aurel Wintner as co-author of what, after I declined, became [WW]. This latter paper contains a positivity theorem: given the m-tuplet densities, under certain regularity conditions on the putative moments of the process, a chaos exists if all the putative probabilities $\mathrm{Prob}\{\mathrm{card}(\gamma \cap S) = m\}$, $S \in \Xi$, $m = 0, 1, 2, \ldots$, are nonnegative. Here, of course, the putative moments and probabilities are expressed by those formulas in the given densities that would describe them if a chaos did exist. The tool used is the joint factorial-moment generating function, the (putative) expectation of the product $\prod_i (i - z_i)^{\mathrm{card}(\gamma \cap S_i)}$ considered as a function of the complex variables z_i and the pairwise disjoint bounded Borel sets S_i. In fact, it suffices to consider this function only for one variable z and one variable S, and to postulate that it be continuous from above in S, regular for $|1 - z| \leq 1 + \delta(S)$, where $\delta(S) > 0$ for all $S \in \Xi$, and completely monotone in S when $z = 1$ (or for each z in $0 < z < 1$.) This fact was very nearly in the authors' hands.

Though [WW] is weak in not providing criteria for the positivity of the necessary probabilities, it is strong in that it proves, under its own sufficient conditions, that a true point process exists. The proof is not a model of clarity but, basically, it reduces the problem to that within a bounded set T where, since $\gamma \cap T$ is a finite set, measure theory on a compact space is directly available.

A true point process is one example of a set-valued random process, or random set. In 1973 David Kendall, [Ke], proved a definitive theorem on such processes. Given a space X and a sufficiently rich class Σ of subsets of X, [Ke] defines a property of X and Σ that resembles compactness (although X need not have a conventional topology). Given then a set function $E\{S\}$ defined for $S \in \Sigma$, if $E\{\cdot\}$ is continuous from above and is completely monotone, then there exists a random set γ such that $E\{S\} = \mathrm{Prob}\{\gamma \cap S = \varnothing\}$. It is further true that, if Σ is a countable ring that covers X and separates its points, and if the random set is locally finite—i.e., if $\gamma \cap S$ is a finite set with probability one, for $S \in \Sigma$ — then no compactness condition is needed. This fact is confirmed by [WW] and is probably deducible directly from [Ke], but the only proof I have is based on the methods of [Mc2].

Anecdotes about Wiener are of course legion. Many of them are transparent adaptations of known absent-minded-professor jokes and are of doubtful validity. One frequently heard complaints, however, from persons whom Wiener knew well that he would pass them on the street or in the hall, staring straight ahead and giving no sign of recognition. This occasionally happened to me. One explanation cites Wiener's obvious near-sightedness, but this explanation will not hold water, as I can prove from personal experience. During my year of work with him I lived in the then "new" Graduate House, what had been the Riverbank Court Hotel across Massachusetts Avenue from the main MIT building. A regular Saturday afternoon event was a handball game with roommate and fellow student Abraham Schwartz. In inclement weather the comfortable way to the handball court led through the halls of the main building. On several occasions during that year our trot through these corridors was interrupted by a hail from far down the hall: "Oh McMillan. . . " in Wiener's best theatrical voice. The handball court then waited while we heard Wiener's latest thoughts about the work in progress. In a semidarkened hallway, from perhaps 100 feet away, near-sighted or not, he could recognize me in shorts, sweatshirt, and sneakers (not the typical student's dress in those days).

During 1938–1939 D. J. Struik lectured on the history of mathematics, a late-afternoon two-hour lecture once a week, not for credit. Attendees came from all over. Wiener was one; he always sat in the front row next the window. Regularly, behind him, sat a silent young man named Slutz. After an absence, Slutz reappeared with a handsome beard, a rarity among students in those days. Slutz also abandoned his habitual seat and sat in the center of the front row. Wiener arrived and sauntered across the front of the room toward his usual place. The new beard caught his eye. He sauntered back to the door and turned to survey the competition from a discreet distance. Then he walked briskly up to Slutz, thrust a welcoming hand forward and introduced himself: "My name's Wiener."

I have only pleasant memories of my personal dealings with Norbert Wiener, from the first days of that class in differential equations. He was always cordial and friendly. I sensed a genuine warmth muted by a faintly Continental formality. He was never patronizing nor was there the slightest suggestion of any master–slave relation between professor and student. He always treated me with full respect.

During the school year of our association, there was little opportunity for more than working meetings. During June and July of 1939, however, I lodged in a boarding house near the Wiener summer place in Tamworth, New Hampshire, and we had more contact. Usually I walked the mile to his place for a working session in the morning, returning to my lodgings for lunch, and possibly repeating these trips in the afternoon. Sometimes he would ferry me in his car. There was a period when his family was not in residence and he took his meals with me.

He was a charming and entertaining companion. Strong as his ego was, he almost never talked in a personal vein. I do remember him reminiscing, amused and self-deprecating, about the unsoldierly PFC Wiener, USA, in service at the Aberdeen Proving Ground during World War I. He and I were bridge partners on a few occasions and he played no better than I! He knew a lot about the folkways, history and dialect of the southern New Hampshire region and enjoyed talking about them. One lunchtime he entertained me with thundering recitations of the poetry of Heine. His humor was subtle, seldom bawdy. He liked limericks — we exchanged only clean ones — and loved a bad pun. We enjoyed exploring science-fiction fantasies, distorting a law of physics and exploring the consequences, or inventing simple ways to harness solar energy.

Wiener never gossiped with me. He spoke with respect of the work of von Neumann and urged me to seek von Neumann out when I went to Princeton. He spoke admiringly of J. D. Tamarkin the man, referring to him as "Jacob Davidovitch," and of his work. He went out of his way to express admiration and respect for Solomon Lefschetz and for his work. He never dropped names. I heard nothing from him about his studies in Cambridge or his work in philosophy and the foundations of mathematics. He seemed always to me as his autobiographical works show him: warm, fond of people, more sensitive than he appeared on the surface, and slightly distant from others more from shyness than from arrogance.

It was inspirational, and discouraging, to watch him work. He did not see me as a competitive threat, but the evidence is that he also worked well with young people who might have been seen as threats. His list of successful collaborations with gifted students is fairly long. He worked hard to get me an appointment for the difficult year 1939–1940, and succeeded. I am in his debt for a career well started. I valued his friendship, and I am pleased to acknowledge this debt in a public way.

BIBLIOGRAPHY

[BF] M. Born and K. Fuchs, *Proc. Royal Soc. London Ser.* A **166** (1938) 391.

[Ca] N. R. Campbell, *Proc. Cambridge Phil. Soc.* **15** (1909) 117–136.

[DMc] G. S. Deem and B. McMillan, *The Wiener program in statistical physics*, [WCW1] 654–671.

[It1] K. Ito, *J. Math. Soc. Japan* **3** (1951) 157–169.

[It2] ——, [WCW1] 513–519.

[Ka] S. Kakutani, *Proc. Nat. Acad. Sci. USA* **36** (1950) 319–323.

[Ke] D. G. Kendall, Foundations of a theory of random sets, in *Stochastic Geometry*, Wiley, London, 1973.

[Ko] A. Kolmogoroff, Grundbegriffe der Wahrscheinlichkeitsrechnungen, *Ergebnisse Math.*, **2** (1933) No. 3.

[Mc1] Brockway McMillan, The calculus of the discrete homogeneous chaos, Thesis, MIT, 1939, unpublished.

[Mc2] ——, Absolutely monotone functions, *Ann. of Math.* **60** (1954), No. 3 467–501.

[MM] J. E. Mayer and M. G. Mayer, *Statistical Mechanics*, Wiley, New York, 1940.

[MMo] J. E. Mayer and E. W. Montroll, *J. Chem. Phys.* **9** (1941) 2.

[S] I. E. Segal, *Trans. Amer. Math. Soc.* **81** (1956), 106–134.

[Sch] W. Schottky, Üeber spontane Stromschwankungen in verscheidenen Elektrizitäetsleitern, *Ann. Physik*, Series 4, **57** (1918), 541–567.

[WCW1] *Norbert Wiener: Collected Works*, Ed. P. Masani, Vol. I, MIT Press, Cambridge, 1976. All but [W2] below are reprinted here.

[W1] Differential-space, *J. Math. Phys.* **2** (1923) 131–174.

[W2] Generalized harmonic analysis, *Acta Math.* **55** (1930) 117–258. Reprinted in *Generalized Harmonic Analysis and Tauberian Theorems*, MIT Press, 1964.

[W3] The homogeneous chaos, *Amer. J. Math.* **60** (1938), 897–936.

[W4] The ergodic theorem. *Duke Math. J.* **5** (1939) 1–18.

[WMc] A new method in statistical mechanics (with B. McMillan), *Abstracts Amer. Math. Soc.* **133**. *Bull. Amer. Math. Soc.* **45** (1939). Paper presented orally in New York (February, 1939).

[WW] The discrete chaos (with A. Wintner). *Amer. J. Math.* **65** (1943) 279–298.

Olga Taussky (Mrs. John Todd) was born in what is now Czechoslovakia and was educated at the University of Vienna, where she received a Ph.D. in 1930 as a student of Philip Furtwängler. She held early positions at the Universities of Göttingen, Vienna, Cambridge, and London; and she was at Bryn Mawr College in 1934–1935 with Emmy Noether. During World War II she was a scientific officer at the Ministry of Aircraft Production in England. Afterwards she joined the National Bureau of Standards and then moved to Caltech where she became a professor in 1971. She has done extensive research in algebraic number theory, matrix theory, and topological algebra.

Some Noncommutativity Methods
in Algebraic Number Theory

OLGA TAUSSKY

This article will not deal with classical Galois theory, nor *K*-theory, nor Langlands theory. It is an autobiographic piece by request, dealing with global aspects. It is split into the following sections:

1. The principal ideal theorem, class field towers, Theorem 94, capitulation.
2. Integral matrices.
3. Conclusion.

LITERATURE

The existence of the two sets:
Reviews in Number Theory, ed. W. J. Le Veque, 6 vols., Amer. Math. Soc., Providence, R.I., 1976
which covers the period 1940–1972, and
Reviews in Number Theory, 1973–1983, ed. Richard K. Guy, 6 vols., Amer. Math. Soc., Providence, R.I., 1984
simplifies my bibliographical tasks.

I will accordingly usually give detailed references only to material issued prior to 1940 and to items which appear specially relevant to my treatment.

1. THE PRINCIPAL IDEAL THEOREM, CLASS FIELD TOWERS, THEOREM 94, CAPITULATION

A most impressive experience happened to me in my student days. A very important problem had been solved, and my thesis teacher was involved too. I became involved too and could not tear myself away from it for years and even returned to it temporarily a few years ago. At that time when my teacher Ph. Furtwängler had not yet found a suitable problem for me the news came through that Emil Artin in Hamburg had reduced the famous principal ideal theorem, at that time only a conjecture, to a problem on finite metabelian groups. Artin had found an explicit correspondence between the ideal class group of the ground field and the Galois group of the Hilbert class field with respect to the ground field. The Hilbert class field is the largest relative abelian and unramified extension of the ground field. The class field of the class field, the second class field, has as Galois group with respect to the ground field a metabelian group, with abelian commutator subgroup, and quotient group with respect to the commutator subgroup a group isomorphic to the class group of the ground field. Only p-groups needed to be considered. It is known that in this case the commutator subgroup can be generated by representatives of the quotient group of the commutator subgroup. This is a case of group extension, and the whole problem comes under the title "transfer of groups."

Artin communicated this to Furtwängler who felt that the group theory problem was in his reach. The reason for this optimism lay in the fact that his previous thesis student Otto Schreier, then in Hamburg, had worked on extension of groups and had developed certain relations which enter into this subject. Furtwängler seemed uniquely suited for proving the relation needed.

Furtwängler was a pioneer in class field theory. He had never met Hilbert, but he had studied Hilbert's work which involved as yet unproved statements and conjectures, some of which Furtwängler disproved. One of the latter was the fact that the class field has class number equal to one. The fact that all ideals of the ground field become principal, the so-called Hauptidealsatz (principal ideal theorem) was now reformulated as a group theoretic relation for nonabelian groups. For many weeks one did not see much of Furtwängler. He was a sick man who could not walk well and he used the snow on the streets as an excuse for staying at home. This was very hard on me for he had not given me a subject to work on, but suggested that I try to refind part of what Artin had done, saying that it was easy, which, of course, it was not. I overworked myself when trying to do it. When Furtwängler realized what he had done to me, he finally sat down and introduced me to the machinery he had developed in the meantime. I had no difficulty understanding this and Furtwängler let me reprove the preparatory relations. By the end of the summer Furtwängler announced that he had proved the principal ideal theorem — via nonabelian p-group work. Although I cannot easily forgive him asking me to reprove part of Artin's ideas and even saying they were not too difficult, I can forgive him for rushing into a proof of the transfer relation which gives the proof of the principal ideal theorem for p-groups as Galois groups. Furtwängler's proof was not liked because it was heavily computational. I defended him when Emmy Noether expressed her feelings about this proof by saying that a first proof gives very much and ought not to be criticized. She reacted in a very friendly manner to my outburst. In due course other proofs emerged, one by Magnus, another one by Iyanaga, a student of Takagi, who had installed himself in Hamburg. There was also Hans Zassenhaus, a pupil of Artin who wrote the first modern book on group theory. Some references are given to others who reproved the principal ideal theorem. Artin published another paper in the *Hamburger Abhandlungen*, vol. 7 (the path breaking one was in vol. 5 — I will never forget this). Artin did indicate that the same method would get information about the subfields of the class field. Furtwängler said that he had now plenty of problems for my thesis. He himself wrote another paper proving the following theorem for $p = 2$:

> Let K be a field with 2-class group type $(2, 2, \dots, 2)$. Then there exists a basis $c_1, \dots, c_n$ for the class group such that each c_i becomes principal in a relatively quadratic unramified extension of K.

He then asked me to generalize this for odd p's starting with $p = 3$. However, it turned out that every p needed another condition, attached to $p - 2$, in such a way that $p = 2$ was no exception, only nicer. I only saw Furtwängler's proof when it was published.

E. Artin had heard about my thesis, possibly the first one written on the new ideas created by him. In a letter, handwritten by him, of which I possess a copy, he inquired about my results. However, later he found this investigation futile — when I met him years later he asked me whether I was still working on these hopeless questions. Furtwängler too had realized this in the meantime and withdrew from them turning to geometry of numbers and found many thesis problems there. However, with no other major problem, no class field colleague in the department, I tried to squeeze more out of these questions, with little success.

Hasse followed Hilbert's example and published a "book" inside the journal *Jahresbericht der Deutschen Mathematiker-Vereinigung* in two parts. Furtwängler checked part of his manuscript, but not all of his papers on class field theory are cited. Since Furtwängler was a member of the Vereinigung and was to receive the "book," Hasse suggested to him to give me the "reprint" he would send to him. Hasse, in his article in the Cassels–Fröhlich book, pointed to the work of Scholz and myself, mentioning the difficulty of our investigation. The only time I saw Furtwängler returning to class field theory was when Takagi visited Vienna and called on Furtwängler. Takagi had by then obtained his famous results characterizing all abelian extensions as class fields. However, even he declined to discuss class field theory. But Takagi did actually listen to my new idea of studying the set of subfields of a given field with respect to which a field is an unramified class field. J. H. Smith generalized my work there. I have also a group-theoretic proof for the fact, proved by Furtwängler by number theory, that a field with class group of order 4 has a class field with cyclic class group. Hence the second class field has class number 1. This theorem had made its way into some group theory books, e.g., W. R. Scott, *Group Theory* (Prentice Hall, 1964).

While I was working on my thesis, a young German, a pupil of Schur, turned up. He was still working on his thesis, I think, which was related to Furtwängler's ideas. I saw him talk to Furtwängler. I gathered that he was very gifted and interested in computing. So when I wanted to find out whether a certain group of order 27 with given relations could be the Galois group of a second class field I wrote to him and asked if he could find a field for which this situation would hold. It did not take long before he supplied me with such a field. It was a cubic subfield of the field product $K_{19}^3 K_{1129}^3$ where K_p^l (p, l prime numbers) is the subfield of degree l of the field of pth roots of unity. His name was Arnold Scholz. In 1930, I went to Königsberg where a meeting of the Deutsche Mathematiker-Vereinigung took place (Hilbert gave a famous lecture on logic there). I gave a lecture on my thesis and met famous people like Noether and Hasse. I recognized Scholz and since our mathematical interests were clearly related (he was a few years older and had already several publications by then) we started a conversation. Soon after I returned to Vienna I had a letter from him suggesting joint work by correspondence, at this time on the field $Q(\sqrt{-3299})$, which has a noncyclic 3-class group of order 27. In connection with this work he wrote an important paper on cubic fields. We added another part to this paper studying the second 3-class groups of certain imaginary quadratic fields. I could be particularly helpful there since my training in number theory had turned out (not with my real preference) more group theoretical than arithmetic. The fields studied were to have 3-class groups of type $(3, 3)$ hence the first class field has its relative Galois group of type $(3, 3)$.

By Hilbert's Theorem 94 at least one ideal class of order 3 becomes principal in each unramified extension field of relative degree 3, a property Scholz called "capitulation," by now generally accepted. In our case it was exactly one class. The pattern of capitulation gave us information on the Galois group of the second 3-class field and even higher ones and it turned out that they came to an end after a finite number.[1] This was information on the

---

[1] I understand that Gold and his student Brink have found one error in our capitulation table, by computer.

so-called "class field tower." Furtwängler had suggested the problem of finding out whether it had to break down after a finite number of steps. I then suggested the name "group tower" for the set of Galois groups with respect to the ground field of these fields.

I further conjectured that the class field tower had to break down because the group tower had to break down. Scholz did not really believe in this. Anyhow this idea was defeated. The first one to do this was Noboru Ito for $p = 3$. However, Magnus was able to show that arbitrarily long group towers can exist. He invented a very ingenious method for this which on its own is very much appreciated. Later Zassenhaus, a guest at Caltech, showed to C. Hobby, my thesis student, the sketch of a very elegant matrix method which led to infinite 2-group towers apart from a small number of exceptions. Serre succeeded for all cases. However, an infinite class field tower was constructed by Golod and Safarevic. At an international congress Safarevic announced that while people some 20 years earlier had tried to prove the breaking up of the class field tower by group theory he saw a way for contradicting this by group theory.

In 1931, I attended a meeting of the Deutsche Mathematiker-Vereinigung for the second time, of course, also hoping to find a position. It was a very interesting meeting, e.g., Gödel was there, and, of great importance for my future. One of my teachers in Vienna, Hans Hahn, spoke to Professor Courant from Göttingen about helping me. It was a very difficult time for young people. My lecture at that meeting was, of course, in class field theory, and they actually needed somebody at that time with knowledge in that field. There were not many young people with a thesis in class field theory and they were trying to publish Hilbert's collected works, with number theory as volume I. The two editors they had working on editing that book had no training of that kind. Hence I was brought to Göttingen not much later.

Now I want to return to Furtwängler once more. I really have very great feelings of gratitude for him. He might have given me a little thesis problem in number theory. It would have spared me from my sufferings. But being introduced to such a profound mountain of great beauty has lifted me up forever. Furtwängler was a very hardworking and talented man. He had many thesis students and found appropriate thesis subjects for all of them. But he also found subjects for the survey articles which were demanded for students who did not write a thesis, but planned to take up teaching in high schools. Algebraic number theory was merely taught in a 2-hour seminar, himself lecturing, no homework. His ill health prevented him from contacts with the steadily growing abstract algebra developments which took place in Germany under the influence of Schur, Artin, Hasse, Noether, van der Waerden, and others and their flourishing schools. Hence my education in this respect was insufficient. One of our teachers in Vienna, a young and enthusiastic man, Karl Menger, a former colleague of O. Schreier in Hamburg, noticed this deficiency. He ran a seminar where abstract algebra was studied. He even invited a former thesis student of Emmy Noether, Heinrich Grell, to visit (he was a former colleague of Nöbeling, who was a thesis student and assistant to Menger) to lecture to us. I recall how excited I was when he introduced left and right ideals in rings. I had never heard about them previously. The books by van der Waerden had not yet appeared.

Furtwängler was also hostile to *p*-adic numbers. I wonder whether he had a dislike for Hensel. He used to say that it was sufficient to use the infinite sequence of congruences which they replace. Furtwängler felt very pleased and honored when he was invited to write the article on "General algebraic number theory" for the new edition of the *Enzyklopädie der mathematischen Wissenschaften*. However, this article was so old fashioned that Hasse and Jehne published a revised edition; Furtwängler had died in the meantime.

So when I came to Göttingen for 1931–1932 not only did I face the hard work of editing Hilbert's work, which contained errors and wrong conjectures, but I was faced with modern abstract algebra of the highest level and had difficulties. There were two very talented young men eagerly awaiting me: One was W. Magnus, a pupil of Dehn in Frankfurt who had worked on infinite groups and his paper on the Freiheitsatz is still very well known, the other H. Ulm, a pupil of Toeplitz who is known for his work on infinite abelian groups. They had not known about class field theory and in addition had little experience in proofreading while Professor Menger made me do quite a bit of such work. In Göttingen I was also asked to attend Courant's course on partial differential equations and grade homework. (Secretly he hoped to win me over to this subject, still the favorite one in the "Courant group." Later, when I was recruited into aerodynamical work in London during WWII, I wished I had learned more about this subject then. I also learned to appreciate its beauty.) However, at that time there was E. Noether, the champion of abstract algebra, waiting for me eagerly. She told me that she and Deuring, her favorite student, had studied class field theory and expressed the hope that her tools of abstract algebra would reprove the current achievements in algebraic number theory. She had frequent visits from Hasse, van der Waerden and, at least once during my stay, E. Artin. She ran a seminar on class field theory in which I lectured too, on cyclic unramified extension fields. There was also Landau who, at least once, made me lecture on Mertens. Emmy was amazed to hear that Hilbert's work contained errors on many levels. She was an editor of part of Dedekind's collected works and felt certain that he made no errors. She discovered that the groups Hilbert associated with the prime ideals in normal fields were already Dedekind's work and insisted that they be renamed as Dedekind–Hilbert groups. She kept on finding more items in Dedekind attributed to others, but once I heard Hasse remarking that she was going too far there. The editing of Hilbert's book was particularly burdensome (although some people told me to make a lifetime employment out of this) since it was a deadline job, to be completed by Springer for Hilbert's 70th birthday. Hilbert was pleased about his papers being published, but had no wish to be involved in the editing. However, he said to me clearly that his work in algebraic number theory had been more satisfying to him than his other contributions.

Since I knew more mathematicians than my two colleagues did, I wrote to people asking if they knew of any errors in the volume and I received helpful replies. I recall particularly Fueter and Speiser in Zurich. But they also wrote that it was wrong to reproduce the Zahlbericht since its presentation was not the best for algebraic number theory; also Schur did not care for reprinting this volume according to what Emmy, who had seen him during the Christmas vacation, reported. I suppose they preferred the treatment of Dickson in *Algebras and their arithmetics*. Emmy made no comments herself, of course, she was very much attached to Hilbert and anyhow there was no question of rewriting Hilbert's Zahlbericht instead of editing it. I heard Emmy frequently shout out: "Dies muss hyperkomplex bewiesen werden". She herself had a very good year at that time. She had reproved and extended Gauss' Principal Genus Theorem by methods one would later call cohomology. Again I heard her saying $(1 - S = 2$ when $S = -1)$ meaning that the operator $1 - S$ leads to squaring when S means "the inverse." She was preparing herself for her 1-hour address at the International Congress at Zurich the same year.

After the Congress I did not see her again until I arrived at Bryn Mawr College in 1934. Among other activities she asked me to present the fundamental items of algebraic number theory to the small class formed by M. Weiss, a group theoretician, Grace Shover, a pupil of MacDuffee (on algebras), Ruth Stauffer, who wrote a thesis under Noether on the integral basis (extending work of Speiser), published in *Amer. J. Math.* **58** (1936) 585–599. The previous year they had studied van der Waerden's book, vol. 1, which had appeared not long before. Now they wanted to study some seminar notes by Hasse. But they had first to revise their number theory. MacDuffee seemed unattached to the big schools like Dickson's, but had worked on algebras on his own. Emmy appreciated him and since she was corresponding with Deuring in Göttingen on the preparation of his book *Algebren* she informed him of MacDuffee's and Grace's thesis publications for his list of references. I met MacDuffee later at meetings of the American Mathematical Society. I will discuss his influence on my work later.

As soon as I started my elementary lectures Emmy flared up and criticized my approaches, which I had taken from Furtwängler, who had taken them from Hilbert. In Göttingen she had not informed me of her feelings. I learned there Emmy's concepts of "cross products" and "factor systems." She cited Artin as saying that the Zahlbericht had held algebraic number theory back for years. Later I was criticized by Mac Lane for not giving Emmy credit in my article on her when referring to her work in Göttingen. He introduced his student Lyndon to cohomology and was led to the well-known book "Eilenberg and Mac Lane".

Although I have used cohomology in some of my work, I have not used it in algebraic number theory, but I am a great admirer of the paper by Brumer and Rosen.

When at Bryn Mawr I traveled frequently with Emmy to Princeton and increased my knowledge in topological algebra there. This helped me finally to break away from my "hopeless work" in class field theory for the time being.

After Bryn Mawr I went almost straight to Cambridge, England where I was to stay for the next two years. There number theory was the work on the Hardy–Wright book and otherwise analytic number theory. However, I gave a voluntary short course during my last term. Heilbronn, of course, was there after his famous achievement concerning the Gauss conjecture. He was interested in working with me on the influence of the Galois group on the ideal class group. However, I had planned to go on with topological algebra on one hand and, on the other hand look for help from Philip Hall, a known expert on p-groups. While B. H. Neumann had a certain interest in my attempts in topological algebra, P. Hall did not know enough class field theory to help with the group-theoretic problems connected with capitulation.

I myself had no chance of teaching algebraic number theory apart from a brief course on class fields, the year after I had left Cambridge, but visited there for one day each week. It was at that time that Hilbert's Theorem 94 started to fascinate me and I hoped to do something about it some day.

I could only teach algebraic number theory again after I was appointed at Caltech in 1957. (I had an odyssey in my academic life for almost 20 years; it included WW II and heavy employment duties otherwise.) Remembering Emmy's dislike for my presentation of it at Bryn Mawr I promised myself to teach it "the modern way." One day the best student in my class came up to me with a small book in his hand. It was H. Mann's book on the subject. Mann was a thesis student of Furtwängler, like myself, and the student wanted me to follow this, actually very nice, book. Bringing up H. Mann gives me the opportunity of mentioning that I possess a write-up by him in which he points out a list of errors in Hilbert's and Furtwängler's work.

I come now to Hilbert's Theorem 94. It concerns unramified relative cyclic extensions of relative degree a prime p. In such a field an ideal class of the ground field, not in the principal class there, becomes principal. It is an "existence" statement.

Although what I am going to describe now happened much later and is still of great interest, I will discuss it now for two reasons: It is still class field theory and further it is connected with my thesis and the work of Furtwängler that preceded and my work with Scholz.

Hilbert had erred about his conjecture concerning the class number of the class field, he had not been able to prove the principal ideal theorem, but he had Theorem 94 with a very elegant proof which informs on the capitulation in the class field. Deuring, in his *Zentralblatt* review of my thesis-attached paper in *J. Reine Angew. Math.*, had pointed out that Theorem 94 had not yet been proved by group theory. One day our group theorist M. Hall produced such a treatment. Zassenhaus then played a role in this. He invited me to give a lecture at a so-called Special Session at a meeting of the American Mathematical Society. (He also invited me to become an editor of his newly formed *Journal of Number Theory*. Some recent number-theoretic work of mine had been appreciated by him.) The lecture was supposed to be of computational nature. I decided to study the joint work with Scholz for $p > 3$. There I noticed two new facts turning up:

(a) a certain commutator which turns up with a power $\sum^{p-1} n^2 \equiv 0(p)$ for $p > 3$;

(b) there are 2 cases to be considered. Each cyclic extension of degree p of a field corresponds to a subgroup of the class group of the ground field. The class which capitulates may be contained in this subgroup or it may not. Hence one has these two cases to consider, a fact which is somehow also contained in the group-theoretic proof.

I had the good fortune of having H. Kisilevsky, a student of Iwasawa, working at Caltech by then. Both of us published papers concerning this problem. Kisilevsky interpreted my observations by cohomology. Somehow, since our work the title Hilbert Theorem 94 has become a title for publications. An example for this is the work of Miyake, who published a chain of such papers. There is also an earlier paper by Iwasawa using cohomology on related material.

Examples of work connected with Theorem 94 are still not plentiful. There is a thesis written under the late Bob Smith in Toronto, by S. H. Chang, concerning related matters. Further, Heilbronn's student Callahan's thesis is connected too.

During my stay in Göttingen, Artin came there to give three lectures, to introduce Herbrand's methods to class field theory. I took careful notes, even adding certain items. They were used in lectures very frequently. Finally, Robert Friedman, a student of Tate, translated them into English and they were incorporated into H. Cohn's book *A classical invitation to algebraic numbers and class fields*.

I will now leave the subject of class field theory in this article, apart from a postscript.

Literature before 1940

E. Artin, Beweis des allgemeinen Reziprozitätsgesetzes, *Abh. Math. Sem. Univ. Hamburg* **5** (1927) 353–363.

——, Idealklassen in Oberkörpern und allgemeines Reziprozitätsgesetz, *Abh. Math. Sem. Univ. Hamburg* **7** (1930) 46–51.

Ph. Furtwängler, Allgemeiner Existenzbeweis für den Klassenkörper eines beliebigen algebraischen Zahlkörpers, *Math. Ann.* **63** (1906) 1–37.

——, Die Reziprozitätsgesetze für Potenzreste mit Primzahlexponenten in algebraischen Zahlkörpern, *Math. Ann.* **67** (1903) 1–50; **72** (1912) 146–386; **74** (1913) 413–429.

——, Beweis des Hauptidealsatzes für die Klassenkörper algebraischer Zahlkörper, *Abh. Math. Sem. Univ. Hamburg* **7** (1930) 14–36.

——, Über das Verhalten der Ideale des Grundkörpers im Klassenkörper, *Monatsh. Math.* **27** (1916) 1–15.

——, Über eine Verschärfung des Hauptidealsatzes, *J. Reine Angew. Math.* **167** (1932) 379–387.

H. Hasse, Bericht über Untersuchungen und Probleme aus der Theorie der algebraischen Zahlkörper, I, Ia, II *Jahresber. d. deutsch. Math. Ver.* **35** (1926) 1–55, **36** (1927), 233–311, Erg. **6** (1930) 1–204.

S. Iyanaga, Zum Beweis des Hauptidealsatzes, *Abh. Math. Sem. Univ. Hamburg* **10** (1934) 349–357.

W. Magnus, Über den Beweis des Hauptidealsatzes, *J. Reine Angew. Math.* **170** (1934) 235–240.

——, Beziehungen zwischen Gruppen und Idealen in einem speziellen Ring, *Math. Ann.* (1935) 259–280.

A. Scholz and O. Taussky, Die Hauptideale der kubischen Klassenkörper imaginärquadratischer Körper: ihre rechnerische Bestimmung und ihr Einfluss auf den Klassenkörperturm, *J. Reine Angew. Math.* **171** (1934) 19–41.

——, Zwei Bemerkungen zum Klassenkörperturm, *J. Reine Angew. Math.* **161** (1929) 201–207.

H. G. Schumann, Zum Beweis des Hauptidealsatzes, *Abh. Math. Sem. Univ. Hamburg* **12** (1938) 42–47 (W. Franz Mitwirkung).

O. Taussky, Über eine Verschärfung des Hauptidealsatzes für algebraische Zahlkörper, *J. Reine Angew. Math.* **168** (1932) 25–27.

——, On unramified class fields, *J. London Math. Soc.* **12** (1937) 85–88.

——, A remark on the class field tower, *J. London Math. Soc.* **12** (1937) 82–85.

Specially Relevant

J. Browkin, On the generalized class field tower, *Bull. Acad. Polon. Sci. Sér. Sci. Math. Astronom. Phys.* **11** (1963) 143–145.

J. W. S. Cassels and A. Fröhlich, eds., Algebraic Number Theory. (Proceedings of the Brighton Instructional Conference, Brighton, 1965), Washington, D.C.: Thompson, 1967. [These Proceedings played a considerable part in restimulating activity in Algebraic Number Theory.]

E. S. Golod and I. R. Safarevic, Über Klassenkörpertürme, [In Russian], *Izv. Akad. Nauk SSSR Sér. Mat.* **28** (1964) 261–272. Also see *Amer. Math. Soc. Transl.* Ser. (2) **48** (1965) 91–102.

C. R. Hobby, The derived series of a finite p-group, *Illinois J. Math.* 5 (1961), 228–233.

N. Ito, A note on p-groups, *Nagoya Math. J.* **1** (1950) 115–116.

K. Iwasawa, A note on the group of units of an algebraic number field, *J. Math. Pures Appl.* (9) **35** (1956) 189–192.

H. Kisilevsky, Some results related to Hilbert's Theorem 94, *J. Number Theory* **2** (1970) 199–206.

R. Schoof, Infinite class field towers, *J. Reine Angew. Math.* 372 (1987) 209–220.

J.-P. Serre, Sur une question d'Olga Taussky, *J. Number Theory* **2** (1970) 235–236.

J. H. Smith, A remark on fields with unramified compositum, *J. London Math. Soc.* (2) **1** (1969) 1–2.

O. Taussky, A remark concerning Hilbert's Theorem 94, *J. Reine Angew. Math.* 239/240 (1970) 435–438.

O. Taussky, Arnold Scholz zum Gedächtnis *Math. Nachr.* **7** (1952) 379–386. (This paper discusses the latest ideas of Scholz which were taken up by Jehne and his school. Note also *J. Reine Angew. Math.* **336** (1982) with Scholz's picture, a page on his life and work, a paper by F.-P. Heider and B. Schmithals "Zur Kapitulation der Idealklassen in unverzweigten primzyklischen Erweiterungen" on pp. 1–25.

A. Weil, Sur la théorie du corps de classes, *J. Math. Soc. Japan* **3** (1951) 1–35.

H. Zassenhaus, *Theory of groups*, second edition, New York: Chelsea, 1958.

Postscript to Section 1

Artin's idea for the proof of the principal ideal theorem and O. Schreier's work used by Furtwängler came under the concept of group and field extensions.

The genus field and group, as studied by Fröhlich, is also relevant. It used to be introduced for abelian fields only. However, Fröhlich describes it as the maximal nonramified extension obtained by composing the given field with absolutely abelian fields.

Literature before 1940

C. G. Latimer and C. C. MacDuffee, A correspondence between classes of ideals and classes of matrices, *Ann. Math.* **74** (1933) 313–316.

A. Scholz, Totale Normenreste, die keine Normen sind, als Erzeuger nicht abelscher Körpererweiterungen, *J. Reine Angew. Math.* **175** (1936) 100–107, II, **182** (1940) 217–234.

O. Schreier, Über die Erweiterung von Gruppen I, *Monatsh. f. Math.* **34** (1926) 165–180. II, *Abh. Math. Sem. Hamburg* **4** (1926) 321–346.

SPECIALLY RELEVANT

G. Cornell and M. Rosen, Group-theoretic constraints on the structure of the class group. *J. Number Theory* **13** (1981) 1–11.

Y. Furuta, The genus field and genus number in algebraic number fields, *Nagoya Math. J.* **29** (1967) 281–285.

G. Gras, Extensions abéliennes non ramifiées de degrée premier d'un corps quadratique, *Bull. Soc. Math. France* **100** (1972) 177–193.

K. Hoechsmann, *l*-extensions, in: *Algebraic Number Theory*, ed. by Cassels and Fröhlich. (Proceedings of the Brighton Instructional Conference, 1965). Washington, D.C.: Thompson, 1967.

H. Koch, Galoissche *Theorie der p-Erweiterungen*; Introduction by I. R. Safarevic. Berlin-New York: Springer and Berlin: VEB Deutscher Verlag der Wissenschatten, 1970. Many references.

B. Mazur and A. Wiles, Class fields of abelian extensions of *Q*, *Invent. Math.* **76** (1984) 179–330.

K. Uchida, On imaginary Galois extension fields with class number one, *Sugaku* **25** (1973) 172–173.

2. INTEGRAL MATRICES

This brings me to my next section, the one of greatest interest to me. Complications dictated by the circumstances of my life have not always allowed me to devote myself to my favorite subject, number theory. My thesis was mainly in group theory. I was rescued for some time by working with Arnold Scholz. The war, WW II, made a numerical analyst of me in Great Britain and later in the USA (with some exceptional breaks through the influence of MacDuffee's work and collaboration with E. C. Dade and M. Newman). In 1957, I started my work at Caltech. Working with thesis students, postgraduates, temporary colleagues like E. C. Dade, Kisilevsky, Estes, Guralnick, P. Morton, P. Hanlon, and other brilliant visitors was terrific. Some of it was number theory. It was noncommutative and concerned integral matrices. H. Cohn permitted me to add an appendix on integral matrices to his 1978 Springer book; I also contributed to Roggenkamp's 1981 volume, Springer Lecture Notes #882.

In M. Newman's book *Integral matrices* (Academic Press 1972, chapter X, 15), he mentions a theorem (obtained jointly with myself) concerning integral circulants: A unimodular circulant of the form AA', where A is a matrix of rational integers, is equal to CC' where C itself is again a unimodular circulant of rational integers (the theorem itself arose from my idea of generalizing the concept of normal basis). Circulants can be looked upon as "group matrices" under the definition given in my article "A note on group matrices." As Newman mentions, the theorem was generalized to statements for more general groups and finally for all groups by Rips, 1973.

Another concept to be mentioned here is the discriminant matrix, the integral symmetric matrix whose determinant is the discriminant of an algebraic number field. The quadratic form associated with this matrix is a "trace form." A paper by myself, entitled "The discriminant matrix of an algebraic number field" showed that this matrix has nonnegative signature. This was used by P. E. Conner and R. Perlis, who wrote the book *A survey of trace forms of algebraic number fields* (World Scientific, 1984).

The fact that the unit of finite order in the group ring of the S_3 does not contain a multiple of the unit element (pointed out by myself) was explained by Takahashi. Reiner felt that the article John Todd and myself wrote on integral matrices of finite order (in 1939, with the war to start any moment and us in London) was a pioneering contribution to integral representations. In his later article on this subject he included a long bibliography.

Now I will show some of my other results. Matrices are mathematical objects which by their own nature are connected with noncommutativity. I am now returning to C. C. MacDuffee, whom I had met in 1935 and whose work had a great influence on me. First of all, I learned from him that Poincaré had introduced a connection between ideals and integral matrices. But this was not taken up very much. At some meeting I met Latimer, who also sent reprints. I observed that the two sets had a common element, a joint paper. This paper is not mentioned in Deuring's book *Algebren*, although I feel that both authors, or at least one of them, would have included it in Emmy's parcel. This is one on which I have spent much time and which has definitely become quite popular. I started this with a paper entitled "On a theorem of Latimer and MacDuffee." The reason why Emmy, and perhaps also Deuring, ignored this paper is that it deals with matrices. While algebraists use matrix algebras they tend to look down on matrices. In my early days I did the same. I now applied the work to the ring of integers $\mathcal{O}$ in an algebraic number field F, assuming it to be of the form $Z[\alpha]$, α assumed the zero of an irreducible monic polynomial $f(x)$ of degree n. The theorem concerns all integral $n \times n$ matrices A for which $f(A)$ is the zero matrix and their division into classes $\{S^{-1}AS\}$ under unimodular integral similarity S. It can be shown that there is a 1-1 correspondence between these matrix classes and the ideal classes in $\mathcal{O}$. Hence the number of matrix classes is finite. The work of Latimer and MacDuffee which is applied to algebras is much more complicated, and MacDuffee's papers are much more complicated for the same reason. I then found an independent proof. I spent extra effort on this since I encountered only people (some of high standards) who thought that there was only one class, i.e., integral matrix roots of $f(x)$ were unimodularly similar. (However, Zassenhaus has done related work on integral representations.) While Latimer and MacDuffee let it go at that, I built up a theory related to this theorem which is much cited and applied. The theorem was even generalized to abstract rings (see D. Estes and R. Guralnick), and Barry Mazur gave me a more modern proof.

I showed that the principal ideal class corresponds to the class of the companion matrix, the inverse ideal class to the class of the transposed matrix.

(A. Fröhlich, in his 1983 *Ergebnisse* book, has a generalization of the latter fact.) On the other hand a matrix is similar to its transpose. This is elaborated in my 1966 *Annalen* paper. Let

$$S^{-1}AS = A'$$

A with irreducible characteristic polynomial. Then S is symmetric and can be expressed as

$$S = (\text{trace } \lambda \alpha_i \alpha_k)$$

where $\lambda \in Q(\alpha)$, where α is a characteristic value of A with characteristic vector $\alpha_1, \ldots, \alpha_n$. The paper contains further results. More generally, since the matrices which enter all classes are all zeros of the same polynomial they must all be similar, but not via unimodular matrices, in general. I connected this question with another concept of MacDuffee, not in the same context. This concerns "Ideal Matrices." I will return to this after discussion of three other items:

(1) I investigated matrix classes which contain symmetric matrices. (This is in connection with D. K. Faddeev's and also E. Bender's work on polynomials with symmetric matrix roots.)

(2) If the maximal order is not generated by a single element, then one can replace the matrix classes by classes of integral representations of the maximal order under unimodular similarity.

(3) When studying matrices one rarely bothers about the numbers which turn up inside the matrix, with the exception of the companion matrix. However, Ochoa in Madrid, Spain, did bother and he found that under circumstances (studied by Rehm) sparse matrices of the following type can occur in a matrix class

$$\begin{bmatrix} 0 & 1 & 0 & \cdots & 0 & 0 \\ 0 & 0 & 1 & \cdots & 0 & 0 \\ & \cdots & & \cdots & & \cdots \\ 0 & 0 & 0 & \cdots & 1 & 0 \\ a_1 & a_2 & a_3 & \cdots & a_{n-1} & a_n \\ b_1 & 0 & 0 & \cdots & 0 & b_2 \end{bmatrix}$$

I now return to the concept of ideal matrix, again for ideals in the maximal order and quite complicated in MacDuffee's work. I begin with the definition.

Let $\mathscr{O}$ have the basis $\omega_1, \ldots, \omega_n$ and let $\alpha_1, \ldots, \alpha_n$ be the basis of an ideal in $\mathscr{O}$ and A an integral matrix for which

$$A \begin{bmatrix} \omega_1 \\ \vdots \\ \omega_n \end{bmatrix} = \begin{bmatrix} \alpha_1 \\ \vdots \\ \alpha_n \end{bmatrix}$$

Then A is called an ideal matrix with respect to the bases chosen. If the bases are altered then A is replaced by UAV where U, V are unimodular. Since A can be transformed into Smith normal form by U's and V's it tells much about the ideal. My paper "Ideal Matrices I" explains how the 'quotient' of two ideal matrices leads to a similarity between different matrix classes. I have by now written four papers on ideal matrices. However, MacDuffee's ideals did more for me. For he showed that mapping the elements of an ideal in the maximal order into the ring $Z^{n \times n}$ (assuming the underlying field of degree n), a principal ideal ring, the ideal matrix turns up as gcrd or gcld of the respective maps.

This led to a paper[2] in which I was able to show that principal or nonprincipal can be replaced by commutativity or noncommutativity. The fact that $Z^{n \times n}$ is a euclidean ring and that there is a procedure for finding gcrd or gcld by computation is of help for numerical work here.

There is a book by C. Chevalley where he studied a more general situation. He assumed his matrices to have entries from a field k, commutative or not, and showed that the arithmetic can be brought down to arithmetic in k. This leads to the study of ideals in the matrix algebra. He points out that the case of k commutative was treated also by V. Korinek.

My work on ideal matrices was picked up by others, the first one by a student of Don Lewis at Ann Arbor, Michigan, Burgie Wagner, in "Ideal matrices and ideal vectors," later by S. K. Bhandari and V. C. Nanda, in "Ideal matrices for relative extensions" and S. K. Bhandari, in "Ideal matrices for Dedekind domains."

In particular they had studied my paper "Integral Matrices II," in which I tried to find information on the ideal matrix for the product of two ideals in the maximal order, in terms of the ideal matrices of the factors. Well, it is not just the product! A factor U, a certain unimodular matrix has to be inserted between A and B. In India they introduced the concept of the "AUB Theorem."

My contribution to integral matrices started to grow steadily after I entered Caltech in 1957. So when I was asked to give a one-hour lecture at an AMS meeting I felt ready to give a lecture entitled "Integral Matrices."

Among my papers on integral matrices I want to point out some subsets.

(α) The first subset is motivated by a theorem for fields.

While any square matrix with elements in a field k can be expressed as the product of two symmetric ones, with elements in k, this is not always true for integral elements, although one factor, say the first one, can be expressed this way.

The question arose: when can both factors be expressed over Z? In a paper dedicated to C. L. Siegel's 70th birthday, by invitation from *Acta Arithmetica*, I solved this for $n = 2$ and received an appreciative letter from Siegel, who was a student of Frobenius and quite devoted to matrix theory. I wrote another paper on this subject, this one at the suggestion and partial advice of my former student E. Bender. This particular problem has led to interesting details.

---

[2] This paper has slight inaccuracies in the presentation.

Let $A = (a_{ik})$ be the matrix to be represented. Then associate with it the quadratic form

$$(1) \qquad a_{12}x^2 + (a_{22} - a_{11})xy - a_{21}y^2$$

However, for the problem considered a second quadratic form turns up. It was shown that it is the square of the first one. They both have as discriminant the discriminant of the matrix A.

It can then be shown that the second form is in a form class whose order is a divisor of 2. This implies that the matrix A has to be in a matrix class whose order is 1, 2, 4. In addition there are some exceptional form classes involved. The first example of this was given by D. Estes and H. Kisilevsky. van der Waerden, who had heard me lecture about this, stated that it would be congruent to (1). But later he told me in two letters that it was the negative square of the first one. His paper received a fairly long review in *Mathematical Reviews* and Cassels from Cambridge, England, asked me to send him a copy of these lectures; they were not yet published, but bound as reports. He said: "but this is Gauss' duplication of a quadratic form," and then he added, "but it is rather weird."

My further work includes three papers with titles:

"Norms from quadratic fields and their relations to noncommuting 2×2 matrices."

(β) Paper II has the subtitle "The principal genus"; Paper III has the subtitle "A link between the 4-rank of the ideal class groups in $Q(\sqrt{m})$ and in $Q(\sqrt{-m})$".

(γ) Let A, B be noncommuting 2×2 integral matrices, with at least one of them, say A, with eigenvalues in $Q(\sqrt{m})$, but not in Q. Then

$$\det(AB - BA) = -\text{norm}\,\lambda, \ \lambda \in Q(\sqrt{m})$$

If both matrices have eigenvalues not in Q, then this can lead to an intersection of norms from two different quadratic fields.

This last theorem was studied by several people and Zassenhaus reproved and elaborated it via cyclic algebras in 1977.

(δ) Another result is:

Let A be a 2×2 integral matrix with eigenvalues in $Q(\sqrt{m})$, m not a square. Let $XAX^{-1} = A'$ where $'$ denotes transpose. Then $-\det X = \text{norm}\,\lambda, \lambda \in Q(\sqrt{m})$.

This is proved in my paper "Ideal matrices 1." Dennis Estes studied (α), (γ), and (δ) via Galois cohomology and a generalized Latimer and MacDuffee correspondence. In connection with my paper "Ideal Matrices III" an application to Galois modules and group matrices was made in the case where the field is normal, with a normal basis also for the integers and even the ideal is to have a normal basis. (The last two conditions are not always satisfied, of course.)

The 1982 Dekker book edited by me contains a review and elaboration of some of the items mentioned here.

There is also a paper from 1979 "A diophantine problem arising out of similarity classes of integral matrices" which was generalized by R. Guralnick in a much-appreciated paper in *J. Number Theory*.

I want to close with mentioning two later papers:

"Some noncommutative methods in algebraic number theory," a paper connected with my 1982 lecture at the symposium honoring Emmy Noether's 100th birthday. It has many references. It also uses central polynomials and is attached to Noether's work on the principal genus.

"Composition of binary integral quadratic forms via integral 2×2 matrices and composition of matrix classes." [Equation (16) in this paper is misleading, but it is explained by the footnote and the equations that follow.] While composition of binary quadratic forms was introduced by Gauss, the matrix approach can be generalized to all dimensions. (This will be mentioned again under problems.)

INTEGRAL MATRICES BEFORE 1940

E. Artin, Zur Arithmetik hyperkomplexer Zahlen, *Abh. Math. Sem. Hamburg* **5** (1928), 261–289.

C. Chevalley, *L'-arithmétique dans les algèbres de matrices*, ASI **323**, (1936), Hermann, Paris.

V. Korinek, Une remarque concernant l'arithmétique des nombres hypercomplexes, *Mém. Soc. Roy. Sci. Bohême* (1931) NR 4, 1–8.

H. Poincaré, Sur un mode nouveau de représentation géométriques des formes quadratiques définies ou indéfinies, *J. École Polytech. Cah.* **47** (1880) 177–245.

SPECIALLY RELEVANT

(apart from items mentioned inside the section)

E. Bender, Classes of matrices and quadratic fields, *Linear Algebra and Appl.* **1** (1968) 195–201.

A. Buccino, Matrix classes and ideal classes. *Illinois J. Math.* **13** (1969) 188–191. (He bases his work on a general integral domain.)

D. Maurer, Invariants of the trace-form of a number field, *Linear and Multilinear Algebra* **6** (1978/1979) 33–36.

M. Newman, Symmetric completions and products of symmetric matrices, *Trans. Amer. Math. Soc.* **186** (1973) 191–210 (1974). (Generalizes Taussky's work on factoring integral matrices.)

W. Plesken and M. Pohst, On maximal finite irreducible subgroups of $GL(n, Z)$ II. The six-dimensional case, *Math. Comput.* **31** (1977) 552–573. (Uses the theorem of Latimer and MacDuffee.)

H. P. Rehm, On Ochoa's special matrices in matrix classes, *Linear Algebra and Appl.* **17** (1977) 181–188.

I. Reiner, Integral representations of cyclic groups of prime order, *Proc. Amer. Math. Soc.* **8** (1957) 142–146. (Uses the theorem of Latimer and MacDuffee.)

J.-P. Serre, L'-invariant de Witt de la forme $Tr(x^2)$, *Comment. Math. Helv.* **59** (1984) 651–676.

O. Taussky, On the similarity transformation between an integral matrix with irreducible characteristic polynomial and its transpose, *Math. Ann.* **166** (1966) 60–63.

D. I. Wallace, Conjugacy classes of hyperbolic matrices in SL(n, Z) and ideal classes in an order, *Trans. Amer. Math. Soc.* **283** (1984) 177–184. (Uses the theorem of Latimer and MacDuffee.)

W. C. Waterhouse, Scaled trace forms over number fields, *Arch. Math.* **47** (1986) 229–231.

3. CONCLUSION

I finish with brief, largely bibliographical, comments on three topics related to my theme and a short list of problems.

GALOIS MODULES

Here Fröhlich and his colleagues Taylor, Bushnell, Ullom, Queyrut,... are leaders. He himself wrote the 1983 Ergebnisse volume *Galois module structure in algebraic number fields.*

In his article in the 1981 Dekker book, *Emmy Noether, A tribute to her Life and Work*, he says that he will concentrate on Galois module structure. He considers her work there as ahead of her time and the developments came 40 years later. He refers to his paper in 1976, "Module conductors and module resolvents." I feel flattered to notice that he refers to a 1956 paper by M. Newman and myself (discussed in chapter 2) and to a much-cited paper by Leopoldt from 1959. As I pointed out earlier, our problem is now settled for all groups. The Durham 1977 *Proceedings* are heavily loaded with relevant material. In his 1976 lecture at the Kyoto number theory conference in honor of Takagi he speaks about Hermitian Galois module structure.

Galois algebras seem to have been introduced by Hasse in 1948. The 1981 thesis of J. Brinkhuis, written under supervision by Fröhlich, refers to this reference. Fröhlich himself has a paper on this and so has Maurer, a paper entitled "... Stickelberger's criterion on Galois algebras and tame ramifications in algebraic number fields."

THE USE OF QUATERNIONS

A number of results are mentioned here:

(1) Venkov, B. On the arithmetic of quaternions (Russian): H. P. Rehm makes use of this paper to give a proof of a famous theorem of Gauss concerning the number of representations of an integer $n > 1, \equiv 1, 2(4)$ as a sum of three squares.

(2) The 1981 Caltech thesis of P. Hanlon studies Rehm's work and finds an application of quaternions to the study of imaginary quadratic ring class groups.

(3) A paper by T. R. Shemanske on ternary quadratic forms and the class number of imaginary quadratic fields.

USE OF CENTRAL POLYNOMIALS

The concept of central polynomials over a field was suggested by Kaplansky, *Amer. Math. Monthly* **77** (1970). Central polynomials were then constructed by E. Formanek, *J. Alg.* **23** (1972) and by Y. P. Razmyslov, *Transl. USSR Izv.* **7** (1973).

The Formanek version was used in O. Taussky, From cyclic algebras of quadratic fields to central polynomials, *J. Austral. Math. Soc.* **28** and Some noncommutative methods in algebraic number theory, *Proc. of Symp. in Honor of Emmy Noether's* 100*th birthday.*

PROBLEMS

Problem I. In the section on Integral Matrices I report mainly on the 2×2 case—however, in my paper on composition of quadratic forms I include the $n \times n$ case. Gauss did not have matrix theory, hence his work on quadratic forms had to stop there. In my paper "From cyclic algebras of quadratic fields to central polynomials" I also study $n > 2$. Although my papers on Ideal Matrices and the Latimer and MacDuffee theorem contain results for $n \geq 2$, my work on factoring an integral 2×2 matrix into the product of two symmetric integral factors has not been generalized so far. Hence I pose this as a problem.

Problem II. In the 1964 Caltech thesis of my student L. L. Foster, the following special case of a problem was studied leading to diophantine problems and other interesting observations:

Given two integral $n \times n$ matrices A, B (may be depending on parameters), in what fields can the eigenvalues of A, B lie (for special values of the parameters)? See *Pac. J. Math.* **18** (1966), 97–110.

Problem III. My work on $\det(AB - BA)$ for A, B integral 2×2 matrices, both with irrational eigenvalues in $Q(\sqrt{m})$, resp. $Q(\sqrt{n})$ leads to a noncommutative statement for the intersection of elements in these fields which are norms. I suggest more work on this.

Robert Osserman received a Ph.D. from Harvard University in 1955, work-ing in Riemann surface theory under the guidance of Lars Ahlfors. He then joined the mathematics department at Stanford University, where his research interests shifted towards differential geometry. In recent years he has made numerous research contributions to the theory of minimal surfaces and other topics in geometry. His elegant book A Survey of Minimal Surfaces *has surely contributed to the geometry renaissance he describes in his article.*

The Geometry Renaissance in America: 1938–1988

ROBERT OSSERMAN

Freeman Dyson, in his 1972 Gibbs lecture to the AMS: "Missed Opportu-nities," and in a 1981 lecture to the Humboldt Foundation: "Unfashionable Pursuits," urges us to look beyond the narrow confines of those subjects and pursuits that happen to be in fashion at a given time. Thinking back to my days as a graduate student I have no trouble in distinguishing what was fash-ionable from what was not, although at the time I would probably have been shocked to hear the word "fashion" used to describe what seemed to be sim-ply the important and exciting areas of research. Certainly Bourbaki was the height of fashion. Also, anything "algebraic," whether topology, geometry, or analysis, had added panache. At the other end of the spectrum there were subjects such as partial differential equations and functions of one complex variable, that had been declared dead for so long that it would be hard even to find mourners.

And somewhere past them, beyond the pale, was differential geometry. It was simply not an option. At least in the case of partial differential equations and complex variables there were faculty members active in those areas, with whom one could write a dissertation. I doubt if there were any Ph.D. theses in differential geometry at Harvard for a period of some 20 years, from the late thirties to the late fifties. During my five years in residence I recall a course in the subject being offered just once. It was given by Ahlfors, and was an excellent course, although not designed to lead to research in the subject. Ahlfors' own work is infused with a deep geometric sense. However, he was not a geometer, in the sense of making contributions to the field; what he did do was to make brilliant use of differential geometry in various parts of function theory. A simple, but penetrating example, was his far-reaching generalization of Schwarz' Lemma. Pick had extracted the geometric content of Schwarz' Lemma by interpreting it as a statement about arc length in the Poincaré metric. Ahlfors showed that it applied much more widely to metrics with certain curvature constraints. At the same time, he de-geometricized the lemma to a certain extent by revealing its roots in partial differential equations. The generality of the method allowed the result to be extended to higher dimensions and to many different classes of mappings. Similarly, Ahlfors' various geometric approaches to Nevanlinna theory made possible the many subsequent higher-dimensional generalizations. But once again, Ahlfors was not primarily a geometer, and none of his Ph.D. students wrote a thesis on differential geometry.

In a broader context, the picture at other major research centers, such as Princeton, Chicago, and MIT was not very different from that at Harvard. At Princeton, Eisenhart had become Dean of the Graduate School in 1933. Between 1938 and his retirement in 1945, his only publications were two introductory textbooks. Veblen's interests were very broad, and included a period of concentration on differential geometry in the twenties during which J. H. C. Whitehead and T. Y. Thomas were his students. But Veblen left Princeton University for the Institute for Advanced Study in 1932, and retired in 1950 at age 70. There was no apparent move to replace either Eisenhart or Veblen with a young geometer. The one locus of geometric activity was Bochner. Like Ahlfors, he was primarily an analyst, but unlike Ahlfors, he had begun in the late forties to work actively in certain areas of geometry, to which he made significant contributions. His Ph.D. students included Rauch in 1947 and Calabi in 1950. However, it should be noted that Rauch's thesis was in analysis. His seminal work in differential geometry — the Rauch comparison theorem and its application to his theorem that a compact manifold whose curvature is close to that of a sphere must be homeomorphic to a sphere—stemmed from a postdoctoral year spent at Zürich, where Heinz Hopf apparently posed the question.

Like Eisenhart at Princeton, there was Struik at MIT and Lane in Chicago — geometers of an earlier generation who during the forties were not doing work destined to have a major impact on the future course of geometry.

Writing on "Fifty years of American Mathematics" for the AMS Semicentennial Publication, G.D. Birkhoff describes the situation in 1938 in these words:

> It must be admitted ... that there are few of our younger men who occupy themselves with algebraic or classical differential geometry, or any other of the geometric questions which seemed most vital fifty years ago.

There is no doubt that the figures who were active in the twenties and thirties and who were to affect most profoundly the future course of geometry were to be found in Europe. First and foremost (in retrospect) there was Élie Cartan in France. In Switzerland, Heinz Hopf, though primarily a topologist, was a powerful force in geometry. Germany maintained a strong geometric tradition, led at the time by Blaschke. There was simply nothing comparable in the United States.

When I say "nothing comparable," I am referring to mathematicians of the stature of Blaschke and Cartan who would classify themselves (and be classified by others) as differential geometers. If one broadens the scope just a bit, one would clearly want to include such major figures as Marston Morse, with his calculus of variations in the large, Hassler Whitney, the inventor of differentiable manifolds and sphere bundles, and Jesse Douglas, winner of one of the first two Fields Medals in 1936, whose solution of Plateau's problem had a geometric component, although it would have to be viewed primarily as analysis. In Morse's case, his application of "Morse theory" to the study of geodesics on a Riemannian manifold would certainly count as a major contribution to differential geometry.

At the Institute for Advanced Study there was also Hermann Weyl. In a curious footnote to the history of geometry, he provided a crucial link between the classical and the general Gauss–Bonnet theorems, after he attended a lecture in 1938 at the Princeton Mathematics Club by the statistician Harold Hotelling. In order to analyze a certain statistical problem, Hotelling wanted formulas for the volumes of tube domains around submanifolds of Euclidean space or of a sphere. Weyl's treatment of the problem led directly to the papers by Allendoerfer and Chern, as well as to later work on submanifolds and integral geometry.

When we look for those core geometers in America whose theorems we still quote, we find them scattered about and isolated. Carl Allendoerfer was at Haverford College and Sumner Myers at the University of Michigan. There was also J. L. Synge, originally at the University of Toronto and in the mid-forties at Ohio State and Carnegie Tech. His well-known geometric results date from the twenties. He later turned to more applied areas, including a period during World War II, when, like Allendoerfer and a number of other mathematicians, he deferred pursuing his own research in order to participate directly in the war effort. Another rather idiosyncratic figure was Herbert Busemann, who was at the Illinois Insitute of Technology and the University of Southern California during the forties, working out his beautiful theories of geometry in the nondifferentiable setting.

If we wish to look beyond the general Bourbaki trend to explain the un-
fashionableness of differential geometry in the forties, we must attribute it
at least in part to the *kind* of geometry that was current at the major institutions in the thirties. Graustein at Harvard, Eisenhart in Princeton, and Lane
at Chicago were not proving the sort of theorems that were destined to be
memorable and influential. Even more, the subject as a whole seemed to have
run out of steam after a surge of fundamental work, including that of Ricci
and Levi–Civita at the turn of the century. There was the much-lamented
"debauch of indices," covering or substituting for geometric content. Furthermore, there was a sense of isolation from the rest of mathematics.

The turnaround can be traced to a series of developments that served first
to renew some of geometry's ties with other fields, and then gradually to move
geometry more and more toward center stage.

First, and perhaps most important, came the development of a global theory, relating geometry to topology. The work of Allendoerfer, Myers, and
Synge already mentioned was almost all in that direction as was that of Hopf,
Cohn–Vossen, Preissmann, and much of Blaschke's. One culmination was
Chern's work on the general Gauss–Bonnet theorem and on characteristic
classes. Still other geometry/topology links arose out of Bochner's vanishing
theorems in the late forties, which in turn had a major impact on algebraic
geometry through Kodaira. Then in the fifties came Rauch's comparison theorem and all the results flowing out of that, in particular the sphere theorems
of Berger and Klingenberg. Both topological and algebraic components were
present in work on Lie groups and their quotients — homogeneous and symmetric spaces — much of which grew out of Cartan's fundamental work. Bott
and Samelson presented Marston Morse in 1958 with a singularly appropriate
65th birthday present consisting of a beautiful application of Morse theory
to the study of symmetric spaces. Like Bott and Samelson, Milnor would be
viewed as primarily a topologist, but with significant geometric interests.

The links between geometry and topology were the central focus of two
series of lectures by Heinz Hopf during visits to the United States: the first
at New York University in 1946 and the second at Stanford University in
1956. Both were written up informally as lecture notes. The Stanford notes,
devoted to the global theory of surfaces, went through several "editions"
and were circulated for years as an underground classic, providing for many
their introduction to differential geometry in the large. In 1983, they were
finally published officially, along with the NYU notes, when Springer–Verlag
was looking for something special to appear as Volume 1000 in their series,
"Lecture Notes in Mathematics."

In another direction, there were the connections with partial differential
equations as exemplified by the work of Philip Hartman and Louis Nirenberg.

But for a single most decisive factor contributing to the rebirth of geometry in America, I would propose the move of Chern to the United States from China in the late forties.

Shiing-shen Chern began his studies in China. He went to Hamburg to work with Blaschke from 1934 to 1936, receiving his doctorate in 1936. He then spent a year in Paris with Élie Cartan before returning to China in 1937. He remained there until the end of 1948, except for two years, 1943–1945, at the Institute for Advanced Study in Princeton. During those two years he did some of his most important work, including his intrinsic proof of the Gauss–Bonnet theorem, and his fundamental paper on characteristic classes, referred to earlier.

In 1949, Chern joined the mathematics department at the University of Chicago. There ensued the first mass production of high-caliber geometry Ph.D.s in United States history, starting with Nomizu in 1953. But Chern's influence was far wider than that. It spread to MIT via Singer, who was a student at Chicago, not at the time a geometer. Singer attended Chern's lectures and then developed his own course at MIT. There he made several converts, including Warren Ambrose and Barrett O'Neill, neither of whom started out in geometry. It was through Chern that André Weil was led to his contributions to the theory of characteristic classes. During the fifties, Chern wrote joint papers with Spanier, Kuiper, Hartman, Wintner, Lashof, Hirzebruch, and Serre. These collaborations accelerated the movement I referred to earlier in which differential geometry was gradually integrated with surrounding areas of mathematics: algebraic topology, algebraic geometry, and partial differential equations.

The same period witnessed one of the more anomalous and yet not insignificant features of the transformation and revitalization of geometry: the notation wars. There was fairly uniform disaffection for the classical coordinate-based notation for vectors, tensors, and forms. It worked well enough for surfaces, where there were fewer indices to manipulate, and where the existence of special types of coordinates allowed significant simplifications. But in higher dimensions the notation could itself be a deterrent to approaching a geometric problem. When Cartan introduced his method of moving frames, it was seen by many as the perfect tool. It was wholeheartedly adopted and promoted by Chern, who took full advantage of its flexibility and advantages, for example in moving up and down from a manifold to a frame bundle or tangent bundle, with perhaps a slight sleight of hand in employing the same "ω_{ij}" in the two contexts.

But the Cartan notation had its own drawbacks. It by no means eliminated the indices. It still involved making arbitrary choices of local frame fields,

forming expressions based on those choices, and then checking behavior under changes of frame field. Finally, one of its strengths — its relative ease of
manipulation in computation involving covariant or exterior derivatives —
had a negative side in a more distant relationship to the underlying geometry.
The much-maligned old-fashioned terminology of gradient, curl, and divergence were thought by many to have been rendered obsolete, as they could
all be subsumed as special cases of the exterior derivative acting on forms of
varying degrees. However, they had (and still have) the advantage of direct
geometric meaning and illuminating physical interpretations. Somewhat in
this spirit, a new notation was invented by Koszul and apparently first introduced in print by Nomizu in the mid-fifties. The fundamental operation
is the covariant derivative of a vector field with respect to a given tangent
vector at a point. The notation is totally independent of local coordinates or
frame fields and is free of indices. It was quickly adopted by a large number
of geometers. However, it too had its drawbacks, including considerable awkwardness for certain types of computations. Thus, unlike most earlier battles
over notation, such as the one in calculus where Leibniz' notation won an almost total victory over Newton's, the outcome here has been a standoff. Just
as one formerly had to learn two modern languages and one dead language,
an aspiring geometer needs now to learn two modern systems of notation
(Cartan and Koszul) and one dead one (coordinates) to be able to read 20th
century papers and books. It may be worth adding that it is not always "just"
a question of notation, since even the content of a theorem may be affected:
proving the existence of a certain kind of frame field is not equivalent to
proving the existence of certain local coordinates.

Before leaving the fifties I should mention one more somewhat isolated,
but important result, which had resonances far beyond its immediate consequences. That was John Nash's embedding theorem. For the first time
one knew that the class of Riemannian manifolds coincided with the class of
submanifolds of Euclidean space with the induced metric. The proof was a
tour de force of original ideas, seemingly coming out of nowhere.

The sixties began with Chern's move from Chicago to Berkeley, which
gradually became a central focus of geometric activity. The decade ended
with the emergence of a new generation of first-rate geometers, including Alan
Weinstein from Berkeley, Jeff Cheeger from Princeton, and Blaine Lawson
from Stanford. In the interim, a powerful force adding to the momentum of
the subject was the appearance of a new generation of books offering modern
presentations and viewpoints, and making use of one of the newer notations.
They almost immediately supplanted the older classics of Eisenhart vintage.
Among them were the hardcover texts by Helgason, Kobayashi and Nomizu,
Bishop and Crittenden, and Sternberg, as well as the no-less-significant soft-
cover lecture notes by Hicks, Berger, Gromoll–Klingenberg–Meyer, and Milnor's notes on Morse Theory. The decade of feverish bookwriting came to a

fitting end at 3:30 a.m. on July 6, 1970, when Michael Spivak finished the preface to Volume II of his *Comprehensive Introduction to Differential Geometry*. In that remarkable book, Spivak takes the reader step-by-step from the origins of differential geometry in the 18th century, through the fundamental papers of Gauss and Riemann, the contributions of Bianchi and Ricci, and into the thicket of the various concepts of a "connection," as seen from the points of view of Levi–Civita, Cartan, Ehresmann, Koszul, and others. In the process, he gives a constructive proof of the invariance of differential geometry under changes of notation by taking one theorem: *a Riemannian manifold with vanishing Riemann curvature tensor is locally isometric to euclidean space*, (referred to as "The Test Case") and providing seven different proofs, from each of seven different viewpoints or notations.

There is no doubt that the spate of new books helped make differential geometry more accessible and interesting to students. But even more important for the health and growth of the subject were the spectacular successes on the research front. The most celebrated was the Atiyah–Singer index theorem — a grand synthesis of analysis, topology, and geometry leading, in particular, to a new way of viewing the Gauss–Bonnet theorem: not as an isolated result, but as one instance of a larger scheme of things.

Other papers were less noted for specific results than for their seminal nature, in some cases laying the foundation for whole new areas of study. Among them are:

1. The 1960 paper on normal and integral currents by Federer and Fleming, which led to the creation of geometric measure theory and to the definitive book on the subject by Federer in 1969.
2. Eells and Sampson's 1964 paper on harmonic mappings of Riemannian manifolds. Although the notion of a harmonic map was not new, this paper was the starting point for the whole future development of the subject.
3. Palais and Smale's 1964 paper on a generalized Morse Theory. This paper, together with others around the same time by each of the authors and earlier ones by Eells, laid the foundation for the study of infinite dimensional manifolds. Lang's book on differentiable manifolds adopted a similar viewpoint and was also influential. The subject was called "global analysis," and found significant applications in the seventies.
4. Kobayashi's 1967 paper introducing an invariant pseudodistance on complex manifolds. Unlike the elaborate machinery used to set up the basis for geometric measure theory and for global analysis, the fundamental idea here is quite elementary and seems almost simpleminded, somewhat like the notion of cobordism. However, the implications have been profound, the latest being unsuspected connections

with Diophantine analysis, described in the 1987 book of Lang on complex hyperbolic spaces.

5. Also in 1967, McKean and Singer's paper on curvature and the eigenvalues of the Laplacian. It played a big role in the subsequent development of that subject.

6. Mostow's rigidity theorem of 1968. It was the first of a whole series in which "the topology determines the geometry." The proof uses and extends Gehring's basic results on higher-dimensional quasiconformal mappings.

7. Also in 1968, Simons' fundamental study of minimal varieties in Riemannian manifolds. This is the first serious account of the subject in full generality, and is chiefly responsible for moving the field of minimal surfaces from a somewhat marginal position to a more central one in differential geometry. There are a number of interesting results in the paper, but the most notable concerns Bernstein's Theorem: *if an n-dimensional minimal hypersurface S in $\mathbb{R}^{n+1}$ has a one-to-one projection onto a hyperplane, then S is itself a hyperplane.* Combining some of the results in his paper with earlier developments using geometric measure theory, Simons proved Bernstein's Theorem for dimensions $n \le 7$. The following year, in 1969, Bombieri, de Giorgi, and Giusti finished the story in a startling fashion: Bernstein's Theorem is false for $n > 7$. Perhaps the only comparable example of a dimensionally-dependent discontinuity in behavior was the founding fact of differential topology: Milnor's exotic 7-sphere, discovered a decade before. There have been various attempts to link the two phenomena, but none have been totally convincing.

I need hardly add that there was a lot more notable work in differential geometry than the sample I have described here. Some old problems were being settled — such as Blaschke's conjecture, by Leon Green, in 1963, and the topology of positively curved complete manifolds by Gromoll and Meyer in 1969 — at the same time as new areas were opening up and a new range of questions being posed.

But it was in the seventies that the field of differential geometry came into full blossom. For the first time, there were whole groups of geometers, rather than one or two isolated individuals, at several universities, most notably Berkeley, SUNY at Stony Brook, and the University of Pennsylvania. It would be hard to even begin to describe the scope of new accomplishments. But it is worth noting that the decade started with Thurston and Yau both doing their graduate work at Berkeley. Thurston went on to Princeton where he inaugurated his monumental study of hyperbolic geometry, and Yau went to Stanford, where his accomplishments included the solution of the Calabi conjecture, a part of the Smith conjecture (together with Meeks, a student of

Lawson at Berkeley) and the positive mass conjecture in relativity (together with R. Schoen — a joint student of Yau and Leon Simon at Stanford).

By the 1980s, the mathematical world was finally ready to award its first Fields Medal ever in differential geometry. Not only one, but two: to both Thurston and Yau.

Another major boost for geometry in America during the seventies was the presence of Gromov at Stony Brook from 1974 to 1980. During that period he wrote his fundamental papers on almost flat manifolds and on bounds for topological types of a manifold with certain curvature and volume constraints. In addition, he did important work on a variety of topics including isoperimetric inequalities, smooth ergodic theory, and scalar curvature (with Lawson). In 1980, he shared with Yau the Veblen prize of the American Mathematical Society. The full title, incidentally, is the "Oswald Veblen Prize in Geometry." It was set up after Veblen's death in 1960. The first seven recipients were all in topology, and it was not till 1976, with Simons and Thurston, that work in geometry proper was deemed award-worthy.

One sign of the burgeoning health of geometry in the eighties can be seen in the growth of regular geometry conferences, such as the Pacific Northwest Geometry Seminar, held three times a year on the West Coast, and the annual Geometry Festival in the East, whose attendance has been growing exponentially. The 1988 Geometry Festival, held at Chapel Hill, North Carolina, had one striking feature: a large proportion of the talks dealt with the construction of specific examples. There was a general feeling, explicitly expressed by Gromoll, that in the past one had been fairly free in making conjectures based on very little concrete evidence, whereas now for the first time we were building a solid basis for our conjectures in the form of examples of manifolds with prescribed topology, curvature of one sort or another, and possibly other geometric constraints, such as diameter or volume bounds. Among the talks at Chapel Hill was one by Gang Tian about his work with Yau on obtaining complete Kähler–Einstein metrics with prescribed Ricci curvature for various noncompact manifolds, and another by Nicolaos Kapouleas presenting his construction of compact and complete surfaces of constant mean curvature in $\mathbb{R}^3$ with prescribed topology. In both of these cases, the proofs involved highly sophisticated uses of partial differential equations.

The eighties had already produced other renowned examples. In 1986, Wente produced his torus of constant mean curvature immersed in $\mathbb{R}^3$, thus answering a question posed by Heinz Hopf in 1951, whether any compact surfaces other than the sphere could be immersed in $\mathbb{R}^3$ with constant mean curvature. (Hopf had proved that starting with a sphere, any such immersion would have to have the standard sphere as its image, and A.D. Alexandrov had shown that a higher genus surface could not be *embedded* in $\mathbb{R}^3$ with constant mean curvature.)

Can we conclude that the pendulum of fashion has now come full swing, from the Bourbaki ideas of generality and structure in the fifties, to the concrete, specific and intuitive in the eighties? And to the degree that it may be true, what does it portend for the future? One of the big unknowns is the impact that computers in general, and computer graphics in particular, will have on the direction and accomplishment of geometric research. The most striking example to date has been the discovery of new families of complete embedded minimal surfaces, with computer graphics playing a significant role. The story (and the pictures) can be found in an article by David Hoffman in the 1987 *Mathematical Intelligencer*. More recently, Hoffman has collaborated with a group of polymer scientists in examining various periodic minimal surfaces and surfaces of constant mean curvature as models for certain interfaces recently revealed by electron microscope photographs. This work appears as the cover article of the August 18, 1988 issue of the journal *Nature*.

Thus, in time for the AMS centennial, differential geometry has recovered not only its links with other parts of mathematics, but its roots in physical reality. It has also entered the realms opened up by the new computer technology. There are now active groups using computer graphics at the University of Massachusetts at Amherst, the University of California at Santa Cruz, Brown University, and Princeton, as well as the new Geometry Supercomputer project, whose goal is to provide a number of mathematicians with high resolution systems, all linked to each other and to a supercomputer at the University of Minnesota. Whether the outcome will be a series of exciting and fundamental new developments, or just a flurry of special cases leading to a renewed cry for a Bourbaki-type clarification and cleansing remains to be seen. That judgement will no doubt be made by the time of the AMS sesquicentennial celebration in 2038.

Postscript. The "geometry" in the title may seem to promise more than the text delivers. I have in fact dealt only with one facet: differential geometry. I did not restrict the title, because I believe that other parts of geometry enjoyed a similar renaissance, but I leave it to others to fill in the details. Even in differential geometry, I do not feel I have done justice to the whole field, but have concentrated on what I know best. In order to compensate at least in part for my own limited knowledge and perspective, I have consulted with a number of people who have offered additional background, comments and suggestions. They are Garrett Birkhoff, Eugene Calabi, Jeff Cheeger, Irving Kaplansky, Blaine Lawson, Cathleen Morawetz, Barrett O'Neill, Halsey Royden, Hans Samelson, James Simons, Isadore Singer, and George Whitehead. My thanks to all of them.

Bibliography

1938 L.V. Ahlfors, An extension of Schwarz's lemma, Trans. Amer. Math. Soc. 43, 359–364.

G.D. Birkhoff, Fifty years of American mathematics, *AMS Semicentennial Publications,* Vol. 2, AMS, New York, pp. 270–315.

1939 H. Hotelling, Tubes and spheres in n-spaces, and a class of statistical problems, Amer. J. Math. 61, 440–460.

H. Weyl, On the volume of tubes, Amer. J. Math. 61, 461–472.

1940 C.B. Allendoerfer, The Euler number of a Riemannian manifold, Amer. J. Math. 62, 243–248.

1941 S.B. Myers, Riemannian manifolds with positive mean curvature, Duke Math. J. 8, 401–404.

1943 C.B. Allendoerfer and A. Weil, The Gauss-Bonnet theorem for Riemannian polyhedra, Trans. Amer. Math. Soc. 53, 101–129.

1944 S.-S. Chern, A simple intrinsic proof of the Gauss-Bonnet formula for closed Riemannian manifolds, Ann. of Math. 45, 747–752.

1945 S.-S. Chern, On the curvatura integra in a Riemannian manifold, Ann. of Math. 45, 674–684.

1946 S. Bochner, Vector fields and Ricci curvature, Bull. Amer. Math. Soc. 52, 776–797.

S.-S. Chern, Characteristic classes of Hermitian manifolds, Ann. of Math. 47, 85–121.

1948 S. Bochner, Curvature and Betti numbers, Ann. of Math. 49, 379–390. (Part II, Vol. 50 (1949), 77–93).

1951 H.E. Rauch, A contribution to differential geometry in the large, Ann. of Math. 54, 38–55.

1953 L. Nirenberg, The Weyl and Minkowski problems in differential geometry in the large, Comm. Pure Appl. Math. 6, 337–394.

1954 K. Nomizu, Invariant affine connections on homogeneous spaces, Amer. J. Math. 76, 33–65.

1955 H. Busemann, *The Geometry of Geodesics,* Academic Press, New York.

1956 J.F. Nash, The imbedding problem for Riemannian manifolds, Ann. of Math. 63, 20–63.

1958 R. Bott and H. Samelson, Applications of the theory of Morse to symmetric spaces, Amer. J. Math. 80, 964–1029.

J. Eells Jr., On the geometry of function spaces, *Symposium de Topologia Algebrica,* Mexico, 303–307.

1960 H. Federer and W.H. Fleming, Normal and integral currents, Ann. of Math. 72, 458–520.

1962 S. Helgason, *Differential Geometry and Symmetric Spaces,* Academic Press, New York.

S. Lang, *Introduction to Differential Manifolds,* Interscience, New York.

1963 L. Green, *Auf Wiedersehensflächen,* Ann. of Math. 78, 289–299.

S. Kobayashi and K. Nomizu, *Foundations of Differential Geometry,* Vol. I, Interscience, New York.

J.W. Milnor, *Lectures on Morse Theory,* Ann. of Math. Studies No. 51, Princeton Univ. Press, Princeton, NJ.

R.S. Palais, Morse theory on Hilbert manifolds, Topology 2, 299–340.

1964 R.L. Bishop and R.J. Crittenden, *Geometry of Manifolds,* Academic Press, New York.

J. Eells, Jr. and H. Sampson, Harmonic mappings of Riemannian manifolds, Amer. J. Math. 86, 109–160.

R.S. Palais and S. Smale, A generalized Morse theory, Bull. Amer. Math. Soc. 70, 165–172.

1965 M. Berger, *Lectures on Geodesics in Riemannian Geometry,* Tata Institute of Fundamental Research, Bombay.

N.J. Hicks, *Notes on Differential Geometry,* Van Nostrand, Princeton, NJ.

1967 S. Kobayashi, Invariant distances on complex manifolds and holomorphic mappings, J. Math. Soc. Japan 19, 460–480.

H. McKean and I.M. Singer, Curvature and eigenvalues of the Laplacian, J. Diff. Geom. 1, 43–69.

1968   D. Gromoll, W. Klingenberg, and W. Meyer, *Riemannsche Geometrie in Grossen,* Springer, Berlin.

G.D. Mostow, Quasi-conformal mappings in n-space and the rigidity of hyperbolic space forms, IHES Publ. Math. 34, 53–104.

J. Simons, Minimal varieties in Riemannian manifolds, Ann. of Math. 88, 62–105.

1969   E. Bombieri, E. de Giorgi, E. Giusti, Minimal cones and the Bernstein's problem, Invent. Math. 243–268.

H. Federer, *Geometric Measure Theory,* Springer, Berlin.

D. Gromoll and W. Meyer, On complete open manifolds of positive curvature, Ann. of Math. 90, 75–90.

S. Kobayashi and K. Nomizu, *Foundations of Differential Geometry,* Volume II, Interscience, New York.

1972   F. Dyson, Missed Opportunities, Bull. Amer. Math. Soc. 78, 635–652.

1973   G.D. Mostow, *Strong Rigidity of Locally Symmetric Spaces,* Ann. of Math. Studies 78, Princeton University Press, Princeton, NJ.

1978   M. Gromov, Manifolds of negative curvature, J. Differential Geom. 13, 223–230.

M. Gromov, Almost flat manifolds, J. Differential Geom. 13, 231–241.

W. Thurston, *The Geometry and Topology of 3-Manifolds,* Lecture Notes, Princeton University.

S.-T. Yau, On the Ricci curvature of compact Kähler manifolds and complex Monge-Ampère equations, I. Comm. Pure Appl. Math. 31, 339–411.

1979   R. Schoen and S.-T. Yau, On the proof of the positive mass conjecture in general relativity, Comm. Math. Phys. 65, 45–76.

1981   M. Gromov, Curvature, diameter and Betti numbers, Comment. Math. Helv. 56, 179–195.

R. Schoen and S.-T. Yau, Proof of the positive mass theorem, II, Comm. Math. Phys. 79, 231–260.

1983 F. Dyson, Unfashionable pursuits, The Mathematical Intelligencer 5, No. 3, 47–54.

H. Hopf, *Differential Geometry in the Large,* Lecture Notes in Mathematics 1000, Springer, Berlin.

1984 W.H. Meeks III and S.-T. Yau, The equivariant loop theorem for three-dimensional manifolds and a review of existence theorems for minimal surfaces, in *The Smith Conjecture,* Academic Press, New York, pp. 153–163.

1986 H.C. Wente, Counterexample to a conjecture of H. Hopf, Pacific J. Math. 121, 193–243.

1987 D. Hoffman, The computer-aided discovery of new embedded minimal surfaces, The Mathematical Intelligencer 9, No. 3, 8–21.

S. Lang, *Introduction to Complex Hyperbolic Spaces,* Springer, New York.

1988 E.L. Thomas, D.M. Anderson, C.S. Henkee, and D. Hoffman, Periodic area-minimizing surfaces in block copolymers, Nature 334, pp. 598–601.

Raoul Bott was born in Budapest. He received his bachelor's and master's degrees in electrical engineering from McGill University, and in 1949 a doctorate in applied mathematics from Carnegie Institute of Technology, where he worked in network theory under the guidance of R. J. Duffin. His conversion to topology was then consummated during a two-year stay at the Institute for Advanced Study. After working for the next eight years at the University of Michigan and again at IAS, he accepted a professorship at Harvard University in 1959. Among his many honors is an AMS Veblen Prize in geometry. He is a member of the National Academy of Sciences and a recipient of the National Medal of Science. This article is the text of an AMS–MAA Invited Address at the Providence Meetings on August 9, 1988.

The Topological Constraints on Analysis

RAOUL BOTT[1]

The other day I mentioned to Erna Alhfors that I was reading some auto-biographical remarks of Jung. "How can you read an autobiography, Raoul," she reproached me. "Don't you know they are just a pack of lies?"

I start with this anecdote to give you fair warning. For, of course, what I have prepared for the occasion are precisely some reminiscences pertaining to my general topic. On balance, I decided that at sentimental occasions such as birthdays, personal lies are still to be preferred to quite impersonal verities.

It has been my honor—involving more work than you might think — to speak in this forum once before and then I talked about the years 1949–1951 at the Institute for Advanced Study in Princeton. This time I would like to talk about my second visit there in 1955–1957. For if I had learned some topology during my first stay, then this second visit was my initiation to a little analysis—granted in a rather algebraic form. But above all I chose this topic because that visit gave me a personal glimpse into a momentous period in the world of mathematics altogether and an especially significant one in the mathematical life of this country.

[1] Work on this article was supported in part under NSF grant DMS 86 05482.

For the start of my story I will take two famous mathematical paradigms of this century: (1) The Maximum Principle and (2) The Brouwer Fixed Point Theorem. Recall here that the Maximum Principle asserts that on a compact space a continuous function must *attain its maximum*. It is from this bedrock that the Birkhoff Minimax and the subsequent Morse, and Lusternick–Schmirelman theories spring, and even when the hypothesis of compactness fails, as it does in all the "harder" calculus variation problems, e.g., Minimal Surface Theory, Yang–Mills Theory, it is still this extension that points the way.

It was this first branch that captivated me during my first stay at Princeton and I cannot let it go by even now without comment. In a sense the "Morse Theory" *quantifies* the maximum principle. If X is a space, the measure of its *topological complexity* was taken by Morse to be its *Poincaré polynomial*, $P_t(X)$. This polynomial is defined by

$$P_t(X) = \sum b_i(X) \cdot t^i,$$

where the $b_i(X)$ are the Betti numbers of the space X. (They — roughly — count the number of holes in X, so that, for instance, $b_0(X)$ measures the number of connected pieces into which X falls.)

Now Morse replaces "maximum" by "extremum" and attaches a measure of complexity $\mu_t^f(Y)$ to every extremum Y of a function f on X. The total complexity of f is then measured by

$$\mathcal{M}_t(f) = \sum_Y \mu_t^f(Y),$$

where Y ranges over the extrema of f. He then shows that under suitable conditions $P_t(X)$ *serves as a lower bound* for $\mathcal{M}_t(f)$. More precisely, Morse writes

$$\mathcal{M}_t(f) - P_t(X) = (1+t)E(t),$$

and then shows that the "error term" $E(t)$ must be a "positive" polynomial in the sense that its coefficients are ≥ 0. In short, Morse's maxim is that "topological complexity of a space goes hand in hand with the extremal complexity of functions on the space." Or to put it most simply, "topological complexity forces extrema."

Let me show you how all this works in an example from analysis which, however, is usually *not* thought of this way in strictly analytical circles. Namely, consider the eigenvalue problem:

$$L(y) \equiv \frac{d^2 y}{dx^2} + q(x)y = \lambda y$$

on the unit interval $0 \leq x \leq 1$ with boundary conditions $y(0) = y(1) = 0$. Now associated to L is its quadratic form:

$$Q(y) = \int_0^1 Ly \, y \, dx = \int (-y'^2 + q(x)y^2) \, dt$$

and you should think of it as a function on the unit sphere S^∞:

$$\int y^2 \, dt = 1,$$

in an appropriate Hilbert space of functions y. For then, as is well known, the extrema of Q on S^∞ correspond to the solutions of (4). Indeed, the Lagrange procedure tells us to look for the extrema of $Q(y) + \lambda \int y^2 \, dt$, and these immediately lead to (4).

But note now that both $Q(y)$ and the constraint $\int y^2 \, dt = 1$ have the symmetry $y \to -y$, so that *properly speaking* the above argument should be applied to the projective space rather than S^∞, and this innocent remark is crucial to us because the topology of $\mathbf{R}P_\infty$ is much, much more complicated than that of the sphere S^∞ at least if we work mod 2. In fact, $\mathbf{R}P_\infty$ then has a "hole" in every dimension so that

$$P_t(\mathbf{R}P_\infty) = 1 + t + t^2 + \cdots,$$

while the *sphere* in an infinite-dimensional *Hilbert space* has no holes whatsoever:

$$P_t(S^\infty) = 1.$$

Hence, if we know *Morse theory* we would immediately expect this problem to have an infinite number of eigenvalues — as of course it does.

But even more is true. Regardless of the degeneracy of the eigenvalues of Q, they are such as to precisely account for the holes in $\mathbf{R}P_\infty$, in short the error term $E(t)$ vanishes *on the nose* for this example, and one speaks of a "perfect Morse function." Indeed if Y is an extremum of Q on $\mathbf{R}P_\infty$, corresponding to an eigenvalue λ of (4) with multiplicity k, then $Y = \mathbf{R}P_k$ and in Morse's way of counting Y is seen to contribute by

$$\mu_t^Q(Y) = P_t(\mathbf{R}P_k) \cdot t^{\lambda_Y}$$
$$= (1 + t + \cdots + t^k)t^{\lambda_Y}$$

with λ_Y equal to the number of eigenvalues less than λ. Hence the μ_ts clearly add up to $P_t(\mathbf{R}P_\infty)$. Q.E.D.

Two comments are now certainly in order:

(1) In some sense the Morse theory or one of its variants is the only way to treat the nonlinear extensions of our problem, although this is often hidden in the actual accounts by analysts.

(2) Secondly, note that the interesting topology in this example was forced by the symmetry, $y \to -y$ of the problem, which was treated here in the most straightforward way imaginable — that is, by simply *dividing by the action of* $\mathbf{Z}_2$. In the more difficult cases, i.e., in the Yang–Mills theory, the extrema are again seen to be created by what the physicists call the group of "gauge transformations," i.e., the symmetries of the problem. However, these now have to be treated in a much more subtle manner.

But so much for my first love in topology. To get on to my topic proper I have to pass on now to my second love, and as you will see, it has, quite characteristically, many similarities to my first one. This second love of mine deals with the fixed point phenomena, and the line I would like to draw here starts with Brouwer, goes on to Lefschetz, and then reappears again, after the deep insight of the 1950s due to the advent of sheaf theory, in the realm of elliptic differential equations. For it is in this guise that Atiyah and I recognized the Lefschetz phenomena again in 1964.

The magnificently simple statement of Brouwer's theorem is of course the following one: "*Any continuous map of the unit disc $D^n \subset \mathbf{R}^n$ into itself has a fixed point.*" I expect everyone here is familiar with this theorem and has used it in some form or another, if not for profound reasons, then at least to amuse some layman with one of its very unlikely consequences in everyday life.

The Lefschetz extension of this phenomenon again uses the same topological invariants of a space as the Morse theory does — but now understood in a more sophisticated sense. Indeed the Betti numbers $b_i(X)$ are really the dimensions of certain vector spaces, $H^i(X)$, "the cohomology spaces" of a space X, and these behave "functorially" with respect to continuous functions.

Precisely, this means that the vector space $H^i(X)$ is not only well defined in terms of X, but *moves* with X in the sense that a continuous function from X to Y,

$$\varphi \colon X \to Y,$$

induces a homomorphism denoted by φ^* — or by $H^i(\varphi)$ by the purists — going in the opposite direction:

$$H^i(Y) \xleftarrow{\varphi^*} H^i(X),$$

and subject to the very simple axiom that $(\varphi \circ \psi)^* = \psi^* \circ \varphi^*$ and that the homomorphism induced by the identity is the identity.

This "simple" and by now quite natural functorial concept due to Eilenberg and Mac Lane is of course also of the 1950s vintage, and its clarifying force on all our mathematical thinking cannot be overestimated. In any case in terms of such a "cohomology functor" $X \rightsquigarrow H^i(X)$ with $H^i(X)$ *finite dimensional* one can then define a Lefschetz number for

$$\varphi \colon X \to X,$$

by the formula

$$\mathscr{L}(\varphi) = \sum (-1)^i \operatorname{trace} H^i(\varphi),$$

which is then some sort of a numerical measure of the extent to which the topology of the space has been perturbed by φ. The Lefschetz theorem now asserts that if $\varphi \colon X \to X$ is a map of the compact polyhedron X into itself, and if $L(\varphi) \neq 0$ *then φ has a fixed point.*

Let me start with a few quite elementary observations concerning this theorem.

(1) When X is the disc, $H^0(X) = \mathbf{R}$ and $H^i(X) = 0$ for $i > 0$. Furthermore, $\varphi^* = 1$. Hence the Lefschetz theorem does extend Brouwer's theorem!

(2) When φ is the identity, then the Lefschetz number is clearly given by

$$\mathscr{L}(I) = \sum(-1)^i \dim H^i(X) = \sum(-1)^i b_i(X).$$

This alternating sum turns out to be much simpler to compute than the Betti numbers individually, and indeed is given by purely counting the number of cells one needs to decompose X! In fact, *this* Lefschetz number turns out to agree with the *oldest* of topological invariants, the Euler number of a polyhedron. That is,

$$\text{Euler}(X) = \sum(-1)^k (\# \text{ of } k - \text{cells in } X).$$

Thirdly, let me remark that the implication $L(\varphi) \neq 0 \Rightarrow \varphi$ has a fixed point, suggests a refinement of the Lefschetz theorem of the sort:

$$\mathscr{L}(\varphi) = \sum_p \mu(p),$$

where p ranges over the fixed point set of φ, and $\mu(p)$ is now some sort of numerical measure of the *local* behavior of φ near p. For instance, if φ is a smooth map of a compact manifold with isolated fixed points $\{p\}$ then the "*Hopf formula*" does just that. That is, under these circumstances,

$$\mathscr{L}(\varphi) = \sum \mu(p)$$

is valid with $\mu(p)$ an integer which measures how often a small sphere about the fixed point p is "wrapped about itself" by φ.

Finally, note that the Lefschetz theorem is not unrelated to the Morse inequalities. Indeed, if φ pushes M a little in the direction of the gradient of a function f on M, then the *fixed points* of φ are precisely the *critical points* of f, and φ now being deformable into the identity, we have $L(\varphi) = L(I)$, so that the Lefschetz number of φ is given by

$$\mathscr{L}(\varphi) = \text{Euler number}(M) = \sum_{df_p=0} \mu(p),$$

while the Morse inequalities yield (after setting $t = -1$):

$$\sum(-1)^i b_i(X) = \sum(-1)^{\lambda p}.$$

And indeed one checks here that in this case the Hopf formula gives $\mu(p) = (-1)^{\lambda p}$ so that these two formulas are identical. In short, the Morse inequalities refine Lefschetz's theorem for maps φ near the identity which are given by small gradient pushes of functions, while the Lefschetz formula isolates that part of the Morse inequalities which is valid for all maps.

Now what I learned at the Institute in these years, 1955–1957, and what had really only just burst upon the mathematical landscape a few years earlier, was that all these "old verities" could be fitted into a much larger framework, and it was my very good fortune that my principal "instructors" in this new lore were two young upstarts called J.-P. Serre and F. Hirzebruch.

The Institute had changed dramatically since 1949. Einstein was no more, von Neumann was gravely ill, C. L. Siegel had left, and Hermann Weyl was commuting to Switzerland. Oppenheimer was strangely altered after the adventures of the McCarthy period, and when I came to pay my respects his whole manner had somehow taken on an "Einsteinian" aura. Of course, my older friends were still quite unchanged: Marston Morse, A. Selberg, and Deane Montgomery — but if my previous stay was dominated by a feeling of awe at the brilliance of the older generation, during this second one, my dominant feeling was one of awe, alas, mixed with envy, at the brilliance of a generation younger or contemporary to my own!

And no wonder. During that time, and largely at Princeton, I met Serre, Thom, Hirzebruch, Atiyah, Singer, Milnor, Moore, Borel, Kostant, Harish-Chandra, James, Adams,..., and I could go on and on. But these people — together with Kodaira and Spencer — and my more or less "personal remedial tutor," Arnold Shapiro, were the ones I had most mathematical contact with.

It was Serre who tried to teach me sheaf theory. "Poor Bott," he would say, whenever he was on the verge of giving up. I had come to Princeton all excited to meet the great topologist, Serre, and, lo and behold, here he was — suddenly an algebraic geometer! But of course the other side of it was that in some sense Serre had brought topology and its techniques to algebraic and analytic geometry and so he could explain it to someone with my interests like nobody else.

Serre is a prime example of what I call a "smart mathematician" — as opposed to a "dumb one." What he knows is so crystal clear in his mind that he can give us lesser mortals the feeling that it is indeed all child's play. He also had, and still has, the infuriating habit of never seeming to work. In public one sees him playing ping pong, chess, or reading the paper — never in the sort of mathematical fog so many of us inhabit most of the time. If one asks him a question, he either knows the answer immediately, and then inside out, or he declines comment. I would ask him: "Well, have you *really thought* about this?" He would say, "How can I *think about it* when I don't know the *answer*!" Added in Proof: Serre disputes this interchange and asserts that it is just one of the many untruths in my "pack of lies." Mrs. Serre comments that from her view "Serre works all the time." And indeed he claims that all his true work is done in his sleep! What can one say? It is an unfair world — as we all know.

But I should return to my subject and tell you a little of what it was that I was being taught. Well, next to what Serre was explaining to me, it was the goings on at the University, namely the activities of the team of Kodaira, Spencer and Hirzebruch, which fascinated me most, and in particular, it was Hirzebruch's "Riemann–Roch" formula which quite took my breath away. So let me try to explain here a little the completely new view that "sheaf theory" brought to our understanding of cohomology, and let me illustrate this with the de Rham theory — which is in any case the most "immediate" concept with deep ramifications in physics, geometry, and topology.

Now "homology" is of course the most intuitive concept in topology. Thus $H_0(X)$ measures the number of connected components into which a space falls, $H_1(X)$ measures how many interesting loops — or 1-dimensional holes — X has and so on. Cohomology was, therefore, at first just "the dual" object to homology. For example, the 1-form

$$\omega = \frac{1}{2\pi} \frac{x\,dy - y\,dx}{x^2 + y^2}$$

defined in $\mathbf{R}^2 - 0$, is dual to the hole in $\mathbf{R}^2$ created by removing the origin, in the sense that the line integral

$$\int_\gamma \omega$$

over any oriented curve in $\mathbf{R}^2 - 0$ measures its winding number around 0.

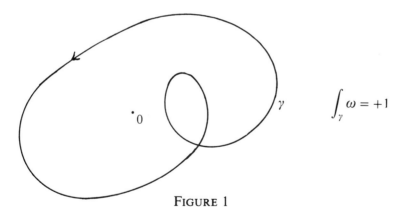

FIGURE 1

In "sheaf theory" (Leray, Cartan, Weil,...) which fashioned such a striking path between topology, analysis, and algebraic geometry, the relation of 1-form to winding numbers is quite *disregarded*. Instead one starts directly from a dual definition of connected component, and then follows this definition to its "logical" or, should I say its "homological" end. Namely, one defines the number of components of a space, as the dimension of the vector space $H^0(X)$ of "locally constant functions" on X. This concept is hopefully quite clear: f is locally constant on X if every point $p \in X$ has a neighborhood U with $f(p) = f(q)$ for all $q = U$.

Now if we are dealing, say, with an open set U in R^n, then the *locally constant C^∞-functions* can already serve to define $H^0(U)$, but these are now clearly characterized by a simple differential equation:

$$f \text{ locally constant} \iff df \equiv 0$$

where

$$df = \sum \frac{\partial f}{\partial x^i} \, dx^i.$$

In short, $H^0(U)$ can now be interpreted as the *vector space of solutions* on U of the differential equation

$$df = 0,$$

and of course one now finds that this equation makes sense in an intrinsic manner on *all C^∞-manifolds*. Furthermore the so-called higher de Rham groups $H^i(X)$ are now seen to be determined through a *"universal principle,"* really, of homological algebra, from H^0, of which the classical de Rham complex procedure is just one special example!

Recall that this so-called de Rham complex of a manifold X, usually notated

$$\Omega^*(X): \Omega^0(X) \overset{d}{\to} \Omega^1(X) \cdots \Omega^n(X)$$

consists of the collection of q-alternating covariant tensors on X, notated Ω^q, so that

$$\Omega^0(X) \equiv C^\infty(X),$$

and to which our d has a canonical "God-given" extension satisfying $d^2 = 0$. In terms of this extended d, the "de Rham" cohomology is given by

$$H^*(X) = \text{Ker } d/\text{Image } d.$$

That is, by

$$H^q(X) = \begin{cases} \text{solutions of } du = 0 \text{ in } \Omega^q \\ \text{modulo the trivial} \\ \text{solutions } u = dv, \; v \text{ in } \Omega^{q-1}. \end{cases}$$

Thus $H^0(X)$ is precisely the space of solutions of $du = 0$, while the others are more mysterious. Now what sheaf theory taught us was that actually all these "higher" cohomology groups could all already be determined from the behavior of just H^0, *but now considered as a function or, rather, functor over all open sets U in X!* Put otherwise, the sheaf

$$U \rightsquigarrow \underline{\mathbf{R}}(U) \equiv \text{locally constant functions on } U,$$

determines everything, and from this vantage point a whole new vista of inquiry appeared, a vista which we have been exploring for the past forty years and which still shows no signs of being exhausted. One new question was of course this: What has this universal cohomological construction to teach us when applied to the sheaf determined by the solutions of general differential equations $Df = 0$?

Now generalization for generalization's sake is always suspect in mathematics. But when the generalization sheds new light on old questions and provides new tools to tackle these old questions — then we have first-rate mathematics. And that was certainly the case in those early 50s, when all the ramifications of the homological point of view were falling into place, primarily in the realm of "complex analytic" and "algebraic" manifolds. The step here is to pass from the operator $df = \frac{\partial f}{\partial x^i} dx^i$ in $\mathbf{R}^n$ to the operator

$$\overline{\partial} f = \frac{\partial f}{\partial \overline{z}_i} d\overline{z}^i \text{ in } C^q.$$

In short, to replace the sheaf of locally constant *functions* $\mathbf{R}(U)$ on $U \subset M$, by the sheaf of locally *holomorphic functions* $O(U)$ on a complex manifold M. The resulting cohomology groups, usually denoted by $H^q(M; O)$, now turned out to be familiar in the classical literature for $q = 0, 1, 2$, and at the same time amenable to quite *new* manipulations inspired by their similarity with the old de Rham theory. It was into these mysteries that my friends were initiating us in Princeton at that time.

For those unfamiliar with the notion of an algebraic and complex manifold, let me recall here the most *fundamental* example of such an animal. That is the "Complex Projective Space": CP_n. It is obtained from $C^n - 0$, by identifying $z = (z_0, \ldots, z_n)$ with $(\lambda z_0, \ldots, \lambda z_n)$, $\lambda \in C^*$, z in $C^n - 0$. This space is compact because we can rescale any $z \in C^n - 0$ to have length $|z|^2 = \sum |z|_i^2$ equal to 1, where CP_n is a quotient space of the $(2n - 1)$-sphere, that is, we have a map

$$\pi \colon S^{2n+1} \to CP_n,$$

with fibers, the circles swept out by λz; $|\lambda| = 1$. (This is the famous Hopf fibering, which by now is well known and fundamental in so many branches of mathematics.)

Classical algebraic geometry really starts with the study of CP_n and all the "algebraic varieties" cut out by a system of polynomials in CP_n, in short, by the spaces $X \subseteq CP_n$ given by

$$z \in X \Leftrightarrow \{f_i(z) = 0\},$$

where $f_i(z)$ is a system of *homogeneous* polynomials in z. For example, the equation

$$z_0^n + z_1^n + z_2^n = 0$$

cuts out in CP_2, the Riemann surface

with $(n-1)(n-2)$ one-dimensional *holes*.

Now the first theorem about the structure sheaf O on such a variety X is really that the new cohomology group $H^i(X;O)$ are "finite-dimensional," and that amongst them one could discover all the "old" invariants of algebraic geometry. In fact, the new *holomorphic* Euler number:

$$\chi(X;O) = \sum(-1)^i \dim H^i(X;O)$$

now turned out to be the so-called "algebraic genus" of X, so that an old and famous inequality of Riemann and Roch could now be re-interpreted as a computation of $\chi(X;O)$ and more generally of $\chi(X;\mathscr{F})$ where $\mathscr{F}$ were certain *twisted* forms of the sheaf O.

This "twisting concept" which replaced ill-defined objects with poles — such as meromorphic functions — by smooth holomorphic "sections of a line bundle L over M" was another insight of this era — originally due to A. Weil, I believe, which was most welcome to topologists. After all, we cut our teeth on the arch *line bundle* of all — the infinite Möbius strip.

Recall here that this line bundle L arises from the strip $-1 \leq x \leq 1$ in $\mathbf{R}_2$ by identifying $(-1, y)$ with $(1, -y)$. This space clearly admits a projection onto the circle S^1, in the guise of the interval $-1 \leq x \leq 1$ where $+1$ has been identified with -1, and this projection π clearly has inverse images $\pi^{-1}(x)$ which are isomorphic to $\mathbf{R}$. (See Figure 2.)

Hence a section s of L, that is, any map

$$s: S^1 \to L, \text{ with } \pi s(x) = x,$$

behaves locally *just* like an $\mathbf{R}$-valued function, but is of course globally very different! Indeed notice that L has an obvious 0-section, s, but it also has a section s_1 which intersects s_0 in one point! But if these were ordinary functions from S^1 to $\mathbf{R}$, they would have to intersect in two points!

Well, in 1954 Kodaira was teaching a course where I started to learn all these things — but in the complex analytic domain and intertwined with complex analysis. And what wonderful lectures those were! Kodaira at the board, a benevolent presence of keen and silent intelligence, printing magnificent symbols and short sentences with his impeccable Japanese hand. Once in a while he would have a look at us — in order to punctuate an argument, — and, after a slight struggle, produce a word or two.

The only problem for me was that these lectures went too fast. Although the immediate impression was one of great calm and of a measured pace, without all that chatter, he could really cover ground!

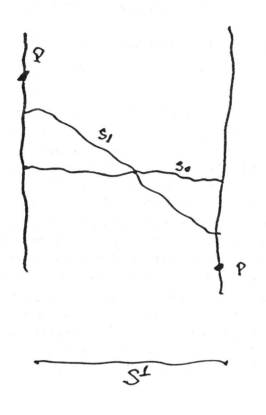

FIGURE 2

But in any case, I did learn the broad outlines of this new view of algebraic geometry. Quite briefly, it turns out that to study the space of meromorphic functions whose poles do not exceed a certain "magnitude" is the same thing as to first build a holomorphic line bundle L over M, and then to study the space of "holomorphic sections $\Gamma_h(L)$" of M. Because L is holomorphic, the operator $\bar{\partial}$ is seen to make sense in this context. Thus the sheaf $O(L)$ of sections $s \in \Gamma(L)$ with $\bar{\partial}s = 0$, makes sense and $\Gamma_h(L)$ is thus re-interpreted as:

$$\Gamma_h(L) \cong H^0(X; O(L)).$$

Also one sees that the notion of "magnitude" or "degree of the total number of poles" is now recoded into the more topological notion of measuring the twist of a line bundle L over X.

For a Riemann surface these reformulations are usually quite transparent. For instance, there the possible twists of a line bundle are measured by an integer $c_1(L)$, and $\Gamma_h(L)$ describes precisely those meromorphic functions whose poles counted with multiplicity add up to $c_1(L)$. Indeed, any such meromorphic function is then the ratio s_1/s_2 of two *holomorphic sections* in $\Gamma(L)$!

In this context the classical Riemann–Roch formula now reads

$$\dim H^0(X; O(L)) - \dim H^1(X; O(L))$$
$$= c_1(L) + 1 - g$$

with $2g$ the number of "1-holes" in X. Note that this is really an existence theorem in disguise. For if $c_1(L) > (1 - g) + 1$ then we must have $H^0(X, O(L)) \geq 2$. In short, there must be a nontrivial meromorphic function s_1/s_2, with total pole "strength" $c_1(L)$! For instance, Riemann–Roch implies that there exists at least one meromorphic function f_P on X with a pole of degree $1 + g$ at a given point P on X! This is a far from obvious fact.

It is this formula which F. Hirzebruch saw how to generalize to arbitrary dimensions, and which — as I said earlier — filled me, and I think all of us, with wonder. Fritz told me once that Hermann Weyl just shook his head in disbelief when Fritz first explained his ambitious conjectures to him.

Let me here at least explain our *wonder* at the nature of his generalization. Indeed, to get anywhere one must realize that the proper measure of the twist of a general line bundle L over M is really an element $c_1(L)$ in $H^2(M; R)$, the *ordinary* second cohomology of M. This granted, let us assume next (only for the sake of brevity) that the tangent bundle of M splits into a direct sum of line bundles

$$T = E_1 + \cdots + E_d.$$

Thus our data X and L furnish us with d elements $x_i = c_1(E_i)$ in $H^2(M; R)$ as well as with additional class $c_1(L)$, describing the twist of L.

Now "Presto" form the expression

$$e^{c_1(L)} \cdot \prod_{i=1}^{d} \frac{x_i}{1 - e^{-x_i}}$$

in $H^*(M)$, then collect the terms of dimension $2m$ and integrate over M to get Hirzebruch's generalization:

$$\sum (-1)^i \dim H^i(M, O(L)) = \int_M e^{c_1(L)} \prod \frac{x_i}{1 - e^{-x_i}}.$$

Thus the innocent $c_1(L)$ and $(1 - g) = \frac{1}{2}$ (Euler number of X) of the original formula had grown into this — at first sight — quite unlikely expression. No wonder Hermann Weyl shook his head.

But Fritz went on undaunted, impetuously using *everything* he had so precociously learned. I like to draw the picture of this proof as in Figure 3.

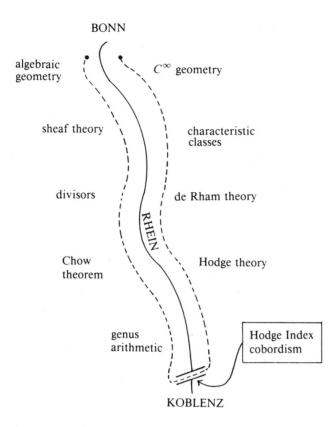

BONN

algebraic
geometry

C^∞ geometry

sheaf theory

characteristic
classes

divisors

de Rham theory

RHEIN

Chow
theorem

Hodge theory

genus
arithmetic

Hodge Index
cobordism

KOBLENZ

FIGURE 3

As you see, he left nothing out.

As my time is limited, I can really say no more about his proof — rather, let me emphasize that this theorem again fits into my general topic of the constraints of topology on analysis, and also establish the theorem more firmly in the line of a fixed point formula. The point is that on the left, the expression

$$\sum (-1)^i \dim H^i(M:\, O(L))$$

is clearly an integer which depends on the complex analytic structure of M. In some sense you should think of it as the dimension of the space of "*virtual solutions of the analytic system of differential equations* $\bar\partial f = 0$ *on* M." On the other hand, on the right-hand side, we find a purely C^∞-invariant of M. It is not purely *topological* in the sense that it doesn't depend only on "purely continuous" concepts. The calculus has entered, but only to the first order, so to speak. Once we know how "twisted" the space of tangents over M is and how twisted L was — all C^∞ concepts — then the right-hand side is determined.

This beautiful theorem refuses to die. We had hardly gotten used to it when Grothendieck produced a magnificently original proof purely in the context of algebraic geometry. Then ten years later or so Atiyah and Singer gave us their general index theorem which generalizes the theorem to all elliptic complexes over M, and again using a new set of ideas. And quite recently we have come to realize that the anomaly calculations of supersymmetric field theory throw new light on this old question. I cannot do justice to all these developments here, but would like to end with some remarks on one of these developments in which I had a personal hand.

Michael Atiyah and I had become very good friends and collaborators by the early sixties and the first proof of his and Iz Singer's index theorem was in fact unveiled in a joint seminar all three of us had started at Harvard during Atiyah's visit in 1963. But the idea of developing a "fixed point formula," to go with the new sheaf theory — and in the context of elliptic complexes — only occurred to us when we met at the Woods Hole Conference of 1964. I think it was a conjecture of Shimura that put us on the right track, as well as some very obscure parts of Eichler's work in number theory.

Recall that as we already saw in the classical case, the Lefschetz number, $L(\varphi)$, was computed in terms of the fixed points by the Hopf formula:

$$\mathscr{L}(\varphi) = \sum_p \mu(p),$$

where the local multiplicities were integers given by the local degree of φ at p. We found the analogous expression in all elliptic situations, but here let me explain it to you only in the complex domain. In this context one of course starts with a *holomorphic* map

$$\varphi\colon M \to M$$

and assumes also that in the general case of a twisting line bundle L, that φ has a lifting, $\hat{\varphi}$, for L over M. Thus we have the diagrams:

$$
\begin{array}{ccc}
L & \xrightarrow{\hat{\varphi}} & L \\
\downarrow & & \downarrow \\
M & \xrightarrow{\hat{\varphi}} & M
\end{array}
$$

Such a lifting induces a homomorphism

$$H^i(\hat{\varphi})\colon\ H^i(M; O(L)) \to H^i(M; O(L)).$$

And so one also has a well-defined holomorphic Lefschetz number

$$\mathscr{L}(\hat{\varphi}) = \sum (-1)^i \operatorname{Trace} H^i(\hat{\varphi}).$$

Now what we did to find the appropriate measure μ_p to go with $\mathscr{L}(\hat{\varphi})$ under the assumption that the fixed points of φ were isolated and nondegenerate. Indeed, under these conditions we found that

$$\mu_p = \frac{\hat{\varphi}(p)}{\det(1 - d\varphi_p)}$$

were

$$d\varphi_p \colon T'_p \to T'_p$$

is the holomorphic differential of φ at p. In short, our version of the Lefschetz formula therefore reads

$$\mathscr{L}(\hat{\varphi}) = \sum \frac{\hat{\varphi}(p)}{\det(1 - d\varphi_p)}, \quad |\varphi(p) = p,$$

so that for the sheaf O itself, for example, when no lifting is needed one has

$$\mathscr{L}(\varphi) = \sum \frac{1}{\det(1 - d\varphi_p)}.$$

We were amazed at first that in this context the "measure" μ_p was actually a complex number, even though in this untwisted case the $\mathscr{L}(\varphi)$ had to be an integer. And Michael recounts in his recollection of that summer that our first attempts to check this formula for elliptic curves with the assembled experts on that subject was a failure! But luckily we didn't believe the experts and pressed on.

I like to think of these formulas as yet another example of the constraints that topology places on analysis. For if we consider the eigenvalues of $d\varphi_p$, say $\lambda_1 \cdots \lambda_d$, then the fixed point formula — read the other way around — puts a remarkable numerical restraint on these eigenvalues:

$$\sum_p \frac{1}{\prod_i (1 - \lambda_i)} = \mathscr{L}(\varphi).$$

Ah! There are so many facets of these formulas I would like to divulge here — but I am afraid my time is up. Still I cannot let you go without at least hinting to you that we are again in an era of new insights which — for instance — have brought about the first refinements of these restrictions on these eigenvalues in the last 25 years. The impetus has this time been from physics, a wonderful byproduct of the newly created vigorous exchange of ideas between our cousins and us.

Thus the much debated physical dreams of string theory have at least had one quite concrete mathematical consequence. The unexpected novelty here was to consider the Lefschetz number in our sense but applied to an *infinite dimensional twisting* bundle L. Thus stimulated by discussions with the topologists P. Landweber, R. Stong, and S. Ochanine, E. Witten predicted that if one twisted $\bar{\partial}$ by a q-dependent bundle which has an "infinite product expansion" of the form:

$$L_q = \bigotimes_{n=1}^{\infty} \Lambda_{q^n} T^* \otimes \Lambda_{q^n} T \otimes S_{q^n}(T^*) \otimes S_{q^n}(T).$$

Then $\chi(M; L_q)$ will have certain quite new "rigidity properties."

I will not even *try* to explain this formula to the uninitiated. Let me simply describe it as the 20th century view of the following formula going back to Gauss and Jacobi:

$$\varphi(\lambda) = \frac{(1+\lambda)}{(1-\lambda)} \prod_{n=1}^{\infty} \frac{(1+\lambda q^n)(1+\lambda^{-1} q^n)}{(1-\lambda q^n)(1-\lambda^{-1} q^n)}$$

for an elliptic function on the torus

$$T_q = C^*/\{q^n\}.$$

I have just returned from a conference on elliptic cohomology where C. Taubes presented his proof of these new conjectures — with a little old man's help from me — which exploits this old Lefschetz formula to the hilt! It is also clear that these physics-inspired ideas are producing a new brand of "elliptic topology" — which someone recently referred to as the beginnings of 21st century K-theory.

In short, I can report to you that once again a quite new point of view has rejuvenated old mathematics — and some old mathematicians — and on this 100th birthday celebration of our Society, who can wish for better news!

Steven L. Kleiman received his Ph.D. from Harvard in 1965, a student of Oscar Zariski and David Mumford. After a time at Columbia and IHES (Paris), he joined the MIT faculty in 1969. He has done extensive research in algebraic geometry. Among his honors have been a NATO Postdoctoral Fellowship, a Sloan Fellowship, and a Guggenheim Fellowship. He is the author of two previous historical articles, on Hilbert's 15th problem and on the Chasles theory of conics.

The Development of Intersection Homology Theory

STEVEN L. KLEIMAN

1. PREFACE

Intersection homology theory is a brilliant new tool: a theory of homology groups for a large class of singular spaces, which satisfies Poincaré duality and the Künneth formula and, if the spaces are (possibly singular) projective algebraic varieties, then also the two Lefschetz theorems. The theory was discovered in 1974 by Mark Goresky and Robert MacPherson. It was an unexpected find, but one highly suited to the study of singular spaces, and it has yielded profound results. Most notably, the Kazhdan–Lusztig conjecture was established via a remarkable bridge between representation theory and intersection homology theory provided by $\mathscr{D}$-module theory. In fact, in 1980, the conjecture, which was a little over a year old, motivated the construction of that bridge, and the bridge in turn led to some far reaching new advances in intersection homology theory. All told, within a decade, the development of intersection homology theory had involved an unprecedented number of very bright and highly creative people. Their work is surely one of the grand mathematical endeavors of the century.

From a broader historical perspective, it is clear that the time was ripe for the discovery of intersection homology theory. Enormous advances had been made in the study of equisingular stratifications of singular spaces during the mid-to-late 1960s. During the early 1970s, various characteristic classes had been found for singular spaces, and there had been several investigations of Poincaré duality on singular spaces, although those investigations were concerned with the degree of failure of Poincaré duality rather than a modification of the homology theory. In addition, about a year and a quarter after the discovery, while the new theory was still undeveloped and virtually unknown, Jeff Cheeger, pursuing an entirely different course of research from that of Goresky and MacPherson, independently discovered an equivalent cohomology theory for essentially the same class of singular spaces: a deRham–Hodge theory corresponding to their combinatorial theory. Furthermore, it is not surprising that there was, during the decade following the discovery of intersection homology theory, a great confluence of topology, algebraic geometry, the theory of differential equations, and representation theory. While those theories had diverged after Riemann, they had converged again on occasion in the hands of Poincaré around 1900, of Lefschetz around 1930, and of others in the 1950s and 1960s.

The present account of the frenetic development of intersection homology theory during the first decade or so after its discovery is intended simply to provide a feeling for who did what, when, where, how, and why, and a feeling for the many interpersonal lines of development. The mathematical discussions are not meant to be outlines or surveys; they are meant to be indications of the philosophy, the aims, the methods, and the material involved. The author has constantly striven to be impartial, historically and technically accurate, and reasonably thorough. Of course, here and there, a delicate line had to be drawn between what to include and what to leave out. The author regrets any errors and oversights.

The present account was based primarily on the rather lengthy literature. There are, first of all, several excellent survey articles [11], [94], [76], and [45]. Of course, it was still necessary to appeal to the original research papers to obtain a more complete and more rounded picture. Unfortunately, some of the historical remarks in print are inaccurate or misleading; their authors seem simply to have been unaware of the whole story.

The present account was also based on numerous interviews: brief interviews with M. Artin, J. Bernstein, R. Crew, D. Kazhdan, J.-L. Verdier, K. Vilonen, and D. Vogan; short interviews with A. Beilinson, J.-L. Brylinski, and S. Zucker; longer interviews with Cheeger, G. Lusztig, L. Saper; and an extended series of interviews with Goresky, D. T. Lê, and MacPherson. In addition, A. Altman, Beilinson, Brylinski, Cheeger, Goresky, B. Kleiman, Lê, Lusztig, J. Lützen, MacPherson, A. Thorup, Verdier, and Zucker read earlier versions of this account and made a number of suggestions, which led to significant improvements. Unfortunately, not everyone who was invited to comment did so. However, it is a pleasure now to thank each and every one of those who did contribute for their invaluable help; the article is all the better for it.

2. Discovery

Intersection homology theory was discovered during the fall of 1974 at the IHES (*Institut des Hautes Études Scientifiques*) in Paris by Mark Goresky and Robert MacPherson. They were seeking a theory of characteristic numbers for complex analytic varieties and other singular spaces. During the preceding four years, the Whitney classes of Dennis Sullivan, the Chern classes of MacPherson, and the Todd classes of Paul Baum, William Fulton and MacPherson had been discovered for such spaces. (The existence of the Chern classes had been conjectured in 1970 by Alexandre Grothendieck and Pierre Deligne, and then the classes had been constructed by MacPherson in a manuscript of July 25, 1972 published in 1974. In 1978, Jean-Paul Brasselet and Marie-Hélène Schwartz proved that the classes correspond under Alexander duality to the cohomology classes that Schwartz had introduced in 1965. The classes are often called the Chern classes of Schwartz and MacPherson.)

All those classes are homology classes, however, and homology classes cannot be multiplied. So Goresky and MacPherson figured, in analogy with the secondary homology operations, that there would be certain "intersectable" homology classes, whose intersection product would be unambiguous modulo certain "indeterminacy" classes.

Goresky was, at the time, writing his Ph.D. thesis under MacPherson's direction on a geometric treatment of cohomology groups, viewing them as the homology groups of a certain kind of cycle. (The thesis was submitted to Brown University in June 1976 and published as [36] in 1981.) By then, they knew why two "geometric cocycles" on a complex analytic variety X can be intersected: one can be made transverse to the other and to each stratum S_α in a Whitney stratification of X.

A *Whitney stratification* is a particularly nice (locally finite) partition of a complex analytic variety X into disjoint locally closed, smooth analytic strata S_α. It satisfies the following 'boundary condition': each closure $\overline{S}_\alpha$ is a union of strata S_β. Also, Whitney's condition (B) holds: if a sequence of points $a_i \in S_\alpha$ and a sequence of points $b_i \in S_\beta$ both approach the same point $b \in S_\beta$, then the limit of the secant lines connecting a_i to b_i lies in the limit of the tangent spaces to S_α at a_i if both limits exist. Consequently, the Thom–Mather isotopy theorem obtains: the stratification is locally topologically trivial along S_β at b. Whitney (1965) proved that given any (locally finite) family of locally closed, smooth analytic subvarieties Y_i of X, whose closures are analytic, there exists a Whitney stratification such that each Y_i is a union of strata.

Goresky and MacPherson found a suitable more general setup in which to intersect geometric cocycles: it is sufficient that X be a piecewise linear space, or pl-space, with a stratification by closed subsets,

$$X = X_n \supset X_{n-1} \supset X_{n-2} \supset X_{n-3} \supset \cdots \supset X_1 \supset X_0,$$

such that: 1. $X_{n-1} = X_{n-2}$;

2. each stratum $X_i - X_{i-1}$ is empty or is a pl-manifold of (pure) topological dimension i along which the normal structure of X is locally trivial (more precisely, each point x of the stratum $X_i - X_{i-1}$ admits a closed neighborhood U, in X, pl-homeomorphic to $B^i \times V$, where B^i is the closed ball of dimension i and V is a compact space with a filtration by closed subsets,

$$V = V_n \supset V_{n-1} \supset \cdots \supset V_i = \mathrm{pt},$$

and the homeomorphism preserves the filtration; that is, it carries $U \cap X_j$ onto $B^i \times V_j$);

3. the closure of each stratum is a union of strata;

4. the largest stratum $X_n - X_{n-2}$ is oriented and dense.

Goresky and MacPherson made several attempts to relax the transversality condition on the cycles by allowing a (piecewise linear and locally finite) i-cycle to deviate from dimensional transversality to X_{n-k} within a tolerance specified by a function $\overline{p}(k)$, which is independent of i; that is, the i-cycle is allowed, for each k, to intersect X_{n-k} in a set of dimension as much as $i - k + \overline{p}(k)$. They called the function $\overline{p}(k)$ the *perversity*. It is required to satisfy these conditions:

$$\overline{p}(2) = 0 \quad \text{and} \quad \overline{p}(k+1) = \begin{cases} \overline{p}(k), & \text{or} \\ \overline{p}(k) + 1. \end{cases}$$

The first condition guarantees that the i-cycle lies mostly in the nonsingular part of X, where it is orientable. The second condition says that the perversity function is nondecreasing and grows no faster than by 1. That condition was not imposed until the summer of 1975 (see below).

All of a sudden one day, Goresky and MacPherson realized that the cycles should be identified by homologies that were allowed to deviate in the same way. Thus they obtained a spectrum of new groups. They called them the "perverse homology" groups, and used that name for about six months. Then Sullivan convinced them to change it, suggesting "Schubert homology" and "intersection homology." The rest is history!

The intersection homology groups $IH_i^{\overline{p}}(X)$ are finitely generated when X is compact. When X is normal, the groups range from the ordinary cohomology groups, where $\overline{p}(k) = 0$ for all k, to the ordinary homology groups, where $\overline{p}(k) = k - 2$ for all k. In addition, the groups possess intersection pairings,

$$IH_i^{\overline{p}}(X) \times IH_j^{\overline{q}}(X) \longrightarrow IH_{i+j-n}^{\overline{p}+\overline{q}}(X),$$

generalizing the usual cup and cap products.

Goresky and MacPherson filled a whole notebook with examples. They felt sure that they were on to something. However, to their dismay, the theory appeared to be tied tightly to the stratification and rather artificial. Then, to see if perchance they had come any further along toward a theory of characteristic numbers, they decided to focus on one characteristic number, the signature. Indeed, in 1970, Sullivan had posed the problem of finding a class of singular spaces with a cobordism invariant signature. The key ingredient here, of course, is Poincaré duality.

Suddenly, they realized that, just as cohomology groups and homology groups are dually paired, so too the intersection homology groups of complementary dimension (i+j=n) and complementary perversity ($\overline{p}(k) + \overline{q}(k) = k - 2$) should be dually paired. They opened the notebook and were astonished to find that the ranks of the complementary groups were indeed always the same. In fact, there was one example where the ranks appeared at first to be different, but they soon located an error in the calculations. There was no doubt about it: Poincaré duality must hold! In particular, Sullivan's problem was clearly solved: if X is compact, of dimension $4l$, and analytic or simply has only even codimensional strata, then the middle perversity group $IH_{2l}^{\overline{m}}(X)$, where $\overline{m}(k) = \lfloor \frac{k-2}{2} \rfloor$, must carry a nondegenerate bilinear form, whose signature is invariant under cobordisms with even codimensional strata (but not under homotopy). It was a magic moment!

After a week or two of very intense effort, Goresky and MacPherson had the essence of the first proof of duality. It was geometric, technical, and messy: they used the Leray spectral sequence of the link fibration over each stratum and then the Mayer–Vietoris sequence to patch. They went to Sullivan and John Morgan, who were also at the IHES, and told them about their discovery. Sullivan for once was dumbfounded. Morgan probably said, "Come on, you can't fool around with the definition of homology." However, Morgan quickly saw the point and used the new ideas to finish Sullivan's program of giving a geometric proof of Poincaré's *Hauptvermutung* [63, p. 1176]; unfortunately, the proof is technical and complicated and has not yet been put in print.

Years passed before Goresky and MacPherson succeeded in writing up and publishing their work. They had not even wanted to start until they had analyzed the invariance of the groups $IH_i^{\overline{p}}(X)$ under restratification, and it was not until the summer of 1975 that that they discovered that the growth condition $\overline{p}(k) \le \overline{p}(k+1) \le \overline{p}(k) + 1$ implies that invariance. Moreover, they did not know what category to work in: differentiable, pl, or topological. When they finally settled on the pl-category, they realized that pl-transversality should say that two pl-chains can be made transverse within each stratum. Clint McCrory, an expert on pl-topology, was at Brown University with them during the academic year 1975–1976, and they asked him about the transversality. He immediately gave them a proof and published it so that they could refer to it.

During the summer of 1976, Goresky and MacPherson struggled with another technical problem. They needed a single chain complex with which to define the intersection homology groups, they needed to be able to move two chains into transverse relative position to intersect them, and they needed to find a dual complex with the same properties. The problem was that all those properties seemed to be technically incompatible. They finally discovered that they had to take the chain complex that is the direct limit over all triangulations to get enough flexibility. They also discovered certain sets $Q(i, p)$ and $L(i, p)$, which are like "perverse skeletons" of the spaces and which allowed them to to prove Poincaré duality without the Leray spectral sequence with coefficients in the intersection homology groups of the fiber in the link fibration.

In addition, Goresky and MacPherson had other serious mathematical projects in progress during those years. Goresky had to write up his thesis. MacPherson was working with Fulton on a literally *revolutionary* new approach to intersection theory in algebraic geometry, [32], [33]. Some other projects involved exciting new ideas in intersection homology theory, which completely captured their attention for months at a time; those ideas will be discussed below.

When Izrail Gelfand visited Paris in the fall of 1976, he met MacPherson and convinced him to write up and publish an announcement of the discovery of intersection homology theory; it [39] appeared in the spring of 1977. With that in print, Goresky and MacPherson felt less pressure to drop everything else, and they did not get back to writing up the detailed treatment until the summer of 1978. Then they worked very hard on the exposition, and, in September 1978, they submitted it for publication. It took almost a year to be refereed and did not appear until 1980 as [40].

3. A FORTUITOUS ENCOUNTER

At a Halloween party near Paris in 1976, Deligne asked MacPherson what he was working on and was told about intersection homology theory. At the time, Deligne was thinking about the Weil conjectures, monodromy, and the hard Lefschetz theorem. He was also thinking about Steven Zucker's work in progress on the variation of Hodge structures over a curve (which eventually appeared in [98]), wondering in particular about how to extend it to higher dimensions. Earlier, Deligne had made significant contributions to the theory of duality of quasi-coherent sheaves (March 1966) and to the formulation and solution of a generalized Riemann–Hilbert problem (fall of

1969). Thus Deligne had been led to the idea of truncating the pushforth of a local system, or locally constant sheaf of vector spaces, on the complement of a divisor with normal crossings on a smooth complex ambient variety X of topological dimension $n = 2d$.

The party was at one of the IHES's large residences, and almost everyone from the institute was there. On a scrap of paper, Deligne wrote down for the first time his celebrated formula,

$$IH_i^{\overline{p}}(X) = H^{2d-i}(\mathbf{IC}_{\overline{p}}^{\cdot}(X)),$$

expressing the intersection homology groups of X, equipped with a suitable stratification by closed sets $\{X_i\}$, as the hypercohomology of the following complex of sheaves:

$$\mathbf{IC}_{\overline{p}}^{\cdot}(X) := \tau_{\leq \overline{p}(2d)} \mathbf{R} i_{2d*} \cdots \tau_{\leq \overline{p}(2)} \mathbf{R} i_{2*} \mathbf{C}_{X-X_2}$$

where $\mathbf{C}_{X-X_2}$ is the complex consisting of the constant sheaf of complex numbers concentrated in degree 0, where i_k is the inclusion of $X - X_{n-k}$ into $X - X_{n-k-1}$, and where $\tau_{\leq k}$ is the truncation functor that kills the stalk cohomology in degree above k. The complex $\mathbf{IC}_{\overline{p}}^{\cdot}(X)$ is, however, well defined only in the 'derived category' — the category constructed out of the category of complexes up to homotopy equivalence, by requiring a map of complexes to be an isomorphism (to possess an inverse) if and only if it induces an isomorphism on the cohomology sheaves.

Deligne asked about a key example, the local intersection homology groups at an isolated singularity. MacPherson responded immediately: they are the homology groups of the link (the retract of a punctured neighborhood) in the bottom half dimensions and 0 in the middle and in the top half dimensions. That answer was exactly what Deligne obtained from his construction. They conjectured that the formula is correct.

Deligne, it seems, had always worked before with a smooth ambient variety and with twisted coefficients. He was rather surprised to learn from MacPherson that there might be a significant theory on a singular space. He could see, however, that his construction would yield cohomology groups that satisfy Poincaré duality because of the Verdier–Borel–Moore duality in the derived category of complexes of sheaves. MacPherson, on the other hand, was surprised at the entrance of the derived category. However, he could see that Deligne's construction might have great technical advantages.

At the time, MacPherson was in the midst of giving a series of lectures on intersection homology theory, and Jean-Louis Verdier was in the audience. Verdier expressed considerable interest in the theory and in Deligne's formula. During the ensuing weeks, he explained more about the derived category and duality to MacPherson.

The next academic year, 1977–78, MacPherson was back at Brown, and Goresky was in his second and final year as a Moore Instructor at MIT.

MacPherson showed Goresky the scrap of paper with Deligne's formula on it and said: "We have to learn derived categories to understand this formula!" In a seminar on intersection homology theory at Brown, they worked out a proof of the formula. The proof was long and messy, involving the derived category of simplicial sheaves and a limit over simplicial subdivisions.

During the following academic year, 1978–1979, Goresky and MacPherson wrote up that proof as part of a first draft of their paper [42], which doubtless is the single most important paper on topological intersection homology theory. However, they were unhappy with that complicated first treatment and decided to streamline it. They made steady progress during the next year, 1979–1980. They found several axiomatic characterizations of $\mathbf{IC}_{\overline{p}}^{\cdot}(X)$ among all complexes in the derived category whose cohomology sheaves are constructible with respect to the given stratification. (A sheaf of $\mathbf{Q}$-vector spaces is called *constructible* with respect to a stratification by closed sets $\{X_i\}$ if its stalks are finite dimensional and its restriction to each stratum $X_i - X_{i-1}$ is locally constant.) They found that the 'constructible derived category' is a "paradise," as Verdier called it: it possesses some two dozen natural properties. However, progress was hampered because Goresky was in Vancouver and MacPherson was in Providence during those years.

The first copy of [42] that was submitted for publication was lost in the mail from Vancouver, and that horrible fact was not discovered for six or eight months. The paper was immediately resubmitted in June 1981. Meanwhile, many people had read the manuscript and offered pages of corrections and suggestions. Their comments were incorporated in a major revision of the paper, which was resubmitted in December 1982. Finally, the paper appeared in print in early 1983, nearly six and a half years after the Halloween party.

In §1 of the paper, Goresky and MacPherson develop the general theory of the constructible derived category. In §2, they study pl-pseudomanifolds X and show how the construction of the $IH_i^{\overline{p}}(X)$ in their first paper [40] actually yields a complex of sheaves. In §3, they develop a first axiomatic characterization $\mathbf{IC}_{\overline{p}}^{\cdot}(X)$ and use it to prove Deligne's formula. In §4 of the paper, Goresky and MacPherson give a second axiomatic characterization of $\mathbf{IC}_{\overline{p}}^{\cdot}(X)$, which they derive from the first. It does not involve the stratification and yields the following remarkable theorem:

THEOREM [42, 4.1]. *The intersection homology groups* $IH_i^{\overline{p}}(X)$ *are topological invariants; in fact, for any homeomorphism* $f : X \to Y$, *the complexes* $\mathbf{IC}_{\overline{p}}^{\cdot}(X)$ *and* $f^* \mathbf{IC}_{\overline{p}}^{\cdot}(Y)$ *are isomorphic in the derived category.*

Earlier, in the summer of 1975, Goresky and MacPherson had figured out that the groups $IH_i^{\overline{p}}(X)$ are independent of the stratification, but they still needed a pl-structure. So, in 1976, Goresky spent some time working with singular chains, but he bumped into an obstacle. About nine years later,

Henry King [59] independently worked out a theory based on singular chains and, without using sheaf theory, he recovered the topological invariance.

In §5, Goresky and MacPherson re-proved using sheaf theory some of the basic properties of the intersection homology groups, such as the existence of the intersection pairing and the validity of Poincaré duality. They also proved some new results, such as the following comparison theorem:

THEOREM (COMPARISON) [42, 5.6.3]. *If X is a complex algebraic variety that is compact, normal, and a local complete intersection, and if $\overline{p}(k) \geq k/2$ for $k \geq 4$, then $IH_i^{\overline{p}}(X) = H_i(X)$ for all i.*

In §6, Goresky and MacPherson proved several theorems about complex algebraic varieties X of (algebraic) dimension d and the middle perversity $\overline{m}(k) = \lfloor \frac{k-2}{2} \rfloor$. This case is particularly important. So, to lighten the notation, set

$$\mathbf{IC}^{\cdot}(X) := \mathbf{IC}_{\overline{m}}^{\cdot}(X), \quad IH_i(X) := IH_i^{\overline{m}}(X), \quad \text{and} \quad IH^i(X) := IH_{2d-i}(X).$$

The first theorem of §6 gives a third and the most important version of the axiomatic characterization of $\mathbf{IC}^{\cdot}(X)$:

THEOREM [42, 6.1]. *Consider the derived category of bounded complexes $\mathbf{K}$ of sheaves such that the cohomology sheaves $\mathbf{H}^i(\mathbf{K})$ are constructible with respect to some Whitney stratification, which depends on $\mathbf{K}$. Then, in this category, there is a unique complex $\mathbf{K}$ satisfying the following conditions:*

(a) (Normalization) *There is a dense open subset U such that $\mathbf{H}^i(\mathbf{K})|U = 0$ for $i \neq 0$ and $\mathbf{H}^0(\mathbf{K})|U = \mathbf{C}_U$.*

(b) (Lower bound) *$\mathbf{H}^i(\mathbf{K}) = 0$ for all $i < 0$.*

(c) (Support) *$\mathrm{codim}(\mathrm{Supp}(\mathbf{H}^i(\mathbf{K}))) > i$ for all $i > 0$.*

(d) (Duality) *$\mathbf{K}$ is isomorphic to its Verdier–Borel–Moore dual $\mathbf{K}^{\check{}}$.*

Condition (d) may be replaced by the following dual condition:

(d˘) (Cosupport) *$\mathrm{codim}(\mathrm{Supp}(\mathbf{H}^i(\mathbf{K}^{\check{}}))) > i$ for all $i > 0$.*

Moreover, $\mathbf{K} = \mathbf{IC}^{\cdot}(X)$.

Goresky and MacPherson used this characterization to prove the following two theorems:

THEOREM (SMALL RESOLUTION) [42, 6.2]. *If a proper algebraic map $f: X \to Y$ is a small resolution, that is, if X is smooth and for all $r > 0$,*

$$\mathrm{codim}\{ y \in Y \mid \dim f^{-1}(y) \geq r \} > 2r,$$

then $IH_i(Y) = IH_i(X) = H_i(X)$; in fact, $\mathbf{R}f_*\mathbf{C}_X = \mathbf{IC}^{\cdot}(Y)$.

THEOREM (KÜNNETH FORMULA) [42, 6.3]. *If X and Y are varieties, then*

$$IH_i(X \times Y) = \bigoplus_{j+k=i} IH_j(X) \otimes IH_k(X).$$

The Künneth formula had already been proved analytically by Jeff Cheeger. Although Goresky and MacPherson referred to Cheeger's article [21] for that proof, the proof did not actually appear explicitly in print before the article's sequel [22, §7.3]. For the unusual story of Cheeger's work, see the beginning of §8.

Later in [45, §A.], Goresky and MacPherson gave two interesting examples concerning small resolutions. In each example, there is a variety Y with two different small resolutions $f_1: X_1 \to Y$ and $f_2: X_2 \to Y$ such that the induced vector space isomorphism between the cohomology rings of X_1 and X_2 is *not* a ring isomorphism.

In [42, §7], Goresky and MacPherson gave a sheaf-theoretic proof of the following theorem, known as the 'Lefschetz hyperplane theorem' or the 'weak Lefschetz theorem':

THEOREM (LEFSCHETZ HYPERPLANE) [42, 7.1]. *If X is a projective variety of (algebraic) dimension d and if H is a general hyperplane, then for all i the inclusion $\alpha: X \cap H \to X$ induces a map*

$$\alpha_*: IH_i(X \cap H) \longrightarrow IH_i(X).$$

Moreover, α_* is bijective for $i < d - 1$ and surjective for $i = d - 1$.

In fact, the theorem is proved not only for the middle perversity $\overline{m}$ but also for any perversity $\overline{p}$ such that $\overline{p}(k) \leq k/2$. Hence, the theorem has the following corollary, whose second assertion results from the comparison theorem stated above:

COROLLARY [42, 7.4.1, 7.4.2]. *If X is normal, then the Gysin map of ordinary cohomology theory $\alpha^*: H^i(X \cap H) \to H^i(X)$ is bijective for $i > d - 1$ and surjective for $i = d - 1$. If X is a normal local complete intersection, then the induced map on the ordinary homology groups $\alpha_*: H_i(X \cap H) \to H_i(X)$ is bijective for $i < d - 1$ and surjective for $i = d - 1$.*

The sheaf-theoretic proof of the Lefschetz hyperplane theorem is like that in [90, XIV 3]. In [42, §7] and in several other places in the literature of intersection homology theory, the latter proof is attributed to Michael Artin. However, Artin says that it is inappropriate to credit the proof to him, because the entire seminar, [90], is a report on joint work and, moreover, that particular proof is due to Grothendieck.

Goresky and MacPherson had learned from Deligne that the sheaf-theoretic proof of the Lefschetz theorem in [90, XIV 3] would carry over to intersection homology theory, and they presented the details in [42, §7]. However, they had already considered the theorem from two other points of view. First, in the summer of 1977, Cheeger and MacPherson had met and conjectured that the related 'hard Lefschetz theorem' and all the other various consequences of Hodge theory should hold for intersection homology theory; for more information about the conjecture, see the beginning of §8. Second, during 1978–1979, Goresky and MacPherson began work on their new stratified Morse theory. That winter, they found they could adapt Thom's Morse-theoretic argument in the nonsingular case to prove the Lefschetz theorem in the singular case. They gave that proof in [43, 5.4]. (René Thom gave his proof in a lecture at Princeton in 1957. It was entered into the public domain in 1959 independently by Raul Bott and by Aldo Andreotti and Theodore Frankel.)

4. THE KAZHDAN–LUSZTIG CONJECTURE

The Kazhdan–Lusztig conjecture grew out of a year of collaboration in Boston starting in the spring of 1978 between David Kazhdan and George Lusztig. Two years earlier, Tony Springer had introduced an important new representation on l-adic étale cohomology groups, of the Weyl group W of a semi-simple algebraic group over a finite field. Kazhdan and Lusztig found a new construction of the representation. Moreover, they allowed the ground field to be $\mathbf{C}$ as well. Indeed, they preferred $\mathbf{C}$ and the classical topology. Their work eventually appeared in their paper [58].

The representation module has two natural bases, and Kazhdan and Lusztig tried to identify the transition matrix. Thus they were led to define some new polynomials $P_{y,w}$ with integer coefficients indexed by the pairs of elements $y, w \in W$, with $y \leq w$, for any Coxeter group W.

Those two bases reminded Kazhdan and Lusztig of the two natural bases of the Grothendieck group of infinite dimensional representations of a complex semi-simple Lie algebra $\mathbf{g}$: the basis formed by the Verma modules M_λ and that by the simple modules L_μ. (By definition, M_λ is the maximal irreducible module with highest weight λ, and L_λ is its unique simple quotient.) Putting aside their work on the Springer resolution, Kazhdan and Lusztig focused on the transition matrix between the M_λ and L_μ. Work by Jens Carsten Jantzen and by Anthony Joseph along with some well-known examples, which indicated that the transition matrix might depend on the topology of the Schubert varieties X_w, the closures of the Bruhat cells B_w, led Kazhdan and Lusztig to formulate the following conjecture. The particular formulation below was taken from Lusztig's paper [72], but the original conjecture appeared in their joint paper [56], which was received for publication on March 11, 1979.

CONJECTURE (KAZHDAN–LUSZTIG) [56, 1.5] [72, (4.4), (4.5)]. *In the Grothendieck group,*

$$L_{-\rho w-\rho} = \sum_{y \leq w} (-1)^{l(w)-l(y)} P_{y,w}(1) M_{-\rho y-\rho}$$

$$M_{\rho w-\rho} = \sum_{w \leq y} P_{w,y}(1) L_{\rho y-\rho}$$

where, as usual, ρ is half the sum of the positive roots, and $l(w) := \dim(X_w)$.

Kazhdan and Lusztig defined the polynomials $P_{y,w}$ by an effective combinatorial procedure, but it is poorly suited for actual computation. However, for restricted Weyl groups of type A_N, Alain Lascoux and Marcel Schützenberger [65] found that the polynomials satisfy some simpler recursion relations determined by the combinatorics, and, using a computer, they worked out some examples. Sergei Gelfand (Izrail Gelfand's son) and MacPherson [35, §5] discussed the Kazhdan–Lusztig algorithm and worked out some examples by hand. Goresky [38], inspired by the latter treatment, implemented the algorithm on a VAX 11 and worked out the cases A_3, A_4, A_5, $B_3 = C_3$, $B_4 = C_4$, D_4, and H_3; the case of A_5 alone took 3 hours of CPU time. In addition, according to Lusztig, Dean Alvis implemented the cases of E_6 and H_4, but the results are too lengthy to print out in full. The study of the polynomials is rather important and has continued. According to MacPherson, recently (1988) Brian Boe, Thomas Enright, and Brad Shelton have generalized the work of Lascoux and Schützenberger to some other types of Weyl groups, and Kazhdan has made the interesting conjecture that $P_{y,w}$ depends only on the partially ordered set of z between y and w.

Kazhdan and Lusztig said [56, top of p. 168] that "$P_{y,w}$ can be regarded as a measure for the failure of local Poincaré duality" on the Schubert variety X_w in a neighborhood of a point of the Bruhat cell B_y. In the appendix, they discussed "some algebraic geometry related to the polynomials," but there they worked exclusively over the algebraic closure of a finite field of characteristic p and used étale cohomology groups with coefficients in the l-adic numbers $\mathbf{Q}_l$, $l \neq p$.

Kazhdan and Lusztig asked Bott about Poincaré duality on a singular space, and Bott sent them to MacPherson. Actually, Lusztig had already learned about intersection homology theory the year before in the spring of 1977 at the University of Warwick, England. At the time, he was on the faculty there. MacPherson came to Warwick and gave a lecture on the theory; after the talk, they discussed it further. Now, Kazhdan, Lusztig, and MacPherson had several discussions in person and by mail. Kazhdan and Lusztig were taken by all the ideas, and at MacPherson's suggestion they wrote to

Deligne. Deligne responded from Paris on April 20, 1979, with a seven-page letter. That letter has often been photocopied and often been cited, because it is the first tangible place where Deligne discussed his sheaf-theoretic approach.

In his letter, Deligne observed that the sheaf-theoretic approach works equally well for a projective variety X over the algebraic closure of a finite field of characteristic p with the étale topology and sheaves of $\mathbf{Q}_l$-vector spaces, $l \neq p$. The strata must be smooth and equidimensional, but it is unnecessary that the normal structure of X be locally trivial in any particular sense along each stratum; it suffices that the stratification be fine enough so that all the sheaves involved are locally constant on each stratum. (In positive characteristic, a Whitney stratification need not exist, and if there is no special hypothesis on the normal structure, then the sheaves $\mathbf{H}^i(\mathbf{IC}^{\cdot}(X))$ need no longer be constructible with respect to a given stratification; nevertheless, the sheaves will be constructible with respect to some finer stratification.)

Deligne stated that Poincaré duality and the Lefschetz fixed-point formula are valid. The latter applies notably to the *Frobenius endomorphism* $\phi_q : X \to X$, which raises the coordinates of a point to the q-th power and is defined when $q := p^e$ is large enough so that the coefficients of a set of equations defining X lie in the field $\mathbf{F}_q$ with q elements. The fixed-points x of ϕ_q are simply the points $x \in X$ with coordinates in $\mathbf{F}_q$, and the formula expresses their number as the alternating sum of the traces of ϕ_q on the $IH^i(X)$.

Deligne said, however, that he could not prove the following statement of *purity*: for every fixed-point x and for every i, the eigenvalues of ϕ_q on the stalk at x of the sheaf $\mathbf{H}^i(\mathbf{IC}^{\cdot}(X))$ are algebraic numbers whose complex conjugates all have absolute value at most $q^{i/2}$. Deligne said that he lacked enough evidence to call the statement a "conjecture", but he did call it a "problem." The problem was solved about fourteen months later by Ofer Gabber. See the beginning of §7.

Deligne noted that if 'purity' holds, then so will the following two theorems, which Kazhdan and Lusztig had asked about. (In the statement of the second theorem, it is implicitly assumed that an isomorphism $\mathbf{Q}_l(1) \cong \mathbf{Q}_l$ has been fixed.) Indeed, given 'purity', then the methods and results of Deligne's second great paper on the Weil conjectures, [29], which was nearly finished at the time, will yield these theorems:

THEOREM (WEIL–E. ARTIN–RIEMANN HYPOTHESIS). *For every i, the eigenvalues of the Frobenius map ϕ_q on $IH^i(X)$ are algebraic numbers whose complex conjugates are all of absolute value $q^{i/2}$.*

THEOREM (HARD LEFSCHETZ). *If* $[H] \in H^2(\mathbf{P}^N)$ *denotes the fundamental class of a hyperplane H in the ambient projective space, then for all i, intersecting i times yields an isomorphism,*

$$(\cap [H])^i \colon IH^{d-i}(X) \xrightarrow{\sim} IH^{d+i}(X) \qquad \text{where } d := \dim(X).$$

Kazhdan and Lusztig then solved the problem of purity directly in case of the Schubert varieties X_w by exploiting the geometry. In fact, they proved the following stronger theorem:

THEOREM [57, 4.2]. *The sheaf* $\mathbf{H}^{2j+1}\mathbf{IC}^{\cdot}(X_w)$ *is zero. On the stalk at a fixed point,* $\mathbf{H}^{2j}\mathbf{IC}^{\cdot}(X)_x$, *the eigenvalues of* ϕ_q *are algebraic numbers whose complex conjugates all have absolute value exactly* q^j.

On the basis of those theorems, Kazhdan and Lusztig then proved their main theorem:

THEOREM [57, 4.3]. *The coefficients of* $P_{y,w}$ *are positive. In fact,*

$$\sum\nolimits_j \dim(\mathbf{H}^{2j}\mathbf{IC}^{\cdot}(X_w)_y)\, q^j = P_{y,w}(q),$$

where the subscript 'y' indicates the stalk at the base point of the Bruhat cell B_y.

5. $\mathscr{D}$-MODULES

By good fortune, the theory that was needed to establish the Kazhdan–Lusztig conjecture was actively being developed at the very same time as the work in intersection homology theory and representation theory, although quite independently. That theory was need as much for its spirit as for its results. The theory is a sophisticated modern theory of linear partial differential equations on a smooth complex algebraic variety X (see for example [5], [6], [68]). It is sometimes called *microlocal analysis*, because it involves analysis on the cotangent bundle T^*X (although the term 'microlocal analysis' is also used more broadly to include more traditional topics in analysis on T^*X). It is sometimes called $\mathscr{D}$-*module theory*, because it involves sheaves of modules $\mathscr{M}$ over the sheaf of rings of holomorphic linear partial differential operators of finite order $\mathscr{D} := \mathscr{D}_X$; these rings are noncommutative, left and right Noetherian, and have finite global homological dimension. It is sometimes called *algebraic analysis* because it involves such algebraic constructions as $\mathrm{Ext}^i_{\mathscr{D}}(\mathscr{M}, \mathscr{N})$. The theory as it is known today grew out of the work done in the 1960s by the school of Mikio Sato in Japan.

During the 1970s, one of the central themes in $\mathscr{D}$-module theory was David Hilbert's twenty-first problem, now called the *Riemann–Hilbert problem*. "This problem," Hilbert [49] wrote, "is as follows: *To show that there always exists a linear differential equation of Fuchsian class with given singular points and monodromic group.*" It is "an important problem, one which very likely Riemann himself may have had in mind." Here Hilbert was, doubtless, thinking of Riemann's 1857 paper on Gauss's hypergeometric equation and of Riemann's 1857 related unfinished manuscript, which was published posthumously in his collected works in 1876.

The hypergeometric equation is of order 2 and has singular points at 0, 1, and ∞, but in the manuscript Riemann began a study of nth order equations with m singular points. Riemann's ingenious idea was to obtain information about the equations and the solutions from the monodromy groups (each group consists of the linear transformations undergone by a basis of solutions as they are analytically continued along closed paths around a singular point). He assumed at the outset that at a singular point x each solution has the form,

$$(z - x)^s[\phi_0 + \phi_1 \log(z - x) + \cdots + \phi_\lambda \log^\lambda(z - x)],$$

where s is some complex number and the ϕ's are meromorphic functions.

Guided by Riemann's paper, Lazarus Fuchs and his students in 1865 took up the study of nth order equations (see [63, p. 724]),

$$y^{(n)} + a_1(z)y^{(n-1)} + \cdots + a_n(z)y = 0.$$

Fuchs showed that for the solutions to have the form described above it is necessary and sufficient that $(z - x)^i a_i(z)$ be holomorphic at x for all i and x. An equation whose coefficients $a_i(z)$ satisfy this condition is said to have *regular singular points* or to be *regular*, although Fuchs used a different term. Fuchs gave special consideration to the class of equations that have at worst regular singular points in the extended complex plane, and so such equations are said to be of *Fuchsian class* or *type*.

The original Riemann–Hilbert problem was given its first complete solution in 1905 by Hilbert himself and by Oliver Kellogg using the theory of integral equations (see [63, p.726]) and in 1913 by George David Birkhoff using a method of successive approximations. Birkhoff added the concepts of a canonical system of differential equations and the equivalence of such systems (and he attacked the case of irregular singular points). The concept of a canonical system is not now present in $\mathscr{D}$-module theory, but, according to Lê Dũng Tráng, it would be good to introduce one and develop it appropriately.

In the fall of 1969, Deligne [27] made a particularly significant advance: he generalized the problem greatly and solved it as follows. Given an open subset U of a smooth complex algebraic variety X of arbitrary dimension d such that the complement $X - U$ is a divisor with normal crossings (that is, locally it is analytically isomorphic to the union of coordinate hyperplanes in the affine d-space) and given a finite dimensional complex representation of the fundamental group $\pi_1(U)$, Deligne constructed a system of differential equations with regular singular points (in an appropriately generalized sense) whose solutions via continuation along paths present the given monodromy. The system is essentially unique. If X is complete (compact), then the equations are algebraic.

Deligne came to the problem from his work on monodromy, in particular that on Picard–Lefschetz theory, which Grothendieck had encouraged between 1967 and 1969 as the next step toward the proof of the remaining Weil conjecture, the Weil–E. Artin–Riemann hypothesis. He drew further inspiration from the work of Michael Atiyah and William Hodge and the work of Grothendieck on the case of the trivial representation and of a number of people on the Gauss–Manin connection (system). The importance of Deligne's contribution to the subject of the Riemann–Hilbert problem cannot be overestimated; it inspired and supported all the subsequent advances.

Around 1977, a definitive generalization of the Riemann–Hilbert problem was formulated. In 1979, that generalization was solved by Zoghman Mebkhout [**79**] and, in 1980, by Masaki Kashiwara [**54**] somewhat differently. Both of those treatments are analytic. In the fall of 1980, Alexandre Beilinson and Joseph Bernstein developed a purely algebraic treatment, which is sufficient for the proof of the Kazhdan–Lusztig conjecture. It is largely analogous to the analytic treatment but is often technically simpler. See [**6**, p. 328, bottom].

To pass to the generalization, first view the monodromy representation in an equivalent form, as a locally constant sheaf of finite dimensional complex vector spaces on U. Then equip X with a Whitney stratification, and let the sheaf be an arbitrary constructible sheaf, or better a bounded complex of sheaves whose cohomology sheaves are constructible.

The definitive generalization does not directly involve any system of differential equations, $AF = 0$, where A is an m by n matrix of linear partial differential operators and F is a vector of meromorphic functions $y(z)$ on X. Rather, it deals with the associated (left) $\mathscr{D}$-module $\mathscr{M}$ defined by a presentation,

$$\mathscr{D}^m \xrightarrow{A^T} \mathscr{D}^n \longrightarrow \mathscr{M} \longrightarrow 0,$$

where A^T denotes the operation of right multiplication with the matrix A. That change is reasonable because applying the functor $\mathrm{Hom}_{\mathscr{D}}(\cdot, \mathscr{O}_X)$ to the presentation yields this exact sequence:

$$0 \longrightarrow \mathrm{Hom}_{\mathscr{D}}(\mathscr{M}, \mathscr{O}_X) \longrightarrow \mathscr{O}_X^n \xrightarrow{A} \mathscr{O}_X^m.$$

So the sheaf of local solutions is $\mathrm{Hom}_{\mathscr{D}}(\mathscr{M}, \mathscr{O}_X)$ and thus depends only on $\mathscr{M}$. There is a further reasonable change: the $\mathscr{D}$-module $\mathscr{M}$ is required to have such a presentation only *locally*. Such an $\mathcal{M}$ is termed *coherent*. (The term is reasonable because $\mathcal{D}$ is left Noetherian.)

The *characteristic variety*, or *singular support,* of a coherent $\mathcal{D}$-module $\mathcal{M}$ is a (reduced) closed subvariety of the cotangent bundle T^*X. It is denoted by $Ch(\mathcal{M})$, or $S.S(\mathcal{M})$, and is defined locally as follows: filter $\mathcal{M}$ by the image of the filtration on $\mathcal{D}^n$ by operator order; then the associated graded module $\mathscr{G}r(\mathcal{M})$ is finitely generated over the associated graded ring $\mathscr{G}r(\mathcal{D})$, and $\mathscr{G}r(\mathcal{D})$ is equal to the direct image on X of the structure sheaf of T^*X; set

$$Ch(\mathcal{M}) := \mathrm{Supp}(\mathscr{G}r(\mathcal{M})).$$

Then each component of $Ch(\mathcal{M})$ has dimension at least d, where $d :=$ $\dim(X)$. In fact, each component comes with a natural multiplicity of appearance, the length of $\mathscr{G}r(\mathcal{M})$ at a general point of the component. The corresponding *characteristic cycle* will also be denoted by $Ch(\mathcal{M})$.

A $\mathcal{D}$-module $\mathcal{M}$ is called *holonomic* if it is coherent and if its characteristic variety $Ch(\mathcal{M})$ is of (pure) dimension d. Then the solution sheaf and its satellites, the sheaves $\mathrm{Ext}^i_{\mathcal{D}}(\mathcal{M},\mathcal{O}_X)$, are constructible with respect to some Whitney stratification.

A holonomic module $\mathcal{M}$ is said to have *regular singular points* or, simply, to be *regular*, if every formal generalized local solution is convergent, that is, if, for every $x \in X$ and every i,

$$\mathrm{Ext}^i_{\mathcal{D}}(\mathcal{M},\mathcal{O}_X)_x = \mathrm{Ext}^i_{\mathcal{D}_x}(\mathcal{M}_x,\widehat{\mathcal{O}}_x),$$

where $\widehat{\mathcal{O}}_x$ is the ring of formal power series at x. Other definitions are also used. In any case, $\mathcal{M}$ is regular if and only its pullback to any (smooth) curve mapping into X is regular. For a curve, the modern concept is equivalent to Fuchs's.

The *dual* of a holonomic $\mathcal{D}$-module $\mathcal{M}$ is, by definition, the $\mathcal{D}$-module,

$$^*\mathcal{M} := \mathscr{H}\mathrm{om}_{\mathcal{O}_X}(\Omega^d_X, \mathrm{Ext}^d_{\mathcal{D}}(\mathcal{M},\mathcal{D})) = \mathrm{Ext}^d_{\mathcal{D}}(\mathcal{M},\mathcal{D}^\Omega),$$

where Ω^d_X is the sheaf of holomorphic d-forms, $d := \dim(X)$, and

$$\mathcal{D}^\Omega := \mathcal{D} \otimes (\Omega^d_X)^{-1} = \mathscr{H}\mathrm{om}_{\mathcal{O}_X}(\Omega^d_X, \mathcal{D}).$$

Then $^*\mathcal{M}$ is holonomic, and $^{**}\mathcal{M} = \mathcal{M}$. If $\mathcal{M}$ is regular, so is $^*\mathcal{M}$. Moreover, $\mathcal{O}_X$ is holonomic (its characteristic variety is the zero-section), and $^*\mathcal{O}_X = \mathcal{O}_X$.

The definitive generalization of the Riemann–Hilbert problem involves bounded complexes $\mathcal{M}$ of $\mathcal{D}$-modules whose cohomology sheaves are regular holonomic $\mathcal{D}$-modules. The duality above, $\mathcal{M} \mapsto {}^*\mathcal{M}$, extends to these complexes, viewed in the derived category. To such a complex $\mathcal{M}$, are associated the following two complexes in the derived category of bounded complexes of sheaves of **C**-vector spaces:

$$\mathbf{Sol}(\mathscr{M}) := \mathbf{RHom}_{\mathscr{D}}(\mathscr{M}, \mathscr{O}_X);$$

$$\mathbf{deR}(\mathscr{M}) := \mathbf{RHom}_{\mathscr{D}}(\mathscr{O}_X, \mathscr{M}).$$

The first complex, $\mathbf{Sol}(\mathscr{M})$, is the complex of generalized solutions; its cohomology sheaves are the solution sheaf and its satellites, $\mathbf{Ext}^i_{\mathscr{D}}(\mathscr{M}, \mathscr{O}_X)$. The second complex, $\mathbf{deR}(\mathscr{M})$, is isomorphic (in the derived category) to the complex,

$$0 \to \mathscr{M} \to \Omega^1_X \otimes_{\mathscr{O}_X} \mathscr{M} \to \cdots \to \Omega^d_X \otimes_{\mathscr{O}_X} \mathscr{M} \to 0,$$

and so it is called the *deRham complex* of $\mathscr{M}$. The two complexes are related through duality and the following two key canonical isomorphisms:

$$\mathbf{Sol}({}^*\mathscr{M}) = \mathbf{deR}(\mathscr{M}) = \mathbf{Sol}(\mathscr{M})^{\check{}},$$

where the '$\check{}$' indicates the Verdier–Borel–Moore dual.

The definitive generalization of the Riemann–Hilbert problem may be stated now. The problem is to prove the following theorem, which describes the nature of the correspondence between a system of differential equations and its solutions:

THEOREM (RIEMANN–HILBERT CORRESPONDENCE) [68], [6]. *Given a bounded complex of sheaves of complex vector spaces, **S**, whose cohomology sheaves are constructible with respect to a fixed Whitney stratification of X, there exists a bounded complex $\mathscr{M}$ of $\mathscr{D}$-modules, unique up to isomorphism in the derived category, such that (1) its cohomology sheaves $\mathscr{H}^i(\mathscr{M})$ are regular holonomic $\mathscr{D}$-modules whose characteristic varieties are contained in the union of the conormal bundles of the strata, and (2) the solution complex $\mathbf{Sol}(\mathscr{M})$ is isomorphic to **S** in the derived category. Moreover, the functor*

$$\mathscr{M} \mapsto \mathbf{Sol}(\mathscr{M}),$$

is an equivalence between the derived categories, which commutes with direct image, inverse image, exterior tensor product, and duality.

The Kazhdan–Lusztig conjecture was proved during the summer and fall of 1980 independently and in essentially the same way by Beilinson and Bernstein in Moscow and by Jean-Luc Brylinski and Kashiwara in Paris. Earlier, in 1971, Bernstein, I. Gelfand, and S. Gelfand had considered a complex semi-simple Lie algebra **g** and constructed a resolution by Verma modules M_λ of the irreducible module L_μ with a positive highest weight μ. In April 1976, George Kempf had given a geometric treatment of the resolution, and Kempf's work provided some initial inspiration for both proofs. Beilinson and Bernstein discussed intersection homology theory with MacPherson during his stay in Moscow for the first six months of 1980. By the middle of September, they had proved the conjecture [2].

Brylinski had become seriously interested in the conjecture in the fall of 1979 and, over the next nine months, he filled in his background. In early June 1980, while reading someone else's notes from a two-day conference that May on $\mathscr{D}$-module theory, he suddenly realized that that theory was the key to proving the conjecture. Shortly afterwards, he attended a lecture of Lê's and told him his ideas. Lê gave him his personal notes from some lectures of Mebkhout and encouraged Brylinski to phone him. Instead of phoning, Brylinski got a hold of Mebkhout's thesis and some articles by Kashiwara and Takahiro Kawai. On July 21, 1980, he wrote up a ten-page program of proof and sent it to a half dozen people; the main problem was to establish the regularity asserted in the following lemma. Soon afterwards, Kashiwara phoned him, saying he wanted to talk about it. They collaborated several times in July and August and, by the middle of September, they had written up a first draft of their proof. The proof was announced in [14] and presented in [15].

The main lemmas used in the proof of the Kazhdan–Lusztig conjecture are these:

LEMMA [94, 3.7, 3.8]. *Let $\mathscr{O}_{\text{triv}}$ denote the (Bernstein–Gelfand–Gelfand) category of representation modules M such that (1) M is finitely generated over the universal enveloping algebra U of the complex semi-simple Lie algebra* **g**, *(2) any $m \in M$ and its translates under the action of the enveloping algebra of a Borel subalgebra form a finite dimensional vector space, and (3) the center of U acts trivially on M. Then the functor $M \mapsto \mathscr{D}_X \otimes M$, where X is the flag manifold, defines an equivalence of the category $\mathscr{O}_{\text{triv}}$ with the category of regular holonomic $\mathscr{D}_X$-modules $\mathscr{M}$ whose characteristic variety is contained in the union of the conormal bundles of the Bruhat cells B_w; the inverse functor is $\mathscr{M} \mapsto \Gamma(X, \mathscr{M})$.*

LEMMA [94, 3.15, 3.16]. *Let $\mathbf{C}_w$ denote the extension by 0 of the constant sheaf on $\mathbf{C}$ on the Bruhat cell B_w. Consider the Verma module $M_w := M_{-\rho y - \rho}$ and its simple quotient $L_w := L_{-\rho y - \rho}$. Set $d := \dim(X)$. Then*

$$\mathbf{deR}(\mathscr{D}_X \otimes M_w) = \mathbf{C}_w[l(w) - d]$$
$$\mathbf{deR}(\mathscr{D}_X \otimes L_w) = \mathbf{IC}^{\cdot}(X_w)[l(w) - d]$$

where the right sides are the shifts down by $l(w) - d$ of the complex consisting of the sheaf $\mathbf{C}_w$ concentrated in degree 0 and of the intersection cohomology complex of the Schubert variety X_w, the closure of B_w.

The second formula of the last lemma is proved by checking the axioms that characterize $\mathbf{IC}^{\cdot}(X_w)$.

The first formula implies by additivity that for any $M \in \mathscr{O}_{\text{triv}}$ the cohomology sheaves of the deRham complex $\mathbf{deR}(\mathscr{D}_X \otimes M)$ are locally constant with finite dimensional stalks on any cell B_w. Hence it is meaningful to consider the "index",

$$\chi_w(M) := \sum_i (-1)^i \dim_{\mathbf{C}} \mathbf{H}^i(\mathbf{deR}(\mathscr{D}_X \otimes M))_w,$$

where the subscript 'w' indicates the stalk at the base point of B_w. For example,

$$\chi_w(M_y) = (-1)^{l(w)-d}\delta_{wy}$$

by the first formula, where δ_{wy} is the Kronecker function. The first formula and additivity now yield the formula,

$$M = \sum_y (-1)^{d-l(y)}\chi_y(M)\,M_y,$$

in the Grothendieck group. Finally, the second formula yields the first formula in the Kazhdan–Lusztig conjecture and, as Kazhdan and Lusztig showed, their second formula is formally equivalent to the first.

6. Perverse sheaves

Beilinson and Bernstein had succeeded in proving the Kazhdan–Lusztig conjecture when Deligne arrived in Moscow in mid-September 1980. The three of them discussed the proof and its implications. There is, they realized, a natural abelian category inside the nonabelian 'constructible derived category' — the derived category of bounded complexes **S** of sheaves of complex vector spaces whose cohomology sheaves $\mathbf{H}^i(\mathbf{S})$ are constructible. It is just the essential image of the category of regular holonomic $\mathscr{D}$-modules $\mathscr{M}$ embedded by the Riemann–Hilbert correspondence, $\mathbf{S} = \mathbf{deR}(\mathscr{M})$. It exists on any smooth complex algebraic variety X. Now, how can this unexpected abelian subcategory be characterized intrinsically?

Ironically, around Easter the year before, 1979, Deligne and Mebkhout had chatted in Paris about the Riemann–Hilbert correspondence. Mebkhout had just established it in his thesis [79], and Lê, then in Stockholm, wrote to Mebkhout and urged him to go and talk to Deligne about it. However, Deligne said politely that while the subject was very interesting, nevertheless it appeared to be far removed from his work in progress on monodromy, pure complexes, and the hard Lefschetz theorem, [29]. That was also the time of Deligne's correspondence with Kazhdan and Lusztig about their conjecture.

In the middle of October 1980, Deligne returned to Paris. MacPherson was there and became excited on hearing about that abelian subcategory; he kept asking Deligne if its existence was not a topological fact. The question had been discussed, according to Beilinson, by Bernstein, Deligne and himself while Deligne was still in Moscow. The time was right, and Deligne soon proved the following theorem, based on those discussions, which characterizes that image category topologically:

THEOREM [**10**, §1]. *Given a bounded complex* **S** *with constructible cohomology sheaves* $\mathbf{H}^i(\mathbf{S})$ *on an arbitrary smooth complex algebraic variety* X, *there exists a regular holonomic* $\mathscr{D}$-*module* $\mathscr{M}$ *such that* $\mathbf{S} \cong \mathbf{deR}(\mathscr{M})$ *in the derived category if and only if both of the following dual conditions are satisfied:*

(i) $\mathbf{H}^i(\mathbf{S}) = 0$ *for* $i < 0$ *and* $\operatorname{codim}(\operatorname{Supp}(\mathbf{H}^i(\mathbf{S}))) \geq i$ *for* $i \geq 0$,

(ĭ) $\mathbf{H}^i(\mathbf{S}^\smallsmile) = 0$ *for* $i < 0$ *and* $\operatorname{codim}(\operatorname{Supp}(\mathbf{H}^i(\mathbf{S}^\smallsmile))) \geq i$ *for* $i \geq 0$,

where $\mathbf{S}^\smallsmile$ *is the Verdier–Borel–Moore dual of* **S**.

Conditions (i) and (ĭ) were not far-fetched; a condition like (i) had appeared in [**90**, XIV 3], and Deligne [**29**, 6.2.13] had generalized the hard Lefschetz theorem to a pure complex **S** satisfying (i) and (ĭ); see §7. The technical aspect of the proof was not that difficult. Indeed, if $\mathbf{S} = \mathbf{deR}(\mathscr{M})$, then $\mathbf{S} = \operatorname{Sol}(\mathscr{M}^*)$ and $\mathbf{S}^\smallsmile = \operatorname{Sol}(\mathscr{M})$ by Mebkhout's local duality theorems [**80**, Thm. 1.1, Chap. III]; hence, (i) and (ĭ) hold by Kashiwara's Thm. 4.1 of [**53**]. Conversely, if (i) and (ĭ) hold, then it can be proved via a 'dévissage' that a complex $\mathscr{M}$ such that $\mathbf{S} \cong \mathbf{deR}(\mathscr{M})$ has cohomology only in degree 0. Independently, according to [**10**, footnote on p. 2], Kashiwara too discovered that theorem.

Deligne had the right perspective, so he proved more of what he, Beilinson, and Bernstein had conjectured together in Moscow. The conditions (i) and (ĭ) of the theorem above define a full abelian subcategory also if X is an algebraic variety in arbitrary characteristic p with the étale topology. The conditions can be modified using an arbitrary perversity so that they still yield a full abelian subcategory. Moreover, unlike arbitrary complexes in the derived category, those **S** that satisfy the modified conditions can be patched together from local data like sheaves. The original conditions (i) and (ĭ) are recovered with the middle perversity. The case of the middle perversity is once again the most useful by far because of the additional theorems that hold in it, such as the next two theorems. It is the only case that will be considered from now on.

Because of all those marvelous properties, everyone calls these special complexes **S** (or sometimes, their shifts by $d := \dim(X)$) *perverse sheaves*. Of course, they are complexes in a derived category and are not sheaves at all. Moreover, they are well behaved and are not perverse at all. Nevertheless, despite some early attempts to change the name 'perverse sheaf', it has stuck.

THEOREM [**3**, 4.3.1(I)]. *The abelian category of perverse sheaves is Noetherian and Artinian: every object has finite length.*

THEOREM [**3**, 4.3.1(II)]. *Let* V *be a smooth, irreducible locally closed subvariety of codimension* c *of* X, *and* **L** *a locally constant sheaf of vector spaces on* V. *Then:*

(1) *There is a unique perverse sheaf* **S** *whose restriction to V is equal to* **L**[$-c$], *which is the complex that consists of* **L** *concentrated in degree c.*

(2) *If* **L** *is the constant sheaf with 1-dimensional stalks, then* **S** *is equal to the shifted intersection homology complex* **IC**$^{\cdot}(\overline{V})[-c]$, *where* $\overline{V}$ *is the closure of V. In general,* **S** *can be constructed from* **L** *by the same process of repeated pushforth and truncation.*

(3) *If* **L** *is an irreducible locally constant sheaf, then* **S** *is a simple perverse sheaf. Conversely, every simple perverse sheaf has this form.*

The perverse sheaf **S** of the last theorem is denoted **IC**$^{\cdot}(\overline{V}, \mathbf{L})[-c]$ and is called the *DGM extension*, or Deligne–Goresky–MacPherson extension, of **L**. It is also called the "twisted intersection cohomology complex with coefficients in **L**". Thus, the family of intersection cohomology complexes was enlarged through twisting and then forever abased, becoming merely the family of simple objects in the magnificent new abelian category of perverse sheaves.

The moment that Deligne told MacPherson the definition of a perverse sheaf, MacPherson realized that some work that he and Goresky had done about three years earlier implied that a perverse sheaf 'specializes' to a perverse sheaf. Indeed, earlier they had thought hard about the way that the intersection cohomology complex specializes. They were rather upset to find that the middle perversity complex did not specialize to the middle perversity complex but to the complex associated to the next larger perversity, which they called the *logarithmic* perversity. Even worse, the logarithmic perversity complex also specialized to the logarithmic complex. The explanation turned out now to be simple: both complexes are perverse sheaves, and the logarithmic complex is in some sense a "terminal" object in the category of perverse sheaves. Goresky and MacPherson's main result in that connection is this:

THEOREM (SPECIALIZATION) [**43**, §6]. *In a 1-parameter family, a perverse sheaf specializes to a perverse sheaf. More precisely, if S is an algebraic curve, $s \in S$ a simple point, $f: X \to S$ a map, $X_s := f^{-1}(s)$ the fiber, and* **S** *a perverse sheaf on $X - X_s$, then the shifted complex of 'nearby cycles'* **R**Ψ_f**S**[-1], *which is supported on X_s, is a perverse sheaf on X. Moreover, the functor* **R**Ψ_f *commutes with Verdier–Borel–Moore duality.*

Goresky and MacPherson used special techniques from stratification theory to construct a neighborhood U of X_s and a continuous retraction $\Psi: U \to X_s$ that is locally trivial over each stratum of X_s. Then they defined **R**Ψ_f**S** by the equation,

$$\mathbf{R}\Psi_f\mathbf{S} := \mathbf{R}\Psi_*\iota_t^*\mathbf{S},$$

where $t \in S$ is a nearby general point and $\iota_t \colon X_t \to X$ is the inclusion. They proved that $\mathbf{R}\Psi_f\mathbf{S}$ is independent of the choice of the stratification and the retraction. Thus $\mathbf{R}\Psi_f\mathbf{S}$ is clearly constructible. They established the support conditions (i) and (ĭ) using their stratified Morse theory.

During the next year, MacPherson told most everyone he met about the specialization theorem. Of course, it has a natural statement, and some people may have thought of it themselves. At any rate, it became well known. It was reproved by Bernard Malgrange, by Kashiwara, and by Bernstein using $\mathscr{D}$-module theory. It was proved in arbitrary characteristic, using Deligne's 1968 algebraic definition of $\mathbf{R}\Psi_f\mathbf{S}$, by Gabber and by Beilinson and Bernstein. Verdier [97] considered the case of specialization to a divisor that is not necessarily principal. At the time, the sheaf $\mathbf{R}\Psi_f\mathbf{S}$ was often (improperly) called the sheaf of 'vanishing cycles'.

The "true" perverse sheaf $\mathbf{R}\Phi_f\mathbf{S}$ of *vanishing cycles* is defined when the perverse sheaf $\mathbf{S}$ is given on all of X. It is defined as the mapping cone over the natural comparison map,

$$\iota_s^*\mathbf{S}[-1] \to \mathbf{R}\Psi_f\mathbf{S}[-1],$$

where $\iota_s \colon X_s \to X$ is the inclusion. Thus it is a measure of the difference between the nearby cycles and the cycles on the special fiber X_s. Deligne conjectured the following remarkable theorem, which enumerates the vanishing cycles:

THEOREM [67, (1.5), (4.1)]. *Choose a local parameter at $s \in S$, and consider the corresponding section df of the cotangent bundle T^*X. Let $\mathscr{M}$ be a regular holonomic $\mathscr{D}$-module such that $\mathbf{S} \cong \mathbf{deR}(\mathscr{M})$, and suppose that the characteristic cycle $Ch(\mathscr{M})$ and the section df have an isolated intersection at a point ξ of T^*X outside the 0-section and lying over a point $x \in X$. Then the support (of every cohomology sheaf) of $\mathbf{R}\Phi_f\mathbf{S}$ is isolated at x, and*

$$\dim(\mathbf{H}^i(\mathbf{R}\Phi_f\mathbf{S})_x) = \begin{cases} mult_\xi(Ch(\mathscr{M}) \cdot [df]), & \textit{if } i = n; \\ 0, & \textit{otherwise.} \end{cases}$$

The assertion about the support of the complex $\mathbf{R}\Phi_f\mathbf{S}$ results directly from the description of the complex given in January 1983 by Lê and Mebkhout, [69, Prop. 2.1]. The formula for the dimension was first proved by Lê at Luminy in July 1983, but that proof required a condition on the restriction of f to the variety $Ch(\mathscr{M})$. In January 1988, Lê [67] eliminated this requirement via a more profound topological analysis inspired by some work that he did with Mitsuyoshi Kato in 1975. Meanwhile, Claude Sabbah (1985) and V.

Ginzburg (1986) gave proofs based on an interesting calculus of 'Lagrangian cycles'. A related proof was sketched earlier (1984) by Alberto Dubson but, according to Lê [**67**, (4.1.3)], he stated a crucial and delicate step without sufficient justification.

In March of 1981, MacPherson went to Moscow and brought along a copy of Deligne's manuscript on perverse sheaves. It turned out that the previous fall Beilinson and Bernstein had worked out an elementary theory of algebraic $\mathscr{D}$-modules and that independently they too had begun to develop the theory of perverse sheaves. When MacPherson mentioned the specialization theorem, Beilinson and Bernstein immediately sat down and came up with their own proof. Then their work became stranded, when all of a sudden Bernstein was granted permission to emigrate. Their theory of algebraic $\mathscr{D}$-modules was later written up and published by Borel *et al.* [**6**].

The two developments of the theory of perverse sheaves were combined by Beilinson, Bernstein, and Deligne, and published in their monograph [**3**], which is the definitive work on perverse sheaves in arbitrary characteristic. It includes the only detailed account of the comparison of the theories in the classical topology and in the étale topology over **C** and the only detailed account of the reduction to the algebraic closure of a finite field. In addition to discussing the theorems already mentioned and some others, which will be considered in the next section, this monograph touches on some more issues of monodromy and vanishing cycles. Parts of the monograph are rather sophisticated and based on some of Gabber's ideas. Gabber should properly have been a fourth co-author, but he declined at the last moment.

MacPherson and Kari Vilonen, another of MacPherson's students, after conversations with Beilinson and Deligne, gave in [**77**] and [**78**] another construction of the category of perverse sheaves on a stratified topological space with only even (real) dimensional strata $X_i - X_{i-1}$. It proceeds recursively, passing from $X - X_i$ to $X - X_{i+1}$. That construction makes the structure of the category more concrete. Previously, a number of other authors had made similar constructions in various special cases — dimension 2, strata with normal crossings, etc. More recently, Beilinson [**1**] gave a short alternative treatment in the general case. Renato Mirollo and Vilonen [**82**] used the construction of MacPherson and Vilonen to extend the results of Bernstein, I. Gelfand, and S. Gelfand about the Cartan matrix of the category $\mathscr{O}_{\mathrm{triv}}$ (see the end of §5) to the category of perverse sheaves on a wide class of complex analytic spaces.

7. PURITY AND DECOMPOSITION

About July 1980, Gabber solved the problem of purity that Deligne posed in his letter to Kazhdan and Lusztig. In fact, he proved more. The precise statement requires some terminology, which was introduced in [**29**, 1.2.1, 1.2.2, 6.2.2, and 6.2.4] and reviewed in [**3**, 5.1.5 and 5.1.8]. An l-adic sheaf **F** on an algebraic variety X defined over the field with q-elements is called *punctually pure of weight w* if, for every n and every fixed-point x of the Frobenius endomorphism $\phi_{q^n} : X \to X$, the eigenvalues of the automorphism $\phi_{q^n}^*$ of $\mathbf{F}_x$ are algebraic numbers whose complex conjugates all have absolute value exactly $(q^n)^{w/2}$.

The sheaf **F** is called *mixed* if it admits a finite filtration whose successive quotients are punctually pure; the weights of the nonzero quotients are called the *punctual weights* of **F**. A complex of l-adic sheaves **S** is called *mixed of weight $\leq w$* if for each i the cohomology sheaf $\mathbf{H}^i(\mathbf{S})$ is mixed with punctual weights $\leq w$. Finally, **S** is called *pure of weight w* if **S** is mixed of weight $\leq w$ and if its Verdier–Borel–Moore dual $\mathbf{S}^\vee$ is mixed of weight $\leq -w$.

Gabber's theorem is this:

THEOREM (PURITY) [**29**, p. 251], [**11**, 3.2], [**3**, 5.3]. *If X is an algebraic variety over the algebraic closure of a finite field, then the intersection homology complex* $\mathbf{IC}^{\cdot}(X)$ *is pure of weight 0; in fact, any DGM extension* $\mathbf{IC}^{\cdot}(\overline{V}, \mathbf{L})[-c]$ *is pure of weight* $-c$.

The theorem shows in particular that there are unexpectedly many pure complexes to which to apply Deligne's theory [**29**].

In the fall of 1980, Gabber and Deligne collaborated to prove some key lemmas about the structure of pure complexes and mixed perverse sheaves and to derive some important consequences. Independently, Beilinson and Bernstein obtained the same results. All the details were presented in the combined treatise [**3**]. The theory is based on Deligne's work on the Weil conjectures [**28**] and [**29**], which in turn is supported by over 3000 pages on étale cohomology theory [**90**], [**91**], on l-adic cohomology theory and L-functions [**92**], and on monodromy [**93**]. Thus these results are some of the deepest theorems in algebraic geometry, if not all of mathematics.

The Weil–E. Artin–Riemann hypothesis and the hard Lefschetz theorem, which were discussed near the end of §4, are two major consequences of the purity theorem. They hold for a projective variety defined over an algebraically closed field; for the Riemann hypothesis, it must be the algebraic closure of a finite field, but for the Lefschetz theorem, it may be arbitrary, it may even be the field of complex numbers **C**! Over **C**, an analytic proof of the Lefschetz theorem, based on a theory of "polarizable Hodge modules" analogous to the theory of pure perverse sheaves, was given by Morihiko Saito [**84**] and [**85**].

One lovely application in intersection theory in algebraic geometry of the hard Lefschetz theorem was made by Fulton and Robert Lazarsfeld; they

used it to give a significantly shorter proof, which moreover is valid in arbitrary characteristic, of the following theorem of Spencer Bloch and David
Gieseker:

THEOREM [31]. *Let X be a projective variety of dimension d, and E an
ample vector bundle of rank e on X. If $e \geq n$, then*

$$\int_X c_n(E) > 0.$$

Doubtless, the single most important consequence of the purity theorem
is the following theorem:

THEOREM (DECOMPOSITION) [11, 3.2.3], [3, 6.2.5]. *If $f: X \to Y$ is a proper
map of varieties in arbitrary characteristic, then $\mathbf{R}f_* \mathbf{IC}(X)$ is a direct sum
of shifts of DGM extensions $\mathbf{IC}(\overline{V}_i, \mathbf{L_i})[-e_i]$, where e_i is not necessarily the
codimension of V_i.*

Indeed, $\mathbf{IC}(X)$ is of 'geometric origin', so it suffices to prove the theorem
when X, Y and f are defined over the algebraic closure of a finite field. Then
$\mathbf{IC}(X)$ is pure by the purity theorem. It therefore follows from Deligne's
main theorem [29, 6.2.3] that $\mathbf{R}f_* \mathbf{IC}(X)$ is pure. Finally, because an eigenvalue of the Frobenius automorphism whose weight is nonzero cannot be
equal to 1, it can be proved that certain Ext^1's must vanish and so the corresponding extensions must split.

The decomposition theorem was conjectured in the spring of 1980 by
Sergei Gelfand and MacPherson [35, 2.10], then proved that fall by Gabber and Deligne and independently by Beilinson and Bernstein. Over **C**, an
analytic proof was given several years later by Morihiko Saito in [84] and
[85]. In fact, more general versions of the theorem are proved in each case:
$\mathbf{IC}(X)$ is replaced by the DGM extension of a locally constant sheaf of a
certain fairly general type.

Some implications of the decomposition theorem are discussed by Goresky
and MacPherson in [41]. In particular, they say in §2 that if the V_i and $\mathbf{L}_i$ are
taken to be irreducible (as they may be), then the summands $\mathbf{IC}(\overline{V}_i, \mathbf{L}_i)[-e_i]$
and their multiplicities of appearance are uniquely determined. However,
the full derived category is not abelian, and the decomposition is in no sense
canonical by itself. On the other hand, Deligne has observed, see [76, §12]
and [64, 4.2, (iii)–(v)], that the decomposition can be made canonical with
respect to a relatively ample sheaf if f is projective.

Sergei Gelfand and MacPherson [35, 2.12] showed that the decomposition theorem yields the main theorem of Kazhdan and Lusztig, which relates
their polynomials to the intersection homology groups of the Schubert varieties (see also [94, 2]). This derivation involves a lovely interpretation of
the Hecke algebra as an algebra of correspondences. Moreover, given the

decomposition theorem over $\mathbf{C}$, the proof involves no reduction to positive characteristic. According to [94, 2.13], similar work was done independently by Beilinson and Bernstein, by Brylinski, and by Lusztig and Vogan [75, 5]. In fact, the latter two authors considered a more general situation, in which the Schubert varieties are replaced by the orbits of the centralizer of an involution. However, all these latter authors used the purity theorem directly rather than applying the decomposition theorem, probably because they were unaware of it at the time. In addition, in [2], Beilinson and Bernstein also treated the case of Verma modules with regular *rational* highest weight, showing that again there is a topological interpretation for the multiplicities in the Jordan–Hölder series in terms of intersection homology groups. Lusztig [73] carried that work further, giving some explicit formulas and applying the results to the classification of the irreducible representations of the finite Chevalley groups; Lusztig's work rests on both the purity theorem and the decomposition theorem.

The decomposition theorem has the following rather useful corollary, which Kazhdan had conjectured in 1979:

COROLLARY [11, 3.2.5]. *If* $f: X \to Y$ *is a resolution of singularities, then* $IH_i(Y)$ *is a direct summand of* $H_i(X)$. *In fact, then* $\mathbf{IC}^{\cdot}(Y)$ *is a direct summand of* $\mathbf{R}f_*\mathbf{Q}_{l,X}$.

Goresky and MacPherson in [41, §A.] gave two examples showing that the direct sum decomposition need not be canonical. Nevertheless, it follows, for instance, that if $H_i(X) = 0$, then $IH_i(Y) = 0$. Thus the odd dimensional intersection homology groups of Y vanish if Y is a Schubert variety or if Y is an orbit closure in the product of a flag manifold with itself; for a proof, see Roy Joshua's paper [51, (3)].

If Y is the toric variety associated to a simplicial d-polytope, then it follows similarly that $IH_i(Y) = 0$ for odd i. Richard Stanley [95] used that fact to prove this: the components h_i of the h-vector of an arbitrary rational d-polytope are nonnegative; in fact, $h_i = \dim(IH_{2(d-i)}(Y))$ for a suitable toric variety Y. Stanley went on to observe that, because of the hard Lefschetz theorem, the vector is unimodal and the generalized Dehn–Sommerville equations are satisfied:

$$1 = h_0 \leq h_1 \leq \ldots \leq h_{[d/2]} \quad \text{and} \quad h_i = h_{d-i}.$$

The equations are obviously also a consequence of Poincaré duality.

Frances Kirwan [60] used the last corollary and the hard Lefschetz theorem to establish a procedure for computing the dimensions of the rational intersection homology groups of the quotient assigned by David Mumford's geometric invariant theory (1965) to a linear action of a complex reductive group on a smooth complex projective variety X. Just before, Kirwan had published a systematic procedure for blowing up X along a sequence of smooth equivariant centers to obtain a variety $\tilde{X}$ such that every semi-stable point of $\tilde{X}$ is stable. Then the quotient of $\tilde{X}$ is a partial desingularization of the quotient of X in which the more serious singularities have been resolved; in fact, the quotient of $\tilde{X}$ is topologically just the ordinary quotient of the open set of semi-stable points $\tilde{X}^{ss}$, and it is everywhere locally isomorphic to the quotient of a smooth variety by a finite group. Hence the intersection homology groups of the latter quotient are equal to its ordinary homology groups. Moreover, they are also equal to the equivariant homology groups of $\tilde{X}^{ss}$, whose dimensions were computed in another of Kirwan's papers. The heart of [60] is a description of the change in the intersection homology groups under the passage to the next successive blow-up. In the sequel [61], Kirwan generalized the work to the case in which X is singular.

Kirwan [62] used the last corollary and Heisuke Hironaka's (1976) equivariant resolution of singularities to treat the rational intersection homology groups of a singular complex projective variety Y with a torus action. The groups are determined by the action of the torus on an arbitrarily small neighborhood of the set of fixed points, and they are given by a generalization of a well-known direct sum formula. Thus Kirwan's results generalize the results of Andre Bialynicki–Birula (1973, 1974) in the case that Y is smooth and the results of James Carrell and Goresky (1983) in the case that Y is singular but its Bialynicki–Birula decomposition is suitably "good." Kirwan also discussed a supplementary treatment using an *equivariant* intersection homology theory. In that discussion, Kirwan referred to the treatments of equivariant intersection homology theory made by Brylinski [13] and Joshua [51]. However, all three treatments of the equivariant theory were, according to MacPherson, developed independently.

Jonathan Fine and Prabhakar Rao [30] used the last corollary to determine the rational intersection homology groups of a complex projective variety Y with an isolated singularity in terms of any desingularization X and its exceptional locus E. They proved that, for all i,

$$B_i(E) = B_i(E^1) - B_i(E^2) + B_i(E^3) - \cdots \qquad \text{where} \quad E^j := \coprod (E_{k_1} \cap \cdots \cap E_{k_j}),$$

In the case that E is a divisor with normal crossings, they went on, by using mixed Hodge theory, to prove a formula for the Betti number $B_i(E) := \dim H^i(E)$, when $i \geq \dim(X)$:

$$B_i(E) = B_i(E^1) - B_i(E^2) + B_i(E^3) - \cdots \qquad \text{where} \quad E^j := \coprod (E_{k_1} \cap \cdots \cap E_{k_j}),$$

where the E_k are the irreducible components of E. Combined, those two results provide a lovely "inclusion–exclusion" formula for the intersection homology Betti numbers of Y in the upper half dimensions. The remaining Betti numbers may be determined by using duality.

Walter Borho and MacPherson in [8, §1] introduced and studied an important case in which the decomposition of the decomposition theorem is in fact canonical. They call a proper map of varieties, $f: X \to Y$, *semi-small* if for all r

$$\text{codim}\{ y \in Y \mid \dim(f^{-1}y) \geq r \} \geq 2r.$$

(Recall from §3 that f is said to be a 'small resolution' if the second inequality is strict and if X is smooth.)

Borho and MacPherson, moreover, weakened the hypothesis in the above corollary on X: it does not have to be smooth but only a *rational homology manifold*; that is, for all $x \in X$,

$$H_r(X, X - x) = \begin{cases} \mathbf{Q}_l, & \text{if } r = 2\dim(X); \\ 0, & \text{otherwise.} \end{cases}$$

It is equivalent, they observe, that $\mathbf{IC}^{\cdot}(X) = \mathbf{Q}_{l,X}$. In this connection, their main result is the following theorem:

THEOREM [**8**, §1]. *Let $f: X \to Y$ be a semi-small proper map of varieties of the same dimension, with X a rational homology manifold. Then $\mathbf{R}f_*\mathbf{Q}_{l,X}$ is a perverse sheaf and, in its decomposition into direct summands, $\mathbf{IC}^{\cdot}(\overline{V}_i, \mathbf{L}_i)[-e_i]$, necessarily $e_i = \text{codim}(V_i)$; that is, the summands are perverse sheaves too. Moreover, the decomposition into isotypical components — the direct sums of all the isomorphic summands — is canonical and, if f is birational, then one of the isotypical components is $\mathbf{IC}^{\cdot}(Y)$.*

Indeed, $\text{codim}(\text{Supp}(\mathbf{H}^r(\mathbf{R}f_*\mathbf{Q}_{l,X}))) \geq r$ for $r \geq 0$ because the map is semi-small, and $\mathbf{R}f_*\mathbf{Q}_{l,X}$ is self-dual because $\mathbf{Q}_{l,X} = \mathbf{IC}^{\cdot}(X)$. Hence, $\mathbf{R}f_*\mathbf{Q}_{l,X}$ is perverse. Hence, so are its direct summands. Since the category of perverse sheaves is abelian, the isotypical decomposition is canonical. Finally, the last assertion is easy to check.

Borho and MacPherson applied the above theorem (or rather the version of it with $\mathbf{Q}_X$ in place of $\mathbf{Q}_{l,X}$) to the (semi-small) Springer resolution $\pi: N' \to N$ of the nilpotent cone N in the dual $\mathbf{g}^*$ of the Lie algebra $\mathbf{g}$ of a connected reductive algebraic group G. They also considered Grothendieck's map $\phi: Y \to \mathbf{g}^*$, which extends π, and they studied the monodromy action of the fundamental group of the open subset of $\mathbf{g}^*$ of regular semi-simple elements (the diagonalizable elements with distinct eigenvalues), recovering Lusztig's construction of Springer's action of the Weyl group W, which is a quotient of the fundamental group, on the fibers $H^*(N'_\xi, \mathbf{Q})$ of $\mathbf{R}\pi_*\mathbf{Q}_{N'}$. Their main result is the following theorem:

THEOREM [**7**], [**8**, §2], [**94**, 4.8, 4.9]. (1) *The nilpotent cone N is a rational homology manifold.*

(2) *There exists a canonical W-stable isotypical decomposition,*

$$\mathbf{R}\pi_*\mathbf{Q}_{N'} = \sum_{(\alpha,\phi)} \mathbf{IC}^{\cdot}(\overline{N}_\alpha, \mathbf{L}_\phi)[-\text{codim}(N_\alpha)] \otimes V_{(\alpha,\phi)},$$

where the N_α are the orbits of G on N, the $\mathbf{L}_\phi$ are all the various locally constant sheaves of 1-dimensional $\mathbf{Q}$-vector spaces on N_α (they are associated to the various irreducible rational characters of the fundamental group of N_α), and $V_{(\alpha,\phi)}$ is a $\mathbf{Q}$-vector space of dimension equal to the multiplicity of ϕ in the locally constant sheaf $(\mathbf{R}^{2\dim(N_\alpha)}\pi_*\mathbf{Q}_{N'})|N_\alpha$.

(3) *The group ring of W is equal to the endomorphism ring of $\mathbf{R}\pi_*\mathbf{Q}_{N'}$ in the category of perverse sheaves. The action of W on the (α,ϕ)-component is of the form $1 \otimes \rho_{(\alpha,\phi)}$, where $\rho_{(\alpha,\phi)}$ is an absolutely irreducible representation of W on $V_{(\alpha,\phi)}$, and every irreducible complex representation of W is obtained in this way.*

In fact, Borho and MacPherson obtain more general results involving parabolic subgroups. In the special case of the general linear group, they obtain a new proof of Lusztig's results on the Green polynomials and the Kostka–Foulkes polynomials.

Assertion (2) above was conjectured by Lusztig [**71**, §3, Conj. 2] after he established the case of the general linear group. The paper was written and available as a preprint in 1980.

Assertion (1) has a curious history. Lusztig recalls discussing it with Deligne in 1974. Lusztig gave a lecture at the IHES in which he mentioned some results in representation theory due to Robert Steinberg. Deligne observed that those results would be explained if (1) holds and, the next day, he had a proof. Seven years later in [**71**, §3, Rem. (a)], Lusztig stated (1), calling it "an unpublished theorem of Deligne" but saying nothing there about how or when Deligne proved it. By the spring of 1981, Borho and MacPherson had proved (2) and (3) in full generality and proved (1) for the general linear group; moreover, using (2) they had reduced (1) to the following lemma, which they conjectured: the trivial representation 1 occurs in the Springer representation on $H^i(N'_\xi, \mathbf{Q})$ with multiplicity 1 if $i = 0$ and 0 otherwise.

Borho and MacPherson announced assertions (2) and (3) in [**7**] but, according to MacPherson, they chose not to discuss (1) in order to keep that Comptes Rendus note sufficiently short. He clearly remembers traveling around Europe, however, lecturing on all three assertions, and asking if (1) was not known. Deligne, at that time, found (1) surprising! At Luminy in July 1981, Borho and MacPherson discussed the lemma with Lusztig. He knew a proof, and so in [**7**, 2.3] they attribute the lemma to him. Lusztig also told them that Deligne had proved (1). Moreover, Lusztig recalls that he had, in fact, proved the lemma as part of his own (unpublished) proof of (1); that proof involved some known properties of the Green polynomials instead of (2). However, since Deligne had no memory whatsoever of having proved (1) and since they did not realize that Lusztig had his own proof, Borho and

MacPherson could feel perfectly comfortable about saying proudly at the beginning of [7, 2.3] that (1) "could have been stated in 1930, but seems to be new."

8. OTHER WORK AND OPEN PROBLEMS

A lot of work has been done on the remarkable relation between L^2-cohomology theory and Hodge theory on the one hand and intersection homology theory on the other. It all began in the winter of 1975–1976 at the State University of New York, Stony Brook, when Cheeger independently found a cohomology theory satisfying Poincaré duality for essentially the same class of spaces as Goresky and MacPherson had considered. Cheeger considered a closed oriented *triangulated* pseudomanifold X. Such an X carries natural piecewise flat metrics. Cheeger formed the L^2-cohomology groups of the incomplete Riemannian manifold U obtained by discarding all the simplices of codimension 2 or more; those are the cohomology groups $H^i_{(2)}(U)$ of the complex of real differential forms ω on U such that

$$\int_U \omega \wedge *\omega < \infty \quad \text{and} \quad \int_U d\omega \wedge *d\omega < \infty.$$

Cheeger found that Poincaré duality could be verified directly or derived formally, in essentially the same way as in the smooth case, from the action of the $*$-operator on the harmonic forms of the associated Hodge theory — in fact, the full Hodge theory holds — given an inductively defined vanishing condition on the middle dimensional L^2-cohomology groups of the links, or given a certain more general '$*$-invariant ideal boundary condition' on the forms. The vanishing condition was later seen to hold whenever X has a stratification by strata of even codimension. The theory automatically also works if X is equipped with any metric that on U is quasi-isomorphic to the previous one; then X is said to have 'conical' or 'conelike' singularities. The theory is invariant under smooth subdivision and, more generally, piecewise smooth equivalence.

In the summer of 1976 at Stony Brook, Cheeger informed Sullivan of his discovery. Cheeger was amazed at Sullivan's response: "You know, Goresky and MacPherson have something like that." Sullivan went on to describe the ideas behind their theory. He suggested that Cheeger had found a deRham–Hodge theory dual to their combinatorial one for the middle perversity, and Cheeger later proved it. So, in particular, Cheeger's L^2-groups are in fact topological invariants. He also observed that Cheeger's 'ideal boundary condition' corresponds to the central condition in Morgan's (unpublished) extension of their theory to a more general class of spaces. He proposed that

Cheeger and MacPherson talk. Within a few weeks, MacPherson, who was on his way to Paris, passed through Stony Brook to talk to Sullivan. He talked to Cheeger as well and was rather surprised to hear about Cheeger's discovery, but he agreed that they must be talking about equivalent theories. He was particularly surprised to hear that there was an L^2-proof of the Künneth formula, because the product of two middle-allowable cycles is seldom middle allowable.

Cheeger's discovery was an extraordinary byproduct of his work on his proof [18], [20] of the Ray–Singer conjecture, which asserts that on a compact Riemannian manifold the analytic torsion and Reidemeister torsion are equal. In an initial attempt to prove it, Cheeger examined the behavior of the spectrum and eigenfunctions of the Laplacian on differential forms on the level surfaces of a Morse function in a neighborhood of a critical value corresponding to a nondegenerate critical point; that level surface has a 'conical' singularity. Engrossed in writing up his proof of the conjecture until October 1977 and, until February 1978, in obtaining local analytic and combinatorial formulas for the signature and total L-class of a pseudomanifold [22], Cheeger did not circulate an announcement of his discovery until the spring of 1978; abridged, it was published in 1979 as [19]. All the details eventually appeared in [21] and [22]. In addition to the first proof of the Künneth formula and the only known explicit local formulas for the L-class, Cheeger's analytic methods in intersection homology theory have yielded a vanishing theorem for the intersection homology groups of a pseudomanifold of positive curvature in the pl-sense [21, pp. 139–40], [23]. Moreover, the general methods themselves have also had significant applications to other theories, including index theory for families of Dirac operators [4], the theory surrounding Witten's global anomaly formula [24], and diffraction theory [26].

In the summer of 1977 in the Cheeger dining room about three miles from the Stony Brook campus, Cheeger and MacPherson talked again. This time they considered not the conical metric of a triangulation but the Kähler metric of a complex projective variety X with nonsingular part U. They conjectured that (i) the L^2-cohomology group $H^i_{(2)}(U)$ is always dual to the intersection homology group $IH_i(X)$ and (ii) the pairing is given by integration. In addition, they conjectured that the various standard consequences of Hodge theory — including the Hodge structure, the primitive decomposition, the hard Lefschetz theorem, and the Hodge index theorem — are valid. Those conjectures were published in [21, §7].

With Goresky's help, the preceding conjectures were developed further and discussed in the joint article [25]. There they observed that, to establish the duality conjecture (i), it suffices to prove that the direct image of the presheaf on U formed of the appropriate L^2-forms of degree i has a 'fine' associated sheaf and that, as i varies, those associated sheaves form a (deRham) complex that satisfies the axioms that characterize $\mathbf{IC}^{\cdot}(X)$; the cohomology groups of the complex are equal to its hypercohomology groups because the sheaves are fine. They conjectured that each class contains a unique harmonic (closed and co-closed) representative and that splitting the harmonic forms into their (p,q)-pieces yields a (pure) Hodge decomposition, compatible with Deligne's mixed Hodge structure on the ordinary cohomology groups of X. They noted that the Hodge decomposition would exist if the metric on U were complete, and they suggested that another approach to constructing a Hodge decomposition of $IH.(X)$ is to construct a complete (Kähler) metric. Moreover, they gave a lot of evidence for the validity of the conjectures. This work of Cheeger, Goresky, and MacPherson has lead to a great deal of work by many people.

Zucker was aware of the work of Cheeger, Goresky, and MacPherson that appears in [21] and [40] when he made the following celebrated conjecture, which first appeared in a 1980 preprint of [99]: if X is the Baily–Borel compactification of the quotient space U of a Hermitian symmetric domain modulo a proper action of an arithmetic group Γ and if U is provided with the natural complete metric, then the L^2-cohomology groups are dual to the (middle) intersection homology groups; the forms may take values in a local system on U of a certain type, and then the intersection homology groups are the hypercohomology groups of the DGM extension of the system. Zucker was led to this conjecture by some examples that he worked out [99, §6] of his general results [99, (3.20) and (5.6)] about the L^2-cohomology groups of an arithmetic quotient of a symmetric space. In the examples, the compactification is obtained by adjoining a finite number of isolated singular points, and Zucker was struck by the values of the local L^2-cohomology groups at these points: they are equal to the singular cohomology groups of the link in the bottom half dimensions and to 0 in the middle and in the top half dimensions. Zucker's work on [99] developed out of an attempt to generalize §12 of [98]. In [99], the L^2-cohomology groups were the objects of initial interest; if they are dual to the intersection homology groups, then they are topological invariants.

Between 1980 and 1987, Zucker's conjecture was proved in various special cases by Zucker himself, by Armand Borel, and by Borel and William Casselman. Finally, in 1987, the general case was proved by Eduard Looijenga [70] and by Leslie Saper and M. Stern [88], [89]. Looijenga uses Mumford's (1975) desingularization of X and the decomposition theorem. Saper and Stern use a more direct method, which they feel will also yield a generalization of the conjecture due to Borel, in which U is an 'equal rank' symmetric space and X is a Satake compactification all of whose real boundary components are equal rank symmetric spaces.

One reason for the great interest in Zucker's conjecture is that it makes it possible to extend the "Langlands program" to cover the important noncompact case, as Zucker indicates in [100]. The program is aimed at relating the L-functions of a Shimura variety, which is a 'model' U_0 of U over a number field, to the automorphic forms associated to the arithmetic group Γ. The forms are directly related to the L^2-cohomology groups. The intersection homology groups, constructed using the étale topology, are compatible with the passage modulo a suitable prime of the number field to positive characteristic, where, it is hoped, the L-functions may be studied; in this connection, also see Kirwan's discussion [62, pp. 396–398]. In the case of Hilbert modular (or Hilbert–Blumenthal) varieties, Brylinski and Labesse [16] did successfully treat the L-functions using intersection homology theory.

The conjectures of Cheeger, Goresky, and MacPherson were also treated with some success in the case that U is the smooth part of a complex projective variety X with isolated singularities. Wu-Chung Hsiang and Vishwambhar Pati [50] gave a proof that $H_{(2)}^i(U)$ is dual to $IH_i(X)$ if X is a normal surface endowed with the induced (Fubini-Study) metric. Saper [86], [87], who was inspired by the case of the Zucker conjecture, constructed a *complete* Kähler metric on U whose L^2-cohomology groups are dual to the intersection homology groups of X. Zucker [101] proved that the corresponding Hodge decomposition is compatible with Deligne's mixed Hodge structure, which, in fact, was proved to be pure by J. H. M. Steenbrink [96], who implicitly used the decomposition theorem, and then by Vicente Navarro Aznar [83], who avoided it. Zucker [101, Rem. (ii), p. 614] notes that the result holds in addition for a Hilbert modular surface, the proof being essentially the same, and that more knowledge about the resolution of the singularities of a Hilbert modular variety of higher dimension will yield the result in the same way in that case as well.

There is other work in the same vein. First, in 1981, Brylinski [10, §3] made the following conjecture: if X is embedded in a smooth variety Y, say with codimension c, and if the regular holonomic $\mathscr{D}$-module $\mathscr{M}$ such that $\mathbf{deR}(\mathscr{M}) = \mathbf{IC}^{\cdot}(X)[c]$ is given the global filtration of Kashiwara and Kawai, then the associated filtration on $\mathbf{deR}(\mathscr{M})$ induces the desired Hodge structure on $IH.(X)$. Second, in a 1985 preprint of [64], János Kollár considered a surjective map $f: X \to Y$ between projective varieties with X smooth, and he related the sheaves $R^i f_* \omega_X$ to certain DGM extensions; then he conjectured a general framework for his results in terms of a corresponding Hodge structure. Third, as mentioned in §7, in July 1983 Saito [84] announced a theory of 'polarizable Hodge modules' analogous to the theory of pure perverse sheaves, and in [85] he provided the details. Zucker's pioneering work [98], which Deligne had in mind when he came up with his pushforth-and-truncate formula, is now perceived as a cornerstone of Saito's theory. Finally, in 1985, Eduardo Cattani, Aroldo Kaplan, and Wilfred Schmid [17] and, in-dependently, Kashiwara and Kawai [55] generalized that work of Zucker's to higher dimensions: they proved that the intersection homology groups of a smooth variety X are dual to the L^2-cohomology groups of the complement U in X of a divisor with normal crossings, with coefficients in a local system underlying a polarizable variation of Hodge structure.

Another major topic of research has been the theory of "canonical trans-forms" of perverse sheaves S; see Luc Illusie's report [52]. The transform $T(S)$ on Y of S on X is defined as $Rq_*(L \otimes Rp^*S)$ where $q: Z \to Y$ and $p: Z \to X$ are maps and L is a local system of rank 1 on Z. If X is a vector bundle, Y the dual bundle, and Z their product, then $T(S)$ is called the *vector Fourier transform*. If Y is a compact parameter space of a family of subvarieties of X and if Z is the total space (or incidence correspon-dence), then $T(S)$ is called the *Radon transform*. The fundamental theory was developed by Brylinski in a 1982 preprint of [12] on the basis of work of Deligne, of Ryoshi Hotta and Kashiwara, of Gérard Laumon, and of Mal-grange. Brylinski also applied the theory to the estimation of trigonometric sums, recovering and extending work of Laumon and Nicholas Katz, and to the study of Springer's representation of the Weyl group via Kashiwara's approach, recovering and extending the results of Springer, of Lusztig, and of Borho and MacPherson.

The transform was used by Laumon [66] to study Langlands' conjecture that there exists a correspondence between the l-adic representations of rank n of the Galois group of the algebraic closure of a finite field and the au-tomorphic forms which are eigenvectors of the Hecke operators on $GL_n(A)$ where A is the ring of adeles. Ivan Mirković and Vilonen [81] used a Radon transformation, which is like the horocycle transform of Gelfand and Graev (1959), to prove the following conjecture of Laumon and Lusztig: let G be a reductive group, S a G-equivariant irreducible perverse sheaf, U a maximal unipotent subgroup, and N the nilpotent cone in the dual of the Lie algebra; then (1) in characteristic zero, S is a character sheaf if and only if its charac-teristic variety lies in $G \times N$, and (2) in arbitrary characteristic, S is a tame character sheaf if and only if the direct image of S on G/U is constructible with respect to the Bruhat cells and is tame. Character sheaves are certain interesting perverse sheaves, which were introduced by Lusztig and studied by him, see [74], and by others as a new way of treating characteristic zero representations of Chevalley groups.

One important open problem is to determine which maps $f: X \to Y$ have a natural associated pair of adjoint maps f_* and f^* on the intersection ho-mology groups. For example, the semi-small resolutions do; see §7. Another important example is the class of placid maps, which was introduced by Goresky and MacPherson in [44] and [42, §4]. By definition, $f: X \to Y$ is *placid* if there exists a stratification of Y such that each stratum S satisfies

$\text{codim}(f^{-1}S) \geq \text{codim}(S)$ (whence equality holds if the map is algebraic). If so, then a map of complexes $f^*\colon \mathbf{IC}^{\cdot}(Y) \to \mathbf{IC}^{\cdot}(X)$ may be defined using generic geometric chains or using Deligne's construction. Virtually every *normally nonsingular* map is placid; those maps were considered earlier in Goresky and MacPherson's paper [42, 5.4] and in Fulton and MacPherson's memoir [34], but they were, in fact, introduced and popularized by MacPherson in many lectures at Brown during the years 1975–1980. To be sure, not every map has such an adjoint pair. An interesting example was given by Goresky and MacPherson in [41, §C.]: it is the blowing-up $f\colon X \to Y$ of the cone Y over a smooth quadric surface in $\mathbf{P}^3$; there exist two small resolutions $g_i\colon Y_i \to Y$ $(i = 1, 2)$ and placid maps $f_i\colon X \to Y_i$ such that $f = g_i f_i$ but $f_1^* g_1^* \neq f_2^* g_2^*$.

A related open problem is to determine which subvarieties X of a variety Y have natural fundamental classes in $IH.(Y)$. Not all do. Indeed, if the graph of a map $f\colon X \to Y$ between compact varieties has a natural fundamental class in $IH.(X \times Y)$, then that class will define a map $f^*\colon IH.(X) \to IH.(Y)$, because by the Künneth formula and Poincaré duality

$$IH.(X \times Y) = IH.(X) \otimes IH.(Y) = IH.(X)^{\check{}} \otimes IH.(Y) = \text{Hom}(IH.(X), IH.(Y)).$$

Nevertheless, it might be that there is a well-defined subspace $A.(X)$ of $IH.(Y)$ that is spanned by all reasonable (though not uniquely determined) fundamental classes. It should contain the duals of the Chern classes in the ordinary cohomology groups of all the algebraic vector bundles on Y, and it should map onto the space of algebraic cycles in the ordinary homology groups. Given any desingularization Y' of Y and embedding of $IH.(Y)$ in $H.(Y')$ coming from the decomposition theorem, $A.(Y)$ should be the trace of $A.(Y')$. Moreover, the intersection pairing on $IH.(Y)$ should restrict to a *nonsingular* pairing on $A.(Y)$. That nonsingularity is unknown even when Y is nonsingular, and in that case it is one of Grothendieck's 'standard conjectures' [48].

The graph of a placid self-map $f\colon X \to X$ is not usually allowable as a cycle for the (middle) intersection homology group; indeed, not even the diagonal itself is. Nevertheless, Goresky and MacPherson [44], [46] proved that these subvarieties carry fundamental classes whose intersection number is equal to the Lefschetz number,

$$IL(f) := \sum_i (-1)^i \text{trace}(f_* | IH_i(X));$$

in other words, the Lefschetz fixed-point formula holds for f. They also observed that the formula holds when f is replaced by a *placid self-correspondence*, a subvariety C of $X \times X$ such that both projections $C \to X$ are placid.

The intersection homology groups with *integer* coefficients of a complex variety do not usually satisfy Poincaré duality. Goresky and Paul Siegel [47] discovered a 'peripheral group', which measures the failure. Remarkably, this group itself admits a nondegenerate linking pairing, and the Witt class of the pairing is a cobordism invariant. According to Goresky and MacPherson, Sylvan Capell and Julius Shaneson are currently (1988) using the invariant to further knot theory.

Finally, there is the problem of developing a reasonable theory of characteristic numbers for singular varieties. Intersection homology theory yields an Euler characteristic and a signature. It also makes it reasonable to expect that every characteristic number will be the same for a variety X and for any small resolution of X. So far, all attempts to lift Chern classes and Whitney classes from ordinary homology groups to intersection homology groups have failed; indeed, Verdier and Goresky gave counterexamples, which were mentioned by Goresky and MacPherson in [46, §A] and explained in detail by Brasselet and Gerardo Gonzales-Sprinberg [9]. On the other hand, Goresky [37] has generalized the theory of Steenrod squares from ordinary cohomology theory to intersection homology theory. While Goresky's theory does not generalize completely, it does make it possible to define in the usual way an intersection homology Wu class, whose Steenrod square is equal to the homology Wu class. Thus, while significant progress has been made, more remains to be done on that problem — the very problem that motivated the discovery of intersection homology theory.

9. REFERENCES

[1] A. Beilinson, *How to glue perverse sheaves*, K-theory, arithmetic, and geometry. Seminar, Moscow University, 1984–1986, Springer Lecture Notes in Math. **1289** (1987), 42-51.

[2] A. Beilinson and J. Bernstein, *Localization of* **g**-*modules*, C. R. Acad. Sci. Paris **292** (1981), 15–18.

[3] A. Beilinson, J. Bernstein, and P. Deligne, *Faiseaux pervers*, in "Analyse et topologie sur les espaces singuliers," Astérisque **100**, Société mathḿatique de france, 1982.

[4] J. M. Bismut and J. Cheeger, *Invariants êta et indices des familles pour des varietés à bord*, C. R. Acad. Sci. Paris **305(I)** (1987), 127–130.

[5] J.-E. Björk, "Rings of Differential Operators," North-Holland Publ. Co., 1979.

[6] A. Borel *et al.*, "Algebraic D-Modules," Perspectives in Math., vol 2, Academic Press, Inc., Orlando, Florida, 1986.

[7] W. Borho and R. MacPherson, *Représentations des groupes de Weyl et homologie d'intersection pour les variétés nilpotentes*, C. R. Acad. Sci. Paris **292** (1981), 707–710.

[8] W. Borho and R. MacPherson, in "Analyse et topologie sur les espaces singuliers," Asté-risque **101–102**, Société mathématique de france, 1982, pp. 23–74.

[9] J.-P. Brasselet and G. Gonzalez-Sprinberg, *Sur l'homologie d'intersection et les classes de Chern des variétés singulières (espaces de thom, examples de J.-L. Verdier et M. Goresky. With an appendix by Verdier*, in "Géométrie algebrique et applications. II (La Rábida, 1984)," Travaux en Cours, no. 23, exposé 2, J.-M. Aroca, T. Sanches-Giralda, J.-L. Vincente (eds.), Hermann, Paris, 1987, pp. 5–14.

[10] J.-L. Brylinski, *Modules holonomes à singularites régulières et filtration de Hodge*, Alge-braic Geometry — La Rábida, Spain 1981, Springer Lecture Notes in Math. **961** (1982), 1–21.

[11] J.-L. Brylinski, *(Co)-homologie d'intersection et faisceaux pervers*, Séminaire Bourbaki **589**, Astérisque **92–93** (1982), 129–158.

[12] J.-L. Brylinski, *Transformations canoniques, dualité projective, théorie de Lefschetz, trans-formations de Fourier et sommes trigonométriques*, Astérisque **140–141** (1986), 3–134.

[13] J.-L. Brylinski, "Equivariant intersection cohomology," Inst. Hautes Études Sci., pre-print, 1986.

[14] J.-L. Brylinski and M. Kashiwara, *Démonstration de la conjecture de Kazhdan and Lusztig sur les modules de Verma*, C. R. Acad. Sci. Paris **291** (1980), 373–376.

[15] J.-L. Brylinski and M. Kashiwara, *Kazhdan–Lusztig conjecture and holonomic systems*, Invent. Math. **64** (1981), 387–410.

[16] J.-L. Brylinski and J.-P. Labesse, *Cohomologie d'intersection et fonctions L de certaines variétès de Shimura*, Ann. Sci. École Norm. Sup. **17** (1984), 361–412.

[17] E. Cattani, A. Kaplan, and W. Schmid, L^2 *and intersection cohomologies for a polarizable variation of Hodge structure*, Invent. Math. **87** (1987), 217–252.

[18] J. Cheeger, *Analytic torsion and Reidemmeister torsion*, Proc. Nat. Acad. Sci. **74(7)** (1977), 2651–2654.

[19] J. Cheeger, *On the spectral geometry of spaces with cone-like singularities*, Proc. Nat. Acad. Sci. **76** (1979), 2103–2106.

[20] J. Cheeger, *Analytic torsion and the heat equation*, Ann. of Math. **109** (1979), 259–322.

[21] J. Cheeger, *On the Hodge theory of Riemannian pseudomanifolds*, in "Geometry of the Laplace operator," Proc. Sympos. Pure Math. **36**, Amer. Math. Soc., Providence, RI, 1980, pp. 91–146.

[22] J. Cheeger, *Spectral geometry of singular Riemannian spaces*, J. Diff. Geom **18** (1983), 575–657.

[23] J. Cheeger, *A vanishing theorem for piecewise constant curvature spaces*, in "Curvature and Topology of Riemannian Manifolds," Proc. (1985) K. Shiohana, T. Sakai, T. Sunada (eds.), Lecture Notes in Math. **1201**, Springer-Verlag, 1986, pp. 333–340.

[24] J. Cheeger, η-invariants, the adiabatic approximation and conical singularities, J. Diff. Geom **26** (1987), 175–221.

[25] J. Cheeger, M. Goresky, and R. MacPherson, L^2-cohomology and intersection homology of singular algebraic varieties, Yau, S. T. (ed.), in "Seminar on differential geometry," Princeton University Press, Princeton, NJ, 1982, pp. 303–340.

[26] J. Cheeger and M. Taylor, On the diffraction of waves by conical singularities I, Comm. Pur. App. Math. **XXV** (1982), 275–331; II, Comm. Pur. App. Math. **XXV** (1982), 487–529.

[27] P. Deligne, Equations différentielles á points singuliers réguliers, Lecture Notes in Math, vol. 163, Springer-Verlag, 1970.

[28] P. Deligne, La conjecture de Weil. I, Publ. Math. IHES **43** (1974), 273–307.

[29] P. Deligne, La conjecture de Weil. II, Publ. Math. IHES **52** (1980), 137–252.

[30] J. Fine and P. Rao, "On intersection homology at isolated singularities," Northeastern University, preprint, March 1988.

[31] W. Fulton and R. Lazarsfeld, Positive polynomials for ample vector bundles, Ann. of Math. **118** (1983), 35–60.

[32] W. Fulton and R. MacPherson, Intersecting cycles on an algebraic variety, in "Real and complex singularities," P. Holm (ed.), Proc. Oslo (1976), Sitjhoff & Noorhoof, 1977, pp. 179–197.

[33] W. Fulton and R. MacPherson, Defining algebraic intersections, Algebraic Geometry — Tromsø, Norway 1977, L. D. Olson (ed.), Springer Lecture Notes in Math. **687** (1978), 1–29.

[34] W. Fulton and R. MacPherson, "Categorical framework for the study of singular spaces," Mem. Amer. Math. Soc. **243**, (1981).

[35] S. Gelfand and R. MacPherson, Verma modules and Schubert cells: a dictionary, Séminaire d'algébre Paul Dubreil and Marie-Paule Malliavin, Springer Lecture Notes in Math. **928** (1982), 1–50.

[36] M. Goresky, Whitney Stratified Chains and Cochains, Trans. Amer. Math. Soc. **267** (1981), 175–196.

[37] M. Goresky, Intersection homology operations, Comment. Math. Helv. **59** (1984), 485–505.

[38] M. Goresky, "Kazhdan-Lusztig Polynomials for Classical Groups," Northeastern University, preprint, 1983.

[39] M. Goresky and R. MacPherson, La dualité de Poincaré pour les espaces singuliers, C. R. Acad. Sci. Paris **184** (1977), 1549–1551.

[40] M. Goresky and R. MacPherson, *Intersection homology theory*, Topology **149** (1980), 155–92.

[41] M. Goresky and R. MacPherson, *On the topology of complex algebraic maps*, Algebraic Geometry — La Rábida, 1981, Springer Lecture Notes in Math. **961** (1982), 119–129.

[42] M. Goresky and R. MacPherson, *Intersection Homology II*, Invent. Math. **71** (1983), 77–129.

[43] M. Goresky and R. MacPherson, *Morse Theory and Intersection Homology*, in "Analyse et topologie sur les espaces singuliers," Astérisque **101–102**, Société mathématique de france, 1982, pp. 135–192.

[44] M. Goresky and R. MacPherson, *Lefschetz fixed point theorem and intersection homology*, A. Borel (ed.), in "Intersection cohomology," Progress in Math. **50**, Birkhäuser Boston, Inc., 1984, pp. 215–220.

[45] M. Goresky and R. MacPherson, *Problems and bibliography on intersection homology*, A. Borel (ed.), in "Intersection Cohomology," Progress in Math. **50**, Birkhäuser Boston, Inc., 1984, pp. 221–233.

[46] M. Goresky and R. MacPherson, *Lefschetz fixed point theorem for intersection homology*, Comment. Math. Helv. **60** (1985), 366–391.

[47] M. Goresky and P. Siegel, *Linking pairings on singular spaces*, Comment. Math. Helv. **58** (1983), 96–110.

[48] A. Grothendieck, *Standard Conjectures on Algebraic Cycles*, in "Algebraic Geometry," Papers Bombaby Colloquium, Oxford Univ. Press, 1969, pp. 193–199.

[49] D. Hilbert, *MATHEMATICAL PROBLEMS*, translated by Dr. Mary Winston Newson, Bull. Amer. Math. Soc. **50** (1902), 437–479.

[50] W.-C. Hsiang and V. Pati, L^2-*cohomology of normal algebraic surfaces, I*, Invent. Math. **81** (1985), 395–412.

[51] R. Joshua, *Vanishing of odd-dimensional intersection cohomology*, Math. Z. **195** (1987), 239–253.

[52] L. Illusie, *Deligne's ℓ-adic Fourier Transform*, Algebraic Geometry — Bowdoin 1985, S. Bloch (ed.), Proc. Symp. Pure Math. **46(2)** (1987), 151–164.

[53] M. Kashiwara, *On the maximally overdetermined systems of linear differential equations. I*, Publ. Res. Inst. Math. Sc. Kyoto **10** (1975), 563–579.

[54] M. Kashiwara, *Faisceaux constructible et systèmes holonomes d'èquations aux dérivées partielles linéaires à points singuliers réguliers*, Sém. Goulaouic-Schwartz (1979–80), Exp. 19.

[55] M. Kashiwara and T. Kawai, *The Poincaré lemma for a variation of polarized Hodge structure.*, Proc. Japan Acad. Ser. A. Math. Sci. **61** (1985), 164–167.

[56] D. Kazhdan and G. Lusztig, *Representations of Coxeter Groups and Hecke Algebras*, Invent. Math. **53** (1979), 165–184.

[57] D. Kazhdan and G. Lusztig, *Schubert varieties and Poincaré duality*, in "Geometry of the Laplace operator," Proc. of Sympos. Pure Math. **36**, Amer. Math. Soc., Providence, RI, 1980, pp. 185–203.

[58] D. Kazhdan and G. Lusztig, *A topological approach to Springer's representations*, Adv. Math. **38** (1980), 222–228.

[59] H. King, *Topological invariance of intersection homology without sheaves*, Topology **20** (1985), 229–234.

[60] F. Kirwan, *Rational intersection cohomology of quotient varieties*, Invent. Math. **86** (1986), 471–505.

[61] F. Kirwan, *Rational intersection cohomology of quotient varieties. II*, Invent. Math. **90** (1987), 153–167.

[62] F. Kirwan, *Intersection homology and torus actions*, J. Amer. Math. Soc. **1** (1988), 385–400.

[63] M. Kline, "Mathematical Thought from Ancient to Modern Times," Oxford Univ. Press, 1972.

[64] J. Kollár, *Higher direct images of dualizing sheaves. II*, Ann. of Math. **124** (1986), 171–202.

[65] A. Lascoux and M. Schützenberger, *Polynômes de Kazhdan–Lusztig pour les Grassmanniennes*, Young tableaux and Schur functors in algebra and geometry, Astérisque **87–88** (1981), 249–266.

[66] G. Laumon, *Correspondance de Langlands géométrique pour les corps de fonctions*, Duke Math. J. **54** (1987), 309–360.

[67] D. T. Lê, "Morsification of $\mathscr{D}$-modules," Université Paris 7 and Northeastern University, preprint, January, 1988.

[68] D. T. Lê and Z. Mebkhout, *Introduction to linear differential systems*, in "Singularities," Proc. of Sympos. Pure Math. **40** Part 2, Amer. Math. Soc., Providence, RI, 1983, pp. 31–63.

[69] D. T. Lê and Z. Mebkhout, *Variétés caractristique et variétés polaires*, C. R. Acad. Sci. Paris **1296** (1983), 129–132.

[70] E. Looijenga, *L^2-cohomology of locally symmetric varieties*, Compositio Math. **67** (1988), 3–20.

[71] G. Lusztig, *Green Polynomials and Singularities of Unipotent Classes*, Adv. Math. **142** (1981), 169–178.

[72] G. Lusztig, *Singularities, character formulas, and a q-analog of weight multiplicities*, in "Analyse et topologie sur les espaces singuliers," Astérisque **101–102**, Société mathématique de france, 1982, pp. 208–229.

[73] G. Lusztig, "Characters of Reductive Groups over a Finite Field," Ann. of Math. Study **107**, Princeton University Press, 1984.

[74] G. Lusztig, *Introduction to Character Sheaves*, Finite Group Representations, Arcata 1986, Proc. Symp. Pure Math. **40** (1987), 165–179.

[75] G. Lusztig and D. Vogan, *Singularities of Closures of K-orbits on Flag Manifolds*, Invent. Math. **71** (1983), 365–379.

[76] R. MacPherson, *Global questions in the topology of singular spaces*, plenary address, in "Proc. of the International Congress of Mathematicians," August 16–24, 1983, Warszawa, PWN–Polish Scientific Publishers, Warszawa, and Elsevier Science Publishers B.V., Amsterdam, 1984, pp. 213–236.

[77] R. MacPherson and K. Vilonen, *Une construction élèmentaire des faisceaux pervers*, C. R. Acad. Sci. Paris **292** (1984), 443–446.

[78] R. MacPherson and K. Vilonen, *Elementary construction of perverse sheaves*, Invent. Math. **84** (403–435), 1986.

[79] Z. Mebkhout, "Cohomologie locale des espaces analytiques complexes," Thèse de Doctorat d'Etat, Université Paris VII, 1979.

[80] Z. Mebkhout, *Théorèms de bidualité locale pour les $\mathscr{D}$-modules holonomes*, Ark. Mat. **120** (1982), 111–122.

[81] I. Mirković and K. Vilonen, *Characteristic varieties of character sheaves*, Invent. Math. **93** (1988), 405–418.

[82] R. Mirollo and K. Vilonen, *Bernstein–Gelfand–Gelfand reciprocity on perverse sheaves*, Ann. Sci. École Norm. Sup. **120** (1987), 311–324.

[83] V. Navarro Aznar, *Sur la théorie de Hodge des variétés algébriques á singularités isolées*, Systèmes Différentiels et Singularités; A. Galligo, M. Granger, Ph. Maisonobe (eds.), Astérisque **130** (1985), 272–307.

[84] M. Saito, *Hodge Structure via Filtered $\mathscr{D}$-modules*, Systèmes Différentiels et Singularités; A. Galligo, M. Granger, Ph. Maisonobe (eds.), Astérisque **130** (1985), 342–351.

[85] M. Saito, "Mixed Hodge Modules," Res. Inst. Math. Sc. Kyoto, preprint no. 585, 1987.

[86] L. Saper, *L^2-cohomology and intersection homology of certain algebraic varieties with isolated singularities*, Invent. Math. **82** (1985), 207–255.

[87] L. Saper, "L^2-cohomology of isolated singularities," preprint, November (1985).

[88] L. Saper and M. Stern, *L^2-cohomology of arithmetic varieties*, Proc. Nat. Acad. Sci. **84** (1987), 5516–5519.

[89] L. Saper and M. Stern, L^2-cohomology of arithmetic varieties, Duke University, preprint, February (1988). To appear in Ann. of Math.

[90] SGA 4, Théorie des Topos et Cohomologie Etales des Schémas. Tome 1, exposés I–IV, Lecture Notes in Math, vol. 269, Springer-Verlag, 1972; Tome 2, exposés V–VIII Lecture Notes in Math, vol. 270, Springer-Verlag, 1972; Tome 3, exposés IX–XIX Lecture Notes in Math, vol. 305, Springer-Verlag, 1973.

[91] SGA $4\frac{1}{2}$, Cohomologie Etale, Lecture Notes in Math, vol. 569, Springer-Verlag, 1977.

[92] SGA 5, Cohomologie l-adique et Fonctions L, Lecture Notes in Math, vol. 589, Springer-Verlag, 1977.

[93] SGA 7 I, Groupes de Monodromie en Géométrie Algébrique, Lecture Notes in Math, vol. 288, Springer-Verlag, 1972; II, Lecture Notes in Math, vol. 340, Springer-Verlag, 1973.

[94] T. Springer, Quelques applications de la cohomologie d'intersection, Séminaire Bourbaki 589, Astérisque 92–93 (1982), 249–274.

[95] R. Stanley, Generalized H-Vectors, Intersection Cohomology of Toric Varieties, and Related Results, Adv. Studies Pure Math. 11 (1987), 187–213.

[96] J. H. M. Steenbrink, Mixed Hodge structures associated with isolated singularities, Singularities, P. Orlik (ed.), Proc. Symp. Pure Math. 40(2) (1983), 513–536.

[97] J.-L. Verdier, Spécialisation de faisceaux et monodromie modérée, in "Analyse et topologie sur les espaces singuliers," Astérisque 101–102, Société mathématique de france, 1982, pp. 135–192.

[98] S. Zucker, Hodge theory with degenerating coefficients: L_2 cohomology in the Poincaré metric, Ann. of Math. 109 (1979), 415–476.

[99] S. Zucker, L^2-cohomology of Warped Products and Arithmetic Groups, Invent. Math. 70 (1982), 169–218.

[100] S. Zucker, L^2-cohomology and intersection homology of locally symmetric varieties, II, Compositio Math. 59 (1986), 339–398.

[101] S. Zucker, The Hodge structure on the intersection homology of varieties with isolated singularities, Duke Math. J. 55 (1987), 603–616.